Advances in Neuroethics

Advances in neuroscience research are bringing to the forefront major benefits and ethical challenges for medicine and society. The ethical concerns related to patients with mental health and neurological conditions, as well as emerging social and philosophical problems created by advances in neuroscience, neurology and neurotechnology are addressed by a specialized and interdisciplinary field called neuroethics.

As neuroscience rapidly evolves, there is a need to define how society ought to move forward with respect to an ever growing range of issues. The ethical, legal and social ramifications of neuroscience, neurotechnology and neurology for research, patient care, and public health are diverse and far-reaching — and are only beginning to be understood.

In this context, the book series "Advances in Neuroethics" addresses how advances in brain sciences can be attended to for the benefit of patients and society at large.

Ariane Lewis • James L. Bernat
Editors

Death Determination by Neurologic Criteria

Areas of Consensus and Controversy

Springer

Editors
Ariane Lewis
NYU Langone Medical Center
New York, NY, USA

James L. Bernat
Dartmouth Geisel School of Medicine
Hanover, NH, USA

ISSN 2522-5677 ISSN 2522-5685 (electronic)
Advances in Neuroethics

ISBN 978-3-031-15946-6 ISBN 978-3-031-15947-3 (eBook)
https://doi.org/10.1007/978-3-031-15947-3

This Springer imprint is published by the registered company Springer Nature Switzerland AG
The registered company address is: Gewerbestrasse 11, 6330 Cham, Switzerland

Contents

Historical Introduction

James L. Bernat and Ariane Lewis

Determination of death by neurologic criteria is now widely accepted around the world as catalogued in the comprehensive report of the World Brain Death Project [1]. Yet, despite this international acceptance, there remain active conceptual, medical, scientific, legal, religious, ethical, and social controversies over it. These contemporary areas of consensus and controversy comprise the subject material of this book.

Death by neurologic criteria was first described in 1959 by French neurologists who called it *le coma dépassé* (beyond coma or irretrievable coma) [2]. After additional cases and commentary were published during the next decade, it gradually began to be accepted more in medical practice. In 1968, the World Medical Assembly noted that the ability to technologically facilitate the circulation of oxygenated blood necessitated a reevaluation of death [3]. Death by neurologic criteria received a major boost in 1968 with the publication of an influential *JAMA* article by an ad hoc committee of Harvard Medical School faculty who delineated the standards for its determination, but misleadingly termed the state "brain death" [4]. Their article also highlighted its dual instrumental value: allowing physicians to withdraw futile chronic ventilator support from hopelessly brain-damaged, ventilator-dependent patients who met the criteria (then believed to be an unlawful act on a living person) and generating multiorgan donors for the emerging practice of deceased donor organ transplantation. However, the *JAMA* article did not provide a rigorous conceptual justification of why these patients were dead. Despite this omission, the article's authority, amplified by its powerful instrumental value, led to a broad

J. L. Bernat (✉)
Department of Neurology, Dartmouth Geisel School of Medicine, Hanover, NH, USA
e-mail: bernat@dartmouth.edu

A. Lewis
Departments of Neurology and Neurosurgery, NYU Langone Medical Center, New York, NY, USA

A. Lewis, J. L. Bernat (eds.), *Death Determination by Neurologic Criteria*, Advances in Neuroethics, https://doi.org/10.1007/978-3-031-15947-3_1

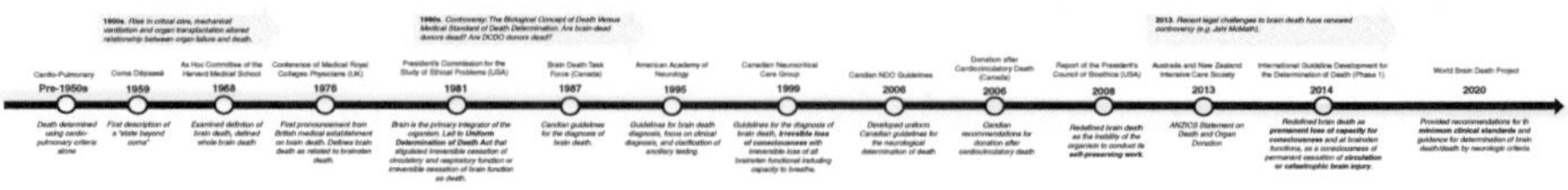

Fig. 1 Timeline depicting significant events in the history of determination of death by neurologic criteria. Reproduced with permission from Zheng K., Sutherland S., Hornby L., Wilson L., Shemie S.D., Sarti A.J. Healthcare professionals' understandings of the definition and determination of death: a scoping review. *Transplant Direct* 2022;8(4):e1309 [19]

expansion of the practice of death by neurologic criteria and to its growing acceptance [5] (see timeline in Fig. 1).

During the 1970s and early 1980s, scholars began to provide justifications for why these patients were dead. The rationales included claims that the patient had sustained the permanent cessation of: (a) those unique characteristics essential to human life; (b) the potential to integrate the body's organs and vital systems; (c) the body's central control mechanism; or (d) the human organism as a whole [6]. Debates over these claims have continued for over 50 years, though, until the past decade, they took place largely within academia. Despite the increasing international acceptance of death by neurologic criteria, conceptual debates over whether such patients are truly dead now have entered the public sphere and penetrated the legal arena through several highly publicized lawsuits that have been covered in the popular press [7].

In 1981, the legal status of death by neurologic criteria in the United States became secured and uniform with the publication of *Defining Death* by the President's Commission for the Study of Ethical Problems in Medicine and Biomedical and Behavioral Research [8]. This book provided a rigorous conceptual justification for death by neurologic criteria, and proposed and defended a uniform model statute of death incorporating it, the Uniform Determination of Death Act, that they urged all states to enact. In conjunction with the American Medical Association, the American Bar Association, and the National Conference of Commissioners on Uniform State Laws (now called the Uniform Law Commission), this effort was successful because the exact language of the statute, or a minor variation of it, was enacted in most states. Similar statutes were developed in other countries.

During the past 40 years, the Uniform Determination of Death Act has largely succeeded in practice in that it: (1) afforded legal grounds for physicians to declare death by neurologic criteria; (2) yielded a reasonable degree of death statute uniformity nationwide; and (3) strengthened the foundation of a popular and successful program of deceased organ donation. Yet physicians and legal scholars have become aware that the Uniform Determination of Death Act is imperfect due to ambiguities of its interpretation [9, 10].

Because of this, in 2020, the Uniform Law Commission empaneled a study committee to examine its alleged deficits, omissions, and ambiguities, on which we both served in an advisory capacity. In 2021, in response to the study committee recommendation, the Uniform Law Commission recommended that the statute should be revised and assigned a drafting committee to rewrite it. This revision procedure entails a multiyear process of drafting and implementation, because as a recommended model statute, states must be convinced to enact it in place of their current death statute [11].

Fig. 2 The international acceptance of death determination by neurologic criteria. Adapted with permission from Lewis A., Bakkar A., Kreiger-Benson E., Kumpfbeck A., Liebman J., Shemie S.D., et al. Determination of death by neurologic criteria around the world. *Neurology* 2020;95(3):e299-e309 [12]

Although death by neurologic criteria is widely accepted in Western society and in many countries in both the Eastern and the developing world [12] (see Fig. 2), certain aspects of its practice continue to generate controversy. Ongoing conceptual and practical disputes include whether it is biologically coherent; whether the whole-brain, brainstem, or higher-brain criterion should be used for its determination; whether the absence of diabetes insipidus (implying continued hypothalamic neurosecretory function) should exclude the determination; whether surrogate consent is necessary to conduct an evaluation for its determination; and whether families of patients should have the authority to overrule or prevent its determination [13, 14]. In a 2018 conference at the Harvard Medical School commemorating the 50th anniversary of the publication of their classic *JAMA* article, invited speakers addressed many of these controversial issues, discussed its legacy over the past half-century, and commented on its future [15]. We are pleased that many of those speakers are contributors to this volume.

The controversies over death by neurologic criteria have an ironic and paradoxical feature. Most practicing neurologists, neurosurgeons, and intensivists, who conduct its evaluation daily in intensive care units around the world, regard it as a quotidian, accepted medical practice and are unaware or unconcerned about its controversies. In contrast, some university philosophy professors have developed a very different and critical attitude about it, emphasizing its lack of biological coherence, its inconsistencies, and other shortcomings. As a result, conferences on death by neurologic criteria in medical settings usually focus on technical refinements to ensure its consistency and accuracy, whereas discussions in universities question whether it should be abandoned altogether [16].

For this volume, we have gathered many of the leading scholars writing about death by neurologic criteria. These authors represent a diverse range of backgrounds and areas of expertise including medicine, law, philosophy, ethics, and religion. The analyses and critiques included here provide the reader with inclusive and thoughtful analyses of the principal areas of controversy and consensus. We have organized the chapters, some of which are paired in a pro-con debate format, into lucidly catalogued sections. The first section on conceptual issues features arguments supporting and rejecting a brain-based death criterion; whether a whole-brain, brainstem, or brain-as-a-whole death criterion is justified; the provisions that belong in a statute on determination of death by neurologic criteria; the impact of preserved hypothalamic neurosecretory function on the whole-brain criterion of death; and the conceptual importance of the distinction between the irreversible and permanent cessation of brain function in assessing death by neurologic criteria. We eschew a discussion of the "higher brain" formulation [17] that would determine death in patients in a vegetative/unaware wakeful state because, despite its popularity in academic circles, it has not been embraced by medical societies or enshrined in laws anywhere in the world. It has been globally rejected because it wrongly classifies living people as dead, such as those in irreversible vegetative states.

The next section on medical issues features discussions of the undesirable variations in death by neurologic criteria practices across institutions, states, and countries; controversies in the determination of death by neurologic criteria in pediatric patients; the necessity to demonstrate absence of brain circulation for its determination; and the role of observation time, body temperature, and repeated evaluations. The section on scientific issues reviews essential research questions and addresses the performance of research after death by neurologic criteria.

The section on legal issues features essays on the content of death statutes; the requirement for consent for testing of death by neurologic criteria (particularly the apnea test); the legal response to objections to its determination on religious, conceptual, or emotional grounds; the view of it as a legal fiction; and the management of the pregnant patient after its determination. The section on religious issues examines Christian, Islamic, and Jewish perspectives on it. The final section on ethical and social issues features essays on public views on death by neurologic criteria, cultural considerations about it in Asia and Africa, the distinction between "determination" and "declaration" of death, management of objections to its determination, and the appropriate role of personal choice in determination of death.

Although a consensus holds that death by neurologic criteria is conceptually and biologically justified and clinically accurate, controversies about this accepted medical practice raise questions that carry a potential to erode public trust [18]. We believe that it is essential for the medical profession to maintain public trust, particularly in a matter as consequential as death determination, which has profound medical, legal, social, and financial consequences. Therefore, these controversies should be squarely addressed with a tolerant and constructive attitude, in a logical, rigorous, and scientifically accurate manner. We hope that by elucidating each debate, this volume will enhance understanding of the diverse perspectives on death by neurologic criteria, and lead, when achievable, to compromise and broadened consensus.

We add a brief explanatory note on terminology, given the variation among authors' choice of words chosen to describe the phenomena under consideration. We tried to maintain uniformity in terminological usage among chapters (see preferred terminology definitions in Table 1), but allowed authors who had used alternative terms consistently in their past writing to continue using them here. We

Table 1 Uniform terminology

- **Activity:** work of cells or groups of cells *(not to be confused with "function")*
- **Ancillary test:** a test to evaluate for brain circulation or electrical activity which can be performed in addition to a clinical evaluation and apnea test *(used in lieu of "confirmatory test," "secondary test," or "supplementary/supplemental test")*
- **Apnea test:** an evaluation to assess the ability to breathe spontaneously in response to hypercarbia and acidosis *(used as "perform an apnea test," used in lieu of "apnea testing")*
- **Brainstem areflexia:** absence of the pupillary, corneal, oculocephalic, oculovestibular, cough, and gag reflexes
- **Brain circulation:** intracranial blood flow *(used in lieu of "brain blood flow," "cerebral blood flow," "cerebral circulation," "intracranial blood flow," or "intracranial circulation")*
- **Cardiopulmonary arrest:** cessation of heartbeat and breathing *(used in lieu of "cardiac arrest," "cardiorespiratory arrest," or "circulatory arrest")*
- **Coma:** eyes-closed state of unconsciousness from which the subject cannot be aroused to wakefulness or awareness
- **Consciousness:** awareness of self and environment with the ability to interact with others
- **Criteria for death:** conditions under which death can be determined *(not to be confused with "definition of death" or "medical standards for death")*
  - **Death by cardiopulmonary criteria:** complete and permanent cessation of circulation and ventilation *(used in lieu of "death by cardiorespiratory criteria" or "death by circulatory criteria")*
  - **Death by neurologic criteria:** complete and permanent cessation of brain function identified by unresponsive coma, brainstem areflexia, and inability to breathe spontaneously *(used in lieu of "brain arrest," "brain death," "brainstem death," "cerebral arrest," "cerebral circulatory arrest," "cerebral death," "death by neurological criteria," "neurologic death," or "neurological death")*
    - **Brainstem criterion:** a formulation of death by neurologic criteria which requires only cessation of function of the brainstem
    - **Whole-brain criterion:** a formulation of death by neurologic criteria which requires loss of function of the entire brain, including the brainstem
- **Declaration of death:** the formal legal identification that death has occurred *(used as "perform a declaration of death" or "declare death," not be confused with determination of death)*
- **Definition of death:** the meaning of the word *death*, as distinguished from *life*
- **Determination of death:** the performance of an evaluation/assessment to see if a patient meets the criteria for death *(used as "perform a determination of death" or "determine death" in lieu of "perform an evaluation/assessment," "evaluate," or "assess," not be confused with declaration of death)*
- **Function:** integrated work within a gland, organ, or organ system, which can be observed or evaluated via laboratory testing, to sustain life *(not to be confused with "activity")*
  - **Irreversible loss of function:** function ceases and cannot resume spontaneously or be restored by medical intervention *(not to be confused with "permanent loss of function")*
  - **Permanent loss of function:** function ceases and will not resume spontaneously or be restored through medical intervention *(not to be confused with "irreversible loss of function")*
- **Medical standards for death:** rules established by an authoritative medical organization which must be utilized to determine death by neurologic criteria *(used in lieu of "tests to determine death," not to be confused with "criteria for death")*
- **Organ support:** interventions to maintain the function of the body, excluding the brain, after determination of death by neurologic criteria *(used in lieu of "somatic support")*
- **Prerequisites for death by neurologic criteria:** the conditions that must be met prior to performance of a determination of death by neurologic criteria *(used in lieu of "preconditions")*
- **Vegetative state/unaware–wakeful state:** a disorder of consciousness with no awareness of self or environment, but with intact wakefulness
- **Ventilation:** mechanical movement of air in the respiratory system *(not to be confused with exchange of gases in the alveoli, "respiration")*
- **Ventilator:** a mechanical device which assists ventilation *(used in lieu of "respirator")*

extended this license to the distinction between "determination" and "declaration" of death which, although the subject of a chapter, is used somewhat differently in the legal chapters. The deficiencies of the old term "brain death" are well recognized, despite its ubiquity. We tried to restrict its usage to its historical context. We chose the term "death by neurologic criteria" as our preference, but recognize that it, too, is not ideal. When possible, we discouraged the use of abbreviations that have become prevalent in scientific writing.

We thank our readers who have chosen to explore the areas of controversy and consensus about death by neurologic criteria and thank each of the authors for contributing their expertise and writing insightful and provocative chapters. We are grateful to Prof. Ralf Jox who, on behalf of the editors of the *Advances in Neuroethics* series, invited us to compose and edit this book. Finally, we are indebted to the staff at Springer Nature for their guidance and assistance in the production of this volume.

References

1. Greer DM, Shemie SD, Lewis A, et al. Determination of brain death/death by neurological criteria: the World Brain Death Project. JAMA. 2020;324:1078–97.
2. Mollaret P, Goulon M. Le coma dépassé. Rev Neurol (Paris). 1959;101:3–15.
3. Gilder SS. Twenty-Second World Medical Assembly. BMJ. 1968;3:493–4.
4. Ad Hoc Committee. A definition of irreversible coma: report of the Ad Hoc Committee of the Harvard Medical School to Examine the Definition of Brain Death. JAMA. 1968;205:337–40.
5. Belkin GS. Brain death and the historical understanding of bioethics. J Hist Med Allied Sci. 2003;58:325–61.
6. Bernat JL, Culver CM, Gert B. On the definition and criterion of death. Ann Intern Med. 1981;94:389–94.
7. Pope TM. Brain death forsaken: growing conflict and new legal challenges. J Legal Med. 2017;37:265–324.
8. President's Commission for the Study of Ethical Problems in Medicine and Biomedical and Behavioral Research. Defining death: medical, legal and ethical issues in the determination of death. Washington, DC: U.S. Government Printing Office; 1981.
9. Lewis A, Bonnie RJ, Pope T, et al. Determination of death by neurologic criteria in the United States: the case for revising the Uniform Determination of Death Act. J Law Med Ethics. 2019;47(suppl 4):9–24.
10. Molina-Pérez A, Bernat JL, Dalle Ave A. Inconsistency between the circulatory and brain criteria of death in the Uniform Determination of Death Act. J Med Philos. (in press).
11. Lewis A. The uniform determination of death act is being revised. Neurocrit Care 2022;36(2):335–8.
12. Lewis A, Bakkar A, Kreiger-Benson E, et al. Determination of death by neurologic criteria around the world. Neurology. 2020;95(3):e299–309.
13. Bernat JL. Controversies in defining and determining death in critical care. Nat Rev Neurol. 2013;9:164–73.
14. Lewis A, Greer D. Current controversies in brain death determination. Nat Rev Neurol. 2017;13:505–9.
15. Defining death: organ transplantation and the 50-year legacy of the Harvard report on brain death. Hastings Center Rep 2018;48(suppl 4):S1–S80.
16. Bernat JL. Whither brain death? Am J Bioeth. 2014;14(8):3–8.

17. Gervais K. Redefining death. New Haven: Yale University Press; 1987.
18. Lewis A, Bernat JL, Blosser S, et al. An interdisciplinary response to contemporary concerns about brain death determination. Neurology. 2018;90:423–6.
19. Zheng K, Sutherland S, Hornby L, Wilson L, Shemie SD, Sarti AJ. Healthcare professionals' understandings of the definition and determination of death: a scoping review. Transplant Direct. 2022;8(4):e1309.

Part I

Conceptual Issues

Arguments Supporting Neurologic Criteria to Determine Death

James L. Bernat

The brain criterion of death ("brain death") occupies a unique niche in the contemporary intersection of medicine and public policy. Since its introduction more than 50 years ago, the brain criterion of death has become increasingly accepted as a standard of death determination by physicians throughout the world, has been codified in public laws in many countries, and has supported a successful program of deceased multiorgan donation [1]. Yet, at the same time, it continues to provoke serious controversies that germinated in academia but now have spread to the lay public [2, 3]. What are the current controversies over the accepted brain criterion of death and why, paradoxically, have they captured public attention at the very moment that the brain criterion of death has achieved its greatest international societal acceptance?

In this chapter, I address these questions by first explaining the conceptual foundation for the brain criterion of death and show how it arose to resolve ambiguities in death determination triggered by technological advances in the support of circulation and respiration. I then update the biophilosophical analysis of death that my colleagues and I first offered over 40 years ago, sequentially exploring the definition, criterion, and tests of death [4]. I analyze the proposed anatomic–physiologic conceptualizations of the brain criterion, known as the whole-brain, brainstem, and higher-brain formulations, and offer critiques and defenses of each. I end with an account of a refined brain criterion, the "brain-as-a-whole" formulation, that resolves some of the current controversies while retaining its essence. Throughout the chapter I cite the conceptual justifications for continued reliance on the brain criterion of death.

J. L. Bernat (✉)
Department of Neurology, Dartmouth Geisel School of Medicine, Hanover, NH, USA
e-mail: bernat@dartmouth.edu

A. Lewis, J. L. Bernat (eds.), *Death Determination by Neurologic Criteria*, Advances in Neuroethics, https://doi.org/10.1007/978-3-031-15947-3_2

1 Medical Standards of Death Determination

To better appreciate the medical context in which the brain criterion of death developed, it is instructive to discuss how physicians traditionally determined death. Typically, physicians established the absence of heartbeat, breathing, responsiveness, and pupillary light reflexes. Because no resuscitative or organ-supportive treatments were available until about 1950, physicians would often delay for a few minutes before declaring death to assure that their findings were valid, namely that the vital functions had completely ceased and would not return. In the nineteenth century, additional tests sometimes were performed to assure the validity of the examination findings, because reversible mimics of death seen in a living patient, such as syncope and hypotension, could be misdiagnosed as death [5].

Prior to the onset of the era of life-sustaining technology, death was a unitary phenomenon. When any of the three classical functions vital for life ceased (circulation, respiration, and brain function), the others quickly and inevitably ceased because of their mutual interdependence. Death was the permanent cessation of the three vital functions. Physicians did not have to consider if one function was paramount or whether a patient was dead in whom only one or two vital functions had ceased because such cases were impossible. In the pretechnological era, the principal medical controversy in death determination centered on the accuracy of the physician's determination, particularly, the certainty that the vital functions had ceased completely.

With the development of tracheal positive-pressure ventilation (TPPV) in the 1950s and cardiopulmonary resuscitation (CPR) in 1960s, the interdependence of the three vital functions was severed forever. Now it became possible for physicians to administer CPR to restart a patient's heartbeat after cardiac arrest and to provide TPPV when breathing ceased. These new therapies were marvelous advances in saving lives. However, they also created a novel problem by permitting the dissociation of vital functions. Prolonged resuscitation after cardiopulmonary arrest was the most common and vexing example. During the interval after cardiopulmonary arrest before CPR could restore heartbeat or before TPPV could restore respiration, all of a patient's brain functions may have permanently ceased as a result of prolonged global oxygen deprivation to brain neurons. Because technology had created an artificial situation in which the three vital functions no longer ceased simultaneously, brain function could have permanently ceased despite resuscitative restoration of heartbeat and circulation, and mechanically produced respiration.

This dissociation of vital functions, impossible in the pretechnological era, meant that death was no longer a unitary phenomenon. Resuscitation technology had produced a new ambiguity in a patient's life state: was such a patient alive or dead? Like a living person, the patient's heart was beating, blood was circulating, and visceral organs continued to function. However, like a dead person, the patient could not respond, could not breathe, and was utterly and permanently immobile and unconscious. Physicians, who were bound by the traditional unitary concept of death, no longer could determine whether such patients were alive or dead because these patients shared features of both states. Consensus on this essential question became impossible until first there could be agreement on what death meant in the new technological era of organ system support.

The existence of this tragic state, clearly an unintended consequence of technology, immediately raised the question as to which of the three vital functions was most essential for human life. During the decade beginning in 1959, neurologists who first described such patients recognized they were unlike any previously described state. Because their coma was of a profundity never encountered, they had lost all circulatory tone and the capacity to breathe, the neurologists claimed that these patients were dead, despite technological support of their respiration and circulation and consequent visceral organ function. These neurologists implicitly ranked the absence of brain function as more essential to human life than the presence of mechanically produced respiration and consequent circulation.

The authors of these early reports provided diagnostic criteria for the determination of "brain death" but usually offered no rigorous conceptual account of why the patients were dead [6]. The clinician-authors presumably relied on their intuition and, perhaps, on their value judgments to assert that "brain-dead" patients were dead. Nevertheless, over the next decades, a societal consensus emerged that these patients were truly dead. This consensus was buttressed by scholars who proposed reasoned rationales arguing that these patients either were biologically dead or that, as a societal matter, they should be treated as if they were dead.

2 Conceptual Analyses of Death

During the mid-1970s and early 1980s, my Dartmouth colleagues, Bernard Gert and Charles Culver, and I were among those scholars who argued why "brain dead" patients were dead using biophilosophical analyses [4]. Our analysis was sequential, proceeding from the conceptual to the tangible, exploring topics that we ordered as assumptions, definition, criterion, and tests. We were influenced by the earlier sequential legal analysis of Alexander Capron and Leon Kass who identified the essential descriptive levels as concepts, standards, criteria, and tests/procedures [7]. Many subsequent scholars have accepted our analytic framework, including some of those who disagree with the substance of our assumptions, definition, or criterion of death. For example, Alan Shewmon, a prominent scholar who completely rejects a brain criterion of death—as he defends in his chapter in this volume—has referred to our analytic framework as "virtually universally accepted" [8].

Our analysis began with a set of assumptions to frame the argument, without which consensus would be impossible. First, we considered death as primarily a biological phenomenon. Although dying and death practices and rituals have important social, legal, anthropological, and religious elements, just as life is fundamentally biological, so is its cessation. We assumed that the concept of death was univocal among higher vertebrate species, that is, when we speak of the death of our pet dog, we mean the same thing by death as we do when we speak of the death of a human. We considered death as an event. Although some deaths culminating a progressive illness do resemble a process, we argued that death is best conceptualized as the event that separates the process of dying from the process of bodily disintegration. Because only two possible states exist—alive and dead—the transition from one to the other is necessarily instantaneous, though the event of death may not be recognized at the instant it occurs and may be able to be determined only in retrospect. Finally, we assumed that death is irreversible. Successful resuscitation of cardiorespiratory arrest means that, earlier, the patient was incipiently dying but was not dead.

The second step in the analysis is the philosophical task of defining death. This task must precede identifying the criterion and tests of death because both depend entirely on first agreeing on exactly what death means in our technological age of organ support. A definition of death requires understanding how we conceptualize and describe the fundamental change that occurs in the transition from alive to dead. Death is a common, nontechnical word that we all use correctly, but whose precise meaning had been rendered ambiguous by the medical technologies of organ support. Identifying the precise definition of death requires establishing the essential concept to which people refer when they say that someone has died. The third step in the analysis is the philosophical and medical task of identifying the criterion of death, that general measurable condition that satisfies the definition by being both necessary and sufficient for death. The criterion is the hierarchical descriptive level that is most suitable for inclusion in a death statute. The final step is the medical task of identifying the tests for death, that set of bedside operations and procedures that show that the criterion has been fulfilled by having no false-positive determinations and minimal false-negative determinations. The specific medical tests evolve over time in response to technologic developments and studies. A perfect analytic system would show a tight correspondence between the definition, criterion, and tests of death. Later, I discuss an imperfection in the relationship between the criterion and tests of death.

3 The Definition of Death

The conceptual foundation for accepting "brain death" as human death rests on the definition and criterion of death. My colleagues and I defined death as the permanent cessation of the organism-as-a-whole. The organism as a whole is a theoretical biology concept developed over a century ago by the biologist Jacques Loeb [9].[1] The organism-as-a-whole is not the whole organism (the sum of its parts), but rather it refers to the essential characteristics of the organism that are greater than the sum of its parts, namely how the intrinsic interrelationship of its parts creates new functions comprising the coherent unity of the human organism that maintain its life and health. Understanding the concept of the organism-as-a-whole requires a brief

[1] Loeb showed how all of an organism's biological processes resulted from their component physical and chemical activities without the need for extra-physical "vitalist" input—a popular prevailing explanation at the time. However, Loeb offered no criteria for delineating the organism as a whole or its essential features.

explanation of three fundamental theoretical biology concepts: emergent functions, hierarchies of organization, and biological mereology. I have explained these concepts elsewhere in greater detail [10].

Higher-level organisms feature hierarchies of organization in which its functional levels are nested in a defined order that generates increasingly complex functions. Nesting is organized such that each level demonstrates a mechanistic property of unity that is not possessed by lower levels and such that wholes at lower levels function as parts at upper levels. While interlevel functional relationships are bidirectional, they are composed in a bottom-up direction, but controlled in a top-down direction. The apex organizational level is the culmination of all lower levels and produces the functions of the organism-as-a-whole. The organism's stratification is functionally integrated such that each stratum contributes to the organism-as-a-whole.

The unique functions produced at each level are called emergent functions because they emerge spontaneously from the ensemble of its parts operating in their natural way. Emergent functions are holistic: they are collective functions of a whole that cannot be localized or reduced to any of its individual parts. They spontaneously self-organize in a way that requires no external controlling agent, but requires only the natural interactions of their component parts. Emergent functions can be seen in many complex physical, chemical, and biological systems. Usually, because the complexity of the systems in which they occur renders the systems incompletely understood, emergent functions cannot be predicted or even comprehended merely by studying their component parts. The emergent functions arising at each level of the organism culminate in the organism-as-a-whole.

Biological mereology is the study of the relationship of parts to a whole or among its parts. Like emergent functions, mereology can be studied in non-biological systems and comprises an evolving branch of mathematics and philosophy. Two biological mereological concepts are relevant here: the distinction between the ontological status of the parts of an organism and its whole, and that the organism's parts serve the whole as its final end and benefactor. The organism's parts cannot survive apart from the whole without external technological support, but the organism may survive the loss of some of its nonessential parts. Thus, many subsystems of a brain-dead patient (e.g., visceral organs) remain alive through the technological support of oxygenation and perfusion despite the death of the organism by the cessation of the organism-as-a-whole.

Scholars have enumerated the characteristics of the organism-as-a-whole. Bonelli and colleagues [11] delineated criteria for life forms: a unity characterized by: (1) dynamics (signs of life)—e.g., metabolism, regeneration, growth, and propagation; (2) integration—the requirement that the life process derives from the mutual interaction of its component parts; (3) coordination—the requirement that the interaction of the component parts is maintained within an order; and (4) immanency—the requirement that these characteristics originate from and are intrinsic to the life form. A life form becomes an organism when it has: (1) completion—an organism is not a component part of another living entity but is itself an intrinsically independent and completed whole; (2) indivisibility—intrinsic unity requires that no organism can be divided into more than one living organism; but if it occurred and the organism survived, the completed organism resides within one of the divided parts; (3) self-reference—the observable life processes of the component parts serve the self-preservation of the whole, even at the expense of the survival of its parts, because the health and survival of the living whole is the primary end; and (4) identity—despite incremental changes in form and the loss or gain of certain component parts (that even could eventually result in the exchange of all its atoms), the living being remains one and the same throughout life.

My colleague, Andrew Huang, and I supplemented this analysis [12] by observing that the organism-as-a-whole has a common ontogeny shared by all organisms. The organism is an anti-entropic entity with processes that promote increasing biological complexity, achieving an integrated wholeness through emergent properties. The characteristics of the organism-as-a-whole vary with the organism's complexity. Species variation in the organism-as-a-whole has an ontological significance. We further distinguished between concept and conception: the former describes the overall principles of a species-wide organism-as-a-whole, whereas the latter describes more precisely the species-specific particular characteristics of the organism-as-a-whole. The species-wide concept of the organism-as-a-whole is manifest by different conceptions that vary as a function of the organism's complexity.

Bonelli and colleagues [11] explained why the cessation of the organism-as-a-whole represents death and, consequently, why brain death is human death. With permanent cessation of brain functions, the organism has lost: (1) immanency because its life processes no longer arise from itself but result from external intensive care support; (2) self-reference because whatever control over the component organ subsystem parts that remains now is directed at the level of the surviving parts and no longer supports the whole; and (3) completeness because its separate component parts no longer relate to each other or constitute a whole. An organism that has permanently lost immanency, totality, completeness, self-reference, and identity no longer functions as a whole and therefore is dead.

Adam Omelianchuk recently proposed a similar concept of the organism-as-a-whole: "An organism as a whole is an enduring, self-directed, and self-moving entity in which its parts derive their identity and function from its internal structure, and it develops by virtue of its own capacities and powers latent in itself according to an information-rich design plan intrinsic to its kind towards a distinctive end or goal." He concluded that "*Death is the end of the organism as a self-moving whole*, meaning there is no more entity that has this active internal structure by which its parts receive their identity and function, no more development or activity towards a kind-distinctive end by virtue of its own capacities and powers." [13].

Notwithstanding these sophisticated analyses, the concept of the organism-as-a-whole remains frustratingly vague and in need of further explication. Yet, despite this shortcoming, the permanent cessation of the organism-as-a-whole has been accepted as the conceptual foundation for death by both the President's Commission for the Study of Ethical Problems in Medicine and Biomedical and Behavioral Research in their work *Defining Death* in 1981 [14] and by the President's Council on Bioethics in their work *Controversies in the Determination of Death* in 2008 [15]. To account for this acceptance, I perceive the presence of a strong intuitive appeal to the cessation of the organism-as-a-whole concept as the foundation of brain death even though the concept remains imprecise [10].[2] Intuition also accounts for the ironic divergence of two opposing trends: the increasing international acceptance of brain death and the persistent misunderstanding of the meaning of brain death by professionals and the lay public that has been recorded in repeated surveys over many years [16]. Although professionals and the lay public may not always understand exactly what brain death means and precisely why it should be considered death, that limitation has not prevented them from accepting it as valid.

4 The Criterion of Death

The criterion of death that shows the permanent cessation of the organism-as-a-whole is the permanent cessation of brain functions. The brain is responsible for the unity and wholeness of the human organism without which there is no organism-as-a-whole. The brain manages our interaction with our environment and operates the control and integrative functions maintaining the unity and health of the human organism. It executes the elegant emergent functions of the organism-as-a-whole, most remarkably, the exquisite yet mysterious capacity for conscious awareness. Obviously, much of the brain-dead human organism remains alive as a consequence of technologic support, but the organism is dead because the organism-as-a-whole has ceased.

Alan Shewmon has shown that the brain's somatic integration capacity, alone, is inadequate to justify its exclusive role as generating the organism-as-a-whole, because the spinal cord and other body systems also perform somatic integration [17]. The President's Council on Bioethics accepted Shewmon's critique but countered that the organism-as-a-whole concept was distinct from the function of somatic integration and did not depend on it.[3] The President's Council argued that the organism-as-a-whole concept turned on the "fundamental vital work of a living organism—the work of self-preservation, achieved through the organism's need-driven commerce with the surrounding world." They stated that total brain failure was a valid criterion of death "… because it is a sign that this organism can no longer engage in the essential work that defines living things" and "is not contingent on the loss of the organism's capacity for somatic integration" [15]. Adam Omelianchuk offered a complementary justification of the brain criterion of death as the end of the organism as a self-moving whole [13]. Andrew Huang and I supplemented these defenses of the brain criterion by adding that the brain is *sui generis* among organs: the conscious awareness it generates can neither be conceivably replaced by a machine nor can it be transplanted successfully [10].

[2] See [10] for a discussion of the different levels of intuition present in people accepting brain death.

[3] A principal motivation for the President's Council report was to address Shewmon's critique of the integration rationale for brain death.

Melissa Moschella criticized Shewmon's bodily integration critique by arguing that an organism is not simply integrated but must be self-integrated. She stated that a true organism "possesses the root capacity for self-integration as evidenced by: (1) possession of the material basis for the capacity for self-integration (the capacity for control of respiration and circulation); or (2) possession of the material basis of the capacity for sentience." She argued that the reason why the brain is the *only* valid criterion for death is that *only* after total brain death can we be certain that a human being has irreversibly lost the material basis of the capacity for self-integration as evidence by *both* (1) the material basis of the capacity to breathe—i.e., the capacity to control the essential vital capacities of circulation and respiration *and* (2) the material basis of the capacity for sentience [18, 19]. Maureen Condic also criticized a simplistic concept of integration and showed that true self-integration was a higher-level function than mere coordination, and the brain required self-integration [20].

The other plausible candidate for a criterion of death that fulfills the definition by being both necessary and sufficient for death is the permanent cessation of circulation. This criterion obviously is sufficient to fulfill the definition because it leads to complete and permanent cessation of brain function. However, I argue that it is unnecessary and that only permanent cessation of brain function is necessary for death. The continued technological support of respiration and, consequently, circulation may be necessary for the support of the whole organism, but is irrelevant to the functioning of the organism-as-a-whole other than by permitting brain function.

An ongoing controversy centers on the anatomic extent of the permanent cessation of brain function that is necessary and sufficient for the cessation of the organism-as-a-whole. Historically, three distinct anatomic–physiologic brain criteria have been proposed over the past half-century, known as the whole-brain, brainstem, and higher-brain formulations [21]. Countries are divided between those that accept the whole-brain and the brainstem formulations, but none favors the higher-brain. Yet, the higher-brain formulation remains popular in some academic circles.

The whole-brain formulation was supported by the President's Commission in 1981 and comprises the basis for the model Uniform Determination of Death Act, from which the majority of American states drafted their death statutes. Its brain criterion stipulates that death requires the "irreversible cessation of all functions of the entire brain including the brainstem" [14]. Many other countries also adopted the whole-brain criterion. The President's Commission cited the original article by my colleagues and me [4] in their justification of the whole-brain criterion. They argued that the whole-brain criterion was chosen because the organism-as-a-whole is distributed throughout the brain and does not reside in any specific part. The whole-brain criterion also serves as a fail-safe mechanism to assure there can be no consciousness, as I have explained elsewhere [21].

The United Kingdom chose the brainstem criterion of death which, through their influence, also has been adopted in several other countries. Christopher Pallis, the principal proponent of the brainstem criterion, has explained convincingly why the

United Kingdom chose it. He pointed out that consciousness requires the brainstem ascending reticular formation, that all motor output and all sensory input (other than vision and olfaction) traverses the brainstem. Further, it is the center for the control of respiration and circulatory tone. Pallis also noted that most of the bedside tests neurologists perform to assess brain death evaluate brainstem function. For these reasons, Pallis referred to the brainstem in this context as "the brain as a whole" [22]. In this volume, Andrew McGee and Dale Gardiner defend the brainstem criterion of death.

The higher-brain formulation (dubbed "neocortical death") was proposed initially by Robert Veatch [23] and later expanded by Karen Gervais [24] and others. It posits that death is best defined as the permanent loss of that which is essential to the nature of man. Supporters of the higher brain formulation argued that only the cerebral hemispheres should count for the life status of the human organism because they control conscious awareness and all higher human behaviors. This formulation captures an important element of the brain death rationale, the essential emergent function of human consciousness, but is a radical redefinition of death. Its strict application would declare dead patients in vegetative states with permanent unconsciousness who breathe spontaneously, and who are considered alive everywhere in the world. For this reason, despite its persisting popularity in some academic circles, it has not been embraced by a single jurisdiction or medical society anywhere in the world.

5 The Tests of Death

The tests for death have been designed by physicians to show that the brain criterion of death has been fulfilled. Medical societies, such as the American Academy of Neurology [25], publish evidence-based, expert-drafted guidelines for testing that are updated periodically to incorporate new outcome studies and developments in technology. The recommended tests should be validated by demonstrating that they correlate with the whole-brain criterion with very high positive and negative predictive values. The tests are fundamentally conducted by bedside physical examination but, when necessary, may be supplemented by ancillary electrodiagnostic and imaging tests whose indications vary internationally [1]. These tests and their controversies are discussed elsewhere in this book.

One essential question that arises in choosing the tests for death and the precise language of a death criterion centers on whether the vital function in question must cease permanently or irreversibly. Irreversible cessation of a vital function means that, once it ceases, it *cannot* be restored with available technology because doing so is impossible (practically, not theoretically). In contrast, permanent cessation of a vital function means that, once it ceases, it *will not* be restored because it will neither restart itself spontaneously ("auto-resuscitation") nor will physicians attempt to restart it with resuscitative interventions. I have further analyzed the relationship between "permanent" and "irreversible" cessation of functions elsewhere [26].

The traditional medical practice of death determination using the circulatory–respiratory criterion requires only the permanent cessation standard. When a physician declares death after cessation of heartbeat, circulation, and respiration in a patient who has a Do-Not-Resuscitate order, they require only that the cessation of those vital functions is permanent. Given the presence of this order, no attempt will be made to resuscitate, and physicians are not required to prove that the cessation of vital functions is irreversible because permanent cessation is sufficient. Other writers concur with my permanent-irreversible distinction and my preference for permanent, rather than irreversible, cessation in death determination using the circulatory–respiratory criterion [27–29]. Similarly, a recent scoping review of controlled organ donation after the determination of death using the circulatory–respiratory criterion concluded that the permanent cessation standard for organ donor death determination has been widely accepted and has accrued "emerging agreement in the donation and transplant community" [30].

In contrast to determination of death using the circulatory–respiratory criterion, brain death determination, from its beginning, has required demonstrating the irreversible cessation of brain function. This irreversibility requirement, however, is an artifact of testing practices because testing for brain death is a retrospective determination. The accepted tests demonstrate that brain functions had ceased earlier as a result of an illness or injury. However, determining death using the circulatory–respiratory criterion typically is prospective. A physician is called to the bedside after a patient's heartbeat and respiration have ceased and determines death in real time by the persistent absence of circulation and respiration. A permanent cessation standard is appropriate for a prospective death determination, whereas an irreversible cessation standard is appropriate for a retrospective death determination. Nevertheless, there is no *prima facie* reason why a permanence standard could not be used for both circulatory–respiratory and brain death determinations when they are performed prospectively. This question and its implications are further analyzed in this volume by Dale Gardiner and Andrew McGee.

Both permanent and irreversible cessation of vital functions are contingent states: permanence is contingent on the decision whether to resuscitate; irreversibility is contingent on the availability of and the choice to use resuscitative technology. Because all plausible biological concepts of death require it to be an irreversible state, a noncongruence exists between the biological concept of death requiring the irreversible cessation of circulatory–respiratory or brain functions and the medical determination of death which requires only their permanent cessation. Elsewhere, I have discussed the historical and societal reasons for and the implications of this confounding circumstance [31]. In recognition of this noncongruence, Alan Shewmon proposed that death is best conceptualized when it is "semantically bisected" into two sequential events: (1) the sociolegal "passing away" or being "deceased" that occurs at the permanent cessation of vital functions; and (2) the ontological–theological "ceasing of the bodily organism" that occurs later at the irreversible cessation of vital functions [8].

6 The Criterion–Test Mismatch and the Brain-as-a-Whole Criterion

An important current controversy centers on the troubling instances of mismatch between the prevailing whole-brain criterion of death and the accepted tests physicians perform to show that the criterion has been fulfilled. Although these instances of mismatch have been reported for decades, concern over them now is increasing. My colleague, Anne Dalle Ave, and I recently reviewed published cases in which physicians noted persisting signs of brain functions on examination after brain death determination [32]. While many, or even most, of these cases undoubtedly resulted from an erroneous preceding brain death determination and thus did not represent valid cases of criterion–test mismatch, one circumstance was reported so frequently that we concluded that it represented a true mismatch.

These frequent mismatch cases were those in which diabetes insipidus was absent, although usually it is present in brain death. The absence of diabetes insipidus suggests the persistence of hypothalamic neurosecretion to the posterior pituitary gland of the precursor molecule of antidiuretic hormone [33].[4] Although these cases have been described in brain death for many years, most neurologists (myself included) either ignored or finessed them with rationalizations such as that the persistence of hypothalamic neurosecretion merely represented a neuronal cellular activity (permissible) and not a brain function (not permissible). However, because this claim becomes harder to defend under more rigorous biological analysis [34], I now accept that hypothalamic neurosecretion counts as a brain function. This topic is discussed in depth in this volume by Alex Manara and Michael Nair-Collins. An alternative solution that sidesteps the conceptual argument has been for medical specialty societies such as the American Academy of Neurology to assert, as a matter of best medical practice, that continued hypothalamic neurosecretory function remains compatible with whole-brain death determination [35].

Anne Dalle Ave and I outlined plausible solutions to reduce the incidence of the criterion–test mismatch and we considered their anticipated benefits and liabilities [36]. The whole-brain criterion could be relaxed to permit the persistence of some brain functions, the tests could be tightened by requiring a neuroimaging demonstration of intracranial circulatory arrest, or both solutions could be implemented. Relaxing the whole-brain criterion to permit certain continued brain functions could be accomplished by accepting the brainstem criterion of death or modifying the whole-brain criterion to the brain-as-a-whole criterion.

[4] The mechanism accounting for the preservation of neurosecretory hypothalamic function may be the separate blood supply to this region, which may remain unaffected by conditions that impair brain perfusion through the principal intracranial arteries [34].

The brain-as-a-whole criterion relates to the whole-brain criterion in a fashion analogous to how the organism-as-a-whole relates to the whole organism. In each case, the "as-a-whole" element attempts to distill the essential features (primarily emergent functions) defining the entity. Thus, the brain-as-a-whole concept denotes that some elements of brain function are more essential to the organism than others, and then attempts to identify them. The brain-as-a-whole criterion lessens the criterion–test mismatch by not requiring the cessation of neurosecretory hypothalamic function that otherwise would be required by the whole-brain criterion. Additionally, it more accurately depicts the varied and often inchoate conceptualizations of the brain criterion held by many physicians as recorded in surveys [37], thus it brings the tests for brain death into greater alignment with its criterion. However, it raises the thorny conceptual question of which brain functions are essential elements of the brain-as-a-whole and why. Accepting it also would require amending death statutes to replace the whole-brain criterion.

The brain-as-a-whole criterion is located on the conceptual line between the whole-brain and brainstem criteria. It requires the brainstem criterion and encompasses much, but not all, of the whole-brain criterion. An important virtue of requiring the absence of brainstem functioning is that it provides a fail-safe protective mechanism, ensuring that any patient who satisfies it cannot possibly retain consciousness,[5] a concept Douglas Gelb called the "shielded-brain formulation" [38]. The bedside test requirements to satisfy the brain-as-a-whole criterion must therefore include the absence of measurable brainstem functions as shown by the loss of the capacity for consciousness, apnea with the loss of the capacity to breathe, the loss of systemic circulatory tone, and the absence of reflexes integrated in the brainstem via the cranial nerves. Omelianchuk and colleagues recently called this concept the neurorespiratory criterion [39]. Elsewhere I have summarized my other recent thoughts on this important but elusive concept [10].

7 Society's Role in Death Determination

Death determination medical practices must be grounded in biology, but also must be acceptable to society. Usually, biological and societal requirements in medicine are compatible, but sometimes precise biological conditions may be modified to serve societal ends. Earlier, I described the noncongruence between the biological requisite of irreversibility for death and the societally permitted medical standards for death determination that allow physicians to declare it earlier at the permanent cessation of vital functions. Societal factors permit physicians to declare death before true irreversibility occurs because little benefit is gained by requiring irreversibility. Moreover, requiring irreversibility produces much harm by delaying

[5] The mechanism by which the absence of brainstem function assures the complete absence of consciousness is the usually irreversible process of lateral displacement and transtentorial herniation of the brainstem that accompanies markedly raised intracranial pressure caused by an expanding supratentorial mass lesion.

death determination for hours, awaiting the moment of demonstrable irreversibility or by requiring the patient to be subjected to invasive maneuvers to prove that cessation is irreversible before physicians can declare death.

Society is better served by permitting physicians to declare death at an earlier moment, once permanent cessation shows it is inevitable, thereby preventing family members from pointless suffering resulting from needless waiting and unnecessary intrusive testing of the patient. This is the moment Shewmon called "passing away" or "deceased" [8]. The biological–societal noncongruence over the timing of death determination has not provoked objections from family members or others but has generated a controversy over whether organ donors after circulatory–respiratory determination of death are truly dead at the moment they are declared dead immediately prior to donation [40, 41]. The claim made by some critics is that prospective organ donors are not dead at the moment they are declared dead because their cessation of circulatory and respiratory functions is not yet irreversible. The justification that the prospective donors are dead at the moment they are declared is that prevailing medical death determination standards require only that donor circulatory and respiratory functions have ceased permanently.

Some scholars have argued that the entire edifice of using the brain criterion of death is merely a societal construct that allows physicians unilaterally to withdraw life-sustaining therapy in hopeless cases of profound brain damage and to facilitate organ donation. These scholars have applied the concept of a "legal fiction," earlier proposed in this context by Robert Taylor [42], to explain why society permits these patients to be declared dead. In this volume, Seema Shah defends the legal fiction concept in detail. The related question of to what extent society should permit each citizen the liberty of personal choice in their criterion of death, originally proposed by Robert Veatch [43, 44], is explained in this volume by Lainie Ross and Christos Lazaridis.

8 Summary

A refined definition and criterion of death became necessary during the mid-twentieth century because the development of cardiopulmonary resuscitation and tracheal positive-pressure ventilation ended the unitary determination of death as the joint cessation of all vital functions. The brain criterion of death was thereafter proposed and accepted quickly by physicians and supported by state laws. The conceptual basis for the brain criterion of death is the permanent cessation of the human organism-as-a-whole: those unique features greater than the sum of the organism's parts, devolving from the intrinsic interrelationship of its parts that create the coherent unity of the human organism. The organism-as-a-whole concept relies on the principles of organizational hierarchies, emergent functions, biological mereology, and self-integration as applied to brain functions. It argues that the brain is responsible for the organism-as-a-whole and acknowledges that many parts of the human organism can remain alive by technological support after the organism-as-a-whole has ceased. This rationale has been endorsed by two prominent US ethics commissions separated by a quarter-century that defined human death and has been cited by numerous scholars. However, this endorsement should not mask the inherent vagueness of the rationale, its reliance on intuition, and the need for its more detailed analysis.

The valid cases of mismatch between the whole-brain criterion of death and the test battery that is generally accepted in the United States could be reduced by modifying the whole-brain criterion of death to become the brain-as-a-whole criterion. Because this is the way many physicians already conceptualize and determine brain death, this change would better align the criterion of death with contemporary medical practice. Death statutes based on the whole-brain criterion would require revision to incorporate this refinement. Yet, the brain-as-a-whole criterion remains in an early conceptual stage and needs further explication with reasoned accounts of which brain functions should be included in the brain-as-a-whole and why.

A peculiar state of noncongruence exists between the biological concept of death as irreversible by definition and the societally accepted medical practice of determining death at the time when the vital functions have ceased permanently, but not yet irreversibly. Although this noncongruence has not been questioned by society, it has triggered disputes over the validity of death determination in organ donors after the circulatory–respiratory determination of death. The reality of the state of death determination noncongruence needs to be more widely recognized. Public laws should formalize the prevailing medical standard of death determination as the permanent cessation of vital functions, irrespective of organ donation candidacy.

References

1. Greer DM, Shemie SD, Lewis A, et al. Determination of brain death/death by neurological criteria: the World Brain Death Project. JAMA. 2020;324:1078–97.
2. Bernat JL. Controversies in defining and determining death in critical care. Nat Rev Neurol. 2013;9:164–73.
3. Lewis A, Greer D. Current controversies in brain death determination. Nat Rev Neurol. 2017;13:505–9.
4. Bernat JL, Culver CM, Gert B. On the definition and criterion of death. Ann Intern Med. 1981;94:389–94.
5. Powner DJ, Ackerman BM, Grenvik A. Medical diagnosis of death in adults: historical contributions to current controversies. Lancet. 1996;348:1219–23.
6. Ad Hoc Committee. A definition of irreversible coma: report of the Ad Hoc Committee of the Harvard Medical School to Examine the Definition of Brain Death. JAMA. 1968;205:337–40.
7. Capron AM, Kass LR. A statutory definition of the standards for determining human death: an appraisal and a proposal. Univ Penn Law Rev. 1972;121:87–118.
8. Shewmon DA. Constructing the death elephant: a synthetic paradigm shift for the definition, criteria, and tests for death. J Med Philos. 2010;35:256–98.
9. Loeb J. The Organism-as-a-whole. New York: G. P. Putnam's Sons; 1916.
10. Bernat JL. Refinements in the organism-as-a-whole rationale for brain death. Linacre Q. 2019;86(4):347–58.
11. Bonelli RM, Prat EH, Bonelli J. Philosophical considerations on brain death and the concept of the organism-as-a-whole. Psychiatr Danub. 2009;21:3–8.

12. Huang AP, Bernat JL. The organism as a whole in an analysis of death. J Med Philos. 2019;44:712–31.
13. Omelianchuk A. Brain death as the end of a human organism as a self-moving whole. J Med Philos. 2021;46:530–60.
14. President's Commission for the Study of Ethical Problems in Medicine and Biomedical and Behavioral Research. Defining Death. Medical, Ethical, and Legal Issues in the Determination of Death. Washington, DC: US Government Printing Office; 1981.
15. President's Council on Bioethics. Controversies in the determination of death: a white paper by the President's Council on Bioethics. Washington, DC; 2008.
16. Bernat JL. Whither brain death? Am J Bioeth. 2014;14(8):3–8.
17. Shewmon DA. The brain and somatic integration: insights into the standard biological rationale for equating 'brain death' with death. J Med Philos. 2001;26:457–78.
18. Moschella M. Deconstructing the brain disconnection—brain death analogy and clarifying the rationale for the neurological criterion of death. J Med Philos. 2016;41(3):279–99.
19. Moschella M. Integrated but not whole? Applying an ontological account of human organismal unity to the brain death debate. Bioethics. 2016;30(6):550–6.
20. Condic ML. Determination of death: a scientific perspective on biological integration. J Med Philos. 2016;41(3):257–78.
21. Bernat JL. How much of the brain must die in brain death? J Clin Ethics. 1992;3:21–6.
22. Pallis C, Harley DH. The ABC of brainstem death. 2nd ed. London: British Medical Journal Publishers; 1996.
23. Veatch RM. The whole brain-oriented concept of death: an outmoded philosophical formulation. J Thanatol. 1975;3:13–30.
24. Gervais K. Redefining death. New Haven: Yale University Press; 1987.
25. Wijdicks EFM, Varelas PN, Gronseth GS, Greer DM. Evidence-based guideline update: determining brain death in adults. Neurology. 2010;74:1911–8.
26. Bernat JL. How the distinction between "irreversible" and "permanent" illuminates circulatory-respiratory death determination. J Med Philos. 2010;35(3):242–55.
27. McGee A, Gardiner D. Permanence can be defended. Bioethics. 2017;31:220–30.
28. Parent B, Turi A. Death's troubled relationship with the law. AMA J Ethics. 2020;22(12):E1055–61.
29. Manara AR, Shemie SD, Large S, et al. Maintaining the permanence principle for death during in situ normothermic regional perfusion for donation after circulatory death organ recovery: a United Kingdom and Canadian proposal. Am J Transplant. 2020;20:2017–25.
30. Murphy N, Weijer C, Smith M, Chandler J, Chambrlain E, Gofton T, et al. Controlled donation after circulatory determination of death: a scoping review of ethical issues, key concepts, and arguments. J Law Med Ethics. 2021;49:418–40.
31. Bernat JL. On noncongruence between the concept and determination of death. Hast Cent Rep. 2013;43(6):25–33.
32. Dalle Ave AL, Bernat JL. Inconsistencies between the criterion and tests for brain death. J Intensive Care Med. 2020;35:772–80.
33. Nair-Collins M, Northrup J, Olcese J. Hypothalamic-pituitary function in brain death: a review. J Intensive Care Med. 2016;31:41–50.
34. Nair-Collins M, Joffe AR. Frequent preservation of neurologic function in brain death and brainstem death entails false-positive misdiagnosis and cerebral perfusion. Am J Bioethics Neurosci. 2021; https://doi.org/10.1080/21507740.2021.1973148.
35. Russell JA, Epstein LG, Greer DM, Kirschen M, Rubin MA, Lewis A. Brain death, the determination of brain death, and members guidance for brain death accommodation requests. Neurology. 2019;92(5):228–32.
36. Bernat JL, Dalle Ave AL. Aligning the criterion and tests for brain death. Cambr Q Healthc Ethics. 2019;28(4):635–41.
37. Joffe AR, Anton NR, Duff JP, deCaen A. A survey of American neurologists about brain death: understanding the conceptual basis and diagnostic tests for brain death. Ann Intensive Care. 2012;2(1):4.

38. Gelb DJ. Building a fence around brain death: the shielded-brain formulation. Neurology. 2021;97:780–4.
39. Omelianchuk A, Bernat J, Caplan A, Greer D, Lazaridis C, Lewis A, et al. Revise the UDDA to align the law with practice through neurorespiratory criteria. Neurology. 2022;98:532–6.
40. Bernat JL. Point: are donors after circulatory death really dead, and does it matter? Yes and yes. Chest. 2010;138:13–6, 18–9.
41. Truog RD. Counterpoint: are donors after circulatory death really dead, and does it matter? No and not really. Chest. 2010;138:16–8.
42. Taylor RM. Reexamining the definition and criterion of death. Semin Neurol. 1997;17:265–70.
43. Veatch RM. The conscience clause: how much individual choice in defining death can our society tolerate? In: Youngner SJ, Arnold RM, Schapiro R, editors. The definition of death: contemporary controversies. Baltimore: John Hopkins University Press; 1999. p. 137–60.
44. Veatch RM, Ross LF. Defining death: the case for choice. Washington, DC: Georgetown University Press; 2016.

Arguments Rejecting Neurologic Criteria to Determine Death

D. Alan Shewmon

1 Introduction

In 1981, Bernat, Culver, and Gert introduced a useful tripartite schema regarding the diagnosis of death [1]. The *concept* is the essence of death, a dictionary definition. The *criterion* is an anatomical specification guaranteeing applicability of the concept (e.g., the Uniform Determination of Death Act [2]). The *tests* are diagnostic protocols to ensure that the criterion is fulfilled in a particular case (e.g., the updated adult and pediatric standards [3, 4]). One cannot develop a criterion for an unspecified concept, nor is there any point in validating tests for a conceptually incoherent criterion. This chapter will examine the correspondence between concepts and criteria of death proposed over the half-century history of "brain death."[1] Tests are discussed in other chapters.

[1] Although the term "death by neurologic criteria" is used throughout this book, it would be oxymoronic to suggest that "death by neurologic criteria" might not be death. Therefore, I shall stick with the historical term "brain death," meaning the condition diagnosed by the updated adult and pediatric standards. The quiet campaign to replace "brain death" or "total brain failure" [5] with "death by neurologic criteria" is an example of semantic engineering, intended to reinforce the quasi-official position that the neurologic diagnosis of death is a settled issue and that scholarly critiques of it do not exist [6–8]. For an extensive bibliography of such critiques, see the online Appendix A of [9].

D. A. Shewmon (✉)
David Geffen School of Medicine at UCLA, Los Angeles, CA, USA
e-mail: ashewmon@mednet.ucla.edu

A. Lewis, J. L. Bernat (eds.), *Death Determination by Neurologic Criteria*, Advances in Neuroethics, https://doi.org/10.1007/978-3-031-15947-3_3

2 Non-ontological Concepts

Denial that death is a real occurrence is usually based on the claim that its conceptual border with life is indistinct, so there is no nonarbitrary point at which "death" can meaningfully be said to happen. The blurry transition zone can be either semantic or physical.

2.1 Semantic Fuzziness: Death as a Cluster Kind

Botkin and Post [10] suggested that

> [l]ife and death may be defined in terms of a cluster of attributes, … death is recognized not through a specific attribute but rather through a group of attributes that may be present or absent in various combinations. (p. 134)

Winston Chiong concurs, arguing that death "eludes definition," because it is a "fuzzy concept" [11, 12]. Chiong builds his argument around two contrasting examples of determinate cases of life: (1) the first few seconds of irreversible circulatory–respiratory arrest before unconsciousness, and (2) "persistent vegetative state" (now preferably called "unresponsive wakefulness syndrome" [13]) with spontaneous breathing. They supposedly illustrate how there is no set of properties both necessary and sufficient for life, and therefore how life is better conceived as a "cluster kind," like Wittgenstein's example of "game," for which no definition can be given that covers all instances of what are called "games" and excludes all non-"games."

I beg to differ. If we follow Wittgenstein's advice (which Chiong favorably quotes)—"*look and see* whether there is anything common to all" [11 (p. 24)] (emphasis in original) – I believe we can look at the world and see, together with the ancients, that some things "move themselves," while all other things are moved only by something else ("movement" in the broad sense of any operation, not just locomotion) [14, 15 (pp. 14–38), 16, 17]. Chiong's determinate cases both exemplify "self-movement," so one need not appeal to a cluster of properties to classify them both as alive.

2.2 Physical Fuzziness: Death as a Process

From the beginning of brain-death history, scholars have debated whether death is a process or an event. In 1968, the World Medical Association asserted that "death is a gradual process at the cellular level" [18], and this concept has been championed up to today [19–30].

Apart from purely linguistic reasons why "death" represents an event and "dying" a process [31], developments over the past century in physics, nonlinear dynamics, and systems biology provide firm theoretical grounds for, and many examples of, dynamical systems with state discontinuity brought about by continuous changes in underlying parameters [32–36].

The physical laws governing relationships among atoms do not exhaustively characterize the fullness of biological reality. There are macroscopic, holistic aspects – global states – which are just as real, are not intelligibly reducible to the elemental level, and exert "top-down" control over the lower levels. These global states are properties of a whole, not of a collection; they are qualitatively distinct and change discontinuously from one into the other. There is no conceptual continuum between unity and multiplicity, or between resistance to entropy and yielding to entropy (*vide infra*).[2]

2.2.1 Criteria

Non-ontological concepts of death correspond to *stipulative* criteria, because the whole purpose of a criterion is to draw a "bright line" between living members of society and dead ones. If the concept is a cluster or a continuum, that line is necessarily arbitrary, and the criterion is essentially a legal status. Stipulative criteria do not claim to identify when death *actually* occurs; they are social constructs based on purely pragmatic considerations.

3 Ontological Concepts

Ontological concepts of death sort into two broad categories corresponding to the components of the perennial mind–body problem. The *psychological* concept identifies the subject of human life and death (the "person") with Descartes' *res cogitans* ("the thinking thing," the mind); human death is the extinction of thinking and self-awareness. It applies uniquely to humans and is compatible with the idea of a surviving nonpersonal human organism. In contrast, the *biological* concept identifies the subject of human life and death (the "person") with the organism (including its cognitive aspects). It is rooted in Aristotle's notion of "substance" and is generally understood as the extinction of the "organism as a whole." It applies to humans in the same way as to all living creatures and is compatible with the idea of a permanently unconscious person.

Although "irreversible apneic unconsciousness" is sometimes presented as a *concept* of death associated with the criterion of irreversible brainstem nonfunction, it is better understood as a stipulative *criterion* for an unspecified concept. The brainstem criterion, championed in the United Kingdom, does not properly correspond, as claimed, with "irreversible apneic unconsciousness" and is insufficient for both psychological and biological concepts (*vide infra*).

[2] Our ability to pinpoint the moment of discontinuity is an epistemic question, which has nothing to do with the fact that, conceptually, death *must* be an event. To deny this amounts to denying that wholes larger than subatomic particles exist and that life is a distinct phenomenon.

3.1 Psychological Concept

The psychological concept of death has had strong advocates all along, being implicitly or explicitly held by around 50% of neurologists, neurosurgeons, and intensivists [37–39]. (An extensive bibliography is provided in the online Appendix B of [9].) As succinctly articulated by Baker and Shemie [40]:

> The capacity for consciousness and self-awareness is uniquely synonymous with human life and personhood, and its absence is necessary and sufficient to identify that death has occurred. (p. 688)

Accordingly, the Harvard Committee's definition of death as "irreversible coma" [41] was an oxymoron, since "coma" is predicable only of live patients.

The Cartesian equating of person with conscious mind is a theme with many contemporary variations, which are of great interest to philosophers, but their distinction is unimportant for brain-death theory. Whether the *living* person is understood as embodied subjective consciousness, or the potential for consciousness [42, 43 (p. 108), 44], or the radical capacity for consciousness [45, 46], or a bundle of cognitive abilities [47], the common denominator for *death* of a person, so understood, is the irreversible loss of consciousness. Whether death is conceived as a loss of personhood, or of personal identity [48], or of "that which is essentially significant to the nature of man" [49 (p. 15)], they are all lost if consciousness is irreversibly lost.

3.1.1 Critique

The main problem with the idea that I *am* my mind – or my mental activities, or my potential for mental activities – is that it does not reflect the universal human experience. We do not experience ourself as a mind connected to a body, but as intrinsically hybrid: mental and corporeal. We say "my thumb," "I have a cold," or "I am dying" – not "my body's thumb," "my body has a cold," or "my body is dying." And especially we say "my mind," implying that my mind is a *component* of "me"; it is not "me." After recovery from coma, I could say "I was in a coma for three days." Such statements are not mere figures of speech but universal expressions of people's self-understanding as corporeal beings. To insist that my body is not an integral constituent of "me" would simply be an artificial forcing of words and concepts to fit some preconceived ontological scheme, not the formulation of an ontological scheme to reflect deeply experienced truths about human nature.

If I am in an irreversible coma from multisystem failure due to end-stage cancer, I am still alive as long as my body is alive. I did not die the moment I lost consciousness. If I suffer a terminal circulatory–respiratory arrest intraoperatively, I die when the possibility of resuscitation is lost, not when the anesthesiologist put me to sleep at the beginning of the operation.

3.1.2 Criterion

The criterion corresponding to the psychological concept is straightforward in principle: partial brain destruction sufficient to eliminate the potential for consciousness. Consciousness is mediated by a diffuse neuronal network involving the neocortex, the thalamus, the white matter tracts connecting them, and the central core of the upper brainstem [50 (p. 31)]. With acquired thalamic and/or cortical lesions, the initial coma typically evolves to unresponsive wakefulness syndrome [13, 50 (pp. 379–436), 51–53]. Congenital absence of cortex, however, does not necessarily preclude consciousness [54, 55]; therefore, we should remain agnostic regarding the necessity of cortex for consciousness even in adults [56, 57].[3]

The criterion associated with the psychological concept has traditionally been called the "higher brain" criterion, but "higher" should be understood functionally rather than anatomically. (Cognitive functions are considered "higher," in the sense of more exalted, than sensitive or vegetative functions.) The anatomically more rostral and phylogenetically "higher" neocortex is an inadequate criterion, because it ignores the possibility of consciousness being mediated by subcortical structures and eliminated by subcortical lesions. Thus, "higher brain death" is not synonymous with so-called "neocortical death" [61–64].

Destruction of the upper brainstem alone, or even the entire brainstem (so-called "brainstem death"), would be an inadequate criterion, because electrical stimulation of the reticular formation rostral to the lesion could potentially produce arousal from coma [65].

The proper criterion for the psychological concept, then, is bilateral thalamic destruction (*caveat* regarding irreversibility, below).

3.2 Biological Concept

The biological concept understands the human being as an organism. In the words of Bernat, Culver, and Gert [1]:

> We define death as the permanent cessation of functioning of the organism as a whole. (p. 390)

They explain that "organism as a whole" does not mean "whole organism," because even a defective organism is physiologically unified if it is alive. I prefer a more ontological focus, because an organism is not a bundle of functions, but a

[3] The advent of functional neuroimaging and computerized electroencephalography has brought to light the presence of covert awareness (cognitive-motor dissociation) in some 15–20% of patients in unresponsive wakefulness syndrome [58–60]. This proportion could well be an underestimate, since those technologies largely depend on language-based testing protocols, so unresponsiveness with diffuse neocortical lesions could conceivably be due to aphasia (inability to understand the commands) rather than inner unawareness.

functioning *thing*. Therefore, to define death as the "cessation of the organism" or "cessation of existence of the organism" captures the deeper essence of their definition.[4]

An important aspect of the biological concept is species-nonspecificity. "When we talk of the death of a man we mean the same thing as we do when we talk of the death of a dog or a cat" [1 (p. 390)]. Another key aspect is that it is not oxymoronic for a living organism, including a human, to be unconscious, even permanently or irreversibly so.

This biological concept was endorsed by the 1981 President's Commission [2]. A slightly different version was proposed by the 2008 President's Council on Bioethics [5]. Others, such as Schrödinger [66] and Korein [23], have emphasized the aspect of endogenous opposition to entropy as the essential feature of life. Still others, such as Maturana and Varela, have focused on the endogenous self-construction of organisms – a process aptly called "autopoiesis" [67–69].

Life is such a rich phenomenon that it is difficult to define in complete generality [70]. Each of the aforementioned aspects is insightful and valid. They are best understood not as competitors for a single necessary and sufficient characteristic, but as complementary. Synthesizing them, we can say that a living organism has the following mutually implicating characteristics (among others, but these suffice for our purpose):

- unity
- anti-entropy
- autopoiesis
- homeostasis
- self-preserving exchange with the environment

Bonelli, Prat, and Bonelli proposed eight criteria for "the specific whole of a living being," which are essentially the above properties described in different terms [71]. It is difficult to conceive of any of these features in the absence of another. Barring examples where the immanent activity is temporarily on hold but its potential remains (e.g., spores, cryopreservation), all living things possess all five features, and no nonliving thing possesses any of them.[5]

Regarding the last item, the qualifier "self-preserving" is important, because mere exchange with the environment occurs with many nonliving things, even a decaying corpse. Environmental commerce differs from the other four properties insofar as its presence is a sufficient but not necessary sign of life, whereas the others are both sufficient and necessary. As the wording indicates, "self-preserving" implies that there is already a living "self" and that this property is about "preserving" that life, not constituting it. Environmental commerce – and more

[4] "As a whole" is already implied, since if the organism exists at all, it is *ipso facto* a unity. "Permanent" and "irreversible" are also implied, since if the organism no longer exists, it cannot bring itself back into existence.

[5] The unity of crystals is structural but not dynamic and emergent; the unity of artifacts is only analogous and extrinsic.

broadly, adaptive interaction with the environment[6] – is a necessary condition for *staying* alive, not for *being* alive [76].

3.2.1 Critique

The main criticism of the biological concept is that it undervalues the mental aspects of human personhood. Some of its endorsers have responded by incorporating capacity for consciousness into their definitions of life and death for humans and higher animals [77 (pp. 290–7), 78, 79]. Such modifications, however, are insufficient for critics committed to a functionalist understanding of personhood, many of whom consider the brain-dead body a living human organism, while insisting that nevertheless it is not a human person. The difference of opinion stems from different fundamental worldviews that no amount of debate is likely to change.

Another version of the same critique is expressed by a subset of those who hold a substantive understanding of personhood,[7] so it is interesting that such unlikely philosophical bedfellows would converge on this same point. I refer to certain scholars of the Aristotelian-Thomistic tradition, which accepts Boethius's definition of person as "an individual substance of a rational nature" [80]; they argue that brain destruction eliminates the capacity for rationality and therefore occasions a substantial change to a new type of living organism with a nonrational substantial form [45, 46, 81]. I was originally of this opinion [82], but have come to agree with those who regard it as functionalism dressed in Thomistic terminology [83, 84 (pp. 141–232), 85].

"Rational nature" is not synonymous with capacity to exercise rationality. An impediment to actuating our most noble and species-specific functions (i.e., mental functions) does not eradicate our nature as rational animals any more than bilateral ocular enucleation eradicates our nature as seeing animals. Human nature is expressed through the human DNA present in every cell throughout the body. That DNA underlies the capacity for endogenous generation (autopoiesis) of every body part, including the brain. That capacity is obviously active in the embryonic stage and is gradually inactivated through epigenetic modifications as development proceeds. In principle, the autopoietic potential inherent in DNA could be reactivated to repair a damaged brain or regrow destroyed parts of a brain or even an entire brain.

That sounds easy to dismiss as mere science fiction, but only a few decades ago genetic engineering was unimaginable. Epigenetic engineering, capable of reactivating selective autopoietic potential latent in DNA is already on the horizon: the potential to regrow amputated limbs is starting to be unlocked in experimental animals [86, 87]. However, the point is not whether epigenetic engineering is

[6] Damiano and Luisi opine that autopoiesis is a necessary but not sufficient criterion for life, since some artificial systems can fulfill the technical definition of autopoiesis but lack the feature of adaptive interaction with the environment (citing [72–74]), which Maturana and Varela call "cognition" [68] (a technical and much broader sense of the term than mental activity) and which Damiano and Luisi consider also necessary for life [75].

[7] Descartes also held a substantive notion of personhood: the person/mind is an immaterial substance. His intellectual progeny (as far as the equation of personhood with mind goes), however, upon abandoning his substance dualism, were left with a functionalist understanding of personhood, generally within a material monist ontology.

easy, or practical, or possible today; it is that our rational nature is diffuse throughout our bodies, not localized in our brains. Although the brain may ground the *capacity* for rationality, it does not ground the *radical capacity*, i.e., the capacity to develop the capacity for rationality inherent in human DNA in the context of a living organism.

This is the answer to Lizza's objection that to attribute potential for rationality to a brain-dead body is based on too "promiscuous" a concept of potentiality [42, 88], as though one might just as well claim that anything at all had a potential for rationality. That does not follow, because most things do not contain human DNA dynamically participating in the life processes of an organism.

Psychological concept advocates might still object that, even if through futuristic epigenetic engineering a new brain could be grown in situ in a brain-destroyed patient, there would be a new person altogether, not a reawakening of the same person. Biological concept advocates would reply that it would be the same person, merely with a new set of memories and personality traits. That disagreement would likely be unresolvable, because the competing fundamental worldviews are irreconcilable. However, either way, that organism would still be *some* human person.

3.2.2 Criterion

Organismal wholeness requires intercommunication among all body parts, and two systems extend to essentially all parts: the nervous system and the circulatory system. We shall now examine the respective candidacies of cessation of brain function and cessation of circulation as the criterion for loss of organismal unity, anti-entropy, autopoiesis, homeostasis, and self-preserving commerce with the environment.

The 1981 President's Commission opined that brain death and circulatory death were merely different clinical contexts for the same physiological state, with the ventilator "masking" that equivalence in the case of brain death [2 (p. 33, 35)]. In hindsight, that theory was so counterfactual that it is difficult to understand how it was taken so seriously at the time. How can a body with all non-neurological systems functioning be said to be in the same physiological state as a circulationless corpse with no systems functioning? If one were to attach that corpse to a ventilator and blow air in and out of its lungs, there would be no deceptive appearance of life, no masking of the death. Vital processes are endogenous, and a mechanical ventilator cannot replace or simulate them.

Thus, the Commission seriously erred in writing that "[r]espiration and circulation in these [brain-dead] bodies may be generated by a ventilator together with intensive medical management" (p. 17), since neither respiration nor circulation is "generated" by the ventilator. The heart has its own intrinsic pacemaker, and respiration (exchange of oxygen and carbon dioxide) is carried out by the pulmonary alveoli and the mitochondria throughout the body.

Moreover, if a brain-dead body and a circulationless corpse were physiologically equivalent, the brain-dead body would undergo decay just as a traditional corpse, though perhaps more slowly.[8] Once the process of postmortem decay had begun, it would progress relentlessly, even if some integrative brainstem function hypothetically were to return. If physiological equivalence were correct, there would be no reason to require that the cessation of brainstem functions be *irreversible*, only that it last long enough for the somatic disintegration to begin down its inexorable path. The requirement of irreversibility obliquely implies that brain death is *not* in fact physiologically equivalent to circulatory death [92].[9]

The true physiological equivalent of a brain-dead body is a brain-disconnected body, since the physiological impact of loss of brain control would be the same regardless whether due to brain destruction or brain disconnection. Clinical experience with both bears this out. Compare any textbook chapter on the intensive-care management of high spinal cord injury with a chapter on the management of brain-dead organ donors: they are essentially identical [93].[10]

Brain–body disconnection can also be functional rather than structural. For example, severe Guillain-Barré syndrome abolishes information transfer into and out of the central nervous system. Apart from inwardly preserved consciousness, its external manifestations are so similar to those of brain death that it can even masquerade as brain death [97–103]. Since neuroendocrine functions are intact in Guillain-Barré syndrome, the physiological comparison can be made exact either by restricting the brain-death cases to those without diabetes insipidus or by supposing that the Guillain-Barré patient also happened to have panhypopituitarism controlled with hormone replacement.

Patients with high spinal cord transection or Guillain-Barré syndrome are obviously alive – not only because they are conscious but also because their bodies are living organisms: unstable and disabled to be sure, but organisms nonetheless. Their physicians and nurses are not caring for a mind/brain imprisoned in "a group of artificially maintained subsystems" [1 (p. 391), 2 (p. 35–6)].[11] Since brain-dead

[8] This is in fact the explanation declared to the court for why Jahi McMath's body was deteriorating during her third and fourth weeks post-arrest: the inevitable decay of a corpse. In fact, the deterioration was iatrogenic, due to total lack of nutrition and vitamins, plus untreated hypothyroidism and hypoadrenalism [89–91].

[9] That the functions required to be absent are "clinical"—namely consciousness, a handful of cranial nerve reflexes, and breathing—implies that the set of somatically integrating functions (which are *not* clinically evident) is assumed to be absent by virtue of anatomical proximity to the brainstem pathways for the various cranial nerve reflexes. Precisely what these mysterious somatically integrative functions are, and what brainstem structures supposedly mediate them, has never been spelled out by proponents of physiological equivalence.

[10] Condic objects that the physiological comparison is not exact, because spinal cord injury patients have intact cranial nerve IX and X function but brain-dead patients do not, and because spinal cord injury patients have a functioning hypothalamus, but many brain-dead patients do not (though many do [94, 95]) [96]. These objections were fully taken into account and the comparison tweaked to an exact physiological equivalence in my original article on the role of spinal shock in the pathophysiology of brain death [93].

[11] Moschella and Condic opine that, in effect, that is *exactly* what such health-care professionals are caring for [96, 104]—an idea that runs counter to the actual clinical experience of medical personnel in neurointensive care units.

patients have the same physiology, it follows that brain-dead bodies are equally integrated wholes, the only difference being that one is conscious and the other not.

Whether an organism is a living whole depends on whether it has endogenous holistic properties, not on whether it requires external assistance to stay alive. We now review the evidence from actual brain-dead bodies and see that many of them manifest the five holistic properties listed above.

Anti-entropy. For decades, a brain-death mantra has been that a proof that brain-dead bodies are disintegrating corpses is the fact that they are so unstable and inevitably succumb to imminent cardiovascular collapse despite all intensive-care measures.[12] Such reasoning would be convincing were it not for two small problems: it is a logical fallacy and its premise is false.

It is a prime example of the classical fallacy of affirming the consequent: "If A, then B; B, therefore A." In this case: "If a body lacks anti-entropy, it will undergo imminent cardiovascular collapse. Brain-dead bodies undergo imminent cardiovascular collapse; therefore, they lack anti-entropy." It is fallacious, because there are other possible, often even likely, explanations for imminent cardiovascular collapse: multisystem damage from the initial etiology [109], secondary multisystem damage from the process of brain herniation [110, 111], and spinal shock [93].

However, the premise of imminent cardiovascular collapse is not even true. By 1998, there had been around 175 reported cases surviving longer than 1 week, some surviving months, and a few surviving years [109], the record being 20½ years [112]. Such cases of "chronic brain death" provide definitive evidence that bodily anti-entropy does not derive from brain function. These patients also exhibited other manifestations of anti-entropy: improving from the initial phase of instability to chronic stability, and overcoming intercurrent illnesses and other complications.

Autopoiesis. These same patients manifested anti-entropy also in the positive sense of teleologically *increasing order* – autopoiesis. They all displayed wound healing. Children supported long enough demonstrated proportional growth (in contrast to the disordered growth of tumors) and rarely even sexual development [109, 113].

Homeostasis. They also possessed homeostasis, because they could go for long periods of time, even at home, without blood tests or adjustments in the composition of tube feedings, and when blood tests were checked from time to time, most if not all items would be in the normal range. Temperature was maintained in the normal or slightly subnormal range with the help of ordinary blankets.

The debate whether hypothalamic function is relevant for the neurologic diagnosis of death is more heated now than ever [7, 9, 95, 114–118]. The answer depends entirely on the concept of death behind the criterion. For the psychological concept, hypothalamic function is completely irrelevant; for the biological concept, it is obviously highly relevant.

---

[12] Commonly cited latencies for this to occur in the great majority are 2–10 days [2 (p. 17)], "several days" [2 (p. 35)], "a few days" [105 (p. 30, 36)], 7 days [23, 96, 106, 107 (p. 117)], and <30 days [108]. Some authors mention that infants and young children can have longer somatic survival times but eventually succumb to the inevitability as well.

Unity. From a metaphysical perspective, it is untenable that the unity of a whole could derive from the activity of just one of its own parts [119]. True unity is a diffusely emergent phenomenon from the mutual interaction of all parts.

The already cited examples of anti-entropy, autopoiesis, and homeostasis also exemplify unity. In addition, some brain-dead bodies exhibit other holistic properties that are difficult to place in one of those other categories [113]: for example, cardiovascular and hormonal stress responses to unanesthetized incision for organ retrieval [120, 121], and successful gestation of a fetus (including even vaginal delivery [122, 123]).

In contrast, no holistic properties are exhibited by

- "a preparation of unintegrated individual subsystems" (Bernat) [124 (p. 48)] or
- "a ventilator [keeping] a heart beating in a corpse" (Daroff) [125 (p. 275)] or
- "a magnificent cell culture" (Wijdicks and Bernat) [125 (p. 276)] or
- an amputated finger perfused in a flask (Masdeu) [125 (p. 146)] or
- "the twitching of a lizard's amputated tail" (Talmudic exegesis of decapitation applied to brain death) [126 (p. 395), 127 (p. 19), 128 (p. 557)] or
- "the thoracic and abdominal contents of a cat [extracted] en bloc and maintained ... with the heartbeat preserved, suspended in a vat with Ringer's solution" (López-Navidad) [129]
- all serious descriptions of the alleged essence of brain-dead patients. Transplant surgeons can attest that no amputated finger or perfused internal organ can be maintained viable for longer than a day or two at most. To compare such body parts to a brain-dead body maintained stably for years at home on a ventilator and tube feedings is the height of absurdity.
- If any comparison is to be made, it is with non-brain-dead dying patients with multisystem failure, comatose, on ventilators, inexorably spiraling downhill toward terminal cardiac arrest. Such patients are universally recognized as alive, yet they have less claim to "organism as a whole" status than some brain-dead patients.
- This is not to say that *every* brain-dead patient is an integrated, living organism. It is quite conceivable that some with massive trauma or diffuse anoxic–ischemic damage are in fact no longer integrated wholes. However, that is not *because* they lack brain function; rather, it is because they suffered supra-critical multisystem damage (including the brain). They will indeed be exceedingly unstable and quickly deteriorate to circulatory–respiratory arrest despite all intensive-care measures. Unfortunately, there is no way to distinguish in practice such "dead brain-dead" patients from "live brain-dead" patients.

A decade ago, I proposed distinguishing various *levels* and *types* of integration [76]. Condic instead proposed a dichotomy between "integration" and mere "coordination," maintaining that the only interaction among body parts worthy of the designation "integration" is mediated by the brain [96]: a conclusion that is both ad hoc and question-begging.

Commerce with the environment. For many of the above reasons, the 2008 President's Council on Bioethics concluded [5]:

> If being alive as a biological organism requires being a whole that is more than the mere sum of its parts, then it would be difficult to deny that the body of a patient with total brain failure can still be alive, at least in some cases. (p. 57)

Nevertheless, the Council (with the dissent of its chairman and another member) opined that the neurologic criterion could be salvaged by appeal to a new, "more compelling account of *wholeness*" than the parts merely "working together in an integrated way." (p. 60, emphasis in original)

> Determining whether an organism remains a *whole* depends on recognizing the persistence or cessation of the fundamental vital *work* of a living organism – the work of self-preservation, achieved through the organism's need-driven commerce with the surrounding world. (p. 60, emphasis in original)

The Council's semantics is hard to follow. Integration literally means "the act of combining or adding parts to make a unified whole." (https://www.thefreedictionary.com/integration). So how could something that is admittedly integrated not be a whole?

In any case, the Council focused on two forms of environmental commerce that it considered jointly necessary for life: consciousness and breathing. Why consciousness was considered a form of "commerce" is unclear – adaptive interaction, to be sure, but consciousness does not exchange any substances with the environment. As for breathing, the Council explained that what was at issue was not actual breathing (which could be substituted by a ventilator) but the "felt need" to breathe, the respiratory drive per se, mediated by the medulla. (The Council's concept of "felt need" was not a conscious feeling but a physiological drive.)

According to the Council, neither irreversible unconsciousness alone nor irreversible apnea alone suffices for death, but their combination suffices. Why this should be was not adequately explained. Consider someone with Ondine's curse, who loses respiratory drive during sleep and is therefore placed on a ventilator before going to bed. They are not dead, despite having no respiratory drive and no consciousness. If that is the case during sleep, why could it not also be the case if it lasted days, weeks, or indefinitely? The Council's proposal has been rightly criticized as ad hoc [118, 130, 131 (p. 72–5)], although it has recently been vigorously defended [17].

Moreover, it ignores non-brain-based forms of environmental commerce, which brain-dead patients exhibit. Gases are exchanged with the environment at the alveolar lining of the lungs, while the ventilator merely substitutes for the diaphragm and intercostal muscles. The gastrointestinal system has a "felt need" to move contents along, to orchestrate the release of digestive enzymes and the opening and closing of sphincters, and to selectively absorb nutrients and exclude wastes. (The gastrointestinal lumen is topologically the "environment.") The gastrointestinal system even has its own intrinsic nervous system, elaborate enough to be dubbed a "second brain" with a "mind of its own" [132, 133]. Brain-dead patients manifest yet other forms of adaptive interaction with the environment, such as the filtering of soluble wastes by the kidneys, the clotting of blood and repairing of breaches in the environmental interface, or the immune response to an environmental invader. Why should the medulla's "felt need" count as relevant, but the gastrointestinal and the immune systems' "felt needs" count as irrelevant? The only conceivable reason is the Council's desire to salvage the neurologic criterion – so its novel rationale is ultimately just question-begging.

In conclusion, regarding the biological criterion: unity, anti-entropy, autopoiesis, homeostasis, and self-preserving environmental commerce are all exhibited by many brain-dead patients. They are emergent properties from the mutual interactions of all body parts, made possible by the endogenous circulation of oxygenated blood. Brain-based behaviors and modulation of physiological functions are important for *staying* alive (in the wild) but not for *being* alive. Therefore, a sufficient criterion for death corresponding to the biological concept is the permanent and irreversible cessation of circulation.

4 Thought Experiments

The brain-death literature is replete with a wide variety of thought experiments intended to prove that the brain-dead body is either not an organism or no longer the body of a person. Although I am sympathetic to this line of argumentation, having based my earlier defense of brain death on it [82], I have come to reject it, for reasons that space does not allow to be reviewed here. In general, "the philosophical method of thought experimentation is inherently circular, starting with a hypothesis and then making up pseudo-evidence to support the hypothesis" [134 (p. 298)]. Regarding brain-death thought experiments in particular, for the psychological concept they are superfluous (since the "higher brain" criterion follows straightforwardly), and for the biological concept they are question-begging and inconclusive (proving neither that a brain-dead body is not an unconscious person nor that it is not an organism). For detailed critiques, see [84 (p. 75–8, 264, 298–9, 507), 118, 131 (p. 80–4), 135–139].

5 Some Key Distinctions

5.1 Sufficiency vs. Necessity

It is generally assumed that the criterion of death should be both necessary and sufficient for instantiating the concept [23, 40, 114, 115, 124, 140–143], although some have argued that criteria that are both necessary and sufficient do not exist [11, 12, 144, 145]. I agree with the latter. For the *biological* concept, the permanent and irreversible absence of circulation is the only sufficient criterion, but it is not a necessary criterion: if one hypothetically took a corpse from the morgue and forcibly pumped oxygenated blood through its vessels, it would still be dead. For the *psychological* concept, bilateral thalamic destruction is sufficient and possibly also necessary (*vide supra*). In the final consequence, the requirement that the criterion be necessary is itself unnecessary. It suffices that (1) the criterion be *sufficient* to instantiate the concept, and (2) it be so broadly applicable that theoretical counterexamples against necessity simply do not occur in practice.

There are two coherent concept–criterion pairings: the psychological concept with the "higher brain" criterion, and the biological concept with the circulatory criterion. (Non-ontological concepts do not logically imply any particular criterion,

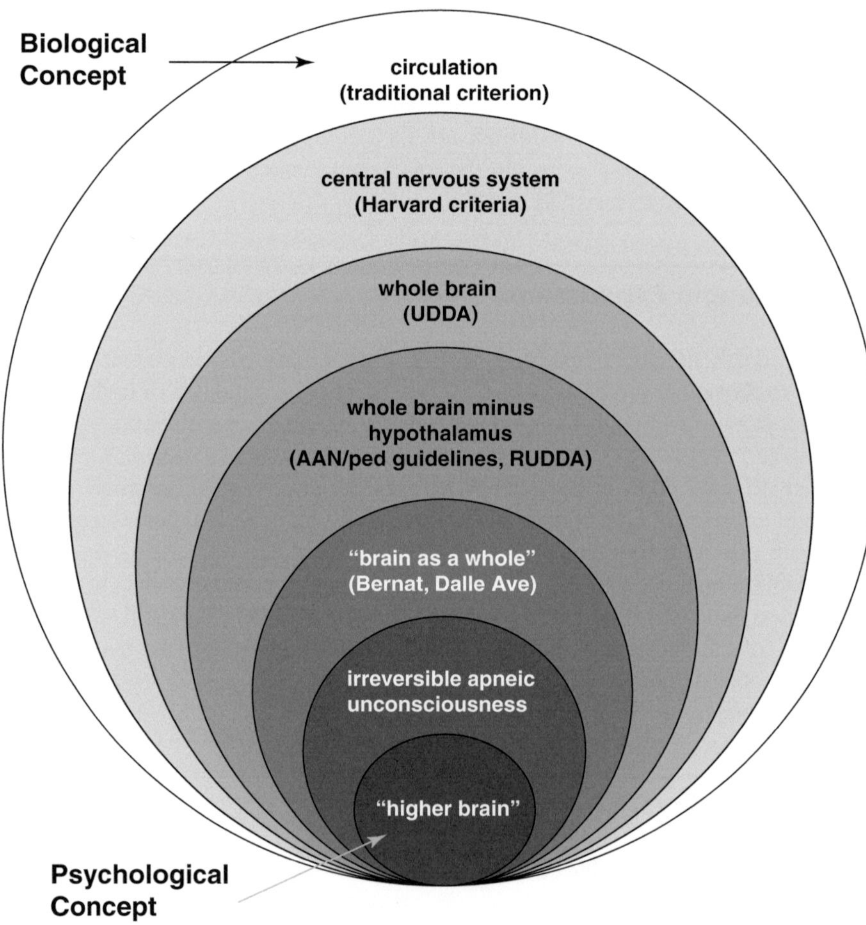

Fig. 1 Hierarchy of criteria for death. Only the outer shell and the inner core correspond coherently to an ontological concept of death. "UDDA": Uniform Determination of Death Act [2]; "AAN/ped guidelines": American Academy of Neurology updated adult guidelines [3] and joint-society updated pediatric guidelines [4]; "RUDDA": revised UDDA as proposed by Lewis and colleagues [116, 117]

and stipulative criteria can be associated with any concept.) All other criteria proposed throughout brain-death history fall anatomically somewhere between these two in concentric layers, like a matryoshka doll (Fig. 1). Criteria beneath the outer shell are insufficient for the biological concept, and criteria beyond the inner core are unnecessary for the psychological concept. (The "brainstem" criterion is omitted, because it was intended to correspond to "irreversible apneic unconsciousness" but does not, being unnecessarily broad for apnea and insufficient for unconsciousness.)

An insufficient criterion entails the possibility of false-positive diagnoses (considering a live patient dead), whereas an unnecessary criterion entails the possibility of false-negative diagnoses (considering a dead patient alive). Clearly, it is ethically much more important that a criterion be sufficient than that it be necessary. Therefore, some who in principle favor an anatomically restricted criterion prefer to stipulate an unnecessarily broad criterion for the sake of diagnostic certainty [115, 145–148]. For the same tutioristic reason (at least implicitly), some who endorse the criterion of "irreversible apneic unconsciousness" favor stipulating a whole-brainstem (the United Kingdom) or a whole-brain criterion [5 (p. 66–7), 17], and Bernat, who favors a "brain-as-a-whole" criterion, prefers to stipulate the "whole-brain" criterion with mandatory blood flow testing [114, 115, 149, 150].

5.2 Structure vs. Function

In the early years of brain-death history, scholars debated whether the neurologic criterion of death should be formulated in terms of destruction [151] or nonfunction [2]. I consider this a nonissue, because as long as the nonfunction is qualified as "irreversible," the two formulations are equivalent, since the only possible basis for irreversible nonfunction is destruction of the structure responsible for the function.

The mere persistence of nonfunction for some specified period of time[13] cannot per se guarantee irreversibility [3, 154]. Therefore, there has been a move to require demonstration of intracranial circulatory arrest (implying destruction) as a source of greater certainty of irreversibility, rather than leave it optional [114, 115, 149, 155]. (Whether current standard tests reliably distinguish no flow from low flow is another question [156–160].)

5.3 Irreversible vs. Permanent

For many years, the terms "irreversible" and "permanent" were treated synonymously, even though their dictionary meanings are modally different: *irreversible* nonfunction means that function *cannot* return, whereas *permanent* nonfunction means that it *will not* return. After circulatory–respiratory arrest, there is a period

[13] For example, 6 h for adults [152], 12 h for children [4], 24 h for hypoxic–ischemic etiology [153], and 48–72 h for post-therapeutic-hypothermia [115].

on the order of minutes when the cessation is permanent but reversible. In most deaths, this interval makes no practical difference, but in the context of organ donation after circulatory determination of death, permanence has been used to ethically justify organ retrieval prior to irreversibility under a reinterpreted dead donor rule, even though the Uniform Determination of Death Act specifies "irreversible" [161, 162].

Bernat offers a number of reasons for a permanence standard in general (for both circulatory and neurologic criteria), most importantly that the onset of permanence can be observed and timed with a precision on the order of seconds, whereas the onset of irreversibility is unobservable and can only be guessed at within a broad time interval [163]. Also, society has traditionally timed death to the onset of permanence (the beginning of grieving and treating the body as a corpse).

I agree with Bernat's arguments for a permanence standard, having previously developed the idea of a semantic bifurcation of the concept of death, which we tend to think of univocally merely because the language we grew up in has the single word "death" [31]. Taking this cue from linguistics, I proposed to distinguish two death-related events: "passing away" (civil or "normative" death [164]), which is observable and occurs at the onset of *permanent* cessation of circulation, and "deanimation" (ontological death), i.e., the loss of the organizational principle conferring unity, anti-entropy, etc., which is unobservable and occurs at the onset of *irreversibility* of cessation of circulation [165].

Regarding irreversibility of brain nonfunction, if the autopoietic potential for brain repair or even regrowth is inherent in human DNA as discussed above, functionalist adherents to the psychological concept may have to abandon the "higher brain" criterion if future epigenetic engineering makes brain destruction reversible.

The idea of irreversibility of circulatory arrest must be understood to include permanence (thereby excluding the hypothetical of the artificially perfused corpse). It could therefore be useful to combine them in the same criterion, so that the irreversibility aspect reflects the ontological reality of death ("deanimation"), while the permanence aspect determines the timing of death ("passing away") for medical, legal, and relational purposes. Such a criterion for the biological concept might be something like:

> the permanent cessation of the functioning of all three systems: (i) circulatory, (ii) respiratory, and (iii) central nervous system, all beyond the physical possibility of resuscitation or restoration. (suggested by Doyen Nguyen, personal communication)

(Although I have been referring throughout to a purely circulatory criterion, specification of all three systems would eliminate the possibility of the onset of permanence being timed prematurely during the seconds between a sudden cardiac arrest and cessation of nervous system functioning.[14]) The distinction

[14] Inclusion of the respiratory system is technically superfluous but motivated by a desire to relate the criterion to the traditional formulation, in which "respiration" is understood as breathing, which is actually more a function of the nervous system than the respiratory system.

between irreversibility and permanence is discussed in more detail elsewhere in this book.

6 Conclusions

Apart from non-ontological concepts and stipulative criteria, there are only two coherent concept–criterion pairs: the psychological concept with a "higher brain" criterion and the biological concept with a circulatory criterion. Every criterion in between results in either false-negative attributions of death for the psychological concept or false-positive attributions for the biological concept.

Given that society will never reach a consensus on fundamental worldviews and corresponding concepts of life and personhood, the only way to respect all deeply held convictions will be for the law to accommodate personal specification of concept and criterion [9, 148, 164].

References

1. Bernat JL, Culver CM, Gert B. On the definition and criterion of death. Ann Intern Med. 1981;94(3):389–94.
2. President's Commission for the Study of Ethical Problems in Medicine and Biomedical and Behavioral Research. Defining Death: Medical, Legal, and Ethical Issues in the Determination of Death. Washington, DC: U.S. Government Printing Office; 1981.
3. Wijdicks EFM, Varelas PN, Gronseth GS, Greer DM. Evidence-based guideline update: determining brain death in adults: report of the Quality Standards Subcommittee of the American Academy of Neurology. Neurology. 2010;74(23):1911–8.
4. Nakagawa TA, Ashwal S, Mathur M, et al. Guidelines for the determination of brain death in infants and children: an update of the 1987 Task Force recommendations. Crit Care Med. 2011;39(9):2139–55.
5. President's Council on Bioethics. Controversies in the Determination of Death. Washington, DC: President's Council on Bioethics; 2008.
6. Wijdicks EFM. Brain death worldwide: accepted fact but no global consensus in diagnostic criteria. Neurology. 2002;58(1):20–5.
7. Russell JA, Epstein LG, Greer DM, et al. Brain death, the determination of brain death, and member guidance for brain death accommodation requests: AAN position statement. Neurology. 2019;92(5):228–32.
8. Greer DM. Determination of brain death. N Engl J Med. 2021;385(27):2554–61.
9. Shewmon DA. Statement in support of revising the Uniform Determination of Death Act and in opposition to a proposed revision. J Med Philos. 2021; Epub ahead of print.
10. Botkin JR, Post SG. Confusion in the determination of death: distinguishing philosophy from physiology. Perspect Biol Med. 1992;36(1):129–38.
11. Chiong W. Brain death without definitions. Hast Cent Rep. 2005;35(6):20–30.
12. Chiong W. What constitutes human death. Against definitions, necessary and sufficient conditions, and determinate boundaries. In: Caplan AL, Arp R, editors. Contemporary Debates in Bioethics. Chichester, UK: Wiley Blackwell; 2014. p. 388–96.
13. Laureys S, Celesia GG, Cohadon F, et al. Unresponsive wakefulness syndrome: a new name for the vegetative state or apallic syndrome. BMC Med. 2010;8:68.
14. Aristotle. Physics.

15. Koren HJ. An Introduction to the Philosophy of Animate Nature. St. Louis, MO: Herder; 1955.
16. Byers S. Life as "self-motion": Descartes and "the Aristotelians" on the soul as the life of the body. Rev Metaphys. 2006;59:723–55.
17. Omelianchuk A. Brain death as the end of a human organism as a self-moving whole. J Med Philos. 2021;46(5):530–60.
18. Gilder SS. Twenty-second World Medical Assembly. Br Med J. 1968;3(5616):493–4.
19. Beecher HK. After the "definition of irreversible coma". N Engl J Med. 1969;281(19):1070–1.
20. Beecher HK. Definitions of "life" and "death" for medical science and practice. Ann N Y Acad Sci. 1970;169(2):471–4.
21. Beecher HK, Dorr HI. The new definition of death. Some opposing views. Int J Clin Pharmacol. 1971;5(2):120–4.
22. Morison RS. Death: process or event? Science. 1971;173(3998):694–8.
23. Korein J. The problem of brain death: development and history. Ann N Y Acad Sci. 1978;315:19–38.
24. Conference of Medical Royal Colleges and their Faculties in the United Kingdom. Diagnosis of death. Br Med J. 1979;1(6159):332.
25. Jennett B. Brain death. Intensive Care Med. 1982;8(1):1–3.
26. Emanuel L. What is wrong with 'dead'. In: Machado C, editor. Brain Death. Proceedings of the Second International Conference on Brain Death, Havana, Cuba, February 27–March 1, 1996. Developments in Neurology, vol. 9. Amsterdam: Elsevier; 1995. p. 47–56.
27. Pallis C. On the brainstem criterion of death. In: Youngner SJ, Arnold RM, Schapiro R, editors. The Definition of Death: Contemporary Controversies. Baltimore, MD: Johns Hopkins University Press; 1999. p. 93–100.
28. Halevy A. Beyond brain death? J Med Philos. 2001;26(5):493–501.
29. Posner JB. Alleged awakenings from prolonged coma and brain death and delivery of live babies from brain-dead mothers do not negate brain death. In: Sánchez Sorondo M, editor. The Signs of Death. The Proceedings of the Working Group 11–12 September 2006. Scripta Varia 110. Vatican City: Pontificia Academia Scientiarum; 2007. p. 116–22.
30. Magnus D. A defense of the dead donor rule. Hast Cent Rep. 2018;48(Suppl 4):S36–S8.
31. Shewmon DA, Shewmon ES. The semiotics of death and its medical implication. In: Machado C, Shewmon DA, editors. Brain Death and Disorders of Consciousness. Advances in Experimental Medicine and Biology, vol. 550. New York: Kluwer Academic/Plenum Publishers; 2004. p. 89–114.
32. Griffiths DJ, Schroeter DF. Introduction to Quantum Mechanics. 3rd ed. Cambridge, UK: Cambridge University Press; 2018.
33. Kuznetsov YA. Elements of Applied Bifurcation Theory. 3rd ed. New York: Springer; 2004.
34. Aderem A. Systems biology: its practice and challenges. Cell. 2005;121(4):511–3.
35. Grocott MP. Integrative physiology and systems biology: reductionism, emergence and causality. Extrem Physiol Med. 2013;2(1):9.
36. Lüttge U. Integrative emergence in contrast to separating modularity in plant biology: views on systems biology with information, signals and memory at scalar levels from molecules to the biosphere. Theor Exp Plant Physiol. 2021;33:1–13.
37. Joffe AR, Anton N. Brain death: understanding of the conceptual basis by pediatric intensivists in Canada. Arch Pediatr Adolesc Med. 2006;160(7):747–52.
38. Joffe AR, Anton N, Mehta V. A survey to determine the understanding of the conceptual basis and diagnostic tests used for brain death by neurosurgeons in Canada. Neurosurgery. 2007;61(5):1039–45; discussion 46–7.
39. Joffe AR, Anton NR, Duff JP, Decaen A. A survey of American neurologists about brain death: understanding the conceptual basis and diagnostic tests for brain death. Ann Intensive Care. 2012;2(1):4.
40. Baker A, Shemie SD. Biophilosophical basis for identifying the death of a person. J Crit Care. 2014;29(4):687–9.

41. Beecher HK, Adams RD, Barger C, et al. A definition of irreversible coma. Report of the Ad Hoc Committee of the Harvard Medical School to Examine the Definition of Brain Death. JAMA. 1968;205(6):337–40.
42. Lizza JP. Potentiality, irreversibility, and death. J Med Philos. 2005;30(1):45–64.
43. Lizza JP. Persons, Humanity, and the Definition of Death. Baltimore: Johns Hopkins University Press; 2006.
44. Lizza JP. On the definition of death. In: Lizza JP, editor. Defining the Beginning and End of Life. Readings on Personal Identity and Bioethics. Baltimore: Johns Hopkins University Press; 2009. p. 533–50.
45. Lee P, Grisez G. Total brain death: a reply to Alan Shewmon. Bioethics. 2012;26(5):275–84.
46. Lee P. Total brain death and the integration of the body required of a human being. J Med Philos. 2016;41(3):300–14.
47. Fletcher J. Indicators of humanhood: a tentative profile of man. Hast Cent Rep. 1972;2(5):1–4.
48. Green MB, Wikler D. Brain death and personal identity. Philos Public Aff. 1980;9(2):105–33.
49. Veatch RM. The whole-brain-oriented concept of death: an outmoded philosophical formulation. J Thanatol. 1975;3(1):13–30.
50. Posner JB, Saper CB, Schiff ND, Plum F. Plum and Posner's Diagnosis of Stupor and Coma. 4th ed. Oxford: Oxford University Press; 2007.
51. Ingvar DH, Brun A, Johansson L, Sammuelsson SM. Survival after severe cerebral anoxia with destruction of the cerebral cortex: the apallic syndrome. Ann N Y Acad Sci. 1978;315:184–214.
52. Kinney HC, Korein J, Panigrahy A, Dikkes P, Goode R. Neuropathological findings in the brain of Karen Ann Quinlan. The role of the thalamus in the persistent vegetative state. N Engl J Med. 1994;330(21):1469–75.
53. Kinney HC, Samuels MA. Neuropathology of the persistent vegetative state. A review. J Neuropathol Exp Neurol. 1994;53(6):548–58.
54. Shewmon DA, Holmes GL, Byrne PA. Consciousness in congenitally decorticate children: "developmental vegetative state" as self-fulfilling prophecy. Dev Med Child Neurol. 1999;41(6):364–74.
55. Merker B. Consciousness without a cerebral cortex: a challenge for neuroscience and medicine. Behav Brain Sci. 2007;30(1):63–81; discussion 81–134.
56. Shewmon DA. The ABC of PVS: problems of definition. In: Machado C, Shewmon DA, editors. Brain Death and Disorders of Consciousness. Advances in Experimental Medicine and Biology, vol. 550. New York: Kluwer Academic/Plenum Publishers; 2004. p. 215–28.
57. Shewmon DA. A critical analysis of conceptual domains of the vegetative state: sorting fact from fancy. NeuroRehabilitation. 2004;19(4):343–7.
58. Schiff ND. Cognitive motor dissociation following severe brain injuries. JAMA Neurol. 2015;72(12):1413–5.
59. Edlow BL, Claassen J, Schiff ND, Greer DM. Recovery from disorders of consciousness: mechanisms, prognosis and emerging therapies. Nat Rev Neurol. 2021;17(3):135–56.
60. Edlow BL, Naccache L. Unmasking covert language processing in the intensive care unit with electroencephalography. Ann Neurol. 2021;89(4):643–5.
61. Brierley JB, Graham DI, Adams JH, Simpsom JA. Neocortical death after cardiac arrest. A clinical, neurophysiological, and neuropathological report of two cases. Lancet. 1971;2(724):560–5.
62. Devettere RJ. Neocortical death and human death. Law Med Health Care. 1990;18(1–2):96–104.
63. Puccetti R. Does anyone survive neocortical death? In: Zaner RM, editor. Death: Beyond Whole-Brain Criteria. Philosophy and Medicine, vol. 31. Dordrecht: Kluwer Academic Publishers; 1988. p. 75–90.
64. Smith DR. Legal issues leading to the notion of neocortical death. In: Zaner RM, editor. Death: Beyond Whole-Brain Criteria. Philosophy and medicine, vol. 31. Dordrecht/Boston: Kluwer Academic Publ; 1988. p. 111–4.

65. Hassler R. Interaction of reticular activating system for vigilance and the truncothalamic and pallidal systems for directing awareness and attention under striatal control. In: Buser PA, Rougeul-Buser A, editors. Cerebral Correlates of Conscious Experience. Amsterdam: Elsevier; 1978. p. 111–29.
66. Schrödinger E. What is Life? The Physical Aspect of the Living Cell. New York: The Macmillan Company; 1946.
67. Varela FJ. Principles of Biological Autonomy. New York: North Holland; 1979.
68. Maturana HR, Varela FJ. Autopoiesis and Cognition. The Realization of the Living. Dordrecht/Boston: D. Reidel; 1980.
69. Maturana HR. The organization of the living: a theory of the living organization. International Journal of Human-Computer Studies. 1999;51(2):149–68.
70. Seifert J. What is Life? The Originality, Irreducibility, and Value of Life. Ginsberg R, editor. Amsterdam: Rodopi; 1997. 163p.
71. Bonelli RM, Prat EH, Bonelli J. Philosophical considerations on brain death and the concept of the organism as a whole. Psychiatr Danub. 2009;21(1):3–8.
72. Bitbol M, Luisi PL. Autopoiesis with or without cognition: defining life at its edge. J R Soc Interface. 2004;1:99–107.
73. Bourgine P, Stewart J. Autopoiesis and cognition. Artif Life. 2004;10(3):327–45.
74. Luisi PL. The Emergence of Life. New York: Cambridge University Press; 2006.
75. Damiano L, Luisi PL. Towards an autopoietic redefinition of life. Orig Life Evol Biosph. 2010;40(2):145–9.
76. Shewmon DA. You only die once: why brain death is not the death of a human being. A reply to Nicholas Tonti-Filippini. Communio. 2012;39:422–94.
77. Gert B, Culver CM, Clouser KD. Bioethics: A Systematic Approach. 2nd ed. New York: Oxford University Press; 2006.
78. Bernat JL. A conceptual justification for brain death. Hast Cent Rep. 2018;48(Suppl 4):S19–21.
79. Huang AP, Bernat JL. The organism as a whole in an analysis of death. J Med Philos. 2019;434(6):712–31.
80. Boethius. Liber de Persona et Duabus Naturis, Chapter 3.
81. Wallace WA. St. Thomas on the beginning and ending of human life. In: Lizza JP, editor. Defining the Beginning and End of Life. Readings on Personal Identity and Bioethics. Baltimore, MD: Johns Hopkins University Press; 2009. p. 469–82.
82. Shewmon DA. The metaphysics of brain death, persistent vegetative state, and dementia. The Thomist. 1985;49(1):24–80.
83. Nguyen D. Why the Thomistic defense of "brain death" is not Thomistic: an analysis from the perspectives of classical philosophy and contemporary biophilosophy. The Thomist. 2018;82:407–46.
84. Nguyen D. The New Definitions of Death for Organ Donation. A Multidisciplinary Analysis from the Perspective of Christian Ethics. Bern: Peter Lang; 2018.
85. Seifert J. "Brain death" is neither human death nor its sign. An answer to Condic, Lee, Moschella and other defenders of brain death definitions and criteria of human death. Aemaet. 2018;7(1):249–358.
86. Muneoka K, Han M, Gardiner DM. Regrowing human limbs. Sci Am. 2008;298(4):56–63.
87. Murugan NJ, Vigran HJ, Miller KA, et al. Acute multidrug delivery via a wearable bioreactor facilitates long-term limb regeneration and functional recovery in adult Xenopus laevis. Sci Adv. 2022;8(4):eabj2164.
88. Lizza JP. Potentiality and human embryos. Bioethics. 2007;21(7):379–85.
89. Shewmon DA. The case of Jahi McMath: a neurologist's view. Hast Cent Rep. 2018;48(Suppl 4):S74–6.
90. Shewmon DA. Truly reconciling the case of Jahi McMath. Neurocrit Care. 2018;29(2):165–70.
91. Shewmon DA, Salamon N. The extraordinary case of Jahi McMath. Perspect Biol Med. 2021;64(4):457–78.
92. Shewmon DA. Brain death: a conclusion in search of a justification. Hast Cent Rep. 2018;48(Suppl 4):S22–5.

93. Shewmon DA. Spinal shock and 'brain death': somatic pathophysiological equivalence and implications for the integrative-unity rationale. Spinal Cord. 1999;37(5):313–24.
94. Nair-Collins M, Northrup J, Olcese J. Hypothalamic-pituitary function in brain death: a review. J Intensive Care Med. 2016;31(1):41–50.
95. Nair-Collins M, Joffe AR. Hypothalamic function in patients diagnosed as brain dead and its practical consequences. In: Swaab DF, Buijs RM, Kreier F, Lucassen PJ, Salehi A, editors. Handbook of Clinical Neurology, vol. 182. Elsevier; 2021. p. 433–46.
96. Condic ML. Determination of death: a scientific perspective on biological integration. J Med Philos. 2016;41(3):257–78.
97. Langendorf FG, Mallin JE, Masdeu JC, Moshe SL, Lipton RB. Fulminant Guillain-Barré syndrome simulating brain death: clinical and electrophysiological findings. Electroencephalogr Clin Neurophysiol. 1986;64:74P.
98. Hassan T, Mumford C. Guillain-Barré syndrome mistaken for brain stem death. Postgrad Med J. 1991;67(785):280–1.
99. Martí-Massó JF, Suárez J, López de Munain A, Carrera N. Clinical signs of brain death simulated by Guillain-Barré syndrome. J Neurol Sci. 1993;120(1):115–7.
100. Bohlega SA, Stigsby B, Haider A, McLean D. Guillain-Barré syndrome with severe demyelination mimicking axonopathy. Muscle Nerve. 1997;20(4):514–6.
101. Vargas F, Hilbert G, Gruson D, Valentino R, Gbikpi-Benissan G, Cardinaud JP. Fulminant Guillain-Barre syndrome mimicking cerebral death: case report and literature review. Intensive Care Med. 2000;26(5):623–7.
102. Tan IL, Ng T, Vucic S. Severe Guillain-Barre syndrome following head trauma. J Clin Neurosci. 2010;17(11):1452–4.
103. Medici C, Gonzalez G, Cerisola A, Scavone C. Locked-in syndrome in three children with Guillain-Barre syndrome. Pediatr Neurol. 2011;45(2):125–8.
104. Moschella M. Deconstructing the brain disconnection-brain death analogy and clarifying the rationale for the neurological criterion of death. J Med Philos. 2016;41(3):279–99.
105. Pallis C, Harley DH. ABC of Brainstem Death. 2nd ed. London: BMJ Publishing Group; 1996.
106. Hung T-P, Chen S-T. Prognosis of deeply comatose patients on ventilators. J Neurol Neurosurg Psychiatry. 1995;58(1):75–80.
107. Wijdicks EFM. Brain Death. 2nd ed. Oxford: Oxford University Press; 2011.
108. Al-Shammri S, Nelson RF, Madavan R, Subramaniam TA, Swaminathan TR. Survival of cardiac function after brain death in patients in Kuwait. Eur Neurol. 2003;49(2):90–3.
109. Shewmon DA. Chronic "brain death". Meta-analysis and conceptual consequences. Neurology. 1998;51(6):1538–45.
110. Yoshida K-I, Ogura Y, Wakasugi C. Myocardial lesions induced after trauma and treatment. Forensic Sci Int. 1992;54(2):181–9.
111. Samuels MA. Cardiopulmonary aspects of acute neurologic diseases. In: Ropper AH, editor. Neurological and Neurosurgical Intensive Care. 3rd ed. New York: Raven Press; 1993. p. 103–19.
112. Repertinger S, Fitzgibbons WP, Omojola MF, Brumback RA. Long survival following bacterial meningitis-associated brain destruction. J Child Neurol. 2006;21(7):591–5.
113. Shewmon DA. The brain and somatic integration: insights into the standard biological rationale for equating "brain death" with death. J Med Philos. 2001;26(5):457–78.
114. Bernat JL, Dalle Ave AL. Aligning the criterion and tests for brain death. Camb Q Healthc Ethics. 2019;28(4):635–41.
115. Dalle Ave AL, Bernat JL. Inconsistencies between the criterion and tests for brain death. J Intensive Care Med. 2020;35(8):772–80.
116. Lewis A, Bonnie RJ, Pope T, et al. Determination of death by neurologic criteria in the United States: the case for revising the Uniform Determination of Death Act. J Law Med Ethics. 2019;47(4 suppl):9–24.
117. Lewis A, Bonnie RJ, Pope T. It's time to revise the Uniform Determination of Death Act. Ann Intern Med. 2020;172(2):143–4.

118. Joffe AR, Khaira G, de Caen AR. The intractable problems with brain death and possible solutions. Philos Ethics Humanit Med. 2021;16(1):11.
119. Accad M. Of wholes and parts: a Thomistic refutation of "brain death". Linacre Q. 2015;82(3):217–34.
120. Fitzgerald RD, Dechtyar I, Templ E, Fridrich P, Lackner FX. Cardiovascular and catecholamine response to surgery in brain-dead organ donors. Anaesthesia. 1995;50(5):388–92.
121. Fitzgerald RD, Dechtyar I, Templ E, Pernerstorfer T, Hackl W, Lackner FX. Endocrine stress reaction to surgery in brain-dead organ donors. Transpl Int. 1996;9(2):102–8.
122. Kinoshita Y, Kamohara H, Kotera A, Sagishima K, Tashiro T, Niimori D. Healthy baby delivered vaginally from a brain-dead mother. Acute Med Surg. 2015;2(3):211–3.
123. Reinhold AK, Kredel M, Markus CK, Kranke P. Vaginal delivery in the 30+4 weeks of pregnancy and organ donation after brain death in early pregnancy. BMJ Case Rep. 2019;12(9):e231601.
124. Bernat JL. The definition, criterion, and statute of death. Semin Neurol. 1984;4(1):45–51.
125. Sánchez Sorondo M, editor. The Signs of Death. The Proceedings of the Working Group 11–12 September 2006. Vatican City: Pontificia Academia Scientiarum; 2007.
126. Tendler MD. Cessation of brain function: ethical implications in terminal care and organ transplant. Ann N Y Acad Sci. 1978;315:394–7.
127. Iserson KV. Death to Dust: What Happens to Dead Bodies? 2nd ed. Tucson, AZ: Galen Press; 2001.
128. Dubois J. Avoiding common pitfalls in the determination of death. Nat Cath Bioeth Q. 2007;(autumn):545–59.
129. López-Navidad A. Chronic "brain death": meta-analysis and conceptual consequences. Neurology. 1999;53(6):1369.
130. Shewmon DA. Brain death: can it be resuscitated? Hast Cent Rep. 2009;39(2):18–24.
131. Miller FG, Truog RD. Death, Dying, and Organ Transplantation. Reconstructing Medical Ethics at the End of Life. Oxford: Oxford University Press; 2012.
132. Gershon MD. The Second Brain. New York: Harper Collins; 1998.
133. Gershon MD. The enteric nervous system: a second brain. Hosp Pract. 1999;34(7):31–2, 5–8, 41–2 passim.
134. Thagard P. Thought experiments considered harmful. Perspect Sci. 2014;22(2):288–305.
135. Shewmon DA. 'Brain-body' disconnection: implications for the theoretical basis of 'brain death'. In: de Mattei R, editor. Finis Vitae. Is Brain Death Still Life? Rome, Italy: Consiglio Nazionale delle Ricerche; Rubbettino; 2006. p. 211–50.
136. Shewmon DA. Mental disconnect: 'Physiological decapitation' as a heuristic for understanding 'brain death'. In: Sánchez Sorondo M, editor. The Signs of Death. The Proceedings of the Working Group 11–12 September 2006. Scripta Varia 110. Vatican City: Pontificia Academia Scientiarum; 2007. p. 292–333.
137. Shewmon DA. On conscious non-organisms, unconscious persons, and bisected person-organisms. APA Newsletters. 2009;9(1):14–8.
138. Verheijde JL, Potts M. Commentary on the concept of brain death within the Catholic bioethical framework. Christ Bioeth. 2010;16(3):246–56.
139. Miller FG, Truog RD. Decapitation and the definition of death. J Med Ethics. 2010;36(10):632–4.
140. Pallis C. Whole-brain death reconsidered—physiological facts and philosophy. J Med Ethics. 1983;9(1):32–7.
141. Pallis C. Brain stem death—the evolution of a concept. Med Leg J. 1987;55(Pt 2):84–107.
142. Wijdicks EFM. The clinical determination of brain death: rational and reliable. Semin Neurol. 2015;35(2):103–4.
143. Youngner SJ, Bartlett ET. Human death and high technology: the failure of the whole-brain formulations. Ann Intern Med. 1983;99(2):252–8.
144. Khushf G. A matter of respect: a defense of the dead donor rule and of a "whole-brain" criterion for determination of death. J Med Philos. 2010;35(3):330–64.
145. Lizza JP. Defining death: beyond biology. Diametros. 2018;55:1–19.

146. Veatch RM. The impending collapse of the whole-brain definition of death [published erratum appears in Hastings Cent Rep 1993;23(6):4]. Hast Cent Rep. 1993;23(4):18–24.
147. Lizza JP. And she's not only merely dead, she's really most sincerely dead [letter]. Hast Cent Rep. 2009;39(5):5–6; author reply 6-7.
148. Lizza JP. In defense of brain death: Replies to Don Marquis, Michael Nair-Collins, Doyen Nguyen, and Laura Sullivan. Diametros. 2018;55:68–90.
149. Bernat JL. On irreversibility as a prerequisite for brain death determination. In: Machado C, Shewmon DA, editors. Brain Death and Disorders of Consciousness. Advances in Experimental Medicine and Biology, vol. 550. New York: Kluwer Academic/Plenum Publishers; 2004. p. 161–7.
150. Bernat JL. Refinements in the organism as a whole rationale for brain death. Linacre Q. 2019;86(4):347–58.
151. Byrne PA, O'Reilly S, Quay PM. Brain death—an opposing viewpoint. JAMA. 1979;242(18):1985–90.
152. Wijdicks EFM. Determining brain death in adults. Neurology. 1995;45(5):1003–11.
153. Shemie SD, Doig C, Dickens B, Byrne P, Wheelock B, Rocker G, et al. Severe brain injury to neurological determination of death: Canadian forum recommendations. CMAJ. 2006;174(6):S1–S12.
154. Shewmon DA. The probability of inevitability: the inherent impossibility of validating criteria for brain death or "irreversibility" through clinical studies. Stat Med. 1987;6(5):535–53.
155. Roberts DJ, MacCulloch KA, Versnick EJ, Hall RI. Should ancillary brain blood flow analyses play a larger role in the neurological determination of death? Can J Anaesth. 2010;57(10):927–35. [Erratum Can J Anaesth 2016;63(9):1116].
156. Joffe AR, Lequier L, Cave D. Specificity of radionuclide brain blood flow testing in brain death: case report and review. J Intensive Care Med. 2010;25(1):53–64.
157. Shewmon DA. False-positive diagnosis of brain death following the Pediatric Guidelines: case report and discussion. J Child Neurol. 2017;32(14):1104–17.
158. Latorre JGS, Schmidt EB, Greer DM. Another pitfall in brain death diagnosis: return of cerebral function after determination of brain death by both clinical and radionuclide cerebral perfusion imaging. Neurocrit Care. 2020;32(3):899–905.
159. Shewmon DA, Salamon N. The MRI of Jahi McMath and its implications for the global ischemic penumbra hypothesis. J Child Neurol. 2021;37(1):35–42.
160. Zuckier LS. Radionuclide evaluation of brain death in the post-McMath era, epilogue and enigmata. J Nucl Med. 2022;63(9):1323–5.
161. Bernat JL. How the distinction between "irreversible" and "permanent" illuminates circulatory-respiratory death determination. J Med Philos. 2010;35(3):242–55.
162. Bernat JL. Conceptual issues in DCDD donor death determination. Hast Cent Rep. 2018;48(Suppl 4):S26–S8.
163. Bernat JL. On noncongruence between the concept and determination of death. Hast Cent Rep. 2013;43(6):25–33.
164. Veatch RM, Ross LF. Defining Death: The Case for Choice. Washington, DC: Georgetown University Press; 2016.
165. Shewmon DA. Constructing the death elephant: a synthetic paradigm shift for the definition, criteria, and tests for death. J Med Philos. 2010;35(3):256–98.

Arguments Supporting the Whole-Brain Criterion

Michael A. Rubin and Ranier Reyes

Loss of the ability to maintain homeostasis after irreversible circulatory–respiratory failure is the traditional medical definition of death, and with such failure, also comes the loss of neurologic function. Historically, a distinction between circulatory–respiratory and neurologic failure was irrelevant, as in the case of catastrophic neurologic failure, circulatory–respiratory collapse inevitably occurred after cerebral herniation and with it came the same loss of homeostasis. With the implementation of advanced organ support, the distinction between the two became relevant. The order of system failure matters, as we can replace some, but not all organ functions with technology. While the same system loss cascade would occur after catastrophic renal injury if it were not for dialysis, there are essential functions of the nervous system that cannot be replaced in the same way our kidneys can. It is that distinction that justifies the organism to be considered dead by neurologic criteria while there is no such determination by renal criteria. Because technology has significantly altered the consequences of organ failure, a change in our understanding of life and death is warranted; the previous paradigm can no longer be applied in all circumstances. Shifting such a paradigm does not mean we are creating a new ontology; rather, we have revealed a truth that has always existed that just never had sufficient relevance to be contemplated and described.

Earlier chapters of this book reviewed the historical context of determination of death by neurologic criteria and arguments for and against its validity. In this chapter, we expand upon the arguments for the validity of death by neurologic criteria by supporting two claims: (1) a definition of death should be based on a biologic foundation and (2) the only clinically reliable and rationally validated way to support a biologic paradigm of death by neurologic criteria is with the whole-brain criterion. Otherwise, the formulation risks being criticized as a socially created concept based

M. A. Rubin (✉) · R. Reyes
UT Southwestern Medical Center, Dallas, TX, USA
e-mail: Michael.Rubin@utsouthwestern.edu; Ranier.Reyes@utsouthwestern.edu

A. Lewis, J. L. Bernat (eds.), *Death Determination by Neurologic Criteria*, Advances in Neuroethics, https://doi.org/10.1007/978-3-031-15947-3_4

on a metaphysical argument or an incomplete assessment of essential biologic function rather than an accurate description of the natural world.

The first argument in support of whole-brain criterion does not require much exploration; the whole-brain criterion is the most widely used and generally accepted so that convention is evidence of its value. While the Harvard Ad Hoc Committee did not offer a justification of the use of neurologic criteria to declare death, the subsequent President's Commission for the Study of Ethical Problems in Medicine and Biomedical and Behavioral Research, the Uniform Determination of Death Act (UDDA), and the American Academy of Neurology (AAN) practice parameters evolved the argument supporting the whole-brain criterion by establishing foundational concepts; it was not until recent history that this understanding was criticized and recast by some as motivated by other interests [1–5]. Perhaps consensus among forbearers led to a lack of perception of need to offer such justification or perhaps they were proceeding based on insufficiently explored intuition. Either way, subsequent scholars and policymakers developed an understanding of the concept. Whether by comfort with convention, an intuition-driven worldview, or rational argument, we have found that surveys of clinicians and the general public, as well as endorsements by numerous medical societies worldwide demonstrate that the whole-brain criterion maintains consensus [6–8]. While tradition is not sufficient to support a position on its own, one cannot ignore that this is the standard against which other viewpoints must argue why they are superior.

1 A Definition of Life

Death is universally accepted as an irreversible transition from the state of being alive to being dead. These are mutually exclusive states, with death being the loss of features that defined something as alive. Consequently, before death by neurologic criteria can be described, the requirements of life must be known. Once those are defined, then the key question can be asked: *does death by neurologic criteria satisfactorily indicate a transition from life to death?* When we argue the whole-brain formulation to be correct, we are really arguing that when certain functions of the central nervous system are irreversibly lost, the criteria of life are no longer met, and the organism is now dead.

It would be incorrect to say that a patient who undergoes cardiac bypass surgery requiring cardioplegia was temporarily dead until revived at the end of their surgery. Similarly, if a patient suffers a circulatory–respiratory arrest and is successfully resuscitated, they likewise were never dead. In both cases, the state in question was shown to be reversible. The time of death is assigned using circulatory–respiratory criteria to the moment when the clinician has determined that function is absent and will not return. An interesting distinction between irreversible and permanent has been proposed because of modern resuscitation techniques that require a judgment as to when efforts ought to be stopped due to proven failure in sustainability. This distinction, which is described in detail elsewhere in this book, allows patient, surrogate, and clinician judgment to determine when permanence is morally sufficient to meet the criteria normally subscribed to irreversible loss of circulatory–respiratory function [9–11].

The property of "living" is defined by features and functions that an organism has and that can be demonstrated. We do not require a pathological examination or the establishment of rigor mortis before we declare death. Such steps might offer greater reassurance, but they deny the cultural and religious norms that require an appropriate legal qualification and burial rights. Consequently, we have determined by consensus what observations are necessary to make a determination of death. Our observation-based declaration leads to the truth that one cannot claim what life *is*, but at most can offer a description of the functions of what a living organism is able *to do*. The current biologic consensus is that living organisms are composed of organic molecules organized into cells, tissues, and organ systems in such a way to utilize energy to maintain homeostasis, respond to the environment, adapt to threats to survival, and complete a generational cycle to reproduce [12, 13].

2 Emergent Properties as an Essential Component of Life

Structural organization allows the ability to compartmentalize and specialize biochemical activities. More importantly, complex organization allows the development of emergent properties: qualities and functions that only exist in the intact organism as a whole [14–18]. Such properties are not localizable in any one place or attributed to a single set of biochemical processes. These properties promote the organism's survival and ability to fulfill the functions of life. Without emergence, the ability to survive is significantly hampered, as evidenced by increasingly complex organisms evolving by natural selection. While "has emergent properties" is not usually listed as a component of life, their presence serves the fulfillment of the other criteria. Specifically, in human beings, consciousness and social behaviors allow protection from threats and the ability to obtain nutrition to maintain homeostasis. A simple thought exercise reveals the necessity of including emergent properties in a definition of life: if one were to place every organ of a human (or any other organism that is composed of more than a single cell) in a jar supported by an adequate milieu to maintain homeostasis, while it is self-evident that these are organic components, the person is no longer a living human organism. Logically speaking, if exclusion of an element of a definition interferes with the integrity of the definition, this element is necessary. When a person fulfills the whole-brain criterion for death, they may maintain some elements of living organisms, but they have lost essential emergent properties, cannot obtain consciousness, cannot respond to the environment, and cannot adapt to new challenges to survival.

The argument of the "organism as a whole" has been supported for decades [2, 13–18]. The "whole" organism possesses specific emergent functions, which are essential for the functioning of the organism and are a product of the collective and unified whole; the permanent cessation of these emergent functions (e.g., control of respiration, circulation, conscious awareness) therefore indicates the organism's death [16]. Centers which are responsible for these emergent properties are distributed throughout various areas within the brain and brainstem; the whole-brain formulation thus argues for the irreversible loss of function of the entire brain, including the neocortex and brainstem, as a necessity for the determination of death [16, 19, 20]. Crucial to the whole-brain criterion is the understanding that "whole" does not equate to "all" [15, 20]. While there is likely to be activity within some neural cells after death, just as there is some retained cardiac myocyte activity after circulatory cessation, this activity no longer has the ability to significantly contribute to the operation of the organism as a whole without support of technology [20].

3 Homeostasis Is Not Enough

A prominent criticism of the whole-brain criterion is that it requires more than the ability to maintain homeostasis, which some critics argue is the *only true criteria* for death. Others argue that this is a "legal-fiction" created to respect the concept of the "dead donor rule" and permit organ recovery [5]. While this argument is coherent and internally consistent, it falls short in that it limits the biologic definition to one component and excludes (or at least, does not include) the presence of essential emergent properties and the ability to respond to the external environment, adapt to challenges to survival, and grow and procreate. One could counter that if one can maintain homeostasis, the other features can be restored, but in the case of irreversible loss of consciousness and the other adaptive features of the nervous system, such is not the case. One cannot exclude a functioning nervous system (or one that may recover to a functional state) in a definition of human life that includes all the aforementioned features of life. As an extension, the criticism that the comparison of death of a human being qualified differently than death in another mammal or a single-cell organism by requiring the emergent properties of consciousness and awareness falls short, as another criterion is not being applied. Rather the claim is consistent: the loss of emergent properties is part of all definitions of death. Such properties are more complex, however, in a human than other organisms, and therefore are easier to identify when they have been lost.

4 Replacing Systemic Integration Does Not Preclude the Importance of the Brain

An early formulation of death by neurologic criteria included the integrative function of the nervous system as grounds for its veracity. In as much as the nervous system controls and influences other systems through direct stimulation and neuroendocrine control, the irreversible loss of neurologic function would lead to system disintegration and cause further organ system failures in a sequential fashion. While this is partially true, supportive therapies can augment these functions and allow other organs to continue for days to months, or perhaps even years [21]. For example, fluid and sodium balance can be managed with intravenous fluids with various concentrations of salt, and loss of cardiovascular tone and cardiac dysrhythmias can be modulated by medications. With improvements in critical care, the whole-brain criterion can no longer claim support from the thought that the nervous system is the keystone on which all other systems depend; however, it does not need to be valid. The inherent functions unique to the nervous system are in themselves requisite to fulfill the qualities of a living human person.

5 The Brainstem Criterion and Why It Falls Short

The argument for death by neurologic criteria based on irreversible loss of all brainstem function uses a reductionist approach based on the premise that death involves two features: (1) the irreversible loss of the capacity for consciousness and (2) the irreversible loss of the capacity to breathe [22]. The irreversible loss of brainstem function, regardless of the mechanism of injury, is believed to produce this state. As the center for respiratory drive, hemodynamic regulation, and arousal and wakefulness, the brainstem is essential for life. Nearly all input and output of the brain passes through the brainstem, making it essential in how we respond to environmental or external stimuli and integrate higher cortical function into survival. Many proponents of the brainstem criterion also argue that most clinical tests utilized in determinations of death by neurologic criteria, using either the whole-brain or brainstem criterion, involve the evaluation of brainstem reflexes; the brainstem criterion is therefore pragmatic and easily testable. Lastly, advocates emphasize that loss of brainstem function ultimately leads to circulatory–respiratory arrest and thus appeases those who define death as irreversible cessation of circulation and respiration [18, 22, 23].

Critics of the brainstem formulation point out ways in which it fails to account for the importance of higher-brain function. It fails to account for conscious awareness and plausible scenarios in which this may be preserved via function of the cerebral cortices and thalami. Bernat proposes a hypothetical patient with extensive, but isolated, brainstem damage who may present with brainstem areflexia yet preserved consciousness; this imaginary scenario portraying the most severe version of a "locked-in" patient illustrates a situation in which death by neurologic criteria may be clinically evident based on bedside clinical testing but arousal is preserved in supratentorial structures [20]. With the advent of invasive neurologic probes to treat movement disorders, one can imagine a future in which invasive technology promotes arousal artificially. In theoretical situations of isolated brainstem damage, this would provide an avenue for preserved cortical function to respond to the surrounding environment and to perform the necessary "work" of the human organism. A practical and essential criticism to the brainstem criterion is that pathologies causing isolated obliteration of brainstem function, but with preservation of cortical function, may not dependably be irreversible. In contrast, supratentorial lesions causing catastrophic elevations in intracranial pressure compressing the brainstem will reliably not be reversible. As we have previously stated, the irreversible loss of function is the *sine qua non* of death; therefore, this uncertainty precludes isolated brainstem dysfunction as a reliable means of declaration.

6 The Higher-Brain Criterion and Why It Falls Short

The higher-brain criterion argues that the primary and distinctive function of the human being includes "the individual's personality, his conscious life, his uniqueness, his capacity for remembering, judging, reasoning, acting, enjoying, worrying, and so on," all of which require an intact cerebral cortex [24]. Therefore, death is the "irreversible loss of integrated function of body and mind" and may be conceptualized as the irreversible loss of higher-brain functions [24, 25]. While some dismiss this approach outright as esoteric philosophy, this view has ancient roots and similar arguments have been made based on the works of Aristotle as well as St. Thomas Aquinas [26]. Whether by a rational understanding of the difference between the material and metaphysical or through a religious or spiritual belief in the eternal soul, there is a component of human psychology that intuitively requires the presence (or a least potential presence) of a conscious mind to see someone as living. In these traditions, death is the moment of separation of the physical body from the incorporeal soul; the separation of the temporary mortal self and the eternal self. The irreversible loss of the human nervous system function would extinguish a person's ability to have access to the connection of mind or soul, and therefore qualify as death.

Robert Veatch made an interesting contribution to the literature in his proposal that the debate regarding the definition of death is really a discourse on the loss of moral and legal standing, rather than the moment that one has switched from one biological category to another [27]. The ability of humans to self-identify, form social relationships, and possess "moral status" is essential to promoting survival of the human organism. This position is persuasive and attractive in many ways. First, it defines death (and life) beyond strictly biological means. Given the societal and cultural implications of the determination of life and death, he argues that the definition of human life and death should not be reduced to somatic and biological measures. There is a reason defining death has such a profound impact outside of scientific medicine and permeates through the realms of law, ethics, and the public. Second, it reiterates the idea that the brain possesses something unique to human existence, the loss of which therefore results in death. The capacity for consciousness in concordance with higher functioning (e.g., reasoning, personality, etc.) is a trait unique not only to humans but to living humans, and is housed in the cerebrum. If death by neurologic criteria accepts the irreversible loss of something within the brain that distinguishes human life from exclusively the existence of cells and tissues, it is reasonable to believe this feature is in fact cognition and the ability to integrate body and mind. This understanding is even more compelling given advances in modern medicine which allow for sustenance of bodily functions.

A practical issue of the higher-brain formulation centers on the examples of anencephalic patients or those in a persistent vegetative state/unaware–wakeful state [19, 20]. These patients, who are breathing and have intact brainstem reflexes, could be labeled as deceased under the higher-brain criterion given they lack the ability for body–mind integration and possession of "full moral standing" [26]. The question is then raised: how much irreversible loss of neocortical function is sufficient to consider someone dead? Are patients with severe and irreversible dementia, who are still breathing and who still have a beating heart, but have profound cortical damage and lack the ability to integrate higher-brain function with the body and external environment, considered deceased? In much of the world today, the majority of cultures would have a difficult time considering these patients, who have long been labeled as alive, now deceased. Another significant concern is the ability to empirically measure or determine death. If death refers to the loss of higher consciousness and an individual's place in moral culture, or loss of personhood, how does one know and quantify when this occurrence takes place [28, 29]? Determinations of death based on loss of higher-brain definitions are therefore individualized and "person-oriented" and strictly non-biological; consequently, specific tests to fulfill this still need be described and validated [20, 28]. Otherwise, determining death becomes highly subjective.

7 Ensuring Certainty Compels the Whole-Brain Criterion

Brainstem and higher-brain formulations are individually important because they each reference critical functions of the human organism. Each, however, is insufficient to stand on its own in satisfying a reliably permanent and measurable definition of death, and therefore a whole-brain formulation is necessary (see Fig. 1). Both the brainstem and cortex are necessary to create the emergent property of consciousness, which is arguably the single feature that makes the brain *sui generis* among organs [16]. The brainstem formulation, while easier to measure with bedside exams, cannot offer guidance on when a radiographically intact cortex or function measured by electroencephalography is moot because such a bedside exam cannot predict when peri-insult edema will recede, or penumbra may recovery adequately to reinstitute reticular activating system function. Likewise, a neurologic insult sufficient to radiographically obliterate cortex but not cause sufficient elevation in intracranial pressure to obviate cranial nerve function or a respiratory drive is not reliably irreversible. This argument has led some to suggest that a determination of death by neurologic criteria should always include a measurement of brain circulation (angiogram, nuclear flow study, or transcranial

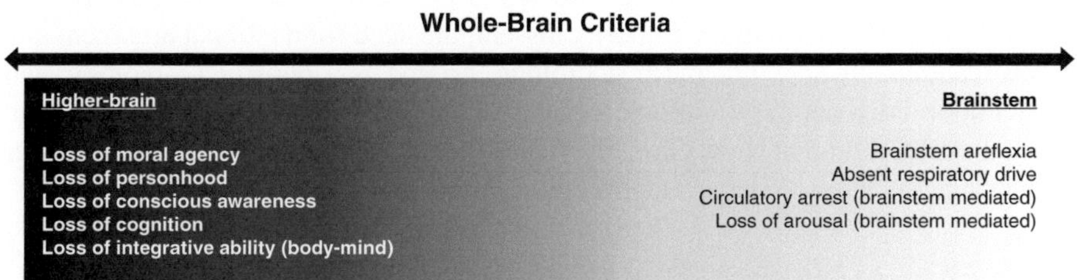

Fig. 1 Potential formulations of death by neurologic criteria

doppler) to demonstrate cerebrovascular arrest, and thereby corroborate that function will not return [30].

The whole-brain criterion offers a higher level of assurance that the damage to the brain is severe enough that function will not return. Requiring loss of all functions of the brain, as described in the whole-brain formulation, has been likened to the Talmudic principle of "building a fence around the Torah" where one assumes a conservative posture outside of an area of uncertainty to ensure that the law or truth is being respected [31]. Emerging science would suggest the growing importance of such reassurance. Investigation into disorders of consciousness have revealed with brain imaging, latent consciousness in severe neurologic injury to the extent that a patient may be able to experience the world in a way that cannot be detected by a clinician [32, 33]. As we cannot ensure what we cannot measure, dysfunction of the whole-brain provides a necessary reassurance that cannot be provided by either the brainstem or higher-brain criterion [34].

In a similar fashion, multiple studies have shown that patients declared dead by neurologic criteria do not always have loss of all cellular architecture [35, 36]. A recent study showed that restoration of brain circulation in a postmortem but intact pig brain allowed return of some cellular activity 4 h after circulatory–respiratory arrest [37]. While opponents would use these data to indicate that death by neurologic criteria is inaccurate and that brain function is still possible, proponents would argue that severe enough injury that leads to clinically indetectable neurologic activity demonstrates that not every cell needs to be lost for the brain to lose its ability to provide the functions necessary to support that the organism is living.

8 Conclusions

Clinicians offering a determination of death by neurologic criteria will inevitably be confronted with questions from patients such as "but why is their heart beating" and "why can the rest of the body be working if they are dead?" While an offering of legal categorization and hospital standard will often suffice, for many patient families, it will not be enough. The clinician themselves ought to establish in their own mind why such a state of being is morally and rationally equivalent to death by circulatory–respiratory criteria. While the brainstem and higher-brain formulations offer significant insight into essential functions of the brain, only the whole-brain criterion supports a valid claim of a transition from life to death in the case of catastrophic neurologic injury. Despite the claim of the minority dissent, the need for a rational argument and moral justification does not mean neurologic criteria is a construct—biology, and most importantly complex biologic organisms, deserve the respect of a thorough rational exploration that shows that life cannot be reduced to a single attribute in our current state of advanced organ supporting and replacing technology.

References

1. Report of the Ad Hoc Committee of the Harvard Medical School to Examine the Definition of Brain Death: a definition of irreversible coma. JAMA 1968;205(6):337–40.
2. President's Commission for the Study of Ethical Problems in Medicine and Biomedical and Behavioral Research Defining Death: Medical, Legal, and Ethical Issues in the Determination of Death. https://repository.library.georgetown.edu/handle/10822/559345. Accessed 23 Feb 2022.
3. The Quality Standards Subcommittee of the American Academy of Neurology. Practice parameters for determining brain death in adults. Neurology. 1995;45(5):1012.
4. Wijdicks EF, Varelas PN, Gronseth GS, Greer DM. Evidence-based guideline update: determining brain death in adults: report of the Quality Standards Subcommittee of the American Academy of Neurology. Neurology. 2010;74(23):1911–8.
5. Truog RD, Miller FG. The meaning of brain death: a different view. JAMA Intern Med. 2014;174(8):1215–6.
6. Joffe AR, Anton NR, Duff JP, Decaen A. A survey of American neurologists about brain death: understanding the conceptual basis and diagnostic tests for brain death. Ann Intensive Care. 2012;2(1):4.
7. Siminoff LA, Burant C, Younger SJ. Death and organ procurement: public beliefs and attitudes. Soc Sci Med. 2004;59(11):2325–34.
8. Lewis A, Bakkar A, Kreiger-Benson E, Kumpfbeck A, Liebman J, Shemie SD, et al. Determination of death by neurologic criteria around the world. Neurology. 2020;95(3):e299–309.
9. Bernat JL. How the distinction between "irreversible" and "permanent" illuminates circulatory-respiratory death determination. J Med Philos. 2010;35(3):242–55.
10. Bernat JL. Controversies in defining and determining death in critical care. Nat Rev Neurol. 2013;9(3):164–73.
11. Bernat JL. Are organ donors after cardiac death really dead? J Clin Ethics. 2006;17(2):122–32.
12. Macklem PT. Emergent phenomena and the secrets of life. J Appl Physiol. 2008;104(6):1844–6.
13. Macklem PT, Seely A. Towards a definition of life. Perspect Biol Med. 2010;53(3):330–40.
14. Bernat JL. Refinements in the 0rganism as a whole rationale for brain death. Linacre Q. 2019;86(4):347–58.
15. Bonelli RM, Prat EH, Bonelli J. Philosophical considerations on brain death and the concept of the organism as a whole. Psychiatr Danub. 2009;21(1):3–8.
16. Huang AP, Bernat JL. The organism as a whole in an analysis of death. J Med Philos. 2019;44(6):712–31.
17. Bernat JL. A conceptual justification for brain death. Hast Cent Rep. 2018;48(Suppl 4):S19–s21.
18. Crane JK. Biological-mereological coincidence. Philos Stud. 2012;161(2):309–25.
19. Bernat JL, Culver CM, Gert B. On the definition and criterion of death. Ann Intern Med. 1981;94(3):389–94.
20. Bernat JL. The biophilosophical basis of whole-brain death. Soc Philos Policy. 2002;19(2):324–42.
21. Shewmon, DA. Statement in support of revising the Uniform Determination of Death Act and in opposition to a proposed revision. J Med Philos. 2021. Online ahead of print.
22. Simpson P, Bates D, Bonner S, Costeloe K, Doyal L, Falvey S, et al. A code of practice for the diagnosis and confirmation of death. Academy of Medical Royal Colleges. https://aomrc.org.uk/wp-content/uploads/2016/04/Code_Practice_Confirmation_Diagnosis_Death_1008-4.pdf. Accessed 24 Feb 2022.
23. Pallis C. ABC of brain stem death. Reappraising death. Br Med J (Clin Res Ed). 1982;285(6352):1409–12.
24. Veatch RM. The impending collapse of the whole-brain definition of death. Hast Cent Rep. 1993;23(4):18–24.

25. Veatch RM. Would a reasonable person now accept the 1968 Harvard brain death report? A short history of brain death. Hast Cent Rep. 2018;48(Suppl 4):S6–s9.
26. Eberl JT. A Thomistic defense of whole-brain death. Linacre Q. 2015;82(3):235–50.
27. Veatch RM. The death of whole-brain death: the plague of the disaggregators, somaticists, and mentalists. J Med Philos. 2005;30(4):353–78.
28. Khushf G. A matter of respect: a defense of the dead donor rule and of a "whole-brain" criterion for determination of death. J Med Philos. 2010;35(3):330–64.
29. Green MB, Wikler D. Brain death and personal identity. Philos Public Aff. 1980;9(2):105–33.
30. Dalle Ave AL, Bernat JL. Inconsistencies between the criterion and tests for brain death. J Intensive Care Med. 2018;35(8):772–80.
31. Gelb DJ. Building a fence around brain death: the shielded-brain formulation. Neurology. 2021;97(16):780–4.
32. Edlow BL, Chatelle C, Spencer CA, et al. Early detection of consciousness in patients with acute severe traumatic brain injury. Brain. 2017;140(9):2399–414.
33. Schiff ND. Cognitive motor dissociation following severe brain injuries. JAMA Neurol. 2015;72(12):1413–5.
34. Verheijde JL, Rady MY, Potts M. Neuroscience and brain death controversies: the elephant in the room. J Relig Health. 2018;57(5):1745–63.
35. Walker AE, Diamond EL, Moseley J. The neuropathological findings in irreversible coma: a critique of the "respirator brain". J Neuropathol Exp Neurol. 1975;34(4):295–323.
36. Wijdicks EF, Pfeifer EA. Neuropathology of brain death in the modern transplant era. Neurology. 2008;70(15):1234–7.
37. Vrselja Z, Daniele SG, Silbereis J, et al. Restoration of brain circulation and cellular functions hours post-mortem. Nature. 2019;568(7752):336–43.

Arguments Supporting the Brainstem Criterion

Dale Gardiner and Andrew McGee

In the debate about whether the whole-brain criterion or the brainstem criterion represents the best formulation of death by neurologic criteria—sometimes called the transatlantic divide due to the opposing conceptual positions in the United States and the United Kingdom—one figure looms large: Professor Christopher Pallis (Fig. 1) [1]. Pallis was a fierce defender of the concept of death by neurologic criteria, but he was also the strongest advocate for the brainstem formulation. Today he would be called an influencer.

Professor Pallis was the Reader Emeritus in Neurology at the Royal Postgraduate Medical School, London and at the Hammersmith Hospital until he retired in 1982. He died in 2005. Pallis came to great prominence in the debate about death by neurologic criteria following a TV show produced by the BBC documentary program Panorama entitled "Transplants—are the donors really dead" [2]. Against strong medical opposition, Panorama proceeded with transmission of this program in the United Kingdom on October 13, 1980 [3]. The program made claims that patients recovered after determinations of death by neurologic criteria and that doctors in the United Kingdom were carrying out determinations of death by neurologic criteria incorrectly and unsafely.

This program provoked a huge outcry by the medical profession and a press conference was organized on Tuesday, November 25, 1980 where Professor Pallis, and others, presented [3]. Pallis reminded the audience that none of the patients shown on the show would have been declared dead by neurologic criteria in any jurisdiction. Nine publications would follow in the *British Medical Journal* over the winter of 1982–1983 in which Pallis defended death by neurologic criteria, the

D. Gardiner (✉)
Adult Critical Care, Nottingham University Hospitals NHS Trust, Nottingham, UK
e-mail: dalegardiner@doctors.net.uk

A. McGee
Queensland University of Technology, Brisbane, Australia
e-mail: a.mcgee@qut.edu.au

A. Lewis, J. L. Bernat (eds.), *Death Determination by Neurologic Criteria*, Advances in Neuroethics, https://doi.org/10.1007/978-3-031-15947-3_5

Fig. 1 Christopher Pallis. Image kindly provided to one of the authors (DG) by Professor Pallis's family

brainstem criterion, and the United Kingdom's approach [4–12]. These papers would be assembled in 1983 as a simple yet era-defining book *The ABC of Brainstem Death* that went through two editions and remained a key resource in many intensive care units until very recently [13, 14].

In this chapter, we outline the legacy of Pallis and his ongoing influence on our understanding of death by neurologic criteria and on the coherence that can be achieved between the concept of death and the standards used to determine it if one accepts the brainstem criterion. We remind readers that the clinical tests for determination of death by neurologic criteria largely examine the brainstem, so that it is the anatomy of the brainstem that unites death by neurologic criteria globally. We argue, like Pallis, that there is a hierarchy of functions that should count when determining human death, and we will show the incoherence of definitions and criteria that insist on the loss of ALL functions of the entire brain, which can only lead to dispute and challenge. We support the standard neurologic teaching that there is no possibility of the cortex being conscious without the brainstem. This is why the entire brain standard is unnecessary. If one accepts the brainstem criterion, then one is better poised to anatomically align definition and criteria to function. A comparison of recent legal cases in the United Kingdom, Canada, and the United States highlights the difficulties that occur if incoherent medical concepts, laws, and practice are allowed to persist. Finally, this chapter answers some criticisms levelled against the brainstem criterion and reflects on the growing evidence base on the importance of the brainstem as the anatomical seat of rudimentary consciousness and the building block for all human consciousness.

1 The Legacy of Christopher Pallis

Apart from the obvious lesson that poorly researched journalism with a sensational and macabre bent will cause societal damage (a lesson seldom heeded in the decades that have followed), Pallis bequeathed to us, the next generation of critical thinkers on the topic of the determination of death, a number of conceptual legacies. First, he taught us that determining death should be a clinical diagnosis. A determination of death is like any other medical diagnosis and is satisfied when certain specified criteria are met. According to Pallis, the best defense against diagnostic errors when applying criteria requires the use of the clinician's brain, not technology – not even some machine. When clinicians are careful and well-trained, the risk of misdiagnosis can be avoided. Second, he taught us that not only is death the loss of biological function but that some functions count more than others. Death is the loss of the capacity for consciousness and breathing, both anatomically located in, and reliant on, a functional brainstem. The loss of brainstem function should, he claimed, be the determinant of human death.

1.1 The Determination of Death is a Clinical Diagnosis

The claim that "death is a clinical diagnosis" reminds us that medicine is concerned with biological processes. Death has not always been a clinical diagnosis determined by doctors. The Hippocratic tradition has been that, as death approached, or seemed to approach, doctors withdrew from patient care and gave way to the family and priests [15].

We would not do justice to Pallis in our analysis if we did not acknowledge his wickedly biting humor [16]. His work is, unexpectedly for scientific papers, often peppered with limericks, some of his own creation, others anonymous.

As Pallis wrote in 1982,

> In the heat of the public controversy about brain death two years ago a limerick was written which summed up the simple wisdom that death is a process:
>
> In our graveyards with winter winds blowing
> There's a great deal of to-ing and fro-ing
> But can it be said
> That the buried are dead
> When their nails and their hair are still growing? [4, p. 1411]

Of course, Pallis would be the first to acknowledge that the apparent growth of hair and nails after death is an artifactual illusion. What Pallis is directly expressing by quoting the limerick is the long-held human fear of being buried alive [17, 18], which in more modern settings is often manifested, like in the Panorama program, in the fear that organ donors are not really dead.

If Pallis is the champion of medical reassurance of the twentieth century, Eugène Bouchut was the champion of the nineteenth century. The Academy of Sciences in Paris had observed that for 50 years in Germany, in an effort to prevent premature burial, the apparently dead had to be placed in stone and unrefrigerated buildings for a number of days (often three) before burial was allowed [19]. Not surprisingly, nobody recovered to life, but families still had to pay for the body to be observed. The Academy of Sciences resolved to offer a cash prize to the physician who successfully made "the diagnosis of death safe, prompt and easy" [19]. The prize was won by Bouchut in 1846 for his Trâité des Signes de la Mort (Treatise on the signs of death) [19]. He made two compelling arguments that were accepted by the Academy. The first was that it should be doctors who determine death and they should be paid for doing so. This would aid public safety in preventing premature burial. His second argument addressed the need for ease and timeliness. A doctor should determine death using four diagnostic criteria. Fascinatingly, three of these are neurologic criteria and only one is cardiovascular:

1. Absent breath.
2. Loss of feeling and movement.
3. Dilation of the pupil.
4. Absent heart sounds by use of a stethoscope for 5 minutes.

1.2 Clinical Diagnoses Have Criteria

Why choose doctors for the societal task of determining death? Doctors, as practitioners of medicine, use and employ the diagnostic process. Diagnoses follow rules which gives them safety and timeliness, while also implicitly allowing for medical advances over time to provide new diagnostic criteria for even greater certainty. This, so society judged, is the best way to determine death. Integral to criteria and associated with any diagnostic process, but perhaps less appreciated by society, are concepts of sensitivity and specificity. In any diagnosis, there must be the possibility of true positives, false positives, true negatives, and false negatives. This is a conceptual truth about the diagnostic process. It accepts the possibility of error, even as it strives for accuracy and improvement over time. The process whereby a rheumatologist diagnoses rheumatoid arthritis follows a similar process to an intensive care doctor making a determination of death by neurologic criteria: a diagnosis is made if the relevant set of criteria are met. We must understand what doctors are doing when they determine death. Doctors are not making a spiritual or familial pronouncement; they are doing what they always do, using criteria to make a diagnosis. Pallis's legacy, that death first and foremost should remain a clinical diagnosis with clear and established criteria, continues to be supported in medical consensus and courts [20, 21].

What may not have sufficiently emerged yet in this chapter is the *pragmatism* inherent to medicine and to the diagnostic process. The Academy of Sciences in Paris recognized a challenge that persists to this day, namely to identify a moment within a continuous biological process of failure and decay that can be determined by doctors and the wider health community as the moment of death, while also being acceptable to lawmakers and society. The moment in question, as the Academy sought, has to be capable of being safely, promptly, and easily identified [19]. These requirements can pull in different directions [22].

The requirement of safety means there can be no coming back to life after death is declared. "Easily" does not mean slipshod in medicine: it means reproducible, clear, and acceptable. Easily identifiable criteria promote safety by increasing accuracy and decreasing interobserver variability. This can still require, as both Bouchut and Pallis advocated, the experience of well-trained doctors. While Bouchut was quick to employ the latest technological aid to assist doctors in determining death, Pallis was more cautious regarding technology. He argued that the determination can and should be a bedside clinical diagnosis and that the best defense against errors was "common sense by experienced and humane physicians [8]." He was therefore against the use of the electroencephalogram (EEG) as a diagnostic aid in determining death by neurologic criteria. Stirring up the transatlantic divide, Pallis stated: "Many American jurors have a touchingly naive faith in the supremacy of machines [12]." The failings of EEG have only become more obvious over time, and it is now difficult to justify their use as an ancillary test [21, 23, 24]. Promptness, or timeliness, is important because the determination of death is not just a medical determination; it has societal, legal, and familial ramifications. The declaration of death by a doctor allows bereavement to formally start, grants family permission to leave the bedside, allows autopsies to commence (very rapidly in warm autopsies), organs to be recovered, and the body to be buried. We may also need an easy way to determine who might inherit under a will using the rules of probate: if both members of a couple die in an accident, we sometimes need to know who died first [25].

Recognizing the tension between safety, ease, and timeliness led the Academy of Sciences to seek a workable medical answer. Rigor mortis as a criterion of death may be safe, but it is not very timely, and according to those concerned about premature burial, not even easy to recognize [26]. The accepted criteria used by doctors to determine death are therefore critical for navigating this tension. What is required are clear diagnostic criteria, which doctors can be trained to recognize so that when a declaration of death is made, the receivers of this pronouncement are accepting.

1.3 Death Is the Loss of Biological Function in the Brain

The clinical criteria, which doctors use to determine death, rely on the identification of loss of biological function in a human being. The alternative perspective might be that a non-biological process determines when human death occurs. An example would be when the soul leaves the body. However, there are no workable criteria for determining when death has occurred in this sense. As we have seen, there has been a societal shift away from priests determining death to doctors. This historical observation should act as a warning to doctors that trust once given can be taken away. A heavy responsibility and duty remain on medicine and science to provide coherent explanations of human death and for death to be determined in a safe and timely manner.

The birth of intensive care in the 1950s only made the job harder [22]. Unlike throughout all of human history, developments in technology meant that brain arrest and circulatory arrest no longer had to coincide. It was now possible for mechanically ventilated patients to permanently lose brain circulation and function, including the ability to breathe spontaneously, yet their hearts continue to beat and other organ and cellular functions to persist. This raised an important conceptual question: did all functions of the body need to cease for death to occur, or only some? This question cannot be answered solely empirically. Human decision sometimes has a role to play in answering such conceptual questions. Empirically, we discovered that brain arrest and circulatory arrest no longer had to coincide. However, that is not the same as *discovering* that patients who had suffered brain arrest were *dead*. A societal decision had to be made to decide whether loss of brain function counted as death. Society decided that people who had suffered "a permanently nonfunctioning brain" were dead [27].

Pallis, and many others then and since, would often use the example of decapitation to defend the societal decision [14, 28]. This was on the basis that most people accept that death has occurred at the moment of decapitation, even if, as would usually occur, the heart continues to beat for a period. Death by neurologic criteria is, by analogy and by loss of function, *physiological* decapitation. This claim has not been without conceptual and physiological challenge and response [29, 30]. Yet recognition of the preeminence of brain function over other bodily functions has only increased as medicine advanced to be able to replace all other organ functions except the brain, e.g., through dialysis, cardiopulmonary bypass for cardiac surgery and extracorporeal membrane oxygenation (ECMO) technology, and transplantation of organs [25]. The decisive point is that, if we replaced your brain with another brain donated by a third party, it would no longer be you who woke up following the operation. This is not true of any other transplanted organ. Although attempts have been made to respond to this point [31], they all fail [22, 28].

1.4 Some Functions in the Brain Count More Than Others

Many nations followed the path of the United States, with definitions and criteria for human death that made determination of death by neurologic criteria additive to the current practice of determination of death by circulatory–respiratory criteria, leading to a conceptual duality: circulatory–respiratory or neurologic [32]. A few nations, most notably the United Kingdom, implemented a unified conception of death: "Whatever the mode of its production, brain death represents the stage at which a patient becomes truly dead" [33]. Under this unified conception, loss of circulatory–respiratory function is only important because of what this signifies to brain function.

If this were the only difference between the US and UK positions, there would be no transatlantic divide, but more a case of "You say tomato, and I say tomahto." The evaluation for death by neurologic criteria would be almost identical in both countries, just with some different flavoring (flavouring). However, this is not the case. The divide is significant, with the United States having legislated that death by neurologic criteria required the loss of *all* functions of the *entire* brain while the United Kingdom identified the loss of brainstem functions as what counted: "It is agreed that permanent functional death of the brainstem constitutes brain death" [34].

It is important to recognize how old the transatlantic divide is. If "[o]ur first problem is to determine the characteristics of a permanently nonfunctioning brain" [27], the Ad Hoc Committee of the Harvard Medical School's answer in 1968 was that "[f]unction is abolished at cerebral, brainstem, and often spinal levels." [27]. Clinically, it was obvious to the Ad Hoc Committee, just as it remains so today, that loss of spinal function was not a requirement for determination of death by neurologic criteria. However, loss of cerebral and brainstem function—anatomically located in the brain—was required. This was supported by the drafters in 1980 of the Uniform Determination of Death Act (UDDA) and approved by the American Medical Association and American Bar Association, where the determination of death by neurologic criteria required "irreversible cessation of all functions of the entire brain, including the brainstem" [32]. Ultimately, this wording made its way into legislation in the majority of states in the United States and in many other countries. Relevant to this chapter's discussion is the prefatory note provided by the drafters defining the "entire brain" and the apparent reasons for this inclusion. The purpose does not appear to be to distinguish the US position from the British brainstem position, but to reject a higher brain criterion for death.

> The 'entire brain' includes the brainstem, as well as the neocortex. The concept of 'entire brain' distinguishes determination of death under the Act from 'neocortical death' or 'persistent vegetative state'. These are not deemed valid medical or legal bases for determining death. [32, p. 3]

In contrast, and as stated above, the UK position is that permanent functional death of the brainstem constitutes the stage at which a patient becomes truly dead, and this was codified in the 1970s. Pallis gave credit for the British support of the brainstem criterion to an earlier individual, Keith Simpson. Simpson was a professor of pathology who, on being asked in 1964 by the Medical Protection Society for a definition of death, proposed the following: "there is life so long as a circulation of oxygenated blood is maintained to live brainstem centres" [14, 35, 36].

So, what is it about the brainstem that led the drafters of the UDDA to specifically identify this part of the brain as part of the entire brain, and led the British to identify permanent loss of this part of the brain as the *sole* criterion for all human death?

The brainstem anatomically houses two essential functional areas of the brain—the *reticular activating system*, an essential component for the generation of consciousness, and the *medulla oblongata* essential for the capacity to breathe. Pallis viewed the loss of consciousness as a reformulation (in terms of modern neurophysiology) of the older cultural concept of the departure of the "conscious soul" from the body and irreversible apnea as the permanent loss of "the breath of life" [14].

So, why privilege consciousness and breathing as the brain functions that count rather than all functions of the entire brain? Many of the arguments used above to privilege brain function over other functions in the body are, in effect, arguments for the privileging of consciousness over other brain functions. They are therefore, in truth, arguments for privileging the brainstem (we explain why below). We have already noted the irreplaceability of the brain. Why? What makes the brain so special that society accepts one can be dead even as other functions in the body can continue? The answer is consciousness. Taking this thought experiment a step further, if one had to choose between losing hormonal function of the brain or consciousness, most would hold to consciousness as the biological function that counts. Just as somatic functions are not equal to brain function, so some brain functions are also not equal—and we find, not unexpectedly, that consciousness reigns supreme.

2 The Coherence that Comes When One Accepts the Brainstem Criterion

Standard textbooks of neurology define consciousness as having two major components: awareness (content) and arousal (wakefulness) [37]. Awareness is identified as a function of the cerebral cortex, whereas arousal relies on a functioning brainstem reticular activating system. Coma, a pathological state marked by deep unconsciousness, typically involves compromised function of the brainstem and other deep brain structures [38]. While it is possible to lose awareness but maintain arousal (e.g., a vegetative state/unaware–wakeful state), loss of brainstem-mediated arousal mechanisms results in simultaneous loss of awareness. Modern textbooks have no difficulty in teaching that intact function of the ascending reticular activating system, with its direct and indirect connections with all levels of the central nervous system, is the basis of, and essential for, any form of consciousness [39, 40]. There is no possibility of the cortex being conscious without the brainstem. This is why the whole-brain criterion is not needed.

Using a computer analogy, the brainstem is like the brain's motherboard. It is what everything else needs to have working so that everything else too can work. However, unlike a computer motherboard, you cannot replace a brainstem. Today, when your brain's motherboard dies, you die. There is no resuscitation possible, and you do not wake up and you do not breathe again.

There is substantial and valid criticism that the insistence in the UDDA on loss of "all functions of the entire brain" does not match clinical practice where patients determined dead by neurologic criteria are observed to have persistent brain hormonal function (e.g., secretion of antidiuretic hormone), as discussed elsewhere in this book [31, 41]. Some have used the UDDA statement "A determination of death must be made in accordance with accepted medical standards" [32] as a kind of get-out-of-jail clause [42]. If ongoing brain hormone function is in accord with accepted medical standards, then it is not necessary to have all functions of the entire brain be absent. However, this does not work as a legal argument, as we shall see below.

While not doubting the sincerity and capability of the American Academy of Neurology (AAN), there is a type of linguistic gymnastics at play in their position statement:

> The AAN endorses the perspective of the UDDA that brain death has occurred when the irreversible loss of all functions of the entire brain including the brainstem has been determined. However, the AAN endorses the belief that preserved neuroendocrine function may be present despite irreversible injury of the cerebral hemispheres and brainstem and is not inconsistent with the whole brain standard of death. [42, p. 230]

The insistence in the UDDA on absence of "all functions of the entire brain" is strange when no such demand is made of circulatory and respiratory functions. Indeed, the heart itself has a hormonal function. It secretes atrial natriuretic peptide (ANP). Ischemia is known to increase ANP release [43]. Nothing is more ischemic than a failing circulatory and respiratory system. And yet, no doctor has ever paused when determining death using circulatory and respiratory criteria to wonder if the heart has ceased secreting ANP. That is because the only function that matters to the doctor (and the patient) is the pumping function. Were it not for "all functions of the entire brain" being locked into the UDDA and many jurisdictions' legislation, it would seem unlikely that the persistence of brain hormonal functions would even be a matter for discussion and nor would it require the AAN to try and endorse a contrary belief. It is not the AAN that is wrong, but the UDDA.

No such challenge applies to the brainstem criterion, where two functions are accepted as being preeminent above all other functions and anatomically located to the brainstem: breathing and consciousness. The persistence of antidiuretic hormone is no different to persisting atrial natriuretic peptide, or heartbeat, or digestion or any other function in the body.

Even more startling is that the diagnostic tests a doctor might use to determine death in a jurisdiction that insists on the loss of "all functions of the entire brain" are not materially different to the tests used in in the United Kingdom, India, or Canada. All that the AAN standards require by way of a clinical evaluation is an examination to neurologically assess for coma, the absence of brainstem reflexes and apnea, i.e., brainstem functions [44]. The impression that the way the patient cohort has their death determined by neurologic criteria in the United Kingdom, India, and Canada is vastly different from the way the patient cohort has their death determined by neurologic criteria in the United States (and jurisdictions modeled on it) is mistaken. Determining death in patients with isolated brainstem lesions in the United Kingdom is rare, perhaps representing an absolute maximum of 2% of all cases and, when it does occur, it is worthy of a case report in the primary UK intensive care journal [45]. One of the authors (DG), a neuro-intensive care specialist, vouches that in his own large UK tertiary hospital neuro-intensive care unit, no patient with isolated brainstem lesions has progressed over the last 15 years to the point of even necessitating the consideration of determining death by neurologic criteria. So, in many jurisdictions, the clinical criteria are agnostic between "entire brain" and "brainstem" and any difference that could be present occurs in only a minority cohort of patients. There is, of course, the requirement in some jurisdictions that ancillary or confirmatory tests be used, many of which can demonstrate absence of brain circulation, establishing entire brain involvement [46]. Yet even this difference is less than it appears given that there is evidence that patients with isolated brainstem lesions lose supratentorial blood flow over time [47].

3 Legal Defense

Some societies are more litigious than others, so the number of legal challenges does not necessarily reflect any deficiency in law or guidance. However, the wording in some jurisdictions' determination of death legislation or guidelines can lead them to be more vulnerable to successful legal challenge.

The legal vulnerability of the UDDA is generated because of the adoption of the (1) "all functions" and (2) "entire brain" (whole-brain) criterion [20]. As stated above, the challenge arises largely because not all functions of the brain may have ceased when death is determined by neurologic criteria. As has been pointed out by many, this violates the requirements of the applicable legislation [20, 31, 48]. Some words in statutes are open to differing interpretations, such as the term "reasonable," which can require a judge to ascertain whether, in the circumstances, any impugned conduct was reasonable or not. Words such as "all" and "entire" are not open to interpretation in the same way. Judges cannot decide, for instance, that "entire" does not really mean "entire," or that "all" means "most" or "some."

Excluding mention of all functions of the entire brain criterion in legal statutes and guidance has not protected Canada and the United Kingdom from legal challenge but it certainly has helped, in two ways. First, it is not vulnerable to the objection that some functions, such as hormonal function, remain after death by neurologic criteria. This objection only arises where the law requires the cessation of all functions of the entire brain. Second, the absence of a statute provides much more flexibility. In the Province of Ontario in Canada, there is no statutory definition of death. When a case was brought to the Ontarian courts, the judge was asked to make a common law ruling [20]. Judge Shaw had no difficulty in finding that the 2006 consensus and expert medical guidelines published in the *Canadian Medical Association Journals* [23] represented "accepted medical practice used by all physicians in not only Ontario but throughout Canada to determine death based on neurologic criteria" [49]. The judge could therefore simply refer to and note the latest medical consensus about the matter raised before her and determine the legal issue accordingly. Judges in the United States, in contrast, are not able simply to find that a determination of death by neurologic criteria was performed in accordance with accepted medical standards and let the matter rest there. They have the additional step of checking that the standards reflect the requirements of the statute. Where current medical practice is out of step with the statutory requirements, there are significant problems and this is precisely what we are seeing in the United States.

In 2015, the Supreme Court of Nevada expressed doubts about whether the AAN standards [44] for determining death by neurologic criteria were authoritative [50]. Following this successful legal challenge in Nevada, the Nevada legislature doubled down and passed legislation stating that the AAN standards are authoritative, and that any subsequent revisions approved by the AAN or its successor organization would also be authoritative [51]. This, however, side-stepped the genuine legal issue that current US clinical practice and AAN guidance do not comply with the legislation [42, 44]. Equally vulnerable to legal challenge are the many jurisdictions that modeled the UDDA in their legislation [20].

The determination of death in the United Kingdom, like in Canada, is also not governed by primary legislation. This makes the legal position in the United Kingdom, too, more flexible because changes in legal guidelines that reflect current medical practice and conceptions of death can be made without the fear of contradicting primary legislation. The United Kingdom is therefore not bedeviled by the problems that we have described with the UDDA in the United States. It is also important to note that the UK courts have specifically accepted the brainstem criterion [20, 52]. In a recent Court of Appeal case in 2020, Patten LJ and King LJ remarked "[t]he courts have, from at least 1992 onwards, accepted the validity of the medical diagnosis arising from an irreversible absence of brainstem function" [53]. Importantly, the recent legal cases have highlighted the importance the common law courts place on the national guidance used in the United Kingdom to determine death [54]. In effect, the common law courts defer to the current applicable national guidance and in that sense confirm their legal force. This is what makes the common law position much more flexible than the position in jurisdictions where the definition of death is governed by legislation, such as the United States.

In a world where experts are increasingly distrusted, our courts and legislators at least still appear to value medical consensus opinion and, when this is challenged, prove supportive. It is just that unlike Canada and the United Kingdom, there are real prospects of successful legal challenge in the United States, given the divergence between medical consensus and the wording of the UDDA. The question is: how do we strengthen the worldwide criterion for the determination of death, so it remains worthy of public trust?

4 The Unfinished Journey to Coherence

Pallis saw clearly that there were two important conceptual steps along the journey to conceptual coherence when determining death [14].

- From classic death ⇨ whole-brain death
- From whole-brain death ⇨ brainstem death

While we can never know for sure, it seems likely that Pallis would be disappointed with how slow the world has been to follow in his steps. Perhaps he might have been pleased with more recent developments, but there is still a long way to go.

In 1998, as inheritors of Pallis's legacy, the United Kingdom Academy of Medical Royal Colleges boldly proclaimed in their Code of Practice for the Diagnosis of Brainstem Death that:

> Death entails the irreversible loss of those essential characteristics which are necessary to the existence of a living human person. Thus, it is recommended that the definition of death should be regarded as 'irreversible loss of the capacity for consciousness, combined with irreversible loss of the capacity to breathe'. The irreversible cessation of brainstem function (brainstem death) whether induced by intra-cranial events or the result of extra-cranial phenomena, such as hypoxia, will produce this clinical state and therefore brainstem death equates with the death of the individual. [55, p. 3]

The successor 2008 UK guidance is very similar in first identifying the functions that count most (breathing and consciousness) and then identifying where one anatomically locates these functions (in the brainstem) [54]. It also endorsed the 1979 unified UK position that ultimately all death, whether from direct brain injury or loss of circulatory or respiratory function, is based on loss of brain function [33, 54].

One may have expected the British brainstem criterion to dominate in Commonwealth nations. However, it is far more mixed than that, almost looking like a lost cause [46]. Australia and New Zealand follow a formulation closely aligned to the UDDA [56]. Likewise, Singapore requires loss of "all functions of the brain" [57]. South Africa leaned heavily on the World Brain Death Project in its recent guidance formulation where previously clinicians were using different available international guidelines, which vary markedly [58]. However, India enacted a law in 1994 which legalized brainstem death [59]. Probably most heartening to Pallis would be that in 2006 Canada moved to the following formulation: "irreversible loss of the capacity for consciousness combined with the irreversible loss of all brainstem functions" [23].

This move in Canada also heralded a distinct shift in the international conceptual debates. Rather than advocating for a criterion focused on "all functions of the entire brain," increasingly proposals are being made for a criterion that emphasizes the functional primacy of "consciousness" and the importance of the brainstem. In

2014, an international consensus development group proposed the following definition for the determination of death:

> Death is the permanent loss of capacity for consciousness and all brainstem functions. [60]

Very similarly, the World Brain Death Project proposed in 2020:

> [Brain death/death by neurologic criteria (BD/DNC)] is defined as the complete and permanent loss of brain function as defined by an unresponsive coma with loss of capacity for consciousness, brainstem reflexes, and the ability to breathe independently. [21]

Although we have emphasized the added flexibility provided when jurisdictions do not have a legislative definition of death, we are not advocating against the use of legislation to define death. If there is legislation already in place, it would seem unlikely the legislation would be repealed rather than amended. However, it is important to be aware of the constraints that legislation can impose when technology in medicine develops rapidly. The legislation can quickly become out-of-date, applicable to technologies and practices that have long since been left behind.

That said, it remains essential that legislation be updated, where problems are caused by outdated wording, to reflect the latest medical position. An example of up-to-date legislation, reflecting the international trend just mentioned, is Nova Scotia. Their legislation may provide some guidance on how the difficulties we have described could be addressed. Nova Scotia used the opportunity in the 2019 Human Organ and Tissue Donation Act to include a definition of death in the legislation [61]. The following statements are made in the Act:

> 2. (g) 'death' means the irreversible cessation of the functioning of the organism as a whole as determined by the irreversible loss of the brain's ability to control and co-ordinate the organism's critical functions;"
>
> "16. The medical tests to demonstrate that death has occurred are those established by the medical profession from time to time. [61]

While an opportunity to make use of the United Kingdom's 2008 and the international consensus development group's 2014 definition was missed in Nova Scotia, the new law did manage to enshrine the crucial point that not all functions in the brain count equally (some are critical) and that it would be the medical profession who would define what functions count as critical by way of their authority to establish the accepted standards of "medical tests to demonstrate that death has occurred."

The UDDA looks increasingly incoherent as we have shown in this chapter; the impact of the UDDA far exceeds the borders of the United States, so the planned revision will have world-wide impact [48, 62]. The responsibility and duty of those who advise on a revision is immense. Table 1 is a summary of the different candidate losses of biological functions the revisers might propose to determine death.

Table 1 Exploring different loss of biological functions as criterion for death

Loss of biological function	Clinical criteria	Criticism/comment
Somatic (whole body) [Ancient and historical criteria]	Rigor Mortis Putrefaction Decapitation	Forensic Historical Not timely (e.g., rigor mortis takes hours, putrefactions days) Indeterminate
Forces in the organism tending to increase entropy irreversibly overcome those that are opposing it [31, 75].	Undefined	Theoretical Impractical Indeterminate
Circulatory–respiratory [76, 77]	Observation period (2–30 min)	Historical Physiologically defined points Reversibility dependent on intention to resuscitate and technology Gray area—ventilated irreversible coma Human centric
Loss of ALL functions of the entire brain [32]	Preconditions Examination brainstem Ancillary investigations (mandatory some jurisdictions)	Physiologically defined points Not all brain functions cease Not-unified to circulatory–respiratory criterion Whole brain Brain centric Human centric
Loss of personhood [Could be used to define those in vegetative states as dead]	Pathology—cortex Ancillary investigations (standard not established)	Inaccurate Residual consciousness Not timely (e.g., vegetative state diagnosed over months) Higher brain Brain centric Human centric
Unresponsive coma with loss of capacity for consciousness, brainstem reflexes, and the ability to breathe independently [21]	Preconditions Examination brainstem Ancillary investigations (mandatory some jurisdictions)	Physiologically defined points Unified Whole brain Brain centric Human centric
"[T]he irreversible cessation of the functioning of the organism as a whole as determined by the irreversible loss of the brain's ability to control and coordinate the organism's critical functions… The medical tests to demonstrate that death has occurred are those established by the medical profession from time to time" [61]	Preconditions Examination brainstem	Critical functions undefined Unified Brain centric Human centric

Table 1 (continued)

Loss of biological function	Clinical criteria	Criticism/comment
Brain injury leading to permanent loss of (a) the capacity for consciousness, (b) the ability to breathe spontaneously, and (c) brainstem reflexes [48]	Preconditions Examination brainstem	Physiologically defined points Not-unified to circulatory–respiratory criterion Brain centric Human centric
Death is the permanent loss of capacity for consciousness and all brainstem functions [60]	Preconditions Examination brainstem	Physiologically defined points Unified Ambiguous if whole brain Brain centric Human centric
"Loss of the capacity for consciousness and the capacity to breathe… The irreversible cessation of brain-stem function… will produce this clinical state" [54]	Preconditions Examination brainstem	Physiologically defined points Unified Brain centric Human centric Consciousness without brainstem?

5 Response to Criticism

As can be seen in Table 1, all the criteria we might choose to accept as a society to determine death are open to criticism. Some are more open to criticism than others. Many of the criteria are vulnerable to the objection that they are brain-centric, and therefore do not accommodate religious and other beliefs [22]. Or they are vulnerable to the criticism that they are human-centric; why, it is asked, should the criterion for death in a human be different than death for a plant or an insect [31]? Such debate is outside the scope of this chapter, but this is discussed elsewhere in this book. The better criteria address the issues raised long ago by the Academy of Science in Paris: how are we to make "the diagnosis of death safe, prompt and easy" [19]? This can be achieved by choosing physiologically defined points that unify the circulatory–respiratory criteria for death with the neurologic criteria for death and, of course, support a hierarchy of brain functions with consciousness at the top and a recognition of the anatomical importance of the brainstem. We consider that the 2008 United Kingdom Code [54] and the 2014 international determination proposal [60] come closest to being the optimal proposal.

There is one standard criticism of death by neurologic criteria that we should dispense with straightaway. This is a criticism of the very idea of a brain-based criterion of death as such, whether it be the whole-brain or the brainstem criterion. On this criticism, influenced by Alan Shewmon, brain-dead people are not dead because the death of the brain does not equate with the loss of the integrated functioning of the organism as a whole. Shewmon correctly showed that many functions in the body are not mediated by, and do not require, a functioning brain, and concluded that a brain-dead person is not dead [29, 63, 64]. No single organ, including the brain, can be the locus of life or death. The death of an *organ* is not the death of an *organism*. There is no difference in principle between the death of a kidney and the death of the brain [31]. We know this because living people can donate a kidney but remain alive. They could even donate a heart and remain alive, at least in principle (and also, for a time, in reality on a heart-lung machine or ECMO). The death of the brain, it is said, is therefore the death only of an organ, not an organism. Let us call this the Loss of Integrated Functioning View.

However, there is a fundamental problem with the Loss of Integrated Functioning View. It can be flipped over to entail an absurdity. Imagine that the whole of the rest of the body has ceased functioning except the brain, which is kept functioning in the body through external support not making use of any of the other organs in the body. On the Loss of Integrated Functioning View, since an organ is not an organism, someone could have lost all integrated functions and so be dead, yet have retained consciousness [65]. This is an unacceptable conclusion, and decisively shows the inadequacy of the Loss of Integrated Functioning View [30]. Death is brain-based.

More challenging for the brainstem criterion is the question whether consciousness might persist without a functioning brainstem. Pallis was very aware of this potential criticism. So much so that he wrote a limerick about it, referring to decapitations by guillotine during the French Revolution:

We knit on, too blasées to ask it:
'Could the tetraparesis just mask it?
When the brainstem is dead
Can the cortex be said
to tick on, in the head, in the basket?' [12, p. 285]

As we already have discussed, standard neurologic textbooks would say the answer to the limerick is no. Consciousness is both arousal and awareness and both functions require a functioning brainstem. This is why the criticism that under the brainstem criterion a quadriplegic apneic patient who suffered an event rendering them in a persistent vegetative state/unaware–wakeful state (and thereby lacking conscious awareness) would be dead [66], is so misguided [67].

A more recent criticism, by Joffe and Nair-Collins, of the brainstem criterion suggests that it is possible that those with isolated brainstem lesions may satisfy clinical criteria for death, in those jurisdictions which allow it, but still retain the capacity for consciousness. This is because there might be preserved viability and function of parts of the meso-pontine tegmentum (the higher part of the brainstem) [41]. This, the authors claim, can be evidenced by preserved alpha/theta activity on the electroencephalogram. We have already highlighted the rarity of determining death in isolated brainstem lesions, but this criticism is not a criticism against the brainstem criterion per se, but our way of knowing if the brainstem has ceased functioning. It concerns our diagnostic tests, not the criterion. Neither Bouchut in 1846, the Ad Hoc Committee in 1969, nor Pallis in the 1980s thought the job was done in their era; they never thought they had succeeded forevermore in proposing a safe, prompt, and easy way to determine death, but only that they had succeeded given the technology and salient medical practices at the time. If it is found that new tests are required in our death determinations, then we are always learning, and this learning drives us to even greater safety.

What we are learning currently is that the brainstem may have a role so vital we have been underplaying it all these years. Limiting its role in generating a capacity for consciousness to mere arousal—a glorified on/off switch for awareness, the consciousness that counts—reflects a view of the brainstem that is becoming outdated. New neuroanatomical work is identifying the brainstem as the rudimentary seat of all consciousness, i.e., awareness as well as arousal. Bjorn Merker's landmark 2006 paper was entitled "Consciousness without a cerebral cortex: A challenge for neuroscience and medicine" [68]. By reflecting on the emotional and orientating reactions of anencephalic children (for which Alan Shewmon above is acknowledged by Merker), and considering neurologic evolutionary development, Merker came to the startling conclusion that it was the brainstem, not the cortex, which fulfilled the primary function of consciousness "matching opportunities with needs in a central motion-stabilized body-world interface organized around an ego-center" [68]. This finding is in keeping with experimental work on mammals in the early part of the twentieth century [69].[1]

Research by Barron and Klein on insect consciousness is discovering that it is in the brainstem that the most basic level of consciousness is found: the capacity for subjective experience [72]. They argue that subjective experience requires the construction of an integrated neural simulation of the agent in space, allowing an egocentric representation of the world to be built. In humans, the midbrain (part of the brainstem) fulfils this role and analogous structures can be found in insect brains [72]. Our overly simple and minimalistic understanding of the role of brainstem-mediated "arousal" is also being challenged [73].

No one is claiming in any of this recent scientific explosion of output that the cortex is unimportant to human consciousness, only that without a functioning brainstem, there cannot be consciousness anywhere in the brain. What is more, some forms of awareness, and not merely arousal, are anatomically located in the brainstem, not merely the cortex. This emphasizes the foundational role of the brainstem as the anatomical seat of rudimentary consciousness. Where this journey will take us, is yet to be discovered, but we hope Pallis would be pleased that the brainstem will no longer be merely a passenger, but this time, will occupy the driver's seat.

6 Conclusion

We have spent so long divided on whether we need a variant of "the entire brain" criterion or should adopt the "brainstem criterion" that we have forgotten there is more that unites us than divides us. Perhaps this joint statement by Alex Manara, an author of the United Kingdom 2008 Code, and Eelco Wijdicks, an author of the 2010 AAN standard, is a start to bridging the transatlantic debate:

[1] Some have even used Merker to ask the question at the opposite end of life—when is the experience of pain developmentally possible in a fetus [70]. This argument is made more compelling by neuroimaging, highlighting the neglected role the brainstem has for nociception and pain processing [71].

> The perceived divide between whole brain and brainstem death is now kept 'alive' only by a minority. It has more to do with emotive concepts rather than hard neurobiological facts, and represents a failure to accept the centrality of the brainstem in defining life or death. [74]

It would be entirely remiss of us to end this chapter without a limerick in dedication to Professor Christopher Pallis. We cannot vouch for his enjoyment, but we hope he would applaud the sentiment.

To you the mad scientist did bray
On a sad and evil day
For my robot your head
Or your body instead
Where would you be would you say?

You would be in your brain, or as Pallis would say: there can be no consciousness without a functioning brainstem. It's the only coherent position to take.

Declaration

Dr. Gardiner is Associate Medical Director—Deceased Organ Donation for NHS Blood and Transplant, the national organ donation organization in the United Kingdom. The views expressed are his own.

References

1. Wijdicks EFM. The transatlantic divide over brain death determination and the debate. Brain. 2012;135(Pt 4):1321–31.
2. Transplants: Are the Donors Really Dead? Panorama. Episode aired Oct 13 1980.
3. Smith T. Medicine and the media. BMJ. 1980;281:1485.
4. Pallis C. ABC of brain stem death. Reappraising death BMJ. 1982;285(6352):1409–12.
5. Pallis C. ABC of brain stem death. From brain death to brain stem death. BMJ. 1982;285(6353):1487–90.
6. Pallis C. ABC of brain stem death. Diagnosis of brain stem death—I. BMJ. 1982;285(6354):1558–60.
7. Pallis C. ABC of brain stem death. Diagnosis of brain stem death—II. BMJ. 1982;285(6355):1641–4.
8. Pallis C. ABC of brain stem death. Pitfalls and safeguards. BMJ. 1982;285(6356):1720–2.
9. Pallis C. ABC of brain stem death. The declaration of death. BMJ. 1983;286(6358):39.
10. Pallis C. ABC of brain stem death. Prognostic significance of a dead brain stem. BMJ. 1983;286(6359):123–4.
11. Pallis C. ABC of brain stem death. The position in the USA and elsewhere. BMJ. 1983;286(6360):209–10.
12. Pallis C. ABC of brain stem death. The arguments about the EEG. BMJ. 1983;286(6361):284–7.
13. Pallis C. ABC of brain stem death. 1st ed. London: Wiley–Blackwell; 1983. p. 40.
14. Pallis C, Harley D. ABC of brainstem death. 2nd ed. London: BMJ Publishing Group; 1996.
15. Powner DJ, Ackerman BM, Grenvik A. Medical diagnosis of death in adults: historical contributions to current controversies. Lancet. 1996;348(9036):1219–23.

16. Lewis P, Goodway D. Christopher Pallis. The Guardian. 2005. https://www.theguardian.com/news/2005/mar/24/guardianobituaries.paullewis.
17. Lear T. Diary account: the last illness and death of General Washington. 1799. https://founders.archives.gov/documents/Washington/06-04-02-0406-0002.
18. Bondeson J. Buried alive: the terrifying history of our most primal fear. New York; London: W. W. Norton & Company; 2002.
19. Bouchut E. Traité des signes de la mort et des moyens de prévenir les enterrements prématurés. Paris: Libraire de L'Académie de médecine; 1849. https://archive.org/details/traitdessignes00bouc.
20. McGee A, Gardiner D. Differences in the definition of brain death and their legal impact on intensive care practice. Anaesthesia. 2019;74(5):569–72.
21. Greer DM, Shemie SD, Lewis A, et al. Determination of brain death/death by neurologic criteria: the world brain death project. JAMA. 2020;324(11):1078–97.
22. Gardiner D, McGee A. Diagnosing death. In: Danbury C, Newdick C, Keene AR, Waldmann C, editors. Law and ethics in intensive care. 2nd ed. Oxford: Oxford University Press; 2020. p. 137–60.
23. Shemie SD. Severe brain injury to neurological determination of death: Canadian forum recommendations. CMAJ. 2006;174(6):S1–12.
24. Webb AC, Samuels OB. Reversible brain death after cardiopulmonary arrest and induced hypothermia. Crit Care Med. 2011;39(6):1538–42.
25. Gardiner D, McGee A, Bernat JL. Permanent brain arrest as the sole criterion of death in systemic circulatory arrest. Anaesthesia. 2020;75(9):1223–8.
26. Tebb W, Vollum EP, Hadwen WR. Premature burial: how it may be prevented. (First Published 1905). Sale J, editor. London: Hesperus Press Limited; 2012.
27. A definition of irreversible coma: report of the ad hoc committee of the Harvard Medical School to Examine the Definition of Brain Death. JAMA 1968;205(6):337–40.
28. Lizza JP. Where's Waldo? The 'decapitation gambit' and the definition of death. J Med Ethics. 2011;37(12):743–6.
29. Shewmon DA. 'Brainstem death,' 'brain death' and death: a critical re-evaluation of the purported equivalence. Issues Law Med. 1998;14(2):125–45.
30. McGee A, Gardiner D, Jansen M. A new defense of brain death as the death of the human organism. J Med Phil. 2023. Epub ahead of print.
31. Miller FG, Truog RD. Death, dying, and organ transplantation: reconstructing medical ethics at the end of life. New York: Oxford University Press; 2012.
32. National Conference of Commissioners on Uniform State Laws. Uniform Determination of Death Act. 1981. https://www.uniformlaws.org/committees/community-home?communitykey=155faf5d-03c2-4027-99ba-ee4c99019d6c&tab=groupdetails.
33. Diagnosis of death. Memorandum issued by the honorary secretary of the Conference of Medical Royal Colleges and their Faculties in the United Kingdom on 15 January 1979. BMJ. 1979;1(6159):332.
34. Diagnosis of brain death. Statement issued by the honorary secretary of the Conference of Medical Royal Colleges and their Faculties in the United Kingdom on 11 October 1976. BMJ. 1976;2(6045):1187–8.
35. Simpson K. The moment of death—a new medico-legal problem. S Afr Med J. 1967;41(46):1188–91.
36. Simpson K. The moment of death. A new medico-legal problem. Acta Anaesthesiol Scand Suppl. 1968;29:361–79.
37. Posner J, Plum F. Pathophysiology of signs and symptoms of coma. In: Plum and Posner's diagnosis of stupor and coma. Oxford: Oxford University Press; 2007. p. 3–37.
38. Purves D, Augustine G, Fitzpatrick D, et al., editors. Cortical states. In: Neuroscience. 6th edn. New York: OUP USA; 2018. p. 609–29.
39. Thömke F. Disturbances of consciousness. In: Urban PP, Caplan LR, editors. Brainstem disorders. Springer; 2016. p. 164–8.

40. Kiernan JA, Rajakumar R. Reticular Formation. In: Barr's the human nervous system: an anatomical viewpoint. 10th ed. Philadelphia: Lippincott Williams and Wilkins; 2013. p. 143–57.
41. Nair-Collins M, Joffe AR. Frequent preservation of neurologic function in brain death and brainstem death entails false-positive misdiagnosis and cerebral perfusion. AJOB Neurosci. 2021:1–14.
42. Russell JA, Epstein LG, Greer DM, et al. Brain death, the determination of brain death, and member guidance for brain death accommodation requests: AAN position statement. Neurology. 2019;92:228–32.
43. Ahmed F, Tabassum N, Rasool S. Regulation of atrial natriuretic peptide (ANP) and its role in blood pressure. Int Curr Pharm J. 2012;1(7):176–9.
44. Wijdicks EFM, Varelas PN, Gronseth GS, Greer DM, American Academy of Neurology. Evidence-based guideline update: determining brain death in adults: report of the Quality Standards Subcommittee of the American Academy of Neurology. Neurology. 2010;74(23):1911–8.
45. Manara A, Varelas P, Smith M. Neurological determination of death in isolated brainstem lesions: a case report to highlight the issues involved. JICS. 2020;21(3):269–73.
46. Wijdicks EFM. Brain death worldwide: accepted fact but no global consensus in diagnostic criteria. Neurology. 2002;58(1):20–5.
47. Varelas PN, Brady P, Rehman M, et al. Primary posterior fossa lesions and preserved supratentorial cerebral blood flow: implications for brain death determination. Neurocrit Care. 2017;27(3):407–14.
48. Omelianchuk A, Bernat J, Caplan A, Greer D, Lazaridis C, Lewis A, et al. Revise the UDDA to align the law with practice through neuro-respiratory criteria. Neurology. 2022;98:532–6.
49. Justice Shaw. McKitty v Hayani (2018) ONSC 4015,[53]. 2018.
50. In Re Guardianship of Hailu 131 Nev, Adv. Op. 2015. https://law.justia.com/cases/nevada/supreme-court/2015/68531.html.
51. Nevada State Legislature. Revises provisions governing the determination of death. (BDR 40–1025) NV AB424. 2017. https://legiscan.com/NV/bill/AB424/2017.
52. Aramesh K, Arima H, Gardiner D, Shah SK. An international legal review of the relationship between brain death and organ transplantation. J Clin Ethics. 2018;29(1):31–42.
53. Lord Justice Patten, Lady Justice King. Re M (Declaration of Death of Child). [2020] EWCA Civ 164. 2020. https://www.bailii.org/ew/cases/EWCA/Civ/2020/164.html.
54. Academy of Medical Royal Colleges. A Code of Practice for the Diagnosis and Confirmation of Death. Academy of Medical Royal Colleges. 2008. https://www.aomrc.org.uk/reports-guidance/ukdec-reports-and-guidance/code-practice-diagnosis-confirmation-death/.
55. Academy of Medical Royal Colleges. A Code of Practice for the Diagnosis of Brain Stem Death. Department of Health; 1998.
56. Australia and New Zealand Intensive Care Society (ANZICS). The statement on death and organ donation. Edition 4.1. 2021. https://www.anzics.com.au/wp-content/uploads/2021/06/ANZICS-Statement-on-Death-and-Organ-Donation-4.1.pdf.
57. Interpretation (Determination and Certification of Death) Regulations—Singapore Statutes Online. 2000. https://sso.agc.gov.sg/SL/IA1965-RG1?DocDate=20040701.
58. Thomson D, Jouber I, De Vasconcellos K, Paruk F, Mokogong S, Mathivha R, et al. South African guidelines on the determination of death. South Afr J Crit Care. 2021;37(1):41–54.
59. Dhanwate AD. Brainstem death: a comprehensive review in Indian perspective. Indian J Crit Care Med. 2014;18(9):596–605.
60. Shemie SD, Hornby L, Baker A, et al. International guideline development for the determination of death. Intensive Care Med. 2014;40(6):788–97.
61. Nova Scotia Legislature. Human Organ and Tissue Donation Act [Internet]. BILL NO. 133. 2019. https://nslegislature.ca/legc/bills/63rd_2nd/3rd_read/b133.htm.
62. Lewis A. The uniform determination of death act is being revised. Neurocrit Care. 2022;36(2):335–8.

63. Shewmon DA. Chronic 'brain death': meta-analysis and conceptual consequences. Neurology. 1998;51(6):1538–45.
64. Shewmon AD. The brain and somatic integration: insights into the standard biological rationale for equating 'brain death' with death. J Med Phil. 2001;26(5):457–78.
65. Chiong W. Brain death without definitions. Hast Cent Rep. 2005;35(6):20–30.
66. Truog RD, Miller FG. Brain death: justifications and critiques. Clin Ethics. 2012;7(3):128–32.
67. Gardiner D, Manara A, Murphy P. Letter in response to: Truog RD and Miller FG. Brain death: justifications and critiques. Clinical Ethics 2012;7: 128-32. Clin Ethics. 2013;8(1):1–1.
68. Merker B. Consciousness without a cerebral cortex: a challenge for neuroscience and medicine. Behav Brain Sci. 2007;30(1):63–81.
69. Staunton H. Arousal and consciousness: effects of total cortical extirpation in the mammal. In: Leisman G, Merrick J, editors. Considering consciousness clinically. Nova Science Publishers, Inc.; 2016. p. 41–6.
70. Derbyshire SWG, Bockman JC. Reconsidering fetal pain. J Med Ethics. 2020;46:3–6.
71. Napadow V, Sciocco R, Henderson LA. Brainstem neuroimaging of nociception and pain circuitries. Pain Rep. 2019;4(4):e745.
72. Barron AB, Klein C. What insects can tell us about the origins of consciousness. Proc Natl Acad Sci USA. 2016;113(18):4900–8.
73. Satpute AB, Kragel PA, Barrett LF, Wager TD, Bianciardi M. Deconstructing arousal into wakeful, autonomic and affective varieties. Neurosci Lett. 2019;693:19–28.
74. Manara A, Varelas P, Wijdicks EF. Brain death in patients with 'isolated' brainstem lesions: a case against controversy. J Neurosurg Anesthesiol. 2019;31(2):171–3.
75. Nair-Collins M. Taking science seriously in the debate on death and organ transplantation. Hast Cent Rep. 2015;45(6):38–48.
76. Lomero M, Gardiner D, Coll E, et al. Donation after circulatory death today: an updated overview of the European landscape. Transpl Int. 2020;33(1):76–88.
77. Dhanani S, Hornby L, van Beinum A, et al. Resumption of cardiac activity after withdrawal of life-sustaining measures. NEJM. 2021;384(4):345–52.

What Provisions Belong in a Statute on the Determination of Death by Neurologic Criteria?

Alexander Morgan Capron

1 Who Should Set the Standards for Determining Death?

Before discussing the topics that ought to be included in a statute on the determination[1] of death by neurologic criteria, one first needs to ask whether a statutory "definition of death" is even necessary. The ad hoc Harvard Medical School committee on "brain death," whose 1968 report focused professional as well as public attention on the subject, concluded that legislation would not be needed in order to give effect to the growing consensus among physicians that ventilator-dependent patients in "irreversible coma" are dead [1].[2]

At that time, the determination that a person had died depended on medical expertise, not on statutes. State laws typically specified that a death certificate must be filled out and filed with the office of vital records by the physician who declared a person dead, but the means for declaring that death had occurred were not spelled out in those laws. Further, then, as now, it was not unusual for death to be

[1] In this chapter, which focuses on the manner in which state law adapted to changes in medical criteria for diagnosing death, the legal standards are described as those for the "determination" of death because that is the wording used in legislation. Prior to 1970, states lacked statutes on this matter but had legislation on the declaration of death; these established the persons authorized to declare death and the procedures they were to follow to record the declaration.

[2] Also in 1968, the National Conference of Commissioners on Uniform State Laws actually went further than the Harvard committee; they unanimously chose not to include *any* standard for determining death in the Uniform Anatomical Gift Act [2], since this "has traditionally been regarded as a question for medical determination and not the proper subject for codification by law" [3, p. 2504]. As one of the drafters explained, the law ought not to "channel medical judgment" on this "much debated question" but should instead leave to the physician in charge of a patient "the burden of determining when life is at an end" in light of the "complex medical circumstances of each case" [4, p. 928].

A. M. Capron (✉)
University of Southern California, Los Angeles, CA, USA
e-mail: acapron@law.usc.edu

A. Lewis, J. L. Bernat (eds.), *Death Determination by Neurologic Criteria*,
Advances in Neuroethics, https://doi.org/10.1007/978-3-031-15947-3_6

determined initially by lay people (such as friends and relatives) based on the very signs on which physicians also rely—the absence of breath or a detectable pulse—even when bodily injuries incompatible with life were not evident.

When the fact or timing of death was at issue in legal actions in the 1960s and early 1970s, courts accepted those traditional "vital signs" (i.e., heartbeat and breathing) as the legal standard. If they needed to cite authority, judges turned to the common law "definition" found in cases in their jurisdiction and summed up in *Black's Law Dictionary* as "a total stoppage of the circulation of the blood, and a cessation of the animal and vital functions consequent thereon, such as respiration, pulsation, etc. [5]." The issue was treated as one of fact, to be resolved by the jury—or by the judge, in cases not tried before a jury—based on the findings and opinions presented by the expert medical witnesses. Thus, it is not surprising that the Harvard committee concluded that judges would likewise defer to physicians who declared dead those comatose, ventilator-supported patients who met the neurologic criteria articulated by the committee. Assuming that judges would modify the common law "definition" to include brain-based determinations of death, the committee members therefore believed that a statute would be needed only if "physicians were unable to agree on the new criteria [1]."

In the decade that followed the Harvard report, physicians achieved the near unanimity that death could be determined as reliably by measuring neurologic functions as by measuring circulatory and respiratory functions. Nonetheless, by 1978, legislatures in 15 states had adopted statutes that superseded the common law definition. Why did this occur? The simple answer is that physicians and their state associations urged legislators to remove the specter that doctors could be held civilly or criminally liable for relying on brain-based criteria when declaring death [6, 7]. The deeper explanation is that the common law did not live up to the Harvard committee's expectations—or at least not quickly or uniformly enough.

1.1 The Problems with Waiting for Judges to Change the Law

The impediments were of four types. First, well into the 1970s, some judges flatly refused to change the common law standard for declaring death. This is hardly surprising. The judiciary's respect for *stare decisis* makes judges inherently conservative. Judges were familiar with physicians applying the circulatory–respiratory criteria when declaring death, whereas physicians transplanting hearts from "brain dead" patients, which then pump away in other patients, seemed mystifying and perhaps suspect. Sanction for such a radical change should therefore come from the legislature, not from the courts, which move incrementally, if at all.

Second, even when judges were willing to entertain the use of the neurologic criteria for declaring death, multiple things needed to occur before the common law would incorporate the new standard. For example, most trial court rulings never come before an appellate court. Thus, even if a trial judge permitted the medical experts in a case to present evidence that neurologic criteria were used to declare death, that would not provide any assurance that a judge presiding at a subsequent trial in that jurisdiction would do the same. Changing the common law standard would require that, having lost the case at trial, the party opposing the use of neurologic criteria appealed the verdict, and the appellate court would have to hold that the trial court had been correct to admit the evidence and allow the jury to conclude that death had occurred despite the continuation of heartbeat and respiration in the mechanically ventilated patient. Further, were the appellate court simply to rule that the physician-witnesses' evidence of "brain death" was admissible but not spell out the basis for determining death, a new legal standard would not have been established. Moreover, to achieve a change in the law across a jurisdiction, the judicial opinion must be issued by its highest court (such as the state's supreme court), which would necessitate the parties continuing the litigation after a ruling by an intermediate appellate court.

Third, judicial opinions are shaped by the facts of the case and the arguments of the opposing parties, who between them may not have reason to call witnesses who represent the full range of professional—much less academic or nonmedical—expertise on the subject before the court; while courts can conduct their own review of scholarly materials, the facts of the case are those found the trial record. Had the Harvard committee taken a deeper look at the subject, it might have concluded—as a contemporary commentator did—that the existing "definition of death" across jurisdictions was not "characterized by uniformity or scientific exactness" and that "inconsistencies and discrepancies" between the legal concepts and medical practice were becoming more apparent [8]. A number of legal cases in the late 1960s and early 1970s added to the confusion, as some courts reiterated the existing "vital signs" standard while others either accepted the irreversible loss of brain functions as a legal standard or, without explicitly modifying the common law, allowed physicians to defend themselves against a malpractice charge by showing that they had correctly used the tests for determining death by neurologic criteria [9].

The fourth problem is of a different sort, since it rests not with the failure of judge-made law to develop clearly, quickly, or uniformly, but rather with its historic deference to physicians. Placing the matter of "defining death" in medical hands seemed unproblematic so long as the standards that physicians used were stable and also familiar and acceptable to the general public. Yet changes of the sort proposed by the Harvard committee made apparent that the concept of death is not simply a technical matter that involves, for example, correlating detectable physical signs (such as total unawareness of, and complete unresponsiveness to, externally applied stimuli) with the presence or absence of activities in, or functions of, particular organs, or predictions about how long such activities and functions can and will continue. Rather, the concept of death involves a judgment about which functions signify the existence of life in a human being and which are salient characteristics of death having occurred. Conclusions on such matters must rest on the best available scientific evidence: a standard that requires findings about a patient's condition that cannot be made with existing technology, or—worse—that uses findings that have been shown to be unreliable would be invalid on its face. Yet the need to update the standard for determining human death is "not only medical but also fundamentally moral and political [10]," which means that scientific evidence and medical expertise are necessary, but not sufficient, to make a legal standard for determining death legitimate.

Ultimately, any standard represents a value judgment concerning what constitutes a basic civic category—being a living member of society. In a democracy, decisions of this sort are made in the legislature, acting within the system of rights and responsibilities established by the constitution.[3] Legislation is crafted by representatives who are accountable to the electorate, and it is scrutinized in hearings which are open to the public and covered by the media. This meant that a range of concerns that were raised by reliance on unfamiliar methods to determine death could be openly addressed, including the possible conflict of interest that the new standard was endorsed by physicians who wished to facilitate organ donation. In contrast to judicial proceedings, legislative hearings provided a means for the rationale for changing the medical criteria to be explained and criticized, not only by physicians and scientists as well as philosophers and theologians but also by members of the public.

1.2 Problems and Success with Legislative Change

Although legislation is the appropriate way to establish standards for the determination of death, especially since the common law responded so slowly and inconsistently to the need to deal with "brain death," the first American statute—drafted by a physician-legislator in Kansas and adopted there in 1970—was controversial. Two commentators praised it in strong terms [12, 13], while another found it to be so clumsy and confusing that legislatures and laymen in general should let common-law judges respond to the changes in evidence presented by physicians [14]. As shown by subsequent model state statutes, careful drafting can avoid the most egregious problems.

Yet, with time, questions have been raised about the conceptual underpinnings of the most widely adopted American statute, the Uniform Determination of Death Act (UDDA), and about whether the criteria and tests used today by most physicians in the United States align with the statutory standards, especially in the context of organ donation. Other chapters in this volume address disagreements about the

[3] The Supreme Court of Washington State, whose legislature had failed to modernize the standard for determining death, followed an ingenious strategy that allowed it both to respond affirmatively to a request to recognize "brain death" and to respect the greater capacity of legislatures to make law on highly technical and value-laden topics. The trial court had ruled that physicians had acted properly in relying on the concept of "brain death" to declare dead 5-year-old Matthew Bowman, who had suffered massive physical injuries, because his brain had ceased functioning. The trial court enjoined the removal of the "extraordinary measures" sustaining Bowman's respiration pending an appeal. The case was set for argument before the supreme court a week later, but one day before that could occur all of Matthew's bodily functions ceased irretrievably. The supreme court decided to rule on the appeal even though his death had rendered the case moot; it "adopted" the Uniform Determination of Death Act, which had been issued by the uniform law commission 2 months earlier [11].

substantive rules under which death may be determined based on physiological (especially neurologic) functions, as well as about subsidiary scientific, legal, religious, and social issues. To set the table for these matters of substance, this chapter discusses how and why various topics ought—or ought not—be addressed in a statute "defining" death.

2 What Objectives Should a Statute Seek?

Some of the basic questions that the drafters of any statute on the determination of death need to address are: what objectives should the statute seek, what principles should guide how it is worded, what should be included in—or excluded from—the statute, and why?

2.1 The Specificity of Public Policy

The reasons for adopting a statute on death are to provide the people who determine death with a clear statement of what constitutes human death in terms that are specific enough to overcome uncertainty about which tests and procedures should be used and to reassure the public that death determinations will be accurate and consistent. Thus, an initial question for legislative drafters concerns level of detail the statute should provide. Four levels of specificity suggested 50 years ago have helped shaped how death is "defined" in American statutes: basic concepts or ideas, general physiological standards, operational criteria for such standards, and the specific tests and procedures that are used to determine when the criteria are met and death has occurred [15].[4]

To explain what it means to die, many people rely on abstract concepts—"departure of the soul" or "irreversible loss of personhood." Such formulations reflect fundamental religious and philosophical views about what it means to transition from being a living human being to a dead body, but their arcane nature offers physicians no guidance for determining when that event has occurred. The statutes under discussion are sometimes described (even in the present chapter) as dealing with the "definition of death." While this term provides convenient shorthand, it is also misleading since the purpose in adopting a statute is not to "redefine" the concepts of life and death. In applying the common law standard, physicians were not "defining" death as the permanent loss of heart and lung functions, but merely using those traditional signs as the means for determining that death had occurred. Modern treatments that prevent the detection of such signs do not necessitate a change in

[4]A three-level schema developed by James Bernat, Charles Culver, and Bernard Gert [16, pp. 390–393] parallels the levels of increasingly specific statutory language presented in the text, below the level of concepts, except that they use the term "definition" in place of "standard." Bernat subsequently added "paradigm" to the top of his sequence; it serves to demarcate the terrain within which the definition/standard is found [17, p. 329].

concepts; they merely require the identification and use of alternative indicators of the same underlying condition.

At the opposite extreme, clinical and laboratory tests and procedures could provide precise means of determining death, but their very specificity would make their inclusion in a statute problematic. To be useful, they must be stated with great care, yet the detail needed would result in an extremely long statute, which is inconsistent with the objective of allowing the public to be easily informed about how the law defines human death. Furthermore, as physicians develop, assess, and adopt new tests and adjust existing procedures in light of experience, either this information would have to be repeatedly incorporated by amending the statute, a task for which legislatures possess neither inclination nor ability, or procedures specified in a statute "might inhibit the development and application of more sophisticated diagnostic methods" [18, p. 357].

Moving up one level, the operational criteria for determining death—such as the absence of pulse, heartbeat, and spontaneous respiration, or a patient's failure to react to painful stimuli or internal need—can be stated more briefly and will be more consistent over time, but the problems inherent in writing diagnostic tests into a statute would also arise were statutes to incorporate the medical criteria for determining death. Legislators are unlikely to possess the ability to weigh the evidence for slightly different ways of expressing the criteria or to know when and how the criteria should be modified in light of new evidence, nor would they want courts to "limit the criteria to a fixed point in the past" by holding that a death determination statute incorporated the medical criteria that were in effect when the statute was adopted [19, p. 776].

It is thus not surprising that the statutes adopted in the United States have been written at the second level, namely, as general physiological standards, which are more concrete than basic concepts, but less technical than criteria or tests. To avoid intractable discord about the standard, James Bernat suggests that discussions begin by specifying the "paradigm of death" being utilized, that is, the "set of assumptions and conditions that frame the argument by making explicit the boundaries of the topic we are discussing, the class of phenomena to which it belongs, and the way in which it should be discussed [17]." An example of "an essential component of [such a] paradigm" would be "the assumption that death is fundamentally a biological (not a social) phenomenon [20]."

Choosing one standard over another entails making philosophical choices, for example, about whether to define death in terms of organ systems, physiological functions, or recognizable human activities, capacities, and conditions, and whether to identify a single underlying capacity or function, the permanent loss of which means that death has occurred, or to recognize several standards as different manifestation of human death. These are the sort of value-based choices that public bodies, such as legislatures, frequently have to make as they set the rules by which members of a society will relate to one another and to the state.

Since the standards found in the UDDA and other modern death determination statutes resemble the format of the traditional common law statements about death, they are likely to seem familiar and comprehensible to members of the public, who should thus be able to relate these standards to whichever conception of death they may hold—for example, that the soul departs the body when breathing ceases forever. Similarly, looking in the direction of greater specificity, physicians and their associates should be able to connect the statutory standards with the criteria on which they will rely in determining death as well as with the tests and procedures they will apply to see whether criteria have been met.

2.2 Avoid the Term "Standards" When "Criteria" Are Intended

The UDDA creates avoidable confusion, however, because immediately after presenting its two substantive standards, it states, "A determination of death must be made in accordance with accepted medical *standards* [2, emphasis added]." In the commentary that accompanied the UDDA, the uniform law commission made clear that, in this sentence, "standards" means the "diagnostic tests and medical procedures" that are accepted by the medical profession, which has the authority to formulate new ones on the basis of "new biomedical knowledge, diagnostic tests, and equipment" [2]. The relationship between standards that are set by the legislature and physicians' agreed-upon means to determine that the standards have been met when death is declared would be clearer if the statute used the phrase "accepted medical criteria and tests" for the latter, instead of "medical standards." (In this context, the term "tests" is not restricted to the procedures applied to measure bodily functions and the like but also specifies the meaning attached to each test result.)

2.3 The Phenomenon of Interest

A death determination statute concerns the death of a human being, not the metaphorical "death" of cells, tissues, or organs, nor the interpersonal or social "death" that occurs when the loss of certain (particularly cognitive) capacities keeps people from functioning fully as members of their family or community. Being clear about the phenomenon of interest is very important for several reasons. First, attaching "death" to a particular organ (such as the brain) implies that organs die, which they do not; organisms die, while organs cease functioning, which in some cases leads to the death of the organism of which they are a part. Second, the use of terms such as "brain death" or "cardiac death" creates the false impression that a number of different types of human death exist. In fact, each standard set forth in death statutes relies on the permanent loss of particular physiologic functions to assess whether the human being as a whole is still alive, since the cessation of each set of functions has implications for the ability of the others to persist. The choice of particular means of assessment depends on the individual circumstances; for example, whether an unconscious patient in an intensive care unit whose respiration is produced by a mechanical ventilator is still alive cannot be determined by applying traditional tests for "vital signs." However, assessing the absence of brain function in such a case does not connote a separate type of death, merely a different way to ascertain whether the same phenomenon exists.

One way to make clear that death is a single phenomenon would be to legislate only a single standard for determining death. The Uniform Brain Death Act (UBDA), a model statute proposed in 1978, which was replaced several years later by the UDDA, stated that, "an individual who has sustained irreversible cessation of all functioning of the brain, including the brain stem, is dead" and required that "reasonable medical standards" be used in making such determinations [21]. What should one make of a stand-alone statute that rests diagnoses of death on the cessation of a single organ's functions? In issuing the UBDA, the uniform law commissioners commented that it did not "preclude" death determinations based on cessation of respiration and circulation or "other legal or medical criteria," which "are practical in cases where artificial life-support systems are not utilized." However, they also negated any implication that such other (unspecified) criteria revealed a separate phenomenon, since, "Even those [other] criteria are indicative of brain death."[5]

The question of whether the permanent cessation of brain functions provides "the unifying concept of human death" [23, p. 270], as some have proposed, is debated elsewhere in this volume. For the moment, one ought to consider the pragmatic reasons why the groups that produced the UDDA—the uniform law commissioners, the President's Commission, the American Bar Association, and the American Medical Association—rejected the approach embodied in the UBDA and instead retained irreversible cessation of circulatory and respiratory functions as an alternative basis for determining death. First, "[t]he overwhelming majority of cases will continue to be determined" under the circulatory–respiratory standard [2]. The newer, brain-based standard was meant to supplement, not to supplant, the older standard. Second, including the circulatory–respiratory standard in the statute allows the law and public perception to change incrementally rather than radically. Continuing to recognize the common law "definition of death" means that should a need arise to explain to an individual's family and others the grounds for declaring a patient dead, in most cases that can be done using familiar terms; indeed, the traditional vital signs are so closely connected to the reality of life and death as perceived by people in everyday life that no explanation is usually necessary, unlike neurologic determinations of death, which remain opaque to many lay persons because patients receiving intensive care typically do not manifest the signs ordinarily associated with death.

---

[5] In 1981, the Law Reform Commission of Canada recommended that the parliament adopt a statute with a single standard for determining death, "an irreversible cessation of all … brain functions," but included the following, not as a comment but as an additional statutory provision:

> (2) The cessation of brain functions can be determined by the prolonged absence of spontaneous cardiac and respiratory functions [22].

This provision explicitly recognizes "prolonged absence" of circulatory (confusingly labeled "cardiac") and respiratory functions, which will continue to be the basis for almost all death determinations, but it also makes the resulting determinations of death indirect and inferential, as discussed in the text.

Further, the continued recognition of the circulatory–respiratory standard in a death determination statute means that when physicians test for the absence of breathing and heartbeat, their findings connect directly to a statutory basis for declaring death, rather than their having to infer that the absence of one set of functions (circulation and respiration) signifies that another set of functions (neurologic) have ceased irreversibly, a conclusion that, had it been reached directly, would have required carrying out a set of complex clinical and laboratory tests for brain function.

The contrast between direct and inferential determinations provides an important reminder that the cessation of the specified bodily systems is a useful way of describing the standards for determining death, not because such cessation is of interest in and of itself, but for what it reveals about the status of the person, as the President's Commission remarked:

> Although absence of breathing and heartbeat may often have been spoken of as "defining" death, review of history and of current medical and popular understanding makes clear that these were merely evidence for the disintegration of the organism as a whole … [25, p. 58].

Being able to detect the collapse of the person as a whole, and to declare death when it is found to exist, is a major objective when drafting and enacting death determination statutes.

3 What Principles Should Guide the Wording?

3.1 Uniformity as to All Persons and for All Purposes

Since being alive or dead is a central attribute in establishing one's civil status in society, the principle of legal equality of all persons means that the same standards for determining death must apply to all. The law on determining death should be uniform, irrespective of individuals' personal characteristics (e.g., sex, race, ethnicity) or the social utility (e.g., organ donation) that is expected to follow a death determination in their case.[6] Of course, the law sometimes gives the same legal status to very different entities. For example, human adults and corporations are

[6]Another type of consistency—uniformity of standards across jurisdictions—is beyond the ability of a single state to achieve, though collectively the states can attempt it. The guiding purpose of the Uniform Law Commission is to achieve uniformity in the law on matters subject to state jurisdiction. Section 2 of the UDDA contains a provision commonly included in model laws from the commission, "**[Uniformity of Construction and Application]**. This Act shall be applied and construed to effectuate its general purpose to make uniform the law with respect to the subject of this Act among states enacting it [2]." With the passage of time, legislative amendments and judicial decisions have reduced the uniformity of state statutes on determining death. Furthermore, a few states have unique statutes; for example, the New Jersey Declaration of Death Act provides that neurological criteria may not be used to determine death when doing so would violate an individual's religious beliefs, so that were such a patient transported from New Jersey to New York across the George Washington Bridge by ambulance the patient could go from being alive to dead at the midpoint in the Hudson River because of differences in the two state's statutes.

both defined as legal "persons" in many context, such as having the right to own property, to enter into enforceable contracts, and to be a party in a legal action. However, when it comes to casting a vote in public elections, a human adult enjoys the right, but a corporate "person" does not. Legal differences of this sort are not problematic, morally or practically.

In the present context, instead of asking "what is death?" one might ask "what difference does it make whether somebody is dead?" [24, p. 629]. As the President's Commission replied, "That question has many answers, most of them familiar to everyone":

> Criminal law (murder v. aggravated assault), tort law (wrongful death), family law (the status of spouse and children), property and estate law, insurance law (payment of life insurance benefits and termination of health insurance payments), and tax law, as well as some actions and culturally determined behaviors of family members, physicians, clerics and undertakers are all initiated by the determination that a death has occurred. Were there good reason for one branch or another of the law or one or another cultural institution to employ a different "definition" of death, logic would not preclude such a step. But in fact, society has found it desirable to employ a single standard for declaring death in all these circumstances and no special-purpose definitions have been seriously advanced [25, p. 60].

Unlike the malleable nature of corporations' legal personhood for different purposes, calling the same person dead for one purpose and alive for another can engender troublesome confusion. Consider what happened in Connecticut, where the legislature placed its neurologic standard for determining death within its anatomical gift act [26]. Although repeated clinical examinations and confirmatory electroencephalograms established that the patient in one litigated case had suffered irreversible loss of total brain function as specified in that statute, her attending physician was unwilling to remove her from the respirator because she was not going to be an organ donor. "Dead for transplantation, but not dead otherwise [27]." Rather than placing separate (and potentially dissimilar) descriptions of death into different statutes, which would imply that each standard is limited to situations covered by a particular statute, a single set of standards for determining death should apply in all circumstances.

3.2 Death's Relationship to Other Bases for Action

Treating death uniformly in all contexts is not, however, inconsistent with recognizing other conditions that can provide a basis for taking the same action as would be triggered by a person's death. Indeed, the original impetus for the formation of the Harvard Medical School committee on death by neurologic criteria was to establish that some long-term comatose patients in the intensive care unit (ICU) were dead, so that removing them from the mechanical ventilators on which their respiration depended would not expose the medical personnel involved to criminal prosecution. Subsequently, courts clarified that physicians who follow instructions from the family members—or incapacitated patients' advance directives—and discontinue support are not committing homicide. In other words, if the question is "when may life-support be withdrawn or withheld?", answering "when the patient is dead" does not preclude other replies, such as "when the patient rejects such interventions."

"Presumed death" statutes provide another example of the precept that, when a reason exists to allow an action associated with the occurrence of death to be taken in the absence of factors that are normally relied on to declare death, the wise choice is not to modify the standards for determining death but instead to recognize a trigger other than death to justify taking the action. These statutes create a presumption that persons who are absent without explanation from their usual place of residence for a prolonged time are dead. The existence of the presumption permits taking certain actions—such as disposition of an absent person's property and remarriage of their spouse—that also occur when a person is declared dead. Such provisions do not modify the standards for determining human death, but create a set of procedures that permit orderly management of the affairs and relations of the absent person while having no effect on the person's status as a living person, should they ever return.

The one area in which a special-purpose definition of death has been proposed is organ transplantation, with the aim of increasing the number of deceased donors. Having a special standard for determining death for organ donors would, however, be problematic for several reasons. First, the status of "prospective organ donor" depends not simply on an individual's physical condition but on choices made by the individual or by others. If a special standard for declaring death were created for organ donors, then designating an ICU patient as a donor, or withdrawing a prior designation, would immediately move that patient from "living" to "dead" or vice versa. Second, a statute that allows physicians to remove organs from bodies of prospective donors when they are "less dead" than nondonors could undermine public confidence in all death determinations because it suggests that the standards (at least for neurologic determinations) are not based on a set of universally applicable factors, but instead can arbitrarily be adjusted for instrumental reasons. Third, any special "definition" of death for transplantation purposes would need to include procedural safeguards regarding the methods for determining death under the special standard for prospective donors; this could then make it appear prudent to extend such procedures to other death determinations, just as the requirement now found in some statutes that neurologic determinations of death be made by two physicians originated with concerns about conflicts-of-interest in determining death in organ donors.

A number of commentators argue that the medical criteria and tests currently used to determine death on neurologic grounds do not fulfill the statutory standards for death and hence that organ donation in such circumstances violates the so-called "dead donor rule" [28, 29]. They argue that the negative implications for organ transplantation could be counteracted by allowing prospective donors to choose to have their organs removed when they are near death rather than dead, as has been suggested for many years [30]. Donation-by-voluntary-euthanasia would represent a major change in current medical ethics standards, even in states that have legalized physician-assisted death. However, at least such an approach would allow the public to confront the merits and risks of such proposals directly—that is, whether the benefits conferred to transplantation justify the harms—which would not be true if physicians decided to apply the neurological standard for determining death more leniently when the deceased is expected to be an organ donor.

3.3 Confine the Delegation of Law-Making Authority

The drafters of American death determination statutes, including the uniform law commissioners, generally accepted the position advocated by American commentators [10, 14], namely, that a good statute sets general standards for determining death rather than enacting specific medical criteria to describe how the standards will be met. Recently, however, Nevada did something akin to the latter. Responding to a court decision [31] that prevented a hospital from declaring a patient dead based on the American Academy of Neurology (AAN) guideline for determining "brain death" [32], the legislature amended its statute [33] to prescribe that death determinations "must be made in accordance with the applicable guidelines" adopted now or in the future by designated medical organizations, including the AAN. The use of such language reduces the risk that the criteria and tests specified in the statute will become obsolete if the legislature later fails to amend the statute as medical knowledge and procedures evolve. However, the amendment creates another, perhaps graver problem. As already noted, death determination statutes typically set a legal standard and make individual physicians responsible for using professionally accepted criteria and tests to comply with the statutory standard. In contrast, Nevada engaged in legislating-medical-criteria-and-tests-by-proxy when it assigned the power to set binding medical criteria to an independent medical body over which the legislature has no control.

The constitutional principle of separation of powers constrains legislators from transferring law-making to another branch of government; however, the nondelegation doctrine is not violated when executive departments or independent administrative agencies are given the task of writing detailed regulations provided that the standards set forth in the legislation give adequate directions regarding the scope and content of such regulations. The UDDA's substantive standards—that is, the irreversible loss of circulatory and respiratory and/or brain functions—fall into the category of legislated standards that aim to confine the actions taken by those who elaborate and apply the specific rules. So long as the relevant medical societies understand that "accepted" medical criteria must align with those standards, or, failing that, watchdogs from the medical profession [34] or from another branch of government (such as the Supreme Court of Nevada [31]) challenge—and even block the use of—any medical criteria that depart from the standards, the strategy adopted by the Nevada legislature seems unnecessary and unwise.

Nonetheless, some states may conclude (as Nevada did) that they need to be more prescriptive about which criteria and tests may be used for determining death in order to prevent medical groups that develop guidelines for determining death from departing from the statutory standards. Legislators in those states could avoid the problem of undue delegation by giving an appropriate state official or body, such as the secretary of health or the state medical board, the power to approve a set of medical criteria and tests for use by physicians within the state that satisfy two conditions.[7] First, the official must determine that criteria and tests are in line with the statutory standards for diagnosing an individual's death—that is, when the tests show that a criterion has been met, the applicable standard for determining death is present. Second, to ensure that criteria and tests are based on scientific evidence and clinical expertise, the official must rely on current, nationally accepted guidelines issued by professional organizations with recognized expertise. A statute could mention the guidelines developed by identified groups such as the AAN as examples of appropriate *sources* on which the state official may draw without creating the problem that arises under the Nevada law of physicians having to diagnose death in accordance with guidelines adopted—and perhaps subsequently modified—by a private body without governmental review and approval.

4 Four Important Terms to Include in a Statute

General principles of sound legislation, such as clarity, brevity, and consistency, apply to the drafting of a statute on determining death. Here we consider certain points peculiar to the subject at hand.

4.1 "An Individual" and "Is Dead"

Reviewing the UDDA's principal provision offers a good entry point to four important building blocks of a good statute on determining death. The first two begin and end the statute's operative sentence: "An individual who has sustained [...] is dead." Describing the person whose status is being evaluated as "an individual" conforms to the now-standard designation of a human being in uniform acts; the term "individual" is favored because "person" is sometimes used by the law for nonhuman entities, such as a corporation.

"Is dead" constitutes a declarative statement of fact about an individual's current civil status. It contains no command that an action be taken, a matter that is left to

[7] If the first section of the statute, which sets forth the physiological standards by which death is determined to have occurred, provides that determinations must be made in accordance with "current nationally accepted medical criteria and tests," a subsequent section that provides definitions for the terms used in the statute should state that this phrase means "one or more sets of criteria and tests (1) that have been approved by the [Secretary of Health] of this State as a means of implementing the standards for determining death set forth in §1, and (2) that are drawn from medical guidelines on the current methods for diagnosing death issued as of the effective date of this [act], or subsequently modified, by nationally recognized sources of technical guidance on this subject, which include, but are not limited to, the American Academy of Neurology, the Society of Critical Care Medicine, the American Academy of Pediatrics, and the Child Neurology Society, [and so forth]."

other statutes that create duties (e.g., for the next of kin to dispose of the body in a respectful fashion) or bestow entitlements (e.g., for heirs to receive their inheritance), and so forth.

The statement that an individual who meets the statutory standards "is dead" avoids the inaccurate, and perhaps disconcerting, implication created by the first model death determination statutes which stated that such people will (or shall) "be considered dead." The latter formulation could be read to indicate that the law treats someone as dead who by some other, perhaps wiser, standard is not dead.

Declaring that an individual "is dead" implies that at some moment prior to the present moment, the individual arrived at "the event that separates the process of dying from the process of disintegration" [27]. The time when that moment occurred—which might, for example, be when a clinical exam revealed that the individual met the statutory standard or later when the necessary repeat testing confirmed that the loss of function was irreversible—need not be specified in the statute on death determination. It may be spelled out in the context of other laws where timing is an important concern (including state laws on death declarations) or it may be left to the clinical skills and judgment of the physician applying accepted medical practices or hospital rules and customs to make the determination of death.

4.2 "Irreversible Cessation"

The center of the same operative sentence in the UDDA also contains two key terms—"irreversible cessation" and "functions"—that have proven to be even more significant and more controverted than the two just examined.

It is understandable that the term "irreversible cessation" has been used in all the model statutes since the first one in 1972 [15]. "Cessation" connotes more than "absence" (the word used in the 1970 Kansas statute); it also suggests that something that was once present has been lost. In some cases, the cessation of a function is temporary, but when a function that is necessary to sustain the life of an organism has ceased permanently—that is, the function is gone forever—the organism is dead. Although neither the uniform law commission's 1980 report on the UDDA [2] nor materials from state legislative adoption of the act define "irreversible," it appears that the term is meant to exclude cases where a cessation should not be thought of as permanent because it is still possible that the function in question could return, either spontaneously or in response to medical interventions.

In the context of neurologic determinations of death, the term "irreversible" is appropriate because certain conditions, which arise either from the manner in which a patient's brain was injured (such as drowning in very cold water, drug intoxication, metabolic abnormalities, etc.) or from medical interventions that are used to protect the brain from further injury (such as drug-induced coma), may give the appearance that neurologic functions have ceased, but with time or medical efforts, either the functions return or the external influence disappears so that the permanence of the cessation of brain functions can be accurately assessed. The statute's use of "irreversible" thus signals to attending physicians the need to proceed cautiously, particularly when performing brain-based determinations of death.

During the first 20 years of statutes "defining" death, "irreversible" seemed like a straightforward modifier, but the term became problematic with the initiation of the so-called "Pittsburgh Protocol" for obtaining organs from patients pronounced dead based on cessation of circulatory and respiratory functions following the elective removal of mechanical ventilation [35]. The problem was that after support has been withdrawn, the period during which asystole is observed is long enough to rule out a spontaneous resumption of circulation and respiration but not long enough to ensure that if resuscitative measures were applied, they would always fail to restore circulatory and respiration functions [36].

As protocols for donation after circulatory determination of death (DCDD) became widely accepted, the statutory requirement "irreversible cessation" came to be "understood in a weaker sense," namely, "cessation of circulatory and respiratory functions under conditions in which those functions cannot return on their own and will not be restored by medical interventions" [37, p. 84]. Drawing on the way in which physicians ordinarily declare death in situations not involving organ donation and on the interchangeable use of the terms "permanent" and "irreversible" in *Defining Death* [25], "which serves as the principal piece of legislative history for the UDDA," commentators next suggested that it is appropriate to read irreversible to mean permanent "because both words designate a condition that is stable and unchanging" [38, p. 964]. In DCDD cases, the cessation of functions will be permanent because the protocol's "no touch" period rules out spontaneous resumption of circulatory and respiratory functions, and the physicians' commitment to the patient and family that cardiopulmonary resuscitation will not be attempted after support has been withdrawn, the cessation will continue forever.

Which adjective is the right one to use in a statute depends on how one regards the issue—as moral, ontological, or practical—and particularly how one conceives of the relationship between "irreversible" (meaning that a function cannot return) and "permanent" (meaning that it will not return). Is a finding of "permanent cessation" a valid "surrogate indicator" for the statutory end-point of irreversibility [20, p. 245]? Or is "irreversible cessation" a step toward the ultimate goal of establishing that the loss of a function is "permanent"? Answers to such questions will differ depending on one's substantive position about the standards for determining death (which are addressed in other chapters in this volume). A main conclusion from the foregoing discussion is that care must be taken when selecting statutory terms, which should probably be defined in the statute so that their intended meaning is conveyed to the physicians who will have to apply them as well as the judges who may be called on to interpret them.

4.3 "Functions"

Statutory standards for determining death are typically framed in terms of certain physiologic "functions." This terminology implicates two important points. The first involves the specification of the functions that will be evaluated when determining death and, if more than one function is specified, how they relate to one another. The choices here relate to one's "paradigm" for death determination (as described above) and to an empirical assessment of the role that particular functions play in supporting organismic homeostasis or in producing other qualities of human life. (Again, these are matters of substance, which are addressed in other chapters in this volume.)

The second point relates to the alternative terms that the choice to use "functions" implicitly reject. It is well understood that some processes persist for a time in organs after they have permanently ceased functioning, and that some of these processes—such as electrical, metabolic, and hormonal activity at the level of tissues and cells (individually and as groups of cells)—may be detected during physicians' clinical and laboratory assessments of possibly deceased persons. Indeed, the more advanced the instrumentation, the more of these signals may be detected. "Unless this cellular activity is organized and directed, however, it cannot contribute to the operation of the organism as a whole" [25, p. 75], and hence is irrelevant in concluding that the organism is dead. In the past, some critics went beyond arguing that signs of cellular "activity" are inconsistent with declaring death on neurologic grounds to claim that until an organ has been destroyed, it is always possible that it might resume functioning [39]. However, the view that the standard for diagnosing death neurologically should be destruction of the brain never gained much traction; while organic destruction has been found when "brain dead" bodies are autopsied, it is clear that cessation of circulation to the brain leads to the loss of brain functions substantially before destruction of brain cells and liquefaction of tissues.

5 Provisions to Leave Out of the Statute

5.1 A "Definition of Death"

The term "defining death" is not only used colloquially but sometimes in serious discussions of the subject, such as those presented by Bernat and his colleagues, where it means the language in the statute that describes what is usually called a standard for determining death. It would, however, be unwise to include in a statute's "Definitions" section, which explains terms used in the statute—both technical terms that may be unfamiliar and common words with a variety of meanings—a definition that begins, "Death is …"? First, the definition would be redundant, since the same language ought to appear in the principal section of the statute, in a statement along the lines of "An individual who has experienced […] is dead." Second, if, to avoid redundancy, the word "death" were placed into the principal section (which is the reason for providing definitions), then it would read, "An individual who has experienced death is dead."

5.2 Scope of Application

It is not unusual for statutes to include a phrase that confines the application of their terminology to the activities directly addressed by that particular statute. Such limitations reflect legislators' reasonable concern that some behavior or category that is defined in a way that is sensible in the context of the statute before them could produce untoward result were that terminology applied in another area to which they had given no thought.

The statements of scope that appear in some of the death determination statutes have been of the opposite sort. For example, the 1975 ABA model law began "for all legal purposes" [40]. Besides being unnecessary—since without this clause, the statute would be presumed to be of general application—the language is confusing. What does "all legal purposes" not encompass? Suits in equity? Likewise, the stated scope of the UBDA ("for medical purposes") [25, p. 117] and the opening phrase of the Kansas statute ("A person will be considered medically ... dead.") [25, p. 127] add nothing to the effect of these laws. Without these phrases, the statutes would govern the actions of any physician who determines death, while the law cannot by fiat make something medically true that conflicts with physical reality or otherwise dictate the laws of nature. Clauses of this sort should not appear in UDDA-type statutes. Nor should determination of death laws be incorporated into the Uniform Anatomical Gift Act (UAGA) or other organ-donation statutes.

5.3 Treatment Termination

The concern of the Harvard committee that unless ventilator-dependent patients in long-term comas could be declared dead, they would fill ICUs might be taken to imply that a statute on death determination ought to contain provisions that would authorize physicians to take certain actions, such as removing patients from scarce ICU beds and the like. In fact, any such actions should be carried out under the authority of, and according to procedures spelled out in, other laws, rather than adding such provisions to a statute on determining death.

6 Conclusion

By 1970, a consensus had begun to emerge in medicine on two points: first, that the presence of heartbeat and breathing in some ventilator-supported, comatose patients with catastrophic brain injuries was an artifact of treatment rather than a sign of life, and second, that death could reliably be diagnosed in such patients through the use of a set of tests of neurologic functions. To legitimize the use of such tests, however, it soon became apparent that courts—which had equated death with the cessation of respiration and circulation—were less suited than legislatures for updating the "definition" of death. While disagreements arose and still exist about which capacities and functions are essential for human life, legislation is most effective if it includes certain provisions and excludes others. First, a conceptual description of what it means to die, such as "loss of personhood," is not suitable. Instead, a statute should state standards for determining death that are clear and understandable to the public but also specific enough to inform medical experts when they formulate the medical criteria and tests that will reveal when the standards have been met. The criteria and tests ought not, however, be included in the statute, as they will need to be revised as medical knowledge and techniques progress, a process that legislators are unlikely to be equipped or inclined to undertake repeatedly. Second, the standards should describe the permanent cessation of physiologic functions that are necessary for life while recognizing that some metabolic, electrical, or other activities can persist in cells or groups of cells after an organ stops contributing to the homeostasis in the organism as a whole. Third, a death determination statute should apply to all persons uniformly; for example, when such a statute is placed in the organ transplant law, the declaration that a potential organ donor has died is negated when the patient is found to be unsuitable to donate an organ. Fourth, treating death uniformly in all contexts does not preclude certain actions, such as withdrawing medical support from patients, on grounds other than a determination of death. Such alternative reasons for acting need to be publicly discussed—and accepted or rejected by courts, legislatures, and professional associations—on their own merits, rather than covertly by applying the usual standards for determining death more loosely in certain situations, as, for example, by allowing some dying persons to become organ donors because they will soon be dead.

References

1. Ad Hoc Committee of the Harvard Medical School to Examine the Definition of Brain Death. A definition of irreversible coma. JAMA 1968;204:337–40.
2. National Conference of Commissioners on Uniform State Laws. Uniform Anatomical Gift Act. Handbook and Proceedings of the Annual Conference. 1968.
3. Sadler AM, Sadler BL, Stason EB. The uniform anatomical gift act: a model for reform. JAMA. 1968;206:2501–6.
4. Stason EB. The Uniform Anatomical Gift Act. The Bus Lawyer. 1968;23:919–29.
5. Black's Law Dictionary, 4th ed. St. Paul, Minn: West Publishing Co.; 1968. ("DEATH. The cessation of life; the ceasing to exist; defined by physicians as a total stoppage of the circulation of the blood, and a cessation of the animal and vital functions consequent thereon, such as respiration, pulsation, etc.)" p. 488.
6. Taylor LF. A statutory definition of death in Kansas. JAMA. 1971;215:296.
7. Baker JC. Liability and the heart transplant. Houston L Rev. 1968;6:85–112.
8. Brickman MJ. Medico-legal problems with the question of death. Calif WL Rev. 1968;5:110–23.
9. Tucker v. Lower, No. 2831 (Richmond, VA, L. & Equity Ct., May 23, 1972).
10. Task Force on Death and Dying of the Institute of Society, Ethics, and the Life Sciences. Refinements in criteria for the determination of death: an appraisal. JAMA 1972; 221:48–53.
11. In Re Bowman, 94 Wn.2d 407. 1980.
12. Curran WJ. Legal and medical death: Kansas takes the first step. New Eng. J. Med. 1971;284:260–1.
13. Mills DH. The Kansas death statute: bold and innovative. New Eng. J. Med. 1971;285:968–9.

14. Kennedy IM. The Kansas statute on death: an appraisal. New Eng. J. Med. 1971;285:946–50.
15. Capron AM, Kass LR. A statutory definition of the standards for determining human death: an appraisal and a proposal. Univ Penn L Rev. 1972;121:87–118.
16. Bernat JL, Culver CM, Gert B. On the definition and criterion of death. Ann Int Med. 1981;94:389–94.
17. Bernat JL. The biophilosophical basis of whole-brain death. Soc Philo Policy. 2002;19:324–42.
18. People v. Eulo, 63 N.Y.2d 341-361. 1984.
19. State v. Guess, 244 Conn. 761-781. 1998.
20. Bernat JL. How the distinction between "irreversible" and "permanent" illuminates circulatory–respiratory death determination. J Med Philo. 2010;35:242–55.
21. National Conference of Commissioners on Uniform State Laws. Uniform Brain Death Act. Handbook and Proceedings of the Annual Conference. 1978.
22. Law Reform Commission of Canada. Report 15: criteria for the determination of death. Ottawa: Law Reform Commission; 1981.
23. Dominguez-Gil B, Ascher N, Capron AM, et al. Expanding controlled donation after the circulatory determination of death: statement from an international collaborative. Intensive Care Med. 2021;47:265–81.
24. Dworkin RB. Death in Context. Ind LJ. 1973;48:623–39.
25. President's Commission for the Study of Ethical Problems in Medicine and Biomedical and Behavioral Research. Defining Death: Medical, Legal and Ethical Issues in the Determination of Death. Washington, D.C.: U.S. Government Printing Office; 1981.
26. 1979 Conn. P.A. 79-556, codified at Conn. Gen. Stat. Ann. §19-139i (West Cum. Supp. 1981), repealed, 1988 Conn. P.A. 88-318, §14.
27. Fabro F. Bacchiocho vs. Johnson Memorial Hospital. Conn Med. 1981;45:267–9.
28. Truog RD, Miller FG. The Dead Donor Rule and Organ Transplantation. New Eng J Med. 2008;259:674–5.
29. Veatch RM. Transplanting hearts after death measured by cardiac criteria: the challenge to the dead donor rule. J. Med. Philo. 2010;35:313–29.
30. Scribner B. Ethical problems of using artificial organs to sustain human life. Trans Am Soc Artif Internal Organs. 1964;10:209–12.
31. Gebreyes v. Prime Healthcare Servs., LLC. 361 P.3d 524. 2015.
32. Wijdicks EFM, Varelas PN, Gronseth GS, Greer DM. Evidence-based guideline update: determining brain death in adults: report of the Quality Standards Subcommittee of the American Academy of Neurology. Neurology. 2010;74(23):1911–8.
33. 2020 Nevada Revised Statutes. §451.007 (Requirements for determination), as amended by A.B. 424, 2017 Statutes of Nevada, Ch. 315. p. 1728–29.
34. Shewmon DA. Brain death or brain dying? J Child Neurol. 2012;27:4–6.
35. DeVita MA, Snyder JV. Development of the University of Pittsburgh Medical Center policy for the care of terminally ill patients who may become organ donors after death following the removal of life support. Kennedy Inst Ethics J. 1993;3:131–43.
36. Lynn J. Are the patients who become organ donors under the Pittsburgh protocol for "non-heart-beating donors" really dead? Kennedy Inst Ethics J. 1993;3:167–78.
37. President's Council on Bioethics. Controversies in the determination of death: a white paper. Washington, DC. www.bioethics.gov. 2008.
38. Bernat JL, Capron AM, Bleck TP, et al. The circulatory-respiratory determination of death in organ donation. Crit Care Med. 2010;38:963–70.
39. Byrne PA, O'Reilly S, Quay PM. Brain death: an opposing viewpoint. JAMA. 1979;242:1985–90.
40. American Bar Association, Feb. 1975 Midyear Meeting. A.B.A. Ann Rpt. 1978;100:231–231.

Preserved Hypothalamic Function Is Not Consistent with the Whole-Brain Criterion for Death

Michael Nair-Collins

The whole-brain criterion for death requires the absence of all functions of the entire brain. This criterion is enshrined in the Uniform Determination of Death Act (UDDA) of the United States, which states that an individual with "irreversible cessation of all functions of the entire brain, including the brainstem" is dead [1].

This criterion should be distinguished from the brainstem criterion, which requires only the cessation of brainstem functions, but not all brain functions [2], as well as the higher-brain criterion, which requires the absence of brain functions necessary for conscious awareness [3–5]. The whole-brain criterion should also be distinguished from the "brain-as-a-whole" criterion, which requires the cessation of *critical* functions of the brain, explicitly allowing some continued brain functions if they are deemed "non-critical" [6, 7]. The brainstem criterion is used in a few nations, while the whole-brain criterion forms the basis for law in most of the world, including the United States; neither the higher-brain nor brain-as-a-whole criteria have been officially enacted in any jurisdiction.

This chapter exclusively addresses the whole-brain criterion, specifically as characterized by the UDDA: "irreversible cessation of all functions of the entire brain, including the brainstem" [1]. Given this concept, it is a matter of valid deductive logic that the preservation of any function of any part of the brain is not consistent with the absence of all functions of the entire brain. The hypothalamus is a part of the brain, and some functions of the hypothalamus, particularly osmoregulation, can continue in some patients declared dead under the whole-brain criterion of death, rendering those declarations of death false-positive misdiagnoses.

In this chapter, I review the literature on this debate, defending the claim—which, one would think, needs no defense—that some brain function is not consistent with absence of all brain function. While there are several related concerns

M. Nair-Collins (✉)
Behavioral Sciences and Social Medicine, Florida State University College of Medicine, Tallahassee, FL, USA
e-mail: michael.nair-collins@med.fsu.edu

A. Lewis, J. L. Bernat (eds.), *Death Determination by Neurologic Criteria*, Advances in Neuroethics, https://doi.org/10.1007/978-3-031-15947-3_7

surrounding the reliability and validity of the determination of death by neurologic criteria (as discussed elsewhere in this book) [8], I maintain a narrow focus on hypothalamic function. Examination of the medical literature on this specific issue reveals broader concerns regarding the role of logic, scientific evidence, and transparency in the determination of death by neurologic criteria.

1 A Note on Terminology

The term "death by neurologic criteria" is used elsewhere in the book as a noun in place of "brain death," following language recommended by the World Brain Death Project [9]. In this chapter I retain "brain death" to refer to the condition in which all functions of the entire brain have ceased irreversibly. The neologism "death by neurologic criteria" incorporates substantive claims that are matters of scholarly dispute, which should not be defined away by stipulation. For example, it is insensible to claim "death by neurologic criteria is not death"; but *whether* brain death is death is a matter of scholarly disagreement. The definition of "death by neurologic criteria" offered by the World Brain Death Project is problematic in additional ways that will be explored in this chapter, in the penultimate section "Demoting the Hypothalamus in the Brain Death Literature."

2 Brain Death Pathophysiology and Diagnostic Tests

Brain death can have a variety of etiologies, but a general pattern in the form of a positive feedback cycle characterizes its basic pathophysiology. As intracranial pressure (ICP) rises, cerebral perfusion decreases, causing damage to neural cells leading to edema, further increase in ICP, further decrease in cerebral perfusion, and so on, until ICP rises above mean arterial pressure (MAP), eventuating in an assumed global loss of brain circulation and hence global cerebral anoxia. This process runs in a rostral-to-caudal direction, with the lower brainstem being the last area to become infarcted and is often accompanied by herniation of the unci of the temporal lobes or the tonsils of the cerebellum.

The accepted diagnostic tests for brain death—also referred to as "medical standards"—are tied to this pathophysiological picture [9, 10]. First, the cause of coma must be known and believed to be severe enough to result in irreversible pathology. This requires the use of neuroimaging, such as a CT scan. Second, potential confounders to further testing must be ruled out, including hypothermia, sedative intoxication, acid–base disturbances, and others, along with a general assessment of patient health, including evaluation of electrolytes and kidney and liver functions. These require laboratory tests.

Third, the patient is evaluated for responsiveness to auditory stimuli and pain and must be unresponsive to all such stimuli. Fourth, a variety of brainstem reflexes are tested, including pupillary response to light, deep cough response to suction, gag reflex, blink response to touching a wisp of cotton to the cornea, and vestibular (eye movement) responses to cold water placed in the ear canals and to brisk movement of the head.

If these findings are all consistent with brain death, then the final evaluation is the apnea test. The patient is hyperoxygenated prior to the test. Oxygen is passively delivered, while the ventilator is disconnected from the patient for a period of 8–10 min as clinicians observe for signs of spontaneous breathing. Arterial carbon dioxide partial pressure is measured before and after the test, and must rise 20 mm Hg from baseline or reach 60 mm Hg, for the apnea test to be considered valid. This test requires arterial blood gas analysis.

Ancillary tests, such as neuroimaging for brain circulation or an electroencephalogram (EEG), are not considered required, but are often used at the physician's discretion, though they are mandatory in some jurisdictions. These accepted diagnostic tests are essentially the same for both the whole-brain criterion of death [9, 10] and the brainstem criterion [2], though there is also variability in practice both internationally and intranationally as discussed elsewhere in this book [9].

In the medical literature, the determination of death by neurologic criteria is uniformly said to be a "clinical diagnosis," meaning that only functions that are observable at the bedside are part of the diagnostic evaluation (e.g., [9, 11]). This is false, as can be seen from the description of the diagnostic tests, which require imaging and laboratory analyses. One might respond that the "core" tests for brain death only include evaluation for unresponsiveness, brainstem areflexia, and apnea. This distinction arbitrarily ignores mandatory aspects of the evaluation, including identifying the cause of coma and ruling out confounders, along with ancillary tests that are often used and can finalize the determination if other parts of the evaluation are equivocal or cannot be performed.

Nonetheless, even granting this arbitrary rejoinder for the sake of the argument, the apnea test—one of the cardinal features of the diagnostic tests for brain death—requires laboratory analysis of arterial blood gases. Brain death is thus just as much of a technological diagnosis as any other in the ICU. It requires a thorough history, many laboratory analyses, neuroimaging, and bedside physical evaluation, combined with ongoing ICU-level monitoring of many physiologic variables; in some cases, the determination also relies on direct measurement of ICP (requiring surgical implantation of the measuring device), along with additional, more advanced imaging, or electrophysiologic analysis. Therefore, it is not a "clinical diagnosis" [9, p. E5]. This claim is a motivated misdescription of how brain death is in fact diagnosed. The relevance of this point will shortly become apparent.

3 Hypothalamic Functions in Patients Declared Dead by Neurologic Criteria

The hypothalamus is a small region at the base and center of the brain and is a component of the diencephalon. Its borders are somewhat indistinct, but is generally considered to be bordered rostrally by the lamina terminalis; caudally by the edge of the mamillary bodies, behind which is found the midbrain; dorsally by the hypothalamic sulcus and above that the thalami; ventrally by the infundibulum, pituitary stalk, and below that the pituitary glands; and laterally by the basal nucleus of Meynert, nucleus accumbens, amygdala, posterior limb of the internal capsule and basis pedunculi, and caudodorsally the subthalamic nucleus [12].

The hypothalamus has been described as "the homeostatic head ganglion" [13, p. 738], for its critical role in multiple homeostatic functions, including osmoregulation (the regulation of osmolarity: the concentration of solutes, of which sodium is the most common, in extracellular fluid), hunger, thirst, sleep–wake cycles, blood pressure control, temperature control, limbic mechanisms, and neuroendocrine and autonomic regulation. The hypothalamus directly controls the posterior pituitary gland through release of vasopressin and oxytocin, and indirectly controls the anterior pituitary gland via hypophysiotropic factors passed through a local blood supply known as the hypophyseal portal system, thereby regulating secretion of adrenocorticotropic hormone, thyroid-stimulating hormone, growth hormone, prolactin, luteinizing hormone, and follicle-stimulating hormone [13].

If all functions of the brain cease, as the whole-brain criterion requires, then the functions of the hypothalamus must cease. As the magnocellular neurons of the supraoptic and paraventricular nuclei in the anterior (supraoptic) region of the hypothalamus become infarcted or damaged, they should stop secreting vasopressin (or antidiuretic hormone), resulting in central diabetes insipidus, which would be easily apparent by the onset of massive hypoosmotic polyuria with hypernatremia. (There are confounders to this, which are mentioned below) [14].

Yet, central diabetes insipidus does not manifest in a large percentage of patients declared dead by neurologic criteria, entailing that hypothalamic osmoregulation remains intact. The largest review of central diabetes insipidus in patients declared dead by neurologic criteria evaluated data from 37 studies and found that 1265/2546 patients (50%, 95% CI [0.478–0.516]) manifested diabetes insipidus [15]. Therefore, the critical brain function of osmoregulation continues in up to half of patients declared dead by neurologic criteria.

Furthermore, the half-life of vasopressin is 15–18 min, and the neural–renal osmoregulatory system is sensitive and rapid, maintaining osmolarity within a narrow, 3% window. The magnocellular neurons are osmoreceptors, directly responsive to their extracellular osmotic environment, regulating the secretion of vasopressin on a minute-to-minute basis [16]. Clearly, they must be receiving ongoing arterial flow as well as venous drainage to perform this function.

It is also worth noting that changes in osmotic pressure local to the magnocellular neurons of the hypothalamus are often insufficient for the cells to reach threshold potential and thus fire an action potential down the axons that traverse the pituitary stalk and terminate in the posterior pituitary gland. There is a secondary osmoreceptive system located in circumventricular areas, including the organum vasculosum of the lamina terminalis and the subfornical organ. These neurons are also osmoreceptors, which supply excitatory, glutamatergic input to the primary osmoreceptors in the hypothalamus [17]. Normal osmoregulation is thus a function of the additive effect of both the primary system located in the hypothalamus and the secondary system located in circumventricular areas, rostrally adjacent to the hypothalamus.

In addition to osmoregulation, neuroendocrine control of anterior pituitary hormones is often not lost either. In a review of 12 studies that included 386 patients, up to 84% did not show central thyroid failure, and in 2 studies that included 24 patients, up to 71% did not show central adrenal failure [14, 16].

The whole-brain criterion for death states that all brain functions must be lost. Since osmoregulation, a brain function, continues in potentially half of patients declared dead by neurologic criteria under the whole-brain criterion for death, it follows that up to half of these declarations of death are false-positive misdiagnoses. Considering the limited data on central thyroid and central adrenal failure, this number of false-positive misdiagnoses potentially rises even further.

4 Demoting the Hypothalamus in the Brain Death Literature: Efforts to Deny the Relevance of Preserved Hypothalamic Function

It has been well known for decades that some patients declared dead by neurologic criteria may have preserved hypothalamic function (e.g., [18]). More recent studies cited above simply attempt to clarify how common this is, but the basic point is (or should be) common knowledge.

However, rather than recommending changes to medical practice based on scientific evidence, in the brain death literature, clinical practice has driven what counts as evidence. Specifically, there have been many attempts to ignore, deny, or minimize the indisputable fact that in some patients declared dead using the whole-brain criterion for death, some brain function continues.

As the UDDA and all extant concepts of brain death rely on the concept of neurologic *function*, a natural target for denying the relevance of hypothalamic functions is to deny that they are functions at all: they are relegated to mere "activities," and therefore their preservation does not run afoul of the whole-brain criterion. For example, the International Guidelines for Determination of Death group allege that "examples of brain function such as the capacity for consciousness … should be distinguished from examples of brain activity such as posterior pituitary antidiuretic hormone release" [11, p. 791]. More recently, the World Brain Death Project has repeated this "function vs. activity" distinction, asserting that "*brain function* refers to the more macro phenomena that are measurable on bedside neurological examination… *brain activity* refers to neuronal cellular micro phenomena recordable by technology" [9, p. E3].

This is an ad hoc definition motivated by the desire to render osmoregulation (and potentially other evidence of brain function) consistent with the determination of death by neurologic criteria. It flies in the face of clinical practice: Physicians assess renal, hepatic, cardiac, and pancreatic function (*inter alia*) with laboratory tests and imaging. Why would the function of the brain, arguably the most complex of all organs, only be assessed by simple bedside evaluation? Furthermore, this would make any findings on the EEG irrelevant to the determination (as, being technology, it can only show "activity"), thus rendering it an unacceptable component of the diagnostic algorithm. However, the medical standards recommend or suggest EEG as an acceptable ancillary test (e.g., [9, 10]), which may be used to finalize the determination in some cases, thus creating an internal contradiction. (The World Brain Death Project has suggested moving away from its routine use in adults except for certain cases such as skull fracture or decompressive craniectomy, thus, still accepting its validity in some cases.) [9, pp. E8–E9]

The apnea test requires laboratory analysis of arterial blood gases as an essential component of the test. If carbon dioxide partial pressure does not sufficiently rise, as measured by laboratory analysis, then the test is not valid. Therefore, one of the core aspects of the diagnostic tests requires laboratory analysis of brain function—specifically, of the medulla's capacity to respond to *a stimulus sufficient to challenge the medulla*—which can only be measured by laboratory analysis of carbon dioxide partial pressure. The "function vs. activity" distinction should thus rule out the apnea test as a valid component of the diagnostic tests as well.

Furthermore, the World Brain Death Project's definitions of "clinical" and "clinical test" are inconsistent. "Clinical" is defined as "Based on direct … observation or examination of the patient," [9, p. E3] while "clinical test" is "A bedside test [which] may include the use of … vital signs monitors." [9, p. E3].

Vital sign monitoring in a modern ICU consists of, at minimum, continuous electrocardiography, continuous photoplethysmography, and an electronic sphygmomanometer, along with a variety of physiologic measurements produced by the mechanical ventilator. These technologies provide information on aspects of patient physiology that are *not directly observable*, such as electrical activity of the heart or peripheral oxygen saturation. This non-directly observable information is detected by technological sensors of various kinds, converted into a signal, and then is altered in accordance with a variety of mathematical and electronic transformations by complex biomedical engineering devices, to finally produce *representations* of that physiologic information, in a form that is interpretable by human observers. The actual physiologic variables measured by these devices are themselves no more "directly clinically observable" than sodium levels in the blood, or cortical activity measured by an EEG—which, notably, is based on the same physical principles as the bedside continuous monitoring electrocardiogram.

Regardless of the motivated definitions discussed above, osmoregulation—the regulation of sodium and free water in the extracellular fluid—is a vital biological function, a function of the brain, necessary for maintenance of homeostasis and the life of the organism [13, 14, 17]. To use the World Brain Death Project's notion of "neurologic function," osmoregulation involves the delivery of "a stimulus to provoke central processing and an efferent response" [9, Supplement 5, p. 20]; or in the words of the President's Commission for the Study of Ethical Problems in Medicine and Biomedical and Behavioral Research (henceforth "President's Commission"), this is cellular activity that "is organized and directed" in maintaining osmolarity, in conjunction with the kidneys, thus rendering it a function and not a mere activity [1, p. 75]. Ad hoc definitions constructed to avoid facts that do not fit professional interests in preserving the status quo in determination of death by neurologic criteria, which create internal inconsistencies in the logic of the tests, and are inconsistent with the actual practice of medicine, do not change those disfavored facts.

A related rejoinder is to assert that only "clinical functions," assessable at the bedside, count. This relies on the false claim that determination of death by neurologic criteria is a clinical diagnosis, which it is not, as reviewed above, and on the additional false claim that normal osmoregulation cannot be observed at the bedside, which it can, through normal urine output. (There are confounders, including severe acute kidney injury, which would cause oliguria, or administration of vasopressin for systemic blood pressure control. However, in a patient without severe acute kidney injury and for whom vasopressin was not administered, normal urine output is a clear sign of brain function [14].)

Another point that has been made since the very early days of the construction of the concept of "brain death" is that not every cell in the brain must die to make the determination [1, 19]. This claim has also been marshalled to avoid acknowledging hypothalamic function in brain death. However, I have not argued that all cells of the brain must die. I have focused specifically on brain *functions*, in accordance with the UDDA and all extant concepts of death. Osmoregulation, subserved by magnocellular nuclei in the hypothalamus with additive input from the circumventricular region, is an organized and directed brain *function* requiring both arterial supply and venous drainage into systemic circulation, participating in a negative feedback system with the kidneys, to maintain a vital physiologic variable within limits necessary for organismic functioning. It is not a mere "cluster of cells" randomly doing nothing of physiological significance.

Arterial supply of the posterior pituitary gland is partially provided by the inferior hypophyseal artery, which branches off the meningohypophyseal trunk of the internal carotid artery. These arteries remain outside the dura until the inferior hypophyseal pierces the dura at the inferior portion of the posterior pituitary gland [14]. This anatomical location is relevant because the inelastic container within which ICP rises is the dura, in addition to the cranium. Therefore, it is plausible that these arteries enjoy some protection from increased ICP, allowing preserved flow while flow to other areas, not protected in this way, would cease. This is yet another reason offered for why preserved hypothalamic function is consistent with the whole-brain criterion for death and the UDDA [20].

However, this does not explain why preserved hypothalamic function is consistent with the whole-brain criterion for death. Neither the UDDA nor the whole-brain criterion make exceptions based on blood supply.

Furthermore, the inferior hypophyseal artery supplies the inferior portion of the posterior pituitary gland, which consists of axons whose cell bodies are located intradurally in the diencephalon, in areas not protected from increased ICP. In addition, the secondary osmoreceptive system in the circumventricular areas is similarly not protected in this way [14]. Therefore, continued arterial supply and venous drainage is required, in areas not protected by the extradural location of the inferior hypophyseal artery, so this does not explain continued osmoregulation anyway.

Another response is that only "critical functions" count in the determination of death by neurologic criteria [21]. This is yet another attempt to carve out "special" brain functions that count while discounting others, in contradiction with the whole-brain criterion of death and the UDDA, thus rendering it irrelevant to the question at hand. The proposal here is to designate some brain functions as "not critical," while insisting only that critical brain functions must be lost to determine death by neurologic criteria, and thus, persistence of any brain function deemed "not critical" would not preclude the determination [6, 7]. Of course, osmoregulation is at the top of the list for demotion to "noncritical" status, thus rendering its preservation allegedly consistent with an accurate diagnosis of brain death. However, this proposal amounts to changing the criterion, from the whole-brain criterion of death that is embodied in the UDDA, to something else, something less than the whole brain. That is not what is at issue here. Furthermore, if anything counts as a critical function, either of the brain or of the organism as a whole, then regulating the chemical composition of the extracellular fluid, a necessary precondition for essentially all cellular functions throughout the organism, surely counts as a critical function. Therefore, osmoregulation cannot be dismissed as a "noncritical function" anyway.

Another, similar move, is to directly claim authority over the criterion itself, so that, rather than law providing the legal standard that physicians are tasked with determining (using diagnostic tests or "medical standards"), some argue that physicians themselves have the authority to define the criterion that is to be identified. Once again, this is irrelevant to the question at hand, which is whether hypothalamic function is consistent *with the whole-brain criterion*, or the UDDA, which it is not.

Wijdicks, for example, has argued that the diagnostic tests, or the medical standards, themselves define the condition that is being diagnosed, rather than being diagnostic tests for a physiologic condition defined by law [22]. If this were true, then, assuming a competently performed examination, a false-positive determination would be impossible in principle: the tests, and the condition being tested for, are one and the same, so a positive result is—by definition—a true positive. If a false-positive determination is impossible in principle, then there is no empirical observation that could refute the claim that "the diagnostic tests have perfect sensitivity and specificity." Without even the possibility of any empirical evidence bearing on this claim, arguably it is not a scientific claim at all, which is inconsistent with medicine's proclaimed commitment to scientific practice [23]. In any case, this amounts to another attempt to change the criterion from the whole-brain criterion for death to something else, and thus is irrelevant.

The first sentence of the UDDA defines the condition—irreversible cessation of all functions of the entire brain—that physicians are tasked with identifying. The second sentence states "Determination of death must be made in accordance with accepted medical standards" [1]. Some suggest that this second sentence gives medicine authority to define the condition being diagnosed. It does not. First, if it did, it would render the first sentence of the UDDA moot, by granting physicians the authority to change the criterion defined in the first sentence. Laws are not written to be self-defeating, to undermine their own authority, nor to render themselves moot. Second, the President's Commission discussed the meaning of each phrase of the UDDA sequentially [1, pp. 72–81]. As for "accepted medical standards" the intent was "to require the use of *diagnostic measures and procedures* that have passed the normal test of scrutiny and adoption by the biomedical community" [1, p. 78; emphasis added]. Thus, the second sentence refers to diagnostic tests used to identify the condition defined in the first sentence, *irreversible cessation of all functions of the entire brain*; it does not grant physicians' authority to ignore the first sentence. Death is not "whatever the medical community says it is."

Finally, two more recent rejoinders have been proposed. Several scholars have questioned whether the hypothalamus is a part of the brain. Lewis, Bonnie, and Pope, in discussing recent lawsuits, wrote "this raised the question of whether the pituitary and hypothalamus are part of the 'entire brain'" [24, p. 143]. In a different article they surmised,

> the authors of the UDDA do not appear to have intended the phrase 'all functions of the entire brain' to encompass functions of the pituitary gland and hypothalamus; in their 188-page report, they mentioned 'coma' 120 times, 'brainstem' 22 times, and 'apnea' 9 times. But not once did the Commission mention any terms to describe pituitary/hypothalamic/hormonal function [25, p. 17].

There are many brain areas that the Commission did not specifically mention; indeed they did not specifically mention *most* areas of the brain. This does not imply they intended "all functions of the entire brain" to mean anything other than *all functions of the entire brain.* This is hardly a convincing argument, either legally or anatomically. Besides, it is unarguable that the hypothalamus is a part of the brain [14].

Finally, both the American Academy of Neurology (AAN) and the World Brain Death Project have simply asserted, in an impressive feat of bold unconcern for logical contradiction, that some brain function is consistent with no brain function; that is, that hypothalamic brain function is consistent with cessation of all functions of the brain. The AAN wrote,

> The AAN endorses the perspective of the UDDA that brain death has occurred when the irreversible loss of all functions of the entire brain including the brainstem has been determined. However, the AAN endorses the belief that preserved neuroendocrine function may be present … and is not inconsistent with the whole brain standard of death [20, p. 230].

Neuroendocrine function is brain function. Preservation of any brain function is inconsistent with "loss of all functions of the entire brain." To assert otherwise is a naked logical contradiction.

The World Brain Death Project made the same claim, although used the term "neuroendocrine activity." However, as repeatedly shown above, osmoregulation is a brain *function*, even under their own definition, and the President's Commission's definition, of "function." Simply *calling* it an "activity" in this context does not change the fact that it is an organized, directed function which maintains, in coordination with the kidneys via a negative feedback process, a vital physiologic parameter within limits necessary for organismic functioning. The World Brain Death Project wrote,

> [Brain death/death by neurologic criteria] is defined as the complete and permanent loss of brain function as defined by an unresponsive coma with loss of capacity for consciousness, brainstem reflexes, and the ability to breathe independently… Persistence of … neuroendocrine activity does not preclude the determination [9, p. E3].

It is to be noted first that the World Brain Death Project has changed the criterion; this is not the whole-brain criterion for death, and is not equivalent to the UDDA. Leaving aside the difference between "permanent" and "irreversible," the first part of the definition closely resembles the whole-brain criterion ("complete and permanent loss of brain function"). However, the second part, which defines the first part, describes the major clinical features of the diagnostic tests (coma, brainstem areflexia, apnea), and is not equivalent to the whole-brain criterion. This definition is similar to the proposal by Wijdicks, where the diagnostic tests define the criterion [22]. The new criterion, labeled with the neologism, "death by neurologic criteria," is a syndrome characterized by unresponsiveness, brainstem areflexia, and apnea. These three characteristics are the "cardinal features" of long-accepted medical standards, or diagnostic tests. As the AAN wrote in its 2010 "Evidence-based guideline update," "the medical standards for the determination of brain death … [consist of, at minimum] 3 clinical findings necessary to confirm irreversible cessation of all functions of the entire brain…: coma (with a known cause), absence of brainstem reflexes, and apnea" [10, p. 1911]. The syndrome characterized by the co-occurrence of these three findings is distinct from the condition of irreversible cessation of all functions of the entire brain. As argued above, the medical community has no authority to change the physiologic condition it is entrusted to identify using diagnostic tests; and even less to do so under the guise of merely clarifying definitions "to ensure consistency" [9, p. E3].

This proposal has the same implication as Wijdicks's [22]. It effectively renders false-positive misdiagnoses impossible by definition, so long as the medical standards are followed competently. Furthermore, this stipulated definition renders hypothalamic function consistent with the determination of death by neurologic criteria merely by definition. Hypothalamic function, especially osmoregulation, is consistent with the syndrome of coma, brainstem areflexia, and apnea, and is not assessed by current medical standards; therefore, its preservation is consistent with these standards. However, we already knew that. The new terminology simply hides the fact that under current practice, patients who do not satisfy the UDDA or the whole-brain criterion for death are routinely declared dead, *specifically by appealing to the UDDA or to the whole-brain criterion for death.*

The World Brain Death Project nonetheless endorsed brain death in terms of the whole-brain criterion, albeit incoherently, by stating in the first part of the definition that brain death is the "complete and permanent loss of brain function." It went on to state that neuroendocrine *activity*, which is actually hypothalamic and circumventricular *function*, is consistent with "complete … loss of brain function" [9, p. E3]. This amounts to the same logical contradiction asserted by the AAN.

5 Concluding Thoughts

Preservation of hypothalamic function in patients declared dead under the UDDA and the whole-brain criterion for death is but one of several concerns surrounding the concept and diagnosis of brain death [8]. However, an in-depth examination of this narrow issue is valuable, as it reveals broader patterns in the medical literature and clinical practices relevant to death determination.

With respect to the preservation of hypothalamic brain function in patients declared dead by neurologic criteria, which explicitly contradicts the UDDA and whole-brain criterion for death, the medical profession has made no changes to its diagnostic practices considering this information since the concept of brain death was developed in the 1960s. Nor has it evinced any transparency about this fact, either with the public, media outlets, or in any of its many "updated" standards for clinicians, who may not be well-conversant with the primary brain death literature and must rely on professional society updates for accurate and scientifically informed guidance for clinical practice. Instead, professional society standards and individuals in both medicine and bioethics have repeatedly ignored, minimized, or denied facts that challenge professional interests in maintaining the status quo in determination of death by neurologic criteria, while continually repeating the false claim that the diagnosis of brain death, relative to the UDDA and the whole-brain criterion for death, is made with near-perfect accuracy (e.g., [10, 20, 22, 26]).

Certainly, it is a weighty responsibility that has been entrusted to the medical profession. The determination of death is unique among all possible medical determinations, and it is associated with profound consequences for the patient, family, and many other interested parties. It is unreasonable to expect perfection in this or any human endeavor; therefore, perfection is not expected. Nonetheless, it *is* reasonable to expect that professional societies, and individual physicians, will be competent, trustworthy, and will follow the law in carrying out such a grave duty. It is also reasonable to expect that diagnostic practices will be informed by scientific knowledge, will be logically coherent, and that, above all, there will be transparency with all stakeholders, including transparency regarding facts that are uncomfortable or that cast doubt on the reliability and validity of accepted practices.

The review of the medical literature on hypothalamic functioning in brain death reveals that these behaviors and standards rightly expected of the medical profession have not characterized its practice with respect to determination of death by neurologic criteria. Instead, the literature is characterized by decades of ad hoc, irrelevant, false, illogical, specious rejoinders, clearly designed to protect the status quo in death determination.

For the medical profession to be worthy of the special trust required to play this important role in society, professional societies and coalitions of such societies, such as the World Brain Death Project, the American Academy of Neurology, and others, must change course. They should stop closing ranks to protect narrow professional interests in maintaining the status quo over the far more important values of truthfulness, scientific credibility, and transparency. They should stop refusing to acknowledge troublesome facts that do not align with the manufactured narrative of brain death as a well-justified, accurately diagnosed medical condition; because it is neither of those things.

Acknowledgments Some of the ideas for this chapter are drawn from an earlier published manuscript [14]. I gratefully acknowledge my coauthor Dr. Ari Joffe for collaboration on this earlier paper, and thank him for permission to rearticulate some of those ideas in the present chapter.

References

1. President's Commission for the Study of Ethical Problems in Medicine and Biomedical and Biobehavioral Research. Defining death: a report on the medical, legal and ethical issues in the determination of death. Washington, DC: U.S. Government Printing Office; 1981.
2. Academy of Medical Royal Colleges. A Code of Practice for the Diagnosis and Confirmation of Death; 2008. http://aomrc.org.uk/wp-content/uploads/2016/04/Code_Practice_Confirmation_Diagnosis_Death_1008-4.pdf.
3. Green MB, Wikler D. Brain death and personal identity. Philos Public Aff. 1980;9(2):105–33.
4. McMahan J. The ethics of killing: problems at the margins of life. Oxford: Oxford University Press; 2002.
5. Lizza JP. Persons, humanity, and the definition of death. Baltimore: The Johns Hopkins University Press; 2006.
6. Bernat JL, Dalle Ave AL. Aligning the criterion and tests for brain death. Camb Q Healthc Ethics. 2019;28(4):635–41.
7. Dalle Ave AL, Bernat JL. Inconsistencies between the criterion and tests for brain death. J Intensive Care Med. 2020;35(8):772–80.
8. Shewmon DA. Statement in support of revising the Uniform Determination of Death Act and in opposition to a proposed revision. J Med Philos. 2021; online ahead of print. https://doi.org/10.1093/jmp/jhab014.
9. Greer DM, Shemie SD, Lewis A, et al. Determination of brain death/death by neurologic criteria. The World Brain Death Project. JAMA. 2020;324(11):1078–97.
10. Wijdicks EF, Varelas PN, Gronseth GS, Greer DM. Evidence-based guideline update: determining brain death in adults: report of the Quality Standards Subcommittee of the American Academy of Neurology. Neurology. 2010;74(23):1911–8.
11. Shemie SD, Hornby L, Baker A, et al. International guideline development for the determination of death. Intensive Care Med. 2014;40(6):788–97.
12. Swaab DF, Buijs RM, Kreier F, Lucassen PJ, Salehi A. Preface. In: Swaab D, Buijs RM, Kreier F, Lucassen PJ, Salehi A, editors. Handbook of clinical neurology vol. 182 (3rd series). The human hypothalamus: neuropsychiatric disorders. Amsterdam: Elsevier; 2021.
13. Blumenfeld H. Neuroanatomy through clinical cases. Sunderland, MA: Sinauer Associates, Inc.; 2002.
14. Nair-Collins M, Joffe AR. Frequent preservation of neurologic function in brain death and brainstem death entails false-positive misdiagnosis and cerebral perfusion. AJOB Neurosci. 2021. Online ahead of print. https://doi.org/10.1080/21507740.2021.1973148.
15. Nair-Collins M, Joffe AR. Hypothalamic function in patients diagnosed as brain dead and its practical consequences. In: Swaab D, Buijs RM, Kreier F, Lucassen PJ, Salehi A, editors. Handbook of clinical neurology vol. 182 (3rd series). The human hypothalamus: neuropsychiatric disorders. Amsterdam: Elsevier; 2021.
16. Nair-Collins M, Northrup J, Olcese J. Hypothalamic-pituitary function in brain death: a review. J Intensive Care Med. 2016;31:41–50.
17. Bourque CW. Central mechanisms of osmosensation and systemic osmoregulation. Nat Rev Neurosci. 2008;9:519–31.

18. Outwater KM, Rockoff MA. Diabetes insipidus accompanying brain death in children. Neurology. 1984;34(9):1243–6.
19. Powner DJ, Snyder JV, Grenvik A. Brain death certification: a review. Crit Care Med. 1977;5(5):230–3.
20. Russell JA, Epstein LG, Greer DM, et al. AAN position statement. Brain death, the determination of brain death, and member guidance for brain death accommodation requests. Neurology. 2019;92:1–5.
21. Bernat JL. The whole-brain concept of death remains optimum public policy. J Law Med Ethics. 2006;34(1):35–43.
22. Wijdicks EF. The case against confirmatory tests for determining brain death in adults. Neurology. 2010;75(1):77–83.
23. Nair-Collins M. Taking science seriously in the debate on death and organ transplantation. Hast Cent Rep. 2015;45:38–48.
24. Lewis A, Bonnie RJ, Pope T. It's time to revise the Uniform Determination of Death Act. Ann Intern Med. 2020;172(2):143–4.
25. Lewis A, Bonnie RJ, Pope T, et al. Determination of death by neurologic criteria in the United States: the case for revising the Uniform Determination of Death Act. J Law Med Ethics. 2019;47(S4):9–24.
26. Magnus DC, Wilfond BS, Caplan AL. Accepting brain death. N Engl J Med. 2014;370(10):891–4.

Preserved Hypothalamic Function Does Not Preclude Determination of Death by Neurologic Criteria

Ian Thomas and Alex R. Manara

Death by neurologic criteria was first proposed as a concept by the Harvard Ad Hoc Committee in the United States of America in 1968 following recognition that with the advent of modern intensive care medicine and the ability to artificially sustain a failing circulatory–respiratory system, there is a clinical state where all brain functions had been irreversibly lost and from which no patient would ever recover [1]. This clinical state had been described previously by Mollaret and Goulon in 1959 and termed "le coma dépassé," a state beyond coma, that is characterized by unresponsive apneic coma, poikilothermia, loss of all brainstem reflexes, and an isoelectric electroencephalogram (EEG) [2]. Clinical and pathological examination of patients in this state of irreversible coma and apnea showed that it is associated with autolysis of the brainstem [3]. The futility of continuing to provide artificial support of the circulatory–respiratory system in the presence of the permanent loss of brain function led the Harvard Ad Hoc Committee to publish its landmark paper calling for death to be confirmed using neurologic criteria as well as the established circulatory–respiratory criteria [1].

Today, there is a legal provision for determination of death by neurologic criteria in about 70% of countries worldwide, predominantly those with deceased donor transplantation programs [4]. Yet inconsistency persists between countries in the definition of, and clinical guidance for, determining death by neurologic criteria with no international consensus as to whether death by neurologic criteria requires loss of functions of the whole brain or just the brainstem (as discussed elsewhere in this book). The World Brain Death Project, a recent international consensus report, attempted to address these inconsistencies. The Project produced a series of recommendations for the minimum clinical standards for the determination of death by

I. Thomas
Southmead Hospital, Bristol, UK
e-mail: Ian.Thomas@nbt.nhs.uk

A. R. Manara (✉)
Southmead Hospital, North Bristol NHS Trust, Bristol, UK

A. Lewis, J. L. Bernat (eds.), *Death Determination by Neurologic Criteria*, Advances in Neuroethics, https://doi.org/10.1007/978-3-031-15947-3_8

neurologic criteria in adults and children, acknowledged the importance of religious, cultural, and legal factors in making the determination, and identified future research questions [5]. The report defines death by neurologic criteria as "the complete and permanent loss of brain function as defined by an unresponsive coma with loss of capacity for consciousness, brainstem reflexes and the ability to breathe independently" [5]. This definition is a clear move away from death being based on an anatomical construct toward one based on loss of defined functions. Specifically, this definition does not require a loss of *all* functions of the entire brain and states that the "persistence of cellular level neuronal and neuroendocrine activity does not preclude the determination" and that "persistence of hormonal regulatory function does not preclude the diagnosis" [5].

In this chapter, we argue that preserved hypothalamic function is consistent with a determination of death by neurologic criteria if a functional definition of death is used, because "death entails the irreversible loss of those essential characteristics which are necessary to the existence of a living human person. Thus, the definition of death should be regarded as the irreversible loss of the capacity for consciousness, combined with irreversible loss of the capacity to breathe" [6].

1 Hypothalamic–Pituitary Function

The hypothalamus is a complex region of the brain and is important in coordinating signals between the nervous system and the endocrine system, primarily via the pituitary gland. Located below the thalamus and posterior to the optic chiasm, it forms the walls and floor of the third ventricle [7]. Lying beneath the hypothalamus is the pituitary fossa which contains the pituitary gland; they are linked via the infundibulum and the pituitary stalk. The pituitary is divided into two glands, the anterior and posterior pituitary gland, that are distinct both embryologically and functionally [8]. The anterior pituitary gland secretes hormones into the systemic circulation via the cavernous sinus. Its function is regulated by inhibitory or releasing factors originating from the hypothalamus that are transmitted to the anterior pituitary via the hypophyseal–portal system, a specialized local blood flow system that runs parallel to the pituitary stalk. In contrast, the posterior pituitary gland contains axons of neurons from the hypothalamus that secrete hormones directly into a capillary plexus that reaches the systemic circulation via the cavernous sinus [7, 8].

The hypothalamus plays a crucial role in maintaining normal endocrine, metabolic, and autonomic function. It incorporates complex and interconnected feedback loops to regulate the release of adrenocorticotrophic hormone (ACTH), thyroid stimulating hormone (TSH), follicle stimulating hormone (FSH), luteinizing hormone (LH), prolactin and growth hormone from the anterior pituitary gland while secreting vasopressin and oxytocin directly into the systemic circulation via the posterior pituitary gland [7, 8]. Disorders of the hypothalamic–pituitary axis will result in distinct clinical syndromes depending on the location and extent of the underlying lesion. As the hypothalamus regulates both endocrine and autonomic function, there is usually a combination of endocrine and neurological disturbance in hypothalamic damage including abnormalities of behavior, appetite, thermoregulation, and hormonal deficiencies. Hypothalamic dysfunction resulting in a failure of vasopressin secretion from the posterior pituitary gland in response to an increase in plasma osmolality or hypovolemia results in central diabetes insipidus. Vasopressin binds to V2 receptors in the distal nephron that control aquaporins in the collecting ducts of the kidneys, so controlling fluid reabsorption and plasma osmolality. Characterized by inappropriate polyuria, central diabetes insipidus can result in dehydration, hypernatremia, and hyperosmolality [9].

2 Hypothalamic–Pituitary Function in Death by Neurologic Criteria

It is well established that hypothalamic function can persist in patients determined dead by neurologic criteria. The clinical manifestations of central diabetes insipidus indicative of hypothalamic dysfunction are readily identified in terms of increased urine output, plasma sodium concentration, and plasma osmolality. A literature review examining hypothalamic–pituitary function in patients determined dead by neurologic criteria concluded that some of the patients who were not polyuric had maintained osmoregulation through some preservation of hypothalamic function [10]. Studies have previously suggested that the incidence of diabetes insipidus in patients who meet the conditions for death by neurologic criteria ranges from 46 to 78% [11, 12]. In a more recent literature review, only 1265 (50%) patients who met the conditions for death by neurologic criteria were found to demonstrate features of central diabetes insipidus [13]. Assessment of residual anterior pituitary function is more difficult and requires direct measurement of hormones released by the anterior pituitary or of those inhibitory or releasing factors produced in the hypothalamus that control anterior pituitary hormonal release. The same review found that there was evidence of preserved anterior pituitary hormones in the peripheral circulation, which are dependent on releasing factors produced in the hypothalamus for their secretion. This suggests the presence of residual hypothalamic–pituitary function, indicating the preservation of a degree of blood flow to the area, although this is often accompanied by peripheral endocrine insufficiency [13]. Additionally, there are reports of patients determined to be dead by neurologic criteria where physiological support has been continued and who have undergone puberty and growth which have been interpreted as further evidence of residual neuroendocrine function controlled by the hypothalamus [14].

Residual hypothalamic–pituitary function may well exist in patients determined dead by neurologic criteria and this has been recognized and accepted by authoritative institutions, yet it is not considered to invalidate the determination of death. This preserved function may be due to the vascular anatomy of the hypothalamus and pituitary gland, providing a potential sanctuary for this region from the adverse effects of raised intracranial pressure (ICP) and consequent ischemia that cause irreversible injury to the remainder of the intracranial contents [5, 10, 15].

Critics, however, have argued that this vascular supply explanation does not account for the fact that for the release of vasopressin to occur from the posterior pituitary gland, additive glutaminergic input from circumventricular (basal forebrain) areas, especially the organum vasularis of the lamina terminalis and the subfornical organ is required and the vascular supply to these areas is not protected from a rise in ICP as is hypothesized for the posterior pituitary gland [16]. Similarly, the blood supply to the hypothalamic nuclei that control the release of hormones from the anterior pituitary gland is not protected from a rise in ICP [16].

3 Death by Neurologic Criteria and Residual Hypothalamic–Pituitary Function

The persistence of residual hypothalamic–pituitary function in patients determined to be dead by neurologic criteria therefore creates a mismatch between the whole-brain criterion of death, which requires the irreversible cessation of *all* functions of the *entire* brain, and how death by neurologic criteria is actually determined in everyday clinical practice [17] where the persistence of neuroendocrine functions is considered to be consistent with determination of death by neurologic criteria. The concept of death by neurologic criteria as proposed by the Harvard Medical Committee [1] has been adopted as law in the United States following the implementation of the Uniform Determination of Death Act (UDDA) in 1981 [18]. The Harvard report required the irreversible cessation of function of the cerebral hemispheres, diencephalon, brainstem, and cerebellum, a so-called "whole-brain" formulation. The UDDA subsequently defined death as "either (1) irreversible cessation of circulatory and respiratory functions, or (2) irreversible cessation of all functions of the entire brain, including the brain stem" and that "a determination of death must be made in accordance with accepted medical standards" [18].

The whole-brain criterion for death has been widely adopted across the United States and endorsed by many professional medical organizations worldwide. However, there is increasing scrutiny about the whole-brain criterion, the definition of death and whether the UDDA itself needs revision [19]. It is recognized that some patients who are defined as dead by neurologic criteria will retain some hypothalamic–pituitary function [15], a position that is inconsistent with a legal definition that requires the "irreversible cessation of all functions of the entire brain." Yet, it is consistent with the legal requirement for making the diagnosis in accordance with accepted medical standards. Thc UDDA may define death, but it makes no attempt to define accepted medical standards. This is left to the professional medical bodies, and the American Academy of Neurology (AAN), in its 1995 standard for determining death by neurologic criteria in adults [20], listed the absence of diabetes insipidus as being compatible with the determination of death, a position that was not addressed further in their 2010 update [21]. International consensus statements and accepted medical standards around the world [5, 22] agree with the AAN standard and allow the determination of death by neurologic criteria despite the presence of some neuroendocrine functions. The whole-brain criterion, therefore, creates a dichotomy in which the legal standard is not in alignment with accepted medical standards, a so-called "legal clinical mismatch" [17]. Furthermore, currently accepted medical standards and technologies cannot categorically demonstrate irreversible loss of *all* brain and brainstem functions and can at best only approximate that legal definition.

An alternative approach to the determination of death by neurologic criteria is adopted in the United Kingdom, where the focus is on the loss of specific brain functions whether secondary to cessation of the circulation or following a devastating brain injury. There is no statutory definition of death in the United Kingdom. Instead, the legal profession has adopted and supported the definition of death, and the standards used to confirm it, as laid down by the Academy of Medical Royal Colleges in its Code of Practice for the Diagnosis and Confirmation of Death of 2008 [6, 23–26]. The Code states that "death entails the irreversible loss of those essential characteristics which are necessary to the existence of a living human person and, thus, the definition of death should be regarded as the irreversible loss of the capacity for consciousness, combined with the irreversible loss of the capacity to breathe. This may be secondary to a wide range of underlying problems in the body, for example, cardiac arrest." [6]. The code also recognizes that all human death is death by neurologic criteria when it states that when determining death by circulatory–respiratory criteria "it is obviously inappropriate to initiate any intervention that has the potential to restore cerebral perfusion after death has been confirmed" [6]. This definition of death is not anatomically based, but rather focused on the loss of brain functions that are judged to be essential to the existence of a living human being. There is, therefore, no requirement that all functions be absent, only a demonstration that there has been irreversible loss of the capacity for consciousness combined with the irreversible loss of the ability to breathe. Indeed, the Code recognizes that while "the body may continue to show signs of biological activity, these have no moral relevance to the declaration of death" [6]. Thus, it is a misconception that the definition of death in the United Kingdom is primarily a brainstem formulation. Instead, it is based on the irreversible loss of those essential functions, a position that has been upheld in law, accepting that some may disagree and challenge this position [23–26].

This functional approach to the definition of death has been further developed by international collaborations seeking to achieve consensus on the scientific, biological, and medical aspects of death in a way that is hoped to supersede international differences, and which may form the basis of more consistent and globally applicable diagnostic criteria [5, 22]. The consensus collaboration with the World Health Organization convened in Montreal in 2012 defined death as occurring "when there is permanent loss of the capacity for consciousness and loss of all brainstem functions. This may result from permanent cessation of circulation and/or after catastrophic brain injury. In the context of death determination, 'permanent' refers to loss of function that cannot resume spontaneously and will not be restored through intervention" [22]. The more recent World Brain Death project also uses a functionally based definition as "the complete and permanent loss of brain function as defined by an unresponsive coma with loss of capacity for consciousness, brainstem reflexes and the ability to breathe independently. This may result from permanent cessation of oxygenated circulation to the brain and/or after devastating brain injury" [5].

These criteria used to determine death are based on loss of specified brain functions rather than anatomically based (cardiac death, whole-brain death or brainstem death) and do not require the absence of all brain functions, acknowledging that "persistence of cellular level neuronal and neuroendocrine activity does not preclude the determination" [5]. These international definitions, and the United Kingdom's definition, offer more clarity in terminology in that there is only one criterion for death, and it is brain-based and can be confirmed using circulatory–respiratory criteria following a circulatory–respiratory arrest or using neurologic criteria following a catastrophic brain injury [27, 28]. None of these standards require the unequivocal demonstration of cessation of all functions of the entire brain. Current accepted medical standards and practice are also more consistent with this functional approach and definitions.

The mismatch between a legal definition of death and accepted medical standards also exists when death is defined as the "irreversible cessation of circulatory and respiratory functions" since, in everyday clinical practice, death is routinely determined using circulatory–respiratory criteria at the point of their permanent cessation. This is the point beyond which the circulation will not return spontaneously and will not be restarted through intervention because a decision has been made not to attempt to do so [29]. However, it is understood that it may still be possible to restore circulatory–respiratory functions through intervention at this point. Also, irreversible cessation of circulatory and respiratory functions is dependent on which technologies are used and/or available to restore these functions. This mismatch does not exist when using a unifying brain-based definition of death. The point of permanent loss of circulatory–respiratory functions is predictive of the irreversible loss of brain function as long as no intervention with the potential to restore brain perfusion is undertaken after the circulatory–respiratory confirmation of death [27, 30]. This point of permanence is reached within 5 min of continuous circulatory–respiratory arrest [31], a point not always compatible with a definition based on irreversible loss of circulatory–respiratory functions. However, it remains the most widely accepted medical standard used in everyday clinical practice when determining death using circulatory–respiratory criteria [32].

Dying is a process, and death is a defined point along that process. Where we choose to place that line between life and death is a decision with significant individual, social, legal, medical, and cultural implications. It determines who is recognized as a person with constitutional and legal rights, who deserves legal entitlements and benefits, and when last wills and testaments become effective [33]. A determination of death also removes any unrealistic expectations the family may have about outcome, giving them a definitive determination of death rather than a prognosis, and allowing them to begin to grieve. The use of neurologic criteria to identify the line between life and death is well established in legal and medical practice. The law and ethics generally defer to medical expertise regarding the standards to determine death by neurologic criteria. This is probably the most reasonable way to manage the process of dying: when consensus professional guidance is followed, there are no false-positive determinations of death [33]. If the whole-brain legal criterion for death were to be strictly adhered to, then 50% of patients with preserved neuroendocrine function who otherwise meet the conditions for death by neurologic criteria could not be declared dead. On the other hand, 100% of patients who meet the accepted medical standards for the determination of death by neurologic criteria will never regain consciousness or breathe independently again, irrespective of whether neuroendocrine function is present or not.

4 The Whole-Brain Criterion vs. the Brainstem Criterion

Some may conclude that the persistence of hypothalamic–pituitary function in patients who have been determined dead by neurologic criteria is consistent with a brainstem criterion for death, but is not consistent with the whole-brain criterion since the hypothalamus is not part of the brainstem. Despite this, and other, apparent differences in the "transatlantic divide" between the two criteria for determining death by neurologic criteria, it is increasingly recognized that in practice the difference is largely one of semantics and the accepted medical standards used to make the determination are largely the same [34, 35]. Irrespective of whether a jurisdiction follows a whole-brain or a brainstem criterion and irrespective of the underlying pathology, both criteria rely on a similar three-stage approach to determine death:

1. Establishing a cause for the clinical state;
2. Excluding reversible causes or confounding factors that could be contributing to the clinical state; and
3. Undertaking a series of clinical tests to confirm the absence of brainstem reflexes and the ability to breathe.

These minimum clinical criteria confirm the permanent loss of the capacity for consciousness and the ability to breathe independently and allow death by neurologic criteria to be confidently determined. The criteria do not, however, demonstrate the "irreversible cessation of all functions of the entire brain." That is practically impossible using currently accepted medical standards and technologies, a situation that perpetuates the legal, ethical, and practical difficulties created by the mismatch between the legal whole-brain requirement for death and the current accepted medical standards used to determine death by neurologic criteria [17]. Potentially, this mismatch not only increases legal challenges to a determination of death by neurologic criteria but may also expose clinicians to accusations of operating outside the law when they follow accepted medical practice in determining death by neurologic criteria. More importantly, it risks undermining public confidence in the determination of death. The adoption of a functional-based definition of death that is based on accepted medical standards will provide greater clarity for both clinicians and the court. Ongoing research on all aspects of accepted medical standards including the clinical examination, modern imaging technologies, and new diagnostic modalities will allow further refinements to the determination. This will maintain confidence and certainty in the determination of death by neurologic criteria, provide stronger evidence of irreversibility and reduce concerns around safety.

In both formulations for death by neurologic criteria, most cases result from infratentorial (brainstem) manifestations of a catastrophic supratentorial (whole brain) event such as an intracerebral hemorrhage, traumatic brain injury, subarachnoid hemorrhage, or hypoxic–ischemic brain injury [36]. It is only in 2–9% of cases that the cause is an isolated posterior fossa lesion or one limited to structures supplied by the posterior cerebral circulation [37, 38]. The few patients with an isolated brainstem lesion who are determined to be dead by neurologic criteria may have persistent supratentorial blood flow initially that is then lost with time [37], meaning that although they initially met only the brainstem criterion, they eventually also met the whole-brain criterion.

The requirement to undertake ancillary investigations is mandatory in some jurisdictions, but not in others [39]. While not mandatory, ancillary investigations assessing electrophysiology or brain blood flow are often undertaken in circumstances where aspects of the clinical testing cannot be performed, when the effects of confounding factors cannot be confidently excluded, or when there is uncertainty about the significance of possible spinally mediated movements [5]. Some have also suggested that ancillary tests should be mandatory when the underlying diagnosis is an isolated posterior fossa lesion due to a hypothetical possibility of sparing of the meso-pontine tegmental reticular formation with the potential for a total apneic, locked-in syndrome mimicking death [40]. This requires, however, knowledge of the boundaries and exact position of the reticular formation and implausible ellipsoid lesions sparing all brainstem nuclei and tracts, none of which are known or seen.

For all these reasons the considerations about persistent neuroendocrine function in the context of determination of death by neurologic criteria are the same irrespective of whether the whole-brain or brainstem criterion is followed, and whether ancillary tests are used or not. If more functional definitions of death consistent with acceptable medical standards are adopted [5, 6, 22], the important consideration is not whether there is any residual neuroendocrine function, but if loss of that function could be a confounding factor contributing to the coma or apnea. Severe metabolic derangement, hypothermia, Addisonian crisis, and myxedema coma can all confound the determination of death by neurologic criteria and can all be caused by hypothalamic dysfunction. The absence of poikilothermia and central diabetes insipidus when hypothalamic–pituitary function persists may therefore be regarded merely as an internal mechanism to control the body temperature and serum sodium at levels that allow neurologic testing. Similarly, the continued secretion of thyrotropin and corticotropin-releasing hormones allows more confident exclusion of either an Addisonian crisis or myxedema coma as contributing to the current clinical state. For those patients where central adrenal or thyroid function was not demonstrable (if tested for), it is biologically implausible to consider that following a sudden and obvious catastrophic intracranial injury in a person who had previously not demonstrated any signs or symptoms of a hyposecretory disorder, the clinical state would be explained by the absence of adrenal or thyroid function in the hours or short days following the event. The presence of residual hypothalamic–pituitary function is not incompatible with a determination of death by neurologic criteria and may even increase confidence in excluding confounding factors that are known to exist as a result of complete failure of the hypothalamic–pituitary axis.

5 Conclusion

There is increasing acknowledgment within academic circles that there is a mismatch between the legal criteria for whole-brain death and the way that death by neurologic criteria is determined using accepted medical standards. The preservation of some degree of neuroendocrine function, indicative of some hypothalamic function, is common in patients who fulfil the conditions for death by neurologic criteria. Whether this mismatch precludes a determination of death appears to be dependent as much on the definition of death, particularly in a legal statute, as it is on the concept of death itself. While making the determination in the presence of retained neuroendocrine function is consistent with internationally accepted medical standards, it may be difficult to reconcile with a whole-brain criterion for death. However, patients who are confirmed dead by neurologic criteria do not ever regain consciousness or breathe again, irrespective of whether neuroendocrine function is present or not.

References

1. A definition of irreversible coma. Report of the ad hoc committee of the Harvard Medical School to examine the definition of brain death. JAMA 1968; 205:337–40.
2. Mollaret P, Goulon M. Le coma dépassé mémoire préliminaire. Rev Neurol. 1959;101:3–15.
3. Mohandas A, Choi SN. Brain death: a clinical and pathologic study. J Neurosurg. 1971;35:211–8.
4. Citerio G, Cypel M, Dobb GJ, et al. Organ donation in adults: a critical care perspective. Intensive Care Med. 2016;42:305–15.
5. Greer DM, Shemie SD, Lewis A, et al. Determination of brain death/death by neurologic criteria: the World Brain Death Project. JAMA. 2020;324:1078–97.
6. Academy of Medical Royal Colleges. A Code of Practice for the Diagnosis and Confirmation of Death. Academy of Medical Royal Colleges. 2008. http://aomrc.org.uk/wp-content/uploads/2016/04/Code_Practice_Confirmation_Diagnosis_Death_1008-4.pdf. Accessed 18 Jan 22.
7. Lechan RM, Toni R. Functional anatomy of the hypothalamus and pituitary. In: Feingold KR, Anawalt B, Boyce A, et al., editors. Endotext. South Dartmouth (MA): MDText.com, Inc.; 2000. https://www.ncbi.nlm.nih.gov/books/NBK279126/.
8. Pop MG, Crivii C, Opincariu I. Anatomy and function of the hypothalamus. In: Baloyannis SJ, Gordeladze JO, editors. Hypothalamus in health and diseases. London: IntechOpen; 2018. https://doi.org/10.5772/intechopen.80728.
9. Kotloff RM, Blosser S, Fulda G, et al. Management of the potential organ donor in the ICU. Crit Care Med. 2015;43(6):1291–325.

10. Nair-Collins M, Northrup J, Olcese J. Hypothalamic-pituitary function in brain death: a review. J Intensive Care Med. 2014;31:41–50.
11. Salim A, Martin M, Brown C, Belzberg H, Rhee P, Demetriades D. Complications of brain death: frequency and impact on organ retrieval. Am Surg. 2006;72:377–81.
12. Gramm HJ, Meinhold H, Bickel U, et al. Acute endocrine failure after brain death? Transplantation. 1992;54:851–7.
13. Nair-Collins M, Joffe A. Hypothalamic function in patients diagnosed as brain dead and its practical consequences. Handb Clin Neurol. 2021;182:433–46.
14. Shewmon DA. Truly reconciling the case of Jahi McMath. Neurocrit Care. 2018;29(2):165–70.
15. Russell J, Epstein L, Greer D, et al. Brain death, the determination of brain death, and member guidance for brain death accommodation requests. AAN position statement. Neurology. 2019;92:228–32.
16. Nair-Collins M, Joffe A. Frequent Preservation of Neurologic Function in Brain Death and Brainstem Death Entails False-Positive Misdiagnosis and Cerebral Perfusion. AJOB Neurosci. 2021. Epub ahead of print.
17. Robbins N, Bernat J. What should we do about the mismatch between the legal criteria for death and how brain death is diagnosed? AMA J Ethics. 2020;22:E1038–46.
18. National Conference of Commissioners on Uniform State Laws. Uniform Determination of Death Act. 1981. http://www.lchc.ucsd.edu/cogn_150/Readings/death_act.pdf. Accessed 18 Jan 22.
19. Lewis A, Bonnie RJ, Pope T. It's time to revise the Uniform Determination of Death Act. Ann Intern Med. 2020;172:143–4.
20. The Quality Standards Subcommittee of the American Academy of Neurology. Practice parameters for determining brain death in adults (summary statement). Neurology. 1995;45:1012–4.
21. Wijdicks EFM, Varelas PN, Gronseth GS, et al. Evidence-based guideline update: determining brain death in adults Report of the Quality Standards Subcommittee of the American Academy of Neurology. Neurology. 2010;74:1911–8.
22. Shemie SD, Hornby L, Baker A, et al. International guideline development for the determination of death. Intensive Care Med. 2014;40:788–97.
23. House of Lords. Airedale NHS trust respondents and Bland appellant [1993] AC 789. 1993. https://www.bailii.org/uk/cases/UKHL/1993/17.html. Accessed 18 Jan 22.
24. High Court of Justice (Family Division). Oxford University NHS Trust and AB (a minor), CD and EF. [2019] EWHC 3516 (Fam). 2019. https://www.bailii.org/ew/cases/EWHC/Fam/2019/3516.html. Accessed 18 Jan 22.
25. England and Wales Court of Appeal (Civil Division) Decisions. Re M (Declaration of Death of Child) [2020] EWCA Civ 164. 2020. https://www.bailii.org/ew/cases/EWCA/Civ/2020/164.html. Accessed 18 Jan 22.
26. High Court of Justice (Family Division). Re A ((a child) [2015] EWHC 443 (Fam). 2015. https://www.bailii.org/ew/cases/EWHC/Fam/2015/443.html. Accessed 18 Jan 22.
27. Manara AR. All human death is brain death: the legacy of the Harvard criteria. Resuscitation. 2019;138:210–2.
28. Shemie S, Gardiner D. Circulatory arrest, brain arrest and death determination. Front Cardiovasc Med. 2018;5:15.
29. Bernat JL, Capron AM, Bleck TP, et al. The circulatory–respiratory determination of death in organ donation. Crit Care Med. 2010;38:972–9.
30. Domínguez-Gil B, Ascher N, Capron AM, et al. Expanding controlled donation after the circulatory determination of death: statement from an international collaborative. Intensive Care Med. 2021;47:265–81.
31. Dhanani S, Hornby L, van Beinum A, et al. Resumption of cardiac activity after withdrawal of life-sustaining measures. N Engl J Med. 2021;384:345–52.
32. Dhanani S, Hornby L, Ward R, Shemie S. Variability in the determination of death after cardiac arrest: a review of guidelines and statements. J Intensive Care Med. 2012;27:238–52.
33. Magnus DC, Wilfond BS, Caplan AL. Accepting brain death. N Engl J Med. 2014;370:891–4.

34. Wijdicks EF. The transatlantic divide over brain death determination and he debate. Brain. 2012;135:1321–31.
35. Manara A, Varelas P, Wijdicks EF. Brain death in patients with "isolated" brainstem lesions: a case against controversy. J Neurosurg Anesthesiol. 2019;31:171–3.
36. Manara A, Varelas P, Smith M. Neurological determination of death in isolated brainstem lesions: a case report to highlight the issues. J Intens Care Soc. 2020;21:269–73.
37. Varelas PN, Brady P, Rehman M, et al. Primary posterior fossa lesions and preserved supratentorial cerebral blood flow: implications for brain death determination. Neurocrit Care. 2017;27:407–14.
38. Ray A, Manara AR, Mortimer AM, Thomas I. Brain herniation on CT is a poor predictor of whether a patient can be confirmed dead using neurological criteria. J Intens Care Soc. 2021. Epub ahead of print.
39. Wahlster S, Wijdicks EF, Patel PV, et al. Brain death declaration: practices and perceptions worldwide. Neurology. 2015;84:1870–9.
40. Walter U, Fernandez-Torre JL, Kirschstein T, et al. When is "brainstem death" brain death? The case for ancillary testing in primary infratentorial brain lesion. Clin Neurophysiol. 2018;129:2451–65.

Does Death Determination by Neurologic Criteria Require Irreversible or Permanent Cessation of Brain Functions?

Andrew McGee and Dale Gardiner

All words have the 'taste' of a profession, a genre, a tendency, a party, a particular work, a particular person, a generation, an age group, the day and hour. Each word tastes of the context and contexts in which it has lived its socially charged life; all words and forms are populated by intentions [1].

—*M.M. Bakhtin, The Dialogic Imagination*

Does death require permanent or irreversible cessation of function? There are different views. This chapter explores those views, focusing ultimately on their application to determination of death by neurologic criteria.

As can be seen in Table 1, at the time when neurologic criteria for the determination death were first proposed, the words "permanent" and "irreversible" were being used interchangeably. It is perhaps only by chance that "irreversible" became the term of legal choice in the Uniform Determination of Death Act (UDDA) and other similar descriptions of death [2].[1]

> An individual who has sustained either (1) irreversible cessation of circulatory and respiratory functions, or (2) irreversible cessation of all functions of the entire brain, the brain stem, is dead. A determination of death must be made in accordance with accepted medical standards [2].

[1] In Australian legislation the wording is "irreversible cessation of circulation in the body" (see McGee and Gardiner [3]). The words "in the body" might have been added to prevent the claim that, when the heart is restarted in the *recipient's* body, circulation is proven not to be irreversible.

A. McGee (✉)
Queensland University of Technology, Brisbane, Australia
e-mail: a.mcgee@qut.edu.au

D. Gardiner
Adult Critical Care, Nottingham University Hospitals NHS Trust, Nottingham, UK
e-mail: dalegardiner@doctors.net.uk

A. Lewis, J. L. Bernat (eds.), *Death Determination by Neurologic Criteria*, Advances in Neuroethics, https://doi.org/10.1007/978-3-031-15947-3_9

Table 1 Historical uses of the terms "irreversible" and "permanent" in landmark determination of death statements from the 1960s to 1980s [4–7]

Document	Example statements of the use of the term "irreversible" and "permanent"
Ad hoc committee of the Harvard Medical School, 1968, the United States	"Our primary purpose is to define irreversible coma as a new criterion for death" "Our first problem is to determine the characteristics of a permanently nonfunctioning brain"
	"We suggest that responsible medical opinion is ready to adopt new criteria for pronouncing death to have occurred in an individual sustaining irreversible coma as a result of permanent brain damage"
Conference of Medical Royal Colleges and their Faculties, 1976, the United Kingdom	"Permanent functional death of the brainstem constitutes brain death"
Conference of Medical Royal Colleges and their Faculties, 1979, the United Kingdom	"Whatever the mode of its production, brain death represents the stage at which a patient becomes truly dead, because by then all functions of the brain have permanently and irreversibly ceased"
President's Commission for the Study of Ethical Problems in Medicine and Biomedical and Behavioral Research, 1981, the United States	"An individual with irreversible cessation of circulatory and respiratory functions is dead" "The accepted standard for determining death has been the permanent absence of respiration and circulation" "Before tissues are removed, the following signs of death… must be present: permanent cessation of the activity of the brain or of the heart" "For most lay people—and in all probability for most physicians as well—the permanent loss of heart and lung function (for example, in an elderly person who has died in his or her sleep) clearly manifests death" "An individual who has sustained either (1) irreversible cessation of circulatory and respiratory functions, or (2) irreversible cessation of all functions of the entire brain, the brainstem, is dead. A determination of death must be made in accordance with accepted medical standards"

The debate about whether "irreversible" or "permanent" is the more appropriate term arose as a direct consequence of the reemergence of organ donation after determination of death by circulatory–respiratory criteria (DCD) in the 2000s. For DCD to result in successful organ transplants, the time after the commencement of circulatory–respiratory arrest must be as short as possible. Every minute that the organs do not have an oxygenated circulation increases warm ischemic damage. The question became: What is the minimum amount of time after circulatory–respiratory arrest that must pass before the donor can be determined to be dead? The generally accepted standard developed in DCD practice worldwide is that the minimum time is 5 min, though some advocate for times as short as 75 s and others favoring periods as long as 30 min [8–11].

The ethical and conceptual challenge is that a person whose circulatory–respiratory function has only been inactive for 5 min can sometimes still have that function restarted by means of cardiopulmonary resuscitation (CPR) [12]. Some therefore hold that in DCD we do not know whether, at just 5 min after circulation has ceased, the cessation is irreversible—as required by the UDDA [13–15]. Even 30 min may not be enough time in the right circumstances and with enough resuscitation effort [12, 16–18]. Indeed, given that we normally would only consider attempting CPR and other resuscitative measures if we think that the brain would not be too damaged from the lack of oxygen, it is actually unknown how long we would have to wait before it was no longer feasible to restart circulation. Restarting circulation enabling good brain function is one thing, but restarting circulation regardless of brain function quality is quite another. Since we normally stop CPR once we know that good function will not be restored to the brain, we just do not know how long we could continue CPR and still eventually recover some circulation.

Whereas "permanent" and "irreversible" might once have been used interchangeably, they have now, in this area of debate at least, taken on very different meanings, where "permanent" is defined as *will not return* and "irreversible" as *cannot return* [19]. Two main arguments support the use of "permanent" in the circulatory–respiratory determination of death. The first is that permanent cessation is the established medical practice standard for determining death [19]. To know that circulatory–respiratory function cannot return, one must either attempt CPR or other forms of resuscitation and fail, or wait a long enough duration for CPR or other forms of resuscitation to always fail. In most modern death determinations, doctors do neither. Most death determinations do not follow an attempt at CPR which fails, and even if they did, the success of CPR is effort and technology dependent as already noted. Nor do doctors necessarily wait a long enough duration to know that CPR would always fail (even when we stop, we could have had a different result if we had carried on for longer, however unlikely) [20]. And this is not even taking into account other resuscitative measures such as extracorporeal cardiopulmonary membrane oxygenation (ECMO), which can begin, in some cases, after conventional CPR fails. Therefore, despite the use of the word "irreversible" in the UDDA and similar instruments, doctors in practice only apply the standard of *permanence*. The UDDA specifically allows for this situation because it provides that a "determination of death must be made in accordance with accepted medical standards [2]." That standard is permanence.

The second argument is that permanent cessation is the meaning of "irreversible" in the UDDA. The point just made (that the UDDA requires death determination to be made in accordance with good medical practice) provides a legal argument that "irreversible" must be interpreted—as a matter of "statutory interpretation," as the lawyers say—to mean permanence, since otherwise it would be difficult to give the wording about determining death in accordance with medical practice any meaning. So, permanent cessation of function will become irreversible cessation of function provided function will not be restored because it will neither return spontaneously nor will it return as a result of medical intervention because resuscitation efforts will not be attempted, these being prohibited [19]. The word "permanent" relies primarily on intent and action to be realized [19], while "irreversible function," at least for advocates of a strong distinction between the two, is function that cannot be restored by any known technology. On this latter view: "'Irreversible' is an absolute and univocal condition that implies impossibility (with currently available technology) and does not rely on intent or action [19, p. 974]." By contrast, for those who reject such a strong distinction between irreversible and permanent, "permanent" records an epistemic limitation that, in the circumstances, has been defanged. When we declare death based on permanent loss of function, it is possible the cessation of functions is biologically irreversible. However, we cannot know for sure without trying resuscitation. This epistemic limitation, however, is irrelevant where trying is itself ruled out. It is defanged because we know all we need to know to declare death in these cases.

Criticism of the modern practice of relying on permanence has been strong and fierce. We will examine one of the most influential and strongest criticisms made in the literature below.

Before we explore the language and meaning of the word "irreversible" in depth, however, we need to preview how the argument for permanence in the determination of death by circulatory–respiratory criteria relates to the determination of death by neurologic criteria. Notwithstanding criticism, the international medical community has found itself advocating for a unified neurologic criterion of death (see Table 2). One is not dead because one's circulation has ceased, but because, when circulation ceases, brain function ceases. If this cessation of brain function is permanent, then, according to the international medical community, death has occurred.

Table 2 Modern determination of death statements which seek to unify determination of death by circulatory–respiratory and neurologic criteria [25–27]

Academy of Medical Royal Colleges, 2008, the United Kingdom	"Death entails the irreversible loss of those essential characteristics which are necessary to the existence of a living human person and, thus, the definition of death should be regarded as the irreversible loss of the capacity for consciousness, combined with irreversible loss of the capacity to breathe … The irreversible cessation of brainstem function whether induced by intra-cranial events or the result of extra-cranial phenomena, such as hypoxia, will produce this clinical state and therefore irreversible cessation of the integrative function of the brain-stem equates with the death of the individual and allows the medical practitioner to diagnose death"
The International Guidelines for Determination of Death phase 1 participants, in collaboration with the World Health Organization, 2014	"Operational definition of human death: Death is the permanent loss of capacity for consciousness and all brainstem functions. This may result from permanent cessation of circulation or catastrophic brain injury. In the context of death determination, 'permanent' refers to loss of function that cannot resume spontaneously and will not be restored through intervention"
World Brain Death Project, 2020	"[Brain death/death by neurologic criteria (BD/DNC)] is defined as the complete and permanent loss of brain function as defined by an unresponsive coma with loss of capacity for consciousness, brainstem reflexes, and the ability to breathe independently. This may result from permanent cessation of oxygenated circulation to the brain and/or after devastating brain injury. Persistence of cellular-level neuronal and neuroendocrine activity does not preclude the determination. In the context of death determination, 'permanent' refers to loss of function that cannot resume spontaneously and will not be restored through intervention"

Does determination of death by neurologic criteria require irreversible or permanent cessation of function? There has been much less debate on this issue compared to the determination of death by circulatory–respiratory criteria. In the determination of death by neurologic criteria, the tests required to confirm death are not carried out until all preconditions are satisfied (e.g., absence of confounding conditions). This can lead to a time gap from when the cessation of the relevant neurologic functions is judged to be irreversible (and so death is strongly suspected to have occurred) to the actual declaration that death has occurred. Typically going from suspicion of death to determination of death through the relevant tests can take many hours to days. This can give the impression that death by neurologic criteria is a *retrospective* determination. Correlatively, since the determination of death by circulatory–respiratory criteria is more temporally immediate to the cessation of function and occurs before neuronal damage is complete, this creates the impression that the determination of death by circulatory–respiratory criteria is a *prospective* determination of what will happen [21–22]. However, the impression regarding the retrospective nature of death by neurologic criteria is misleading because it is not conceptually sustainable, as we will show in more detail below. For reasons we shall see, we can only suspect that death has occurred if we mean "permanent cessation of function" by "death." Suffice it to say at this point that technological and medical advances such as therapeutic decompressive craniectomy, deep brain stimulation probes, and the BrainEx machine, which restored some cellular and synaptic activity in pig brains 4 h after decapitation, all demonstrate that even when determining death by neurologic criteria, it remains the case that it is the intention to resuscitate and treat (or not to do so) that remains paramount, and this decision precedes a determination of death by neurologic criteria. As with the use of circulatory–respiratory criteria, the use of neurologic criteria relies on the accepted medical standard of permanence, not irreversibility [23].

Our aim in this chapter is not to defend a unified, brain-based, definition of death, but instead to defend the medical community's endorsement of permanence rather than irreversibility as the necessary precondition to accurate determination of death by both circulatory–respiratory and neurologic criteria.

But can permanence really be defended? As critics claim, it is obvious that death is irreversible and reliance on the alternative criterion of permanent cessation is "little more than [a] medical charade" [24]. We turn now to the criticisms.

1 Arguments from Language: The Meaning of Irreversible

Death was irreversible, he suspected, and he began to think he was going to lose.
—Joseph Heller, Catch-22

Don Marquis, in a well-cited paper in the *Hastings Center Report*, claimed that reversibility is a dispositional property and that, at the time death is declared, this dispositional property still obtains [28]. People declared dead for whom resuscitative measures are not appropriate (such as those who have do-not-resuscitate orders) are therefore not known to be dead at the point at which death is declared. This claim has been endorsed by other critics, such as Ari Joffe, Michael Nair-Collins, Franklin Miller, and Robert Truog [13–15, 29, 30].

In this section of the chapter, we argue that some of these criticisms can be answered, and that the focus on irreversibility, if interpreted to exclude rules about what is appropriate, is beset by problems that are equally as serious as those that critics claim beset the focus on permanence. Adopting permanence instead of irreversibility (or construing "irreversible" and "irreversibility" to include rules about whether it is permitted to attempt resuscitation) as the prerequisite threshold for accurate death determination is rational and defensible.

What does "irreversible" mean? Defenders of determination of death by circulatory–respiratory criteria have claimed that "irreversible" need not entail that a person can only be declared dead if they cannot, as a matter of fact, be resuscitated by human effort [19, 31, 32]. The claim is that it can *also* mean that a person can be determined dead if circulation and respiration have ceased and resuscitation is not ethically appropriate. This allows practitioners to declare death much earlier than would be the case where resuscitative measures are applicable—on the basis that, without such resuscitative measures, the cessation of circulation and respiration will be permanent.

Don Marquis disagrees with these defenders of determination of death by circulatory–respiratory criteria. In developing his influential criticisms in the context of DCD, Marquis discusses separately what he calls "the appeal to permanence" and "the appeal to a norm" [28].[2] However, we treat these together because they are related. The reason why the cessation of circulation and respiration is permanent for those endorsing DCD protocols or, more generally, protocols applying to those with a do-not-resuscitate order, is that there is a norm in place that precludes the use of resuscitative measures. Because this norm exists, those who declare death under such protocols believe that they do not need to wait for a second period of time to pass, once the possibility of auto-resuscitation has passed, before declaring death. For that second period of time is *only* required for those cases where resuscitation may be attempted, and which may therefore bring the patient back—but resuscitation is inapplicable if there is do-not-resuscitate order. We have all the knowledge we need to declare death.

We should therefore understand the appeal to permanence as partly relying on what Marquis calls "the appeal to a norm" (we say "partly" because it is *also* reliant on auto-resuscitation no longer being possible at the point of death declaration). Aside from its reference to the impossibility of auto-resuscitation, where death is declared for patients with a do-not-resuscitate order, "irreversible" means "normatively irreversible," in the sense that, say, it is not possible to reverse a legal decision if one has no power to reverse it. However, this, Marquis claims, is clearly not what is meant by the word "irreversible" when speaking of the cessation of circulation in a patient's body [28, pp. 27–30]. Marquis instead insists that what we mean is that it is not *physiologically* possible to restart circulation, and only when we know that this is so can we declare death *knowing* it to have occurred.

[2] Marquis, "Are DCD Donors Dead?", 26, 27.

To explain this point, Marquis claims that "reversible" refers to what he calls a "dispositional property" that has a corresponding "occurrent" manifestation. To illustrate "dispositional" and "occurrent" with a simple example, a sugar cube is "disposed" to dissolve when put in water. It manifests this disposition when it *actually* dissolves in the water—its disposition is then "occurrent."

The correlative term for "reversible"—"irreversible"—means that the entity in question has no such dispositional property. We can perhaps bring out the force of Marquis's criticisms of DCD death declaration if we start with examples where the relevant dispositional property is absent. Consider the term "non-combustible," an example of our own but one which nicely illustrates Marquis's point.[3] A non-combustible substance is one that is fire-resistant. That property of being fire resistant is an inherent property of the entity in question, part of its physical nature. It would be absurd to say that a substance that *does* have the relevant dispositional property of being combustible is fire resistant when there is a rule about keeping the substance away from fire to prevent it from being ignited. The rule exists precisely because the substance is combustible, and so *cannot* mean that the substance is non-combustible. When the substance is in fact ignited, the property of being combustible becomes occurrent, or realized. However, if it is never ignited, all that this means is that the dispositional property of being combustible is never realized, or never becomes occurrent. It does not mean that it does not *have* the property of being combustible.

Other examples given by Marquis himself include the properties of being breakable and soluble [28]. A rule against breaking a china cup, or against dissolving a ring in *aqua regia*, does not mean that the china cup is non-breakable, or that the ring is insoluble [28].[4] It is precisely because the cup is breakable and the ring is soluble that we have these norms in the first place. Marquis concludes that "in these contexts, in which moral norms apply, ethical interpretations of these dispositional terms seem incorrect" [28, p. 27]. By analogy with these terms, Marquis claims that "reversible" and "irreversible" are dispositional properties. For the purpose of determining death, they refer to whether, as a matter of fact, a person's circulation can be physiologically restored.

We should note, however, a difference here that Marquis ignores. The terms "non-combustible," "insoluble," and "non-fragile" never have normative meanings (i.e., they never embody rules about what we are allowed to do or prohibited from doing). These terms are *never* used in a normative sense. We cannot refer to a legal prohibition on the use of some combustible material as making this material "non-combustible." This fact is partly what makes Marquis's claims here seem so compelling. In contrast, "irreversible" clearly has both a dispositional and a normative meaning. We can refer to President Obama's decision at the end of his Presidency to

[3] The term "flammable" is mentioned by Marquis, as a dispositional term, but the term "non-combustible" is that which mirrors the term "irreversible," and the example we give is our own because we think it brings out his point more clearly. Marquis's own example to mirror "irreversible" is "non-fragile." These differences are not relevant to the issues discussed.

[4] Marquis,, "Are DCD Donors Dead?", 27.

commute the Chelsea Manning sentence as "irreversible", and so as not capable of being undone by the incoming President. "Irreversible" clearly has this normative meaning, and not merely a dispositional meaning. Of course, these are separate meanings of "irreversible." However, the point for now is that there are two meanings of "irreversible," whereas there is only one meaning of "non-combustible" or "insoluble." It is therefore too quick for Marquis to conclude, from his dispositional analysis of these other terms alone, that the "assumption" that "irreversible" can be given a normative meaning "does not seem to be true" [28].[5] For the moot question is whether there is a legitimate basis on which we can read "irreversible" as having its normative meaning, and the appeal to terms that are *entirely* dispositional, and do not bear any normative meaning at all, does not answer *this* particular question. It begs it.

Consider reversible T-shirts. These are T-shirts that can be worn inside out or back to front, without anyone else noticing a difference; the labels of the T-shirt are removed and the seams are stitched in such a way that they are not showing, whichever way the T-shirt is worn. Does this mean that so-called "non-reversible" T-shirts are really non-reversible? Physically speaking, all T-shirts can clearly be worn inside out or back to front, or reversed, and so *any* T-shirt is reversible. However, there is a social norm about not wearing one's clothes with the labels or seams showing, which explains why only a subset of these are called "reversible" T-shirts, rather than all T-shirts being so. Now, are we to say here that, in the case of T-shirts, they have the dispositional property of being reversible, and so all T-shirts should be called reversible, and we therefore should not have a special class of so-called reversible T-shirts? This seems to us to be absurd, but why isn't Marquis committed to this claim?

Of course, Marquis can reply that people are not T-shirts and "irreversible" must bear its physiological meaning, but this argument cannot be established solely by reference to these other dispositional terms that do not have a second, normative sense in the way that "irreversible" does. On the contrary, the points about these other dispositional terms already presuppose that he has *independently* established that it is not legitimate to give "irreversible" a partly normative meaning when describing death [32].[6]

We must therefore turn to the question of whether it is legitimate ever to give "irreversible" its normative sense when speaking of whether someone is dead, or whether we must always mean it in its dispositional sense.

---

[5] Marquis here refers to John Robertson's assumption, who is an early proponent of the view Marquis is criticizing, but for convenience we leave that wording out here.

[6] Elsewhere we claim that "irreversible" might be more like "inoperable," "irreparable," and "irretrievable" than like "insoluble," inasmuch as it carries an intrinsic reference to someone acting on intentions (see McGee and Gardiner [32]).

1.1 Can "Irreversible" Legitimately Be Given a Normative Construal?

So, *is* it ever legitimate to give "irreversible" its normative sense when speaking of whether someone is dead? At first glance, there seems to be a very strong argument against giving it this sense. Whether someone is dead must surely depend entirely on their physical state, and not on decisions made which affect that physical state or prevent actions capable of changing that physical state. Death is a physiological state, and anyone in that same state must therefore be dead. As Marquis puts it, "if an individual is dead in virtue of being in state S, then all other individuals in state S are also dead" [28]. It cannot be the case that some people are in state S and known to be dead, while others are in state S and alive or not known to be dead. Yet permanence advocates seem to be committed to precisely this possibility, to the extent that they seem to accept that two people can be in state S, yet one is known to be dead (if there is a valid do-not-resuscitate order which applies to them and it is between 2 and 5 min after asystole), while the other is *not* known to be dead because there is no such order.

Consider now the following case. This case is imaginary, but we will later present a real-life case that, in our view, highlights the same issues. Suppose today that our practice is to declare death in a person after about an hour following asystole, when, let us suppose,[7] we can be confident that neither CPR nor any other resuscitative technology such as ECMO could work—even if it is not appropriate to try it. Suppose this has been accepted and a standard practice for decades. Imagine now that new technology is announced that is capable of restarting circulation in some people after a downtime[8] of 2 days but works best in people under the age of 30. Nevertheless, even though circulation can be restarted after 2 days, the condition to which such people can be restored means that they would not have a sufficient quality of life to make it appropriate to use this technology. On Marquis's dispositional account, it seems that nobody now would be known to be dead until after the point at which even this new technology would definitely fail to restart circulatory–respiratory function, that is, until at least the end of the 2-day period. This would be so even though it is not appropriate to use it on anyone (the quality of life to which people could be restored being too low).

Would Marquis's view be the *only* plausible view to take of this case? Let us consider all the possible alternative views available and how they may impact on our interpretation of the meaning of "irreversibility" and "permanence." These are:

1. We say that the case is a mere thought experiment that bears no relation to the current situation and practice, and so is irrelevant to his criticism of DCD protocols, or protocols based on cessation of brain function;

[7] In reality, nobody actually knows when the first point of irreversibility is reached, but we can leave that complication aside here for now.

[8] "Downtime" refers to the period of time that the heart has stopped before any resuscitative effort has been attempted.

2. We insist that people really *would* be dead (and so known to be dead) only after the 2 days, and could no longer be known to be dead prior to the end of that period, thereby maintaining the dispositional account;
3. We claim that, if this or any other such imagined scenario should occur, our concept of death would change at that point, and so people today known to be dead after a few hours would tomorrow not be known to be dead until after 2 days, but "dead" would be indexed to what is possible given the new technology, and so would have a slightly different application than it has now, given only our current technology;
4. We concede that we might have different standards of death, depending on the category of patient, so that it remains appropriate to consider people to whom this technology is not applicable to be dead, and only those to whom it may be applicable to potentially be still alive.

Marquis's view, given the commitments of his dispositional account of reversibility, is most closely aligned with view 2, but he might adopt one of the other views. Let us look at each possibility in more detail.

View 1: This is a mere thought experiment having no bearing on current DCD practice and protocols.

The first possible view would, in our view, be weak. It is a standard philosophical practice to examine the implications of an idea by examining imaginary cases to see whether those implications are acceptable. Imaginary cases may be better than real cases, since we can vary the possible range of facts more to work out what we really believe, or how our beliefs would change with a particular variation—and the implications of any such change in beliefs for what we currently believe under the status quo. Since, on current practice, a distinction is drawn between the time at which death is determined for patients for whom resuscitation is appropriate and those for whom it is not, our thought experiment is relevant, for it tests how far Marquis and those who endorse his view may be willing to go to defend his criticism of the standard practice, and to defend his own position and his dispositional account of "irreversible". Our claim is that the dispositional account commits him to the view that we have discovered that people are now not known to be dead until after 2 days. Our thought experiment can be used to test whether this is an acceptable conclusion, or at least the only rational conclusion, and one which Marquis himself would accept—or whether there may be another, equally rational view to take.

It is also worth briefly noting here that in a recent English case,[9] the court decided to permit the cryopreservation of a 14-year-old girl who died from cancer; the prospect of such technology becoming realized is not so fanciful as to make courts reject applications from minors who seek judicial approval to have themselves cryopreserved in cases where there is disagreement between the minor's parents about

[9] *JS* [2016] EWHC 2859 (Fam).

cryopreservation. The questions we raise in our imaginary case are equally raised by this technology and need to be faced. We return to this case later.

View 2: If this technology succeeded, we really would have discovered that people were not dead at the time we had originally assumed.

On the second view, if the technology succeeds in restarting circulation in people who were otherwise (thought to be) dead an hour after asystole, we have discovered that people were not dead at the time we originally assumed they were. This second possible view is plausible. The question is whether it is the *only* plausible view. Why can't we say that, since nobody can be revived with this new technology to a condition that makes reviving them appropriate, we shall continue to consider people to be dead at the time we always have? Why would it be "a medical charade" to declare these people to be dead at the time we currently declare them to be, and then change our practices when the technology is developed to such an extent that it is appropriate (because worthwhile) to use it on those patients for whom it was formerly considered to be inappropriate? Suppose we develop the technology further, so that the quality of life to which people can be restored is good, but that the technology still works better on people under age 30—we do not try it on people over age 60 because it is deemed unlikely to work and, even if it did, it will not restore them to a worthwhile state. Why should an external factor that is inapplicable to a 60-year-old woman (the fact that it can restart circulation in those under 30) make a physiological difference to this 60-year-old woman?

A general problem with this second possible view is that it seems to entail that we are never in a position really to know when someone is dead, because new technology may be invented that enables us to restart circulation much later than we currently can, or currently believe to be possible. Although this is only an epistemic limit under this second possible view (there is a fact of the matter about when someone is dead, and it may be that we just don't know, as yet, when that point is), Marquis's point against those who rely on *normative* irreversibility is that the donor (in DCD) is not known to be dead when organ recovery proceeds, and this point applies to his own view (a version of this second possible view). On the logic of this view, we may well be engaging in many practices on people who are not known to be dead, including burial and autopsies, at the time we declare them to be dead. If this is right, then this undermines the criticism that, in current death determination practice, we may be engaging in other practices (such as organ donation) when the patient is not known to be dead. For on this view, we never truly know the point at which anyone is dead (save after putrefaction and decomposition have set in) because new technology capable of restarting circulation at times much further after asystole than is currently possible may be invented. Note that, on this view, it is not possible for Marquis to say that these people are known to be dead, given *current* technology (but not given any *future* technology), as that is possible view 3, which we will discuss shortly (under possible view 2, we are instead discovering that people were not known to be dead when we thought they were).

Another problem with this second view is that it ignores a different possibility. Instead of saying that we have discovered people are not known to be dead until after 2 days have passed, we may instead say that we have discovered ways of bringing people back from the dead. Return to our imaginary case, and suppose we take the option of saying that we have discovered ways of bringing people back from the dead after 2 days. It follows that we could still consider people to be dead whether we use the technology on them or not. Those on whom we use the technology would be dead but brought back to life. And those on whom we do not use the technology would also be dead but would not be brought back to life. And we may adopt a whole host of new rules for this kind of case.[10] On this position, DCD candidates or people declared dead based on the permanent cessation of brain function, on neither of whom it is appropriate to use the technology, would be dead, and so organ retrieval from them would not violate the dead donor rule.

What is it that would stop us from adopting the option of saying that, in our imaginary case, we have discovered ways of bringing certain people back from death? It seems to us that no fact of the matter could restrain us from adopting this option. Only external constraints—the implications for practical matters such as the disposition of property under a will, the status of marriages, and the concept of bigamy—would have a say about which is the better option out of the two possible ways of proceeding (saying we bring people back from death or saying that we have discovered people are not dead at the time we thought they were). Furthermore, choosing which is the *better* of the options is itself a normative exercise, and this might undermine the claim that normative considerations have no place in declaring death—we return to this criticism again later.

These two difficulties, then, perhaps undermine some of the criticisms of the other options (views 3 and 4 we turn to next) that someone inclined to adopt view 2 would make.

View 3: If this technology were discovered, our concept of death would shift at that point, but, at present, it is merely a logical possibility we can ignore.

Consider now a third possible view. On this view, what counts as "death" is indexed to what is possible given *current* technology. To say, as Marquis does, that "death is, as a matter of fact, irreversible" contemplates the *logical* possibility that new technology could emerge that allows us to restart circulation much later than we are currently able to do with our existing technology. However, the restriction to *factual* (rather than logical) irreversibility considers the concession of the logical possibility to be irrelevant, because death by circulatory–respiratory criteria is irreversible once *current* technology is no longer able to reverse the cessation of circulation. Suppose, then, that we adopt view 3. Returning to our imaginary case, this

[10] To give one example, we could reject Marquis's view that, if death were reversible, a woman who married after her husband had died and before he had been brought back from the dead, would be guilty of bigamy once he is brought back (p. 28). We may instead refuse to count this as bigamy, on the basis that the husband had truly died, and was dead when the woman remarried.

means that, prior to the technology being developed, people really were dead after about an hour, whereas, once the new technology is used, people are now not dead—hence not known to be dead—until after 2 days. However, "dead" has a slightly different meaning[11] in each case on this possible view. Because death, on this view, is indexed to what is possible given current technology, and because we are not denying (on this view) that people who were declared dead after an hour *really were* dead after an hour (on the basis that the new technology had yet to be invented), death is instantiated by different paradigms of irreversibility as technology improves and, like different colors used as paradigms to illustrate the meaning of color words, "dead" in each case necessarily has a slightly different meaning. (This would be akin to having a concept of red before having the concept of magenta, and then introducing the new concept of magenta when we decide to distinguish between red and the shade we now call "magenta." Prior to introducing the concept of magenta, what we now call magenta would simply have been called red even if we could discriminate between shades. Similarly, prior to introducing this new paradigm of irreversibility, what we are now calling reversible would beforehand have been called irreversible.)

Could Marquis and those influenced by him choose this third possible view? If they choose this view, they face the same objection Marquis raises against those who rely on normative irreversibility (permanence). In that objection against normative irreversibility, Marquis pointed out that two people could be in exactly the same physiological state, but one person (on whom resuscitative measures remain appropriate) could be alive, while the other one (on whom resuscitative measures are not appropriate) would be dead.[12] In addition, Marquis said that this consequence of normative irreversibility "is unacceptable" (p. 29). However, accepting view 3 leads to an equivalent difficulty. It means that someone could be in the same physiological state today and tomorrow yet be dead (and be known to be dead) today and not be dead (nor known to be dead) tomorrow. (Remember that, on this view, what counts as "irreversible" depends on the technology that exists at the time). That being so, what is the objection to those who choose to adopt permanence in death determination protocols, and who thereby interpret "irreversible" normatively? True, "dead," under view 3, now has a slightly different application, since it applies now to paradigms of irreversibility that were not previously in existence (it not being possible to revive someone after the end of 2 days, rather than it not being possible to revive someone after the end of the 1-h mark). However, this is precisely the claim that is made now by those who endorse permanence: when determining death in those patients with a do-not-resuscitate order, we do not need to wait for a second period of time to pass, where that time is only necessary to rule out the

[11] Alternatively, it might be said that "dead" does not have a slightly different meaning because it means what it always means: the irreversible cessation of circulation (or brain function). Rather, death might be instantiated at later points in the future, given future technology, to the points at which it is currently instantiated given existing technology. We can accept this alternative analysis here as well, as nothing turns on the analysis we choose.

[12] Marquis, "Are DCD Donors Dead?", 29.

possibility of reversal via resuscitation, given that resuscitative measures do not apply. On the permanence view, we arguably already operate with a two-tiered understanding of death, one tier of which applies to those people for whom resuscitation is appropriate, and the other of which applies to those for whom it is not.[13] The only difference between the permanence view, and view 3, is that the permanence understanding applies contemporaneously (we therefore adopt different paradigms of "irreversibility" at the same time), whereas the different paradigms of "irreversibility" under view 3 apply *across* time, rather than at the same time. However, we see no reason why this difference should be relevant.

So, as with view 2, it appears that the opting for view 3 also undermines the critique of normative irreversibility (permanence).

View 4: We can have different standards of death, depending on the category of patient we are dealing with. The permanence view.

In our imaginary case, we discover new technology that can restart circulation after 2 days following mechanical asystole. Prior to this, we could only restart circulation after about an hour from mechanical asystole. View 3, just discussed, accounts for this by claiming that what counts as death is always indexed to what it is possible to do, given our current technology. On that view, prior to the development of the technology, someone really was dead and known to be dead after an hour from mechanical asystole, since it was not at that time possible to reverse the cessation of circulation after an hour. However, once the technology developed, anyone now in the physiological state of asystole after an hour would no longer be known to be dead, since technology might be used to restart circulation for up to 2 days following mechanical asystole. We pointed out that this means that someone could be in the same physiological state today and tomorrow yet be dead (and be known to be dead) today and not be dead (nor known to be dead) tomorrow. We claimed that this seems to be an equivalent problem to that pointed out by Marquis and followers, where A and B could be in the same physiological state now and A be dead because it is not appropriate to try to reverse the cessation of circulation while B is alive (or not known to be dead) because it is appropriate to try to reverse the cessation of circulation. If that is right, then the same criticism could be leveled at proponents of view 3 as is leveled against advocates of permanence or normative irreversibility in current death declaration practice. This might mean that Marquis and followers should retreat to option 2. However, we have seen that there are problems with this option too. Which option to choose depends ultimately on how serious we consider these problems to be. Returning to our imaginary case, given that

[13] For reasons we shall see later, everyone, including Joffe, adopts permanence even as they claim to endorse irreversibility (assuming these concepts are distinct—they are not if "permanence" means normative irreversibility). This is because even when we attempt CPR or other resuscitative measures and fail, we do not know whether we could have succeeded if we had tried for longer. The main reason we do not try for longer is that the brain would be too damaged for the efforts to be worthwhile. However, that is a different point to the point about whether we know we have reached biological irreversibility.

level of function to which we could restore someone after 2 days is acceptable only in people under age 30, we might claim that, although theoretically possible to restore circulation in a 60-year-old, it is not appropriate to try to do so, and so these people should continue to be considered dead. We see no reason to think it differs from current practice for death determination in those for whom resuscitation is not appropriate.

If we can have different paradigms of irreversibility and there is no difference whether that is synchronic or diachronic, view 3 leads us to view 4: Marquis should concede that we might have different standards of death, depending on the category of patient, so that it remains appropriate to consider people to whom this technology is not applicable to be dead, and only those to whom it may be applicable to potentially be still alive. Why can't we choose view 4? View 4 is effectively the position of those who endorse normative irreversibility or permanence, now.

It is important to consider what the remaining objection to this option might be. The objection is that death is a biological, and so a physiological, phenomenon, a "matter of fact" and therefore normative considerations of the kind appealed to under view 4 (which we were led to in considering the full implications of view 3) cannot enter into the issue. However, this reply ignores the point we made earlier in this chapter when discussing view 2; that is, that what counts as someone's having died can become an open question when technological advances are made. We need only consider here what we might say if cryopreservation technology does allow us to revive people in the future. Marquis and his followers can object, of course, to our discussion of cryopreservation on the basis that we do not yet know if we will ever be able to revive such people. However, as noted earlier, this does not prevent us from exploring the logic of his position by imagining what would be the case if we succeeded and revived a cryopreserved person for the first time, and others who endorse his position have taken up a position in respect of this case [30].

What, then, would be the analysis applicable to cryopreservation? If cryopreserved people could have their circulation restarted in the future, do we say these people have been brought back to life, or do we assume instead that they were never really dead? These are not factual but conceptual questions to be determined if, indeed, the technology does become a success. Our point, however, can still be made. If the technology does become a success, and the question of whether we say these people were dead, or were never dead, is raised, the issue about which option to choose (were they dead and brought back to life, or were they never dead) is no longer purely biological, nor purely a matter of fact. Furthermore, the definition of death, if it is tied to what counts as "irreversible" given the then current technology, is no longer strictly biological, but carries an intrinsic reference to human capabilities ("irreversible" in that case would be like "irreparable" rather than "insoluble") [32]. As we have noted, this opens the door to normative considerations forming part of our understanding of what it is possible to do to a patient, including the ones we currently adopt in death declaration practice in people with a do-not-resuscitate order. For example, our decision about what it is *better* to say would partly appeal to other normative considerations such as rules about bigamy, burials, autopsies, the administration of wills, etc. However, if that is so, why can't we also allow normative considerations to partly determine what is better to say of those people for whom it is not ethically appropriate to apply CPR or current forms of resuscitation such as ECMO? Where is the fundamental difference between these cases?

There is also a wider sense in which normative considerations intrude into our determination of when someone is dead. Consider the choice between view 2 and view 4. Under view 2, someone is not dead unless they are not revivable given *any* technology, current *or* future. Under view 4, someone is dead if they are not revivable given *current* technology, even if they *would* be revivable given *future* technology. The choice between these two views is not based on a matter of fact but is normative in the wide sense of fixing the meaning of the term "death." Furthermore, whichever view is chosen, it will then be true that we are committed to saying that "irreversible" *must* or *should* mean "irreversible given *any* technology" (if we choose view 2) or "irreversible" *must* or *should* mean "irreversible given *current* technology" (if we choose view 3 or 4). As the words "must" and "should" imply, this recommendation is a normative one. It may not be *ethically* normative—although there are grounds, indicated in the previous paragraph, for thinking that some ethical considerations inform the recommendation we advance—but it is at least *conceptually* normative in so far as it is a recommendation about how we should apply the concepts of death and irreversibility. Since Marquis and followers would *themselves*, in recommending or promoting one of these options, be relying on normative considerations, this undermines their criticisms of those who *also* rely on normative considerations when they say that a person to whom CPR and other resuscitative efforts are inapplicable is dead after the possibility of auto-resuscitation has passed.

2 Arguments from ECMO and the Reality of Permanence

[S]hall we say to them they are dead?; or should we not rather speak of different meanings of the word 'dead' and distinguish between say, 'heart-dead' and 'dead' in some other way?
—F. Waismann, The Principles of Linguistic Philosophy (written between 1929–1936) [33].

We have discussed the extent to which normative considerations can enter even the accounts of those who favor strict irreversibility, noting that this fact (that normative considerations can enter these accounts) seems to undermine criticisms of normative irreversibility. One immediate difficulty with the analogy with non-combustible properties, and dispositional properties such as the property of being dissolvable in *aqua regia*, is that, in the case of human beings, the point at which the dispositional property of being reversible will no longer obtain depends on the specific physiology in the patient (no two patients are ever exactly the same, unlike a sugar cube in water), and on the technology used. In reality, a dispositional account of reversibility presupposes certain background conditions in order to ascertain the point at which the loss of function is no longer reversible. However, these background conditions mean that this point will vary, depending on: (a) whether we are attempting resuscitation or not, (b) the technology that is widespread in the country concerned, (c) the resuscitation technique used, and (d) physiology and pathology of the patient.

To make this point very clear, we can consider the current medical practice that, once again, is challenging our concept of irreversibility: extracorporeal cardiopulmonary resuscitation (ECPR). This uses an extracorporeal cardiopulmonary membrane oxygenation (ECMO) technology which, over the last decades, has developed to be more portable and easier to instigate in patients following unexpected cardiac arrest. While outcomes are debated, its use and availability is growing [12, 17]. The question for today's medical practitioner evaluating a patient with circulatory–respiratory arrest is: (a) whether to attempt resuscitation or not (e.g., do-not-resuscitate order, other injuries); (b) whether ECPR is available or could be made available; (c) whether the patient should have standard CPR, with higher effort and sustained CPR (e.g., it is typical in younger patients to sustain the attempt for longer before "calling it"), or have ECPR; and (d) how the physiology and pathology of the patient would impact on the above decisions. In each of these decisions, the claim of irreversibility cannot escape the normative elements of intention, decision, technology available, and the often-unknowable elements of individual patient physiology and pathology.

The problem faced by Marquis and others who agree with his argument is this. Suppose we say that a person first becomes irreversible only when the very best technology, like ECPR, would not achieve reversal. We could *theoretically* carry out ECPR on every single patient prior to determining death. In practice, we would never dream of doing so—it not being *appropriate*, in many cases. This is because the ischemic damage to neurologic function would be too severe to achieve a recovery compatible with a patient's values, wishes, or beliefs. The limits of modern ECPR are yet to be fully elucidated [12] but historic animal work suggests the limit for restoring some neurologic function is very long—hours at least [34–37]. However, if mere return of circulation (rather than *good* function) is the goal desired—which when considering death by circulatory–respiratory criteria indicates the patient is still alive—this historic work suggests ECPR (and other resuscitative efforts) can restore circulation way beyond the point at which we normally declare someone to be dead under modern death determination practice [20, 38].

So, on a dispositional account, why aren't we committed to indexing the time of death to when someone would be dead if ECPR had been used but the circulation could not be restored? If we rely on the fact that it is not appropriate to use ECPR on a great many patients, then normative considerations are feeding into the point at which we consider the cessation of circulation to be irreversible—and if normative considerations can come in here, then why can't they come in at the point of auto-resuscitation no longer being possible, as permanence supporters would claim? It would not be appropriate to commence ECPR on an elderly patient with a do-not-resuscitate order in a nursing home once their heart stops. It is not clear what the objection can be to determining death at the point just beyond which auto-resuscitation would no longer be possible, once we accept that all forms of resuscitation are not permitted in the case of those with a do-not-resuscitate order. If we rely, however, on what it is appropriate to do, then we have not exploited the dispositional potential of the patient in the way that we could have done, notwithstanding that it would not have been appropriate to exploit that dispositional potential. It means that we cannot legitimately declare this elderly patient dead because we do not actually know the point at which their circulation has truly irreversibly ceased. This seems to be an unacceptable consequence of the dispositional view.

A standard move in response to this type of problem is to claim that "it does not matter" that death is declared early because "nothing of any ethical significance" is done to the patient [14]. In contrast, in organ donation contexts—where this debate has been played out—it has been said that "lethal acts" will be performed [14]. The problem with this standard move is that there are other contexts where such "lethal acts" would also be performed (prior to the putative point of irreversibility), such as in warm autopsies. Are we to stop these practices too [32]? And there are many other contexts where the time of death *is* ethically and legally significant. One such context is precisely the one involving the elderly patient: if we know, at the time their heart stops, that they are not *really* dead and could theoretically be revived up to many hours later with ECPR, then their loved ones are being falsely told that they have died. Why isn't this an ethical problem [22]?[14]

3 Irreversibility and Death by Neurologic Criteria

To die: to sleep–
No more—and by a sleep to say we end.
—William Shakespeare, Hamlet Act III, Scene 1

As we indicated in the introduction, determination of death by neurologic criteria has largely avoided the "irreversible" or "permanent" debate. This is changing as new therapies and technological advances make questions on intent to resuscitate just as relevant as in determination of death by circulatory–respiratory criteria. While the mode of resuscitation and treatment may vary, current and future medical practice impacts determination of death by neurologic criteria in similar ways to the alternative views of "irreversible" for death by circulatory–respiratory criteria we outlined above (see Table 3).

Is then a brain resuscitable after a determination of death by neurologic criteria just as a heart is resuscitable after a determination of death by circulatory–respiratory criteria? There is every reason to suspect that it is. In a postmortem case series, Wijdicks and Pfeifer examined the brains of patients following a determination of death by neurologic criteria. They concluded that "No distinctive neuropathologic features were apparent in our series of patients with brain death. Neuronal ischemic changes were frequently profound, but mild changes were present in a third of the examined hemispheres and in half of the brainstems… Neuropathologic

[14] 14. See Gardiner, McGee, and Bernat for other examples, including determining inheritance under a will, which cannot rely on irreversibility but requires permanence [20].

Table 3 How the possibility of current and future medical practice similarly impact the way we might view the meaning of "irreversible" in the determination of death by both circulatory–respiratory and neurologic criteria[a]

View	Explanation
View 1: This is a mere thought experiment having no bearing on current practice and protocols	Thought experiments are standard philosophical practice to help examine the implications of an idea Some of the example medical practices are possible now
View 2: If this technology succeeded, we really would have discovered that people were not dead at the time we had originally assumed	Raises the general problem that it seems to entail that we are never in a position really to know when someone is dead Raises the option to say that we have discovered ways of bringing people back from the dead
View 3: If this technology were discovered, our concept of death would shift at that point, but, at present, it is merely a logical possibility we can ignore	On this view, what counts as "death" is indexed to what is possible given *current* technology Death is instantiated by different paradigms of irreversibility as technology improves. Someone could be in the same physiological state today and tomorrow, yet be dead today and not be dead tomorrow. The definition of death, if it is tied to what counts as "irreversible" given the then current technology, is no longer strictly biological, but carries an intrinsic reference to human capabilities
View 4: We can have different standards of death, depending on the category of patient we are dealing with. The permanence view	It remains appropriate to consider people to whom this medical practice is not appropriate to be dead, and only those to whom it may be appropriate to potentially be still alive Recognizes and accepts an intrinsic reference to human intention and capabilities

[a]Examples of current and future medical practices that might allow for the return of function beyond the time when death by circulatory–respiratory criteria would have been determined by accepted medical standards include cardiopulmonary resuscitation (for patients with do-not-resuscitate orders), sustained standard cardiopulmonary resuscitation, extracorporeal cardiopulmonary resuscitation, and cryopreservation. Examples of current and future medical practices that might allow for the return of function beyond the time when death by neurologic criteria would have been determined by accepted medical standards include therapeutic decompressive craniectomy, BrainEx, and future technologies

examination is therefore not diagnostic of brain death" [39]. While neurologic function may have ceased in the brain for the determination of death by neurologic criteria to be made, there is no pathological reason to suppose that, with enough effort, it could not be theoretically restorable—at least for hours or even days.

Therapeutic decompressive craniectomy is a surgical intervention which removes part of the skull in patients with severe brain swelling in an attempt to reduce life-threatening intracranial pressure. Decompressive craniectomy can be lifesaving, though debate persists as to the quality of outcome, making the decision to use it nuanced [40–43]. Given that neuropathologic changes are not universal in patients determined to be dead by neurologic criteria, it is entirely feasible that were decompressive craniectomy performed in such patients, return of function might occur, reversing their determination of death. An example from the circulatory–respiratory criteria debate is the patient with a do-not-resuscitate order who is resuscitated more than 5 min after cardiac arrest by a clinician who is unaware the order exists [14, 44, 45]. We now have a similar example case using death by neurologic criteria (although harder to imagine actually happening as decompressive craniectomy requires a whole theatre team) where an unsuspecting neurosurgeon carries out a decompressive craniectomy minutes or even hours after the determination of death. Clearly, the other medical doctors had ruled out decompressive craniectomy as a treatment option prior to their determination of death. They could have based this decision on the availability of decompressive craniectomy in their institution, a judgment on the physiology and pathology of the patient, and the likely impact of this intervention on achieving an outcome consistent with the values, wishes, and beliefs of the patient. If the only gateway to reversibility is an intervention, but that intervention is not appropriate, then we know all there is to know to determine death [32]. This is the permanence standard.

We mentioned above that our imaginary case may be vulnerable to the objection that it is merely a hypothetical case, and that we would later discuss a real-life case. We have done this with ECPR, but there is another interesting real-life case that parallels our imaginary case. Using a custom-made extracorporeal pulsatile–perfusion system and a hemoglobin-based cryoprotective perfusate, scientists and clinicians in Yale School of Medicine restored some cellular functions (specifically vascular and glial responsiveness to pharmacological and immunogenic interventions) in pig brains 4 h after decapitation in a food production slaughterhouse [23]. They also observed spontaneous synaptic activity and active cerebral metabolism during this period. The scientists named their technology, BrainEx.[15]

While a completely different mechanism to what might lead to death by neurologic criteria, some debate arose about whether this new technology undermines the concept of death by neurologic criteria [46–49]. From the above discussions in this chapter, it can be seen that even if BrainEx or a future technology could restore consciousness in a decapitated head, a decision will need to be made about whether to use this technology. BrainEx is to brain function as ECPR is to cardiac arrest. Just as we do not need to await the point at which ECPR could not succeed in order to determine death—because that would be applying a time frame that is inapplicable to this category of patients—so we would not need to await the point at which BrainEx could no longer work before we could determine death on the basis of brain arrest [22]. The BrainEx experiment does not mean we now need to wait a minimum of 4 h to determine death

[15]The authors distinguished between restoring brain function and cellular activity in the brain, being careful to describe the activity as postmortem activity. However, one reason for this is that they were very careful to emphasize that consciousness was not restored (which might imply the authors regarded awareness or consciousness as the criterion of life, which is controversial). Regardless of how the authors themselves describe the results of the study, it is reasonable to regard them as having restored life at least in a minimal sense, but we can in any event imagine that, in future, a greater level of function is restored.

by neurologic criteria. BrainEx technology is discussed further elsewhere in this book.[16]

It is essential to understand, then, that the mere existence of these possibilities, imagined or real, does not throw any doubt on the utility of our current concept of death, which caters perfectly well for the demands of current practice. A relevant factor in this claim is that, in current practice, no person ever recovers from a proper determination of death by neurologic criteria. However, our current concept of death as it stands cannot be expected to budget for every imaginable case in advance, and it makes sense to explore, to a limited degree, the options we have for dealing with new advances in technology enabling us to resuscitate people we cannot resuscitate now. A successor of BrainEx that could restore brain function 2 days after the time at which we currently consider it impossible to bring someone back might call for such a decision to revise the concept, for example (to say either that we have discovered people are not dead when we thought they were, or we have discovered a way to bring people back from death). This decision from that point then determines whether, in a particular case, someone has had their life saved, or has been brought back to life from having been dead. In addition, whatever decision society takes, it remains true that, as with all our previous examples, doctors and families will invariably choose not to use this technology on *each and every* patient because function might be too limited or might not reach an acceptable quality, making it inappropriate to use. We may in that case continue to say that those on whom we do not use the technology are dead at the time we have always declared them to be. The mere introduction of new technologies, then, does not imply that patients who were determined dead before this technology was invented, or who are declared dead at the current applicable timeframes when the technology is not appropriate to use, are no longer known to be dead at the time of that death determination.

Returning to death by neurologic criteria, while we may once have been tempted to consider death by neurologic criteria a retrospective diagnosis, in contrast to a prospective determination of death by circulatory–respiratory criteria, this temptation should be resisted. Medical practice and technological advances raise the same issues of intention and decision for death by neurologic criteria as are raised in the case of death by circulatory–respiratory criteria [22]. A strictly biological concept of "irreversible"—at least if that does not take into account what we consider it appropriate to do to patients—appears poorly positioned to respond to some of the challenges in these shifting technologies. Table 4 shows how the impression of retrospectivity regarding death by neurologic criteria is misleading and, in light of our arguments above, no longer conceptually sustainable.

---

[16] The same point applies if recent reports about researching the possibility of restoring function to brain-dead people are realized. See https://www.thestatesman.com/lifestyle/health/biotechnology-company-attempts-reverse-state-brain-death-1503044180.html and https://urldefense.com/v3/__https:/www.labroots.com/trending/neuroscience/6401/brain-death-reversed__;!!NVzLfOphnbDXSw!WhN4amr7M17AUI80LPgdOLwl7scACNOBhrhgi8-u-ytXRD0txGb4v-kkb0kkJuqd6Q$

Table 4 Application of "permanent" and "irreversible" perspectives to the typical steps for the determination of death by accepted medical standards

Death by circulatory–respiratory criteria (A prospective determination)	Death by neurologic criteria (The retrospective belief)	Death by neurologic criteria (The prospective reality)
1. Circulatory–respiratory arrest leading to cessation of circulatory and respiratory and neurologic function	1. Severe devastating brain injury in an apneic and comatose patient who has absent brainstem reflexes	1. Severe devastating brain injury in an apneic and comatose patient who has absent brainstem reflexes
2. Decision made not to attempt resuscitation or that further attempts are futile (permanence) 3. Five minutes elapse 4. **Determination of death by accepted medical standard** 5. Restoration of circulatory and respiratory and neurologic function still theoretically possible (permanence) 6. In the absence of resuscitation, organ and cellular injury ensue 7. Restoration of circulatory and respiratory and neurologic function are no longer possible (irreversible)	2. Brain arrest—cessation of neurologic function 3. Neurologic injury judged to be "irreversible" **4. Death suspected** 5. Testing takes time to organize and carry out 6. Even more brain injury **7. Determination of death by accepted medical standard**	2. Brain arrest—cessation of neurologic function 3. Decision made that further attempts at resuscitation or treatment are futile (permanence) **4. Death suspected** 5. Testing takes time to organize and carry out 6. Even more brain injury 7. **Determination of death by accepted medical standard** 8. Restoration of neurologic function still theoretically possible (permanence) 9. In the absence of resuscitation, organ and cellular injury ensue 10. Restoration of neurologic function is no longer possible (irreversible)

4 Conclusion: The Source of Disagreement: Two Rival Conceptions of Death

In our view, the source of the disagreement between those who insist that death requires irreversibility and those who defend permanence is that the former tend to see death as an event, like a flash of lightning, after which there is no way back for the patient. Franklin Miller and Robert Truog, who are sympathetic to the criticisms by Marquis and who have extensively criticized death determination practice, have claimed that death is an event, and that the event is the "moment…when the forces tending to increase entropy irreversibly overcome those that are opposing it…" [15, pp. 70–71]. This is, however, a recommendation that death be understood as an event, and that the event in question be identified with this moment of entropy overcoming the forces that resist it. While it is certainly a plausible recommendation, it is no less a stipulation than the recommendation of those who claim that death by neurologic criteria is death.

It is possible, however, to understand death (and not merely dying) as a process, or a question of degree. For example, all sides in this debate agree that it is not necessary to wait until putrefaction has set in before we can determine there is irreversible loss of function. It is also agreed by all sides that not every cell in the body need have died in order to claim that the organism has died. Yet, in a perfectly innocuous sense, we can claim that a person is "more dead" when putrefaction has set in than they are at the time rigor mortis begins to occur, and even more dead again weeks after putrefaction. If we accept that being dead is a matter of degree, it makes more sense to say of a group of patients for whom resuscitative measures are not even appropriate that the degree of cessation of bodily function shall be considered sufficient for a valid determination of death in their case. It is true that, on this view, a theoretically resuscitable patient can be in the same physiological state as a potential organ donor, yet in one further resuscitative measures are appropriate and in the donor they are not.

Where death is construed as a process rather than as an event, this situation is entirely plausible and expected, since resuscitation, if *successful*, would involve changing the physiological state of a person from the state they are currently in to one they were in earlier (see Fig. 1). However, the mere *possibility* of changing the physiology is not itself such a change. Further, for reasons we have already explored, the situation where one person can be in the same physiological state as another, yet one be alive and the other dead, can also arise with irreversibility at the point at which new medical practice or technology extends the time at which someone can be brought back, so this objection to permanence is not sufficiently strong if the alternative to permanence is irreversibility. We have also seen above that irreversibility is not essential to our concept of death.

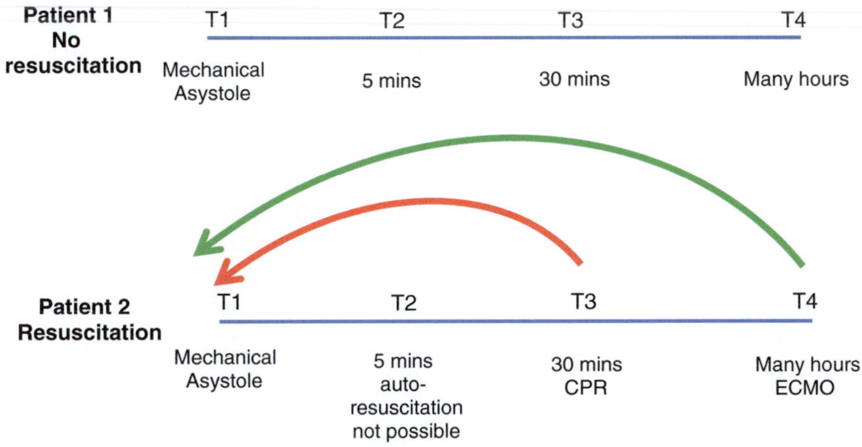

Fig. 1 Gradual loss of functions after mechanical asystole. Illustrating the role of resuscitation and how death is a process

Once we accept that death is a more process than an event, we can say that, during the first few hours of death, death is, in some cases, potentially reversible. Only after many hours does it become irreversible. However, since there is an element of human choice about what we shall call death, society has in practice adopted a compromise between two extremes: the extreme of the layperson (and resuscitation researchers such as Parnia) who claim that someone with a cardiac arrest "died three times" on the way to hospital, and the irreversibility proponents such as Marquis, Joffe, and Nair-Collins who, to be strictly consistent, would require many hours to pass before we could truthfully know that circulation could not be restarted [18, 28–30]. It is this that allows us to wait only 5 min when we know that patients with a do-not-resuscitate order will not be resuscitated, while also not allowing us to say that a person with *no* do-not-resuscitate order is dead at 5 min: the adoption of permanence is a rule about what it is appropriate to do to the patient, and its application is context dependent (in truth, even the irreversibility proponent will declare death before many hours with people who do not have a do-not-resuscitate order).

A final point should be emphasized. Permanence is a defeasible concept, which means permanence obtains *unless* something else occurs. We can compare the logic of declaration by permanence to a vicar's declaration that a couple is now husband and wife in a marriage ceremony. When does the couple become married? When the ceremony is over. However, if the marriage is not consummated, then the marriage is void from the beginning, not merely from the time consummation fails to occur. It is the same with permanence. If resuscitation were attempted and succeeded after a declaration of death, the death declaration would be invalid from the time of the declaration, not the time the resuscitation succeeded. Permanence advocates are not committed to the claim that someone is raised from the dead when the death declaration is overturned—it would not be permanent cessation if the patient were brought back [17]. Yet if resuscitation is not attempted, or attempted and fails, the person is dead from the time of death declaration, not from the time resuscitation is attempted and fails.

Why does permanence have this strange logic? Consider, by analogy, two living-room lights controlled by an automatic dimmer switch. In one room, the light becomes gradually dimmer so that, at some stage, we can uncontestably describe the light in the room as "dim." In the other room, someone overrides the automatic control and turns the light back up again. It is no longer dim in this second room. However, it *was* dim. The light has been brought back from the state of being dim to that of being bright. However, we would neither conclude: (a) that it never *was* dim, nor (b) that the light in the first room never became dim because it, too, *could* have been made brighter by someone overriding the automatic control to turn it up.

This is where dimness, and death, differ. With the dim room, if we bring back the bright lights, the room is no longer dim. In contrast, with the dead person, if we bring back that person through resuscitation, the person never was dead in the first place. Why is there this difference? Death is *sui generis* here. We never say that no object can be called "red" unless it is irreversibly so. Even with the related concept of extinction, bringing back the woolly mammoth would not mean that the woolly mammoth had never gone extinct. Few predicates require irreversibility as a precondition of their application.

We believe that the difference can be explained by the fact that the requirement of irreversibility reflects an ethical rule in standard cases: everything should be done to try to bring a patient back. People have been dying for 200,000 years but genuine possibilities of reversal (as opposed to early fantasies) only became real in the 1700s. The idea of reversing death is a recent achievement, and we made the decision that reversing what used to be sufficient for death should not be called reversing *death* but instead should mean that death has not occurred. The reason for this is understandable. We do not want to give up on people prematurely, and we do not want to be buried before we are dead. However, this norm is not appropriate in all contexts—e.g., in patients with do-not-resuscitate orders. This is why this debate has arisen: there is a tension between two conflicting practical requirements, and permanence is the perfect solution to it.

Declaration and Acknowledgement Dr. Gardiner is Associate Medical Director—Deceased Organ Donation for NHS Blood and Transplant, the national organ donation organization in the United Kingdom. The views expressed are his own. The authors are very grateful for the helpful comments of Franklin G Miller on an earlier draft of this chapter.

References

1. Bakhtin MM. In: Holquist M, editor. The dialogic imagination: four essays. Texas: University of Texas Press; 1981.
2. National Conference of Commissioners on Uniform State Laws. Uniform Determination of Death Act. 1981. https://www.uniformlaws.org/committees/community-home?communitykey=155faf5d-03c2-4027-99ba-ee4c99019d6c&tab=groupdetails.
3. McGee A, Gardiner D. Differences in the definition of brain death and their legal impact on intensive care practice. Anaesthesia. 2019;74(5):569–72.
4. A definition of irreversible coma: report of the ad hoc committee of the Harvard Medical School to Examine the Definition of Brain Death. JAMA 1968;205(6):337–40.
5. Diagnosis of brain death. Statement issued by the honorary secretary of the Conference of Medical Royal Colleges and their Faculties in the United Kingdom on 11 October 1976. Br Med J. 1976;2(6045):1187–8.
6. Diagnosis of death. Memorandum issued by the honorary secretary of the Conference of Medical Royal Colleges and their Faculties in the United Kingdom on 15 January 1979. Br Med J. 1979;1(6159):332.
7. President's Commission for the Study of Ethical Problems in Medicine and Biomedical and Behavioral Research. Defining Death: Medical, Legal and Ethical Issues in the Determination of Death. U.S. Government Printing Office Washington, DC; 1981. https://repository.library.georgetown.edu/bitstream/handle/10822/559345/defining_death.pdf.
8. Dhanani S, Hornby L, van Beinum A, et al. Resumption of cardiac activity after withdrawal of life-sustaining measures. N Engl J Med. 2021;384(4):345–52.
9. Boucek MM, Mashburn C, Dunn SM, et al. Pediatric heart transplantation after declaration of cardiocirculatory death. N Engl J Med. 2008;359(7):709–14.
10. Lomero M, Gardiner D, Coll E, et al. Donation after circulatory death today: an updated overview of the European landscape. Transpl Int. 2020;33(1):76–88.

11. Domínguez-Gil B, Ascher N, Capron AM, et al. Expanding controlled donation after the circulatory determination of death: statement from an international collaborative. Intensive Care Med. 2021;47(3):265–81.
12. Guy A, Kawano T, Besserer F, et al. The relationship between no-flow interval and survival with favourable neurological outcome in out-of-hospital cardiac arrest: implications for outcomes and ECPR eligibility. Resuscitation. 2020;155:219–25.
13. Joffe AR, Carcillo J, Anton N, et al. Donation after cardiocirculatory death: a call for a moratorium pending full public disclosure and fully informed consent. Philos Ethics Humanit Med. 2011;6:17.
14. Truog RD, Miller FG. Counterpoint: are donors after circulatory death really dead, and does it matter? No and not really. Chest. 2010;138(1):16–8; discussion 18-19.
15. Miller FG, Truog RD. Death, dying, and organ transplantation: reconstructing medical ethics at the end of life. Oxford University Press; 2011.
16. Goldberger ZD, Chan PS, Berg RA, et al. Duration of resuscitation efforts and survival after in-hospital cardiac arrest: an observational study. Lancet. 2012;380(9852):1473–81.
17. Inoue A, Hifumi T, Sakamoto T, Kuroda Y. Extracorporeal cardiopulmonary resuscitation for out-of-hospital cardiac arrest in adult patients. J Am Heart Assoc. 2020;9(7):e015291.
18. Parnia S. The Lazarus effect: the science that is rewriting the boundaries between life and death. Rider; 2013.
19. Bernat JL, Capron AM, Bleck TP, et al. The circulatory-respiratory determination of death in organ donation. Crit Care Med. 2010;38(3):963–70.
20. Gardiner D, Housley G, Shaw D. Diagnosis of death in modern hospital practice. In: Leisman G, Merrick J, editors. Considering consciousness clinically. Nova Science Publishers, Inc.; 2016. p. 93–77.
21. Bernat JL. On noncongruence between the concept and determination of death. Hast Cent Rep. 2013;43(6):25–33.
22. Gardiner D, McGee A, Bernat JL. Permanent brain arrest as the sole criterion of death in systemic circulatory arrest. Anaesthesia. 2020;75(9):1223–8.
23. Vrselja Z, Daniele SG, Silbereis J, et al. Restoration of brain circulation and cellular functions hours post-mortem. Nature. 2019;568(7752):336–43.
24. Truog RD, Miller FG, Halpern SD. The dead-donor rule and the future of organ donation. N Engl J Med. 2013;369(14):1287–9.
25. Academy of Medical Royal Colleges. A Code of Practice for the Diagnosis and Confirmation of Death. Academy of Medical Royal Colleges. 2008. https://www.aomrc.org.uk/reports-guidance/ukdec-reports-and-guidance/code-practice-diagnosis-confirmation-death/.
26. Shemie SD, Hornby L, Baker A, et al. International guideline development for the determination of death. Intensive Care Med. 2014;40(6):788–97.
27. Greer DM, Shemie SD, Lewis A, et al. Determination of brain death/death by neurologic criteria: the world brain death project. JAMA. 2020;324(11):1078–97.
28. Marquis D. Are DCD donors dead? Hast Cent Rep. 2010;40(3):24–31.
29. Nair-Collins M. Taking science seriously in the debate on death and organ transplantation. Hast Cent Rep. 2015;45(6):38–48.
30. Joffe A. DCDD donors are not dead. Hast Cent Rep. 2018;48(Suppl 4):S29–32.
31. McGee A, Gardiner D. Permanence can be defended. Bioethics. 2017;31(3):220–30.
32. McGee A, Gardiner D. Donation after the circulatory determination of death: some responses to recent criticisms. J Med Philos. 2018;43(2):211–40.
33. Waismann F. In: Harri R, editor. The principles of linguistic philosophy. 2nd ed. Palgrave Macmillan; 1997.
34. Kountz WB. Revival of human hearts. Ann Intern Med. 1936;10(3):330–6.
35. Hinzen DH, Müller U, Sobotka P, Gebert E, Lang R, Hirsch H. Metabolism and function of dog's brain recovering from longtime ischemia. Am J Phys. 1972;223(5):1158–64.
36. Merkle RC. The technical feasibility of cryonics. Med Hypotheses. 1992;39(1):6–16.
37. DeVita MA. The death watch: certifying death using cardiac criteria. Prog Transplant. 2001;11(1):58–66.

38. Rittenberger JC, Menegazzi JJ, Callaway CW. Association of delay to first intervention with return of spontaneous circulation in a swine model of cardiac arrest. Resuscitation. 2007;73(1):154–60.
39. Wijdicks EFM, Pfeifer EA. Neuropathology of brain death in the modern transplant era. Neurology. 2008;70(15):1234–7.
40. Mohan Rajwani K, Crocker M, Moynihan B. Decompressive craniectomy for the treatment of malignant middle cerebral artery infarction. Br J Neurosurg. 2017;31(4):401–9.
41. Cooper DJ, Rosenfeld JV, Murray L, et al. Decompressive craniectomy in diffuse traumatic brain injury. N Engl J Med. 2011;364(16):1493–502.
42. Zeiler F, Trickey K, Hornby L, Shemie S, Lo B, Teitelbaum J. Mechanism of death after early decompressive craniectomy in traumatic brain injury. Trauma. 2018;20(3):175–82.
43. Hutchinson PJ, Kolias AG, Timofeev IS, et al. Trial of decompressive craniectomy for traumatic intracranial hypertension. N Engl J Med. 2016;375(12):1119–30.
44. NCEPOD. Time to intervene? A review of patients who underwent cardiopulmonary resuscitation as a result of an in-hospital cardiorespiratory arrest. 2012. https://www.ncepod.org.uk/2012report1/downloads/CAP_fullreport.pdf.
45. McGee A, Gardiner D. The Papworth donation-after-circulatory-death heart technique and its challenge to the permanence standard. J Heart Lung Transplant. 2017;S1053-2498(17):31949–6.
46. Youngner S, Hyun I. Pig experiment challenges assumptions around brain damage in people. Nature. 2019;568(7752):302–4.
47. Dunphy S. A new study in pigs challenges the definition of "brain dead". European Scientist. 2019. https://www.europeanscientist.com/en/public-health/new-study-in-pigs-challenges-the-definition-of-brain-dead/.
48. Busl KM. When is dead really dead? Study on pig brains reinforces that death is a vast gray area. The Conversation. 2019. http://theconversation.com/when-is-dead-really-dead-study-on-pig-brains-reinforces-that-death-is-a-vast-gray-area-115750.
49. Bernat JL, Delmonico FL. Restoring activity of pig brain cells after death does not invalidate the determination of death by neurologic criteria or undermine the propriety of organ donation after death. Transplantation. 2019;103(7):1295–7.

Part II

Medical Issues

Intra/International Variability in the Determination of Death by Neurologic Criteria

Ali Daneshmand and David Greer

One might think that a medical determination of death by neurologic criteria would not be subject to variability in any regard, because it is a medical and legal determination of death that has been utilized for many decades, both in the United States and throughout the world. However, as with many things in medicine and law, there is both intra- and international variability across a number of domains including the concept of death by neurologic criteria; laws on declaration of death by neurologic criteria; the required credentials for the clinical examiner; the technique to perform the clinical evaluation and apnea testing; the indications for, and selection of, ancillary testing; determination of death by neurologic criteria in children; and communication, documentation, discontinuation of organ support and organ donation after the determination. This variability raises concerns about the accuracy and validity of determinations of death by neurologic criteria, which could be detrimental to the medical field and the public's trust in this core medical determination. In this chapter, we analyze different types of variability in the determination of death by neurologic criteria and explain the steps that are being taken to reduce variability.

1 The Concept of Death by Neurologic Criteria

As described elsewhere in this book, there are two ways to conceptualize death by neurologic criteria: (1) the whole-brain criterion and (2) the brainstem criterion. The United States, and most other countries, utilize the whole-brain criterion, which defines death by neurologic criteria as the irreversible loss of all functions of the entire brain, including the brainstem, and implies that both infratentorial and supratentorial structures are affected. The United Kingdom and many Commonwealth

A. Daneshmand · D. Greer (✉)
Department of Neurology, Boston University School of Medicine, Boston Medical Center, Boston, MA, USA
e-mail: Ali.Daneshmand@bmc.org; dgreer@bu.edu

A. Lewis, J. L. Bernat (eds.), *Death Determination by Neurologic Criteria*, Advances in Neuroethics, https://doi.org/10.1007/978-3-031-15947-3_10

countries, on the other hand, use the brainstem criterion, in which irreversible loss of all brainstem functions is sufficient for death by neurologic criteria [1].

In clinical practice, both criteria embrace a three-step approach to the determination: first, establishing a cause for the catastrophic brain injury and determining that the injury is permanent; second, excluding confounders and reversible causes for the neurologic state; and third, confirming the patient is comatose and has absence of all brainstem reflexes and breathing capacity using apnea testing. Because most injuries to the brain that lead to death by neurologic criteria are supratentorial and progress to involve loss of function infratentorially, there is usually no practical difference between these ways to conceptualize death by neurologic criteria [2]. However, questions arise when considering how death should be determined using the whole-brain criterion in patients with a primary injury to the posterior fossa, as they may have persistent brain circulation and/or electrical activity in supratentorial structures. Varelas et al. elucidated this concern by reporting cases of patients with primary posterior fossa injury that satisfied the clinical standards for death by neurologic criteria, but had supratentorial brain circulation but later developed secondary brain edema and hydrocephalus, and ultimately were determined dead using the whole-brain criterion [3]. However, it is unclear whether this is always the case. The potential for reversibility of the condition should give clinicians great pause when considering determination of death by neurologic criteria in a patient who has suffered a primary brainstem or other posterior fossa insult.

2 Laws on Death by Neurologic Criteria in the United States

In 1968, an ad hoc committee at Harvard Medical School introduced the first medical standard for death by neurologic criteria [4]. The committee believed that statutory changes to acknowledge death by neurologic criteria as legal death were not needed unless there was controversy within the medical community about the adoption of neurologic criteria to declare death. A decade later, due to social and legal consequences of this new criterion for death, President Carter tasked the President's Commission for the Study of Ethical Problems in Medicine and Biomedical and Behavioral Research to evaluate the definition of death. Through a process involving the American Bar Association, the American Medical Association, the National Conference of Commissioners on Uniform State Laws, and several religious officials, the Commission created the Uniform Determination of Death Act (UDDA) [5]. The UDDA codified determination of death in the setting of "irreversible cessation of function of the entire brain, including the brainstem," according to "acceptable medical standards."

Since the creation of the UDDA, all 50 states have incorporated death by neurologic criteria into their judicial or statutory definitions. However, only 36 states adopted the complete language of the UDDA into their statutes [6] and there are inconsistencies in a number of areas across the country.

One area of variability is in both the use and interpretation of the phrase "acceptable medical standards." This was exemplified by a Nevada Supreme Court ruling in 2015, prompting the State of Nevada to revise their UDDA to stipulate that the determination of death by neurologic criteria should be based on the 2010 standards for determination of death by neurologic criteria in adults published by the American Academy of Neurology (AAN) and the 2011 standards for determination of death by neurologic criteria in pediatric patients written by the Society of Critical Care Medicine, American Academy of Pediatrics, and the Child Neurology Society (SCCM/AAP/CNS), or subsequent standards [7–9].

A second area of variability in laws about death determination is in the management of objections to death by neurologic criteria. The state of New Jersey allows family objection to determination of death by neurologic criteria based on religious or moral grounds [10]. California and New York law allow "reasonable accommodation" of religious and moral objections to determination of death by neurologic criteria, while Illinois indicates that a patient's religious beliefs must be taken into account for documentation of time of death [11]. The AAN issued a position statement in 2019 regarding accommodation, which provides a framework for states and hospitals to develop a formal stance on management of objections, which are arising more and more in recent years [12–14].

International variability in the content of laws on the declaration of death by neurologic criteria is discussed elsewhere in this book.

3 Variability in Institutional Standards on Determination of Death by Neurologic Criteria in Adults in the United States

Variation exists in the United States because each hospital is responsible for developing policies for determination of death by neurologic criteria. In a 2008 study comparing standards for determination of death by neurologic criteria among the top 50 neurology and neurosurgery programs in the United States (according to US News and World Report) against the 1995 AAN standards for determination of death by neurologic criteria in adults, variability was measured according to five domains: number and qualifications of examiners, prerequisites for determination, clinical examination, apnea testing, and ancillary testing [15]. The authors found that in 71% of standards, multiple evaluations were required, and distinct and separate clinicians were required to conduct repeat testing in 44% of standards. Eighty-nine percent of standards noted a minimum temperature for the evaluation, but this varied from 32 to 36 °C. For the clinical evaluation, the lowest concordance with the AAN standard was for evaluating the absence of pain above the foramen magnum (42%) and the absence of spontaneous respirations prior to initiating apnea testing (27%). Guidance regarding apnea testing also demonstrated relatively poor concordance with the AAN standard, with only 66% of standards specifying the need for arterial blood gas sampling prior to testing, 39% requiring a specific $PaCO_2$ level prior to testing, and 76% requiring preoxygenation. Finally, guidance regarding specific circumstances to pursue ancillary testing was included in only 66% of standards.

In a 2015 follow-up study of the "top 50" neuroscience centers in the United States, the authors assessed the same five domains as in 2008 to determine whether improvements had been made since the interval update to the AAN standard in 2010 [16]. Seventy-six percent of institutions had updated their standards on determination of death by neurologic criteria by that time. Ninety-four percent of institutional standards required the absence of hypothermia (compared to 89% previously). Compliance in the specifics of the clinical evaluation also improved, with the absence of pain above the foramen magnum required in 53% (from 42%), absence of a jaw jerk reflex in 24% (from 18%), and absence of spontaneous respirations in 47% (from 27%). The most significant improvements were related to apnea testing and ancillary testing. Fifty-three percent of standards that required two evaluations stipulated that there was a need for a waiting period between them.

In 2016, an expanded analysis was done of the standards on death by neurologic criteria from 492 individual hospital or system systems in the United States [17]. The areas of greatest difference from the AAN standard included prerequisites for death by neurologic criteria testing, clinical examination of the lower brainstem, apnea testing, and ancillary testing. Exclusion of hypotension and hypothermia prior to the evaluation was specified in only 56% and 79% of standards, respectively. Only 83% of standards required identification of the cause of the patient's neurologic state before the evaluation. Although the 2010 AAN standard noted that ancillary testing is only necessary for death by neurologic criteria determination when clinical and apnea testing cannot be safely or fully completed and interpreted, ancillary testing was mandated in 7% of standards [7]. Clear guidance for ancillary testing indications, timing, and performance were described in only 64% of standards.

4 Variability in Institutional Standards on Determination of Death by Neurologic Criteria in Children in the United States

There is also variability between pediatric institutional standards for determination of death by neurologic criteria throughout the United States as discussed in detail elsewhere in this book. In a recent study, death by neurologic criteria standards were obtained from pediatric institutions in the United States via organ procurement organizations [1]. The standards were assessed with respect to general procedures, prerequisites, clinical examination, apnea testing, and ancillary testing, similar to the adult studies discussed above.

Of the 118 standards developed or revised after 2011, 97% required identification of the mechanism of irreversible brain injury and 67% required an observation period after the brain injury before death by neurologic criteria evaluation. The majority of standards required prerequisites consistent with the pediatric standards published by the SCCM/AAP/CNS such as the absence of hypotension (94%), hypothermia (97%), and metabolic derangements (92%). In respect to the clinical examination, 91% required a lack of responsiveness, 93% no response to noxious stimuli, and 99% the loss of brainstem reflexes. Eighty-four percent of standards required two apnea tests, in accordance with the SCCM/AAP/CNS standards. $PaCO_2$ targets were consistent with the SCCM/AAP/CNS standards in 64% of institutional standards. Fifteen percent of pediatric standards required ancillary testing for all patients, and 15% allowed ancillary studies that are not validated in the pediatric population.

5 Variability in National Standards on Determination of Death by Neurologic Criteria around the World

A subsequent study by Lewis et al. reviewed standards for determination of death by neurologic criteria around the world [18]. Of the 197 countries in the world, contact was made with representatives from 136 countries (69% of the world); representatives from 83 of these countries indicated that they had a national standard for death by neurologic criteria (42% of the world). Notable findings included that 18% of standards referred to "brainstem death"; different rules for multiple age groups was common; the number of examiners was usually 2 (93%), with multiple different areas of expertise specified; and there was variability for the prerequisites, clinical evaluation, apnea testing, and ancillary testing, similar to the aforementioned findings in the studies from the United States. Most international standards (82%) required a known etiology of the neurologic catastrophe, but only 30% required brain imaging. Wide variability and insufficiencies existed in stipulated examination requirements; for example, only 82% of standards stipulated the need for absence of the gag reflex, 79% the cough reflex, and 74% the oculocephalic reflex. Ninety-one percent of standards provided guidance for apnea testing, but most (76%) provided a target for the final $PaCO_2$, ranging from 50 to 60 mmHg. A surprisingly high percentage (28%) of standards required ancillary testing for determination of death by neurologic criteria, with most mentioning EEG (71%), but also digital subtraction angiography in 59% and transcranial doppler ultrasound in 55%. Additionally, instructions on performance and interpretation of ancillary testing are also variable and at times inconsistent, which could cause an error in declaration of death by neurologic criteria [19].

Table 1 reviews the variability in the medical standards for determination of death by neurologic criteria in the United States [17], Europe [20], Asia [21], and Latin America [22].

Table 1 Comparison of standards in the United States [17], Europe [20], Asia [21], and Latin America [22] on determination of death by neurologic criteria

Criteria	Component	United States (492 hospitals) (%)	Europe (28 countries) (%)	Asia (24 countries) (%)	Latin America (15 countries) (%)
Number of exams required	One exam	13	18	14[a]	37[a]
	Two or more exams	87	82	86[a]	63[a]
Prerequisites for clinical exam	Established cause	83	96	88	73
	Absence of hypotension	56	64	37	40
	Absence of hypothermia	79	96	25	40
	Absence of electrolyte/acid–base/endocrine disturbance	71	54	71	67
	Absence of muscle relaxants/paralytics	94	100	63	60
Specifics of clinical exam	Absence of pupillary reflex	93	96	87	87
	Absence of corneal reflex	90	100	83	80
	Absence of oculocephalic reflex	88	96	79	60
	Absence of oculovestibular reflex	89	96	87	80
	Absence of gag reflex	87	79	79	67
	Absence of cough reflex	79	100	75	67
Specifics of apnea testing	Apnea testing is required	97	100	87	87
	Preoxygenation specified	79	NA	79	53
	Arterial blood gas prior to testing	66	82	54	53
	Target $PaCO_2$ or pH specified	84	86	79	67
	Suspension of test with hemodynamic instability	63	71	47	47
Requirement for ancillary testing	Required in all patients	7	50	21	13
	Inability to complete clinical evaluation	51	61	47	38
	Inconclusive apnea test	48	50	47	46
	Presence of drugs that could depress the central nervous system	32	57	21	38

NA data not available

[a] Only 53% of Latin American protocols and 58% of Asian protocols specified the number of exams

6 Qualifications for the Examiner

Given the importance of determinations of death, one might think that both the law and medical standards would be prescriptive about who can perform the evaluation. However, this is not routinely the case. In the United States, for example, only Florida and Virginia require a clinician with expertise in the field of critical care or neuroscience to make a determination of death by neurologic criteria [6].

Surveys and reviews of standards have demonstrated variability in the level of experience and specialization of the recommended examiner(s) (including neurology, neurosurgery, critical care, or simply a "licensed physician") [23].

Although there is clearly value in having trainees and others learn the entire detailed process of determination of death by neurologic criteria from beginning to end, this must be done with strict and direct supervision by a physician with appropriate expertise. The examiner needs to have experience in both the determination of death by neurologic criteria and in the recognition of potential pitfalls [24]. For example, a neurologist with a primarily outpatient practice may be uncomfortable performing a determination of death by neurologic criteria every few years when called upon to do so, whereas a medical or surgical intensivist who is doing coma examinations in the intensive care setting every day has appropriate comfort and expertise in completing the evaluation, particularly as pertains to performance of the apnea test, which could lead to hemodynamic compromise.

7 Communication, Documentation, Discontinuation of Organ Support and Organ Donation

In addition to variability in the evaluation for death by neurologic criteria, there is variability in communication and documentation about death by neurologic criteria, as well as discontinuation of organ support and organ donation following determination of death by neurologic criteria. In the aforementioned international study of national standards on determination of death by neurologic criteria, 50 countries (64%) did not specifically address communication with a patient's family before or after determination of death by neurologic criteria [18]. On the other hand, nine countries (12%) required physicians to inform families before discontinuation of organ support. In 45 countries (58%), the specifics of the time of death were defined in the standard: time of death was the time of completion of the clinical exam in 30 countries, the time of interpretation of ancillary testing in four countries, the time of performing ancillary testing in one country, and other protocols listed a variety of other times. Interestingly, no standards in this study designated the time of arterial blood gas sampling or resulting as the time of death.

The standards from 36 countries (46%) included guidance on discontinuation of organ support after determination of death by neurologic criteria. Five countries allowed "immediate discontinuation" of organ support after determination of death by neurologic criteria. Five countries counseled consideration of a patient's religious beliefs when planning the time to discontinue organ support, five to consider the family's objection to discontinuation of organ support, three recommended waiting for a "reasonably respectful period" before discontinuing organ support, two to consider a patient's "moral beliefs," and one with specific guidance surrounding death by neurologic criteria in pregnancy.

Thirty-four countries' standards for determination of death by neurologic criteria mentioned organ donation, of which 19 provided specific guidance.

8 Rectifying Variability

8.1 Intranational (the United States)

At the time of this writing, the AAN is leading an effort to combine guidance for determination of death by neurologic criteria for adults and children, as many of the practices and procedures are shared. Of course, there are notable differences, such as norms for blood pressure based on age, as well changes in cranial physiology before and after closure of fontanelles and sutures, and thus there will be essential "carve out" sections to deal with these issues in particular. However, the hope of such a combined document is to harmonize practice wherever possible, thus reducing confusion and inconsistent practice, hopefully leading to a highly stringent process. It remains to be seen how this will impact institutional standards.

In an attempt to rectify the legal variability in death determination in the United States, a Drafting Committee has been convened by the Uniform Law Commission to revise the UDDA [25]. However, it remains unclear what language the new version will contain, how and whether it will be implemented state by state, and whether the revised UDDA will improve variability.

8.2 International

The World Brain Death Project (WBDP), an international consensus statement on death by neurologic criteria, was published in 2020 [26]. Prior to this, there was no global consensus regarding death by neurologic criteria determination. The WBDP provides specific and detailed guidance for clinical and apnea testing ("Minimum Clinical Determination of Brain Death/Death by Neurologic Criteria") and ancillary testing ("Beyond Minimum Clinical Determination of Brain Death/Death by Neurologic Criteria"), including unapproved tests such as computed tomographic angiography. It remains to be seen whether countries around the world will ensure their standards are consistent with the guidance in the WBDP.

As determination of death by neurologic criteria requires consistency and accuracy, it is important to recognize the multiple ways to ensure examiners have proper training and expertise. Options include simulation courses, online training experiences, and national courses, such as those offered by the AAN and the Neurocritical Care Society [27–29]. Some institutions have established brain death "champions," a core group of practitioners who are well-versed in death by neurologic criteria determination and potential pitfalls, and who do the bulk of determinations for the sake of consistency and practice excellence.

9 Conclusions

Although there is variability in the determination of death by neurologic criteria, both in the United States and worldwide, much work has been done to reduce unevenness and to harmonize standards and practice. We emphasize that much of the research that has been performed to date has evaluated *standards* at the hospital and national levels, but very few have looked at the bedside *practice* of determination of death by neurologic criteria, which may be better (or worse) than that stipulated in standards. Although some variability in practice is likely acceptable—for example, there can be some nuance to evaluation technique—it is necessary to adhere to core minimum standards so that there are no erroneous determinations of death by neurologic criteria. This is truly one of the few areas in medicine where there is no room for error; if there is any doubt as to whether a patient is dead or not, clinicians must err on the conservative side, *not* declaring death until there is irrefutable and consistent evidence to support the determination. Efforts to reduce variability should continue in earnest, ensuring not only stringent, consistent standards at the institutional, national, and international levels but also sound practice at the bedside. These efforts will help to maintain the public's trust in the process, leading to fewer legal challenges and strife for families dealing with the tragic loss of a loved one.

References

1. Francoeur C, Weiss MJ, Macdonald JM, et al. Variability in pediatric brain death determination protocols in the United States. Neurology. 2021;97(3):e310–9.
2. Shemie SD. The pathophysiology of brain death and care of the potential organ donor. In Resuscitation and stabilization of the critically ill child. London: Springer; 2009. p. 1–13.
3. Varelas PN, Brady P, Rehman M, et al. Primary posterior fossa lesions and preserved supratentorial cerebral blood flow: implications for brain death determination. Neurocrit Care. 2017;27(3):407–14.
4. Beecher HK. A definition of irreversible coma: report of the Ad Hoc Committee of the Harvard Medical School to examine the definition of brain death. JAMA. 1968;205(6):337–40.
5. President's Commission for the Study of Ethical Problems in Medicine and Biomedical and Behavioral Research. Defining death: medical, legal, and ethical issues in the determination of death. Washington DC: U.S. Government Printing Office; 1981. p. 73.
6. Lewis A, Cahn-Fuller K, Caplan A. Shouldn't dead be dead?: the search for a uniform definition of death. J Law Med Ethics. 2017;45(1):112–28.
7. Wijdicks EF, Varelas PN, Gronseth GS, Greer DM. Evidence-based guideline update: determining brain death in adults: report of the Quality Standards Subcommittee of the American Academy of Neurology. Neurology. 2010;74(23):1911–8.
8. Nakagawa TA, Ashwal S, Mathur M, Mysore M. Society of Critical Care Medicine. Guidelines for the determination of brain death in infants and children: an update of the 1987 Task Force recommendations. Pediatrics. 2011;128(3):e720–40.

9. Lewis A. Contentious ethical and legal aspects of determination of brain death. Semin Neurol. 2018;38(05):576–82.
10. Son RG, Setta SM. Frequency of use of the religious exemption in New Jersey cases of determination of brain death. BMC Med Ethics. 2018;19(1):1–6.
11. Lewis A, Varelas P, Greer D. Prolonging support after brain death: when families ask for more. Neurocrit Care. 2016;24(3):481–7.
12. Russell JA, Epstein LG, Greer DM, et al. Brain death, the determination of brain death, and member guidance for brain death accommodation requests: AAN position statement. Neurology. 2019;92(5):228–32.
13. Lewis A, Greer D. Current controversies in brain death determination. Nat Rev Neurol. 2017;13(8):505–9.
14. Shewmon DA. Truly reconciling the case of Jahi McMath. Neurocrit Care. 2018;29(2):165–70.
15. Greer DM, Varelas PN, Haque S, Wijdicks EF. Variability of brain death determination guidelines in leading US neurologic institutions. Neurology. 2008;70(4):284–9.
16. Wang HH, Varelas PN, Henderson GV, Wijdicks EF, Greer DM. Improving uniformity in brain death determination policies over time. Neurology. 2017;88(6):562–8.
17. Greer DM, Wang HH, Robinson JD, et al. Variability of brain death policies in the United States. JAMA Neurol. 2016;73(2):213–8.
18. Lewis A, Bakkar A, Kreiger-Benson E, Kumpfbeck A, Liebman J, Shemie SD, et al. Determination of death by neurologic criteria around the world. Neurology. 2020;95(3):e299–309.
19. Lewis A, Liebman J, Kreiger-Benson E, Kumpfbeck A, Bakkar A, Shemie SD, et al. Ancillary testing for determination of death by neurologic criteria around the world. Neurocrit Care. 2021;34(2):473–84.
20. Citerio G, Crippa IA, Bronco A, Vargiolu A, Smith M. Variability in brain death determination in Europe: looking for a solution. Neurocrit Care. 2014;21(3):376–82.
21. Lewis A, Liebman J, Bakkar A, Kreiger-Benson E, Kumpfbeck A, Shemie SD, et al. Determination of brain death/death by neurologic criteria in countries in Asia and the Pacific. J Clin Neurol. 2020;16(3):480.
22. Lewis A, Kreiger-Benson E, Kumpfbeck A, et al. Determination of death by neurologic criteria in Latin American and Caribbean countries. Clin Neurol Neurosurg. 2020;197:105953.
23. Wahlster S, Wijdicks EF, Patel PV, et al. Brain death declaration: practices and perceptions worldwide. Neurology. 2015;84(18):1870–9.
24. Busl KM, Greer DM. Pitfalls in the diagnosis of brain death. Neurocrit Care. 2009;11(2):276–87.
25. Lewis A. The uniform determination of death act is being revised. Neurocrit Care. 2022;31:1–4.
26. Greer DM, Shemie SD, Lewis A, et al. Determination of brain death/death by neurologic criteria: the World Brain Death Project. JAMA. 2020;324(11):1078–97.
27. MacDougall BJ, Robinson JD, Kappus L, Sudikoff SN, Greer DM. Simulation-based training in brain death determination. Neurocrit Care. 2014;21(3):383–91.
28. Hocker S, Schumacher D, Mandrekar J, Wijdicks EF. Testing confounders in brain death determination: a new simulation model. Neurocrit Care. 2015;23(3):401–8.
29. NCS Courses. Brain death toolkit. Neurocritical Care Society; 2019. https://www.pathlms.com/ncs-ondemand/courses/1223. Accessed 10 Jan 2022.

Overcoming Challenges in the Determination of Death by Neurologic Criteria in Pediatric Patients

Matthew P. Kirschen and Wynne Morrison

1 Introduction

The Uniform Determination of Death Act is the legal foundation for declaration of death by neurologic criteria in infants and children in the United States. The Uniform Determination of Death Act states that "An individual who has sustained either (1) irreversible cessation of circulatory and respiratory functions, or (2) irreversible cessation of all functions of the entire brain, including the brain stem, is dead. A determination of death must be made in accordance with accepted medical standards [1]." The initial clinical standards for the diagnosis of death by neurologic criteria were provided by an ad hoc committee at Harvard Medical School [2]. This was followed by a Report of the Medical Consultants on the Diagnosis of Death to the Presidential Commission for the Study of Ethical Problems in Medicine and Biomedical and Behavioral Research in 1981 [3, 4]. However, these reports did not provide guidance on the determination of death by neurologic criteria in children younger than 5 years of age. In 1987, the American Academy of Pediatrics' Task Force for the Determination of Brain Death in Children provided the first standards for determination of death by neurologic criteria in children [5]. These standards were updated in 2011 by a multidisciplinary committee from the Society of Critical Care Medicine, the Section on Critical Care of the American Academy of Pediatrics, and the Child Neurology Society [6]. The goal of this update was to promote accuracy and uniformity in pediatric determination of death by neurologic criteria and documentation. More recently, the World Brain Death Project provided minimum criteria for determination of death by neurologic criteria in children that are largely concordant with the 2011 standard [7]. In the United States, the 2011 standard is the

M. P. Kirschen (✉) · W. Morrison
Departments of Anesthesiology and Critical Care Medicine, Neurology, and Pediatrics, Children's Hospital of Philadelphia, Perelman School of Medicine at the University of Pennsylvania, Philadelphia, PA, USA
e-mail: kirschenm@chop.edu; morrisonw@chop.edu

A. Lewis, J. L. Bernat (eds.), *Death Determination by Neurologic Criteria*, Advances in Neuroethics, https://doi.org/10.1007/978-3-031-15947-3_11

accepted medical standard for determination of death by neurologic criteria in children [6, 8].

Even in the 1987 pediatric standard, the authors recognized the challenges of determination of death by neurologic criteria in children, and infants in particular [5]. In this chapter, we review the epidemiology of death by neurologic criteria in children, clinical challenges in pediatric determination of death by neurologic criteria, and systems issues in pediatric determination of death by neurologic criteria including variability in standards between institutions, management of objections to death by neurologic criteria, and strategies to address catastrophic brain injury and death by neurologic criteria with families in a pediatric environment.

2 Epidemiology of Death by Neurologic Criteria in Pediatrics

Because of the need for mechanical ventilation, children evaluated for death by neurologic criteria are cared for exclusively in intensive care units (ICUs), primarily pediatric ICUs (PICUs), but occasionally neonatal ICUs (NICUs) or cardiac ICUs (CICUs). Neurologic injury or disease causing a poor neurologic prognosis is the most common contributor to death in pediatric ICUs, following either determination of death by neurologic criteria or withdrawal of technological support [9–11]. Thirteen to twenty-one percent of patients who die in PICUs in the United States are declared dead by neurologic criteria [9, 10, 12]. Similar percentages of pediatric patients are declared dead by neurologic criteria in other countries including Canada, China, United Kingdom, Spain, Turkey, Argentina, Saudi Arabia, and Brazil [11, 13–19]. The largest study on pediatric death by neurologic criteria in the United States examined a national multicenter database over a 5-year period and found that the most common etiologies of brain injury leading to death by neurologic criteria was hypoxic-ischemic injury from either cardiac arrest or shock and/or respiratory failure (which accounted for nearly two-thirds of cases) or trauma [12]. Nearly 85% of patients declared dead by neurologic criteria were neurologically normal prior to their inciting event. Interestingly, compared to patients declared dead by circulatory-respiratory criteria, patients declared dead by neurologic criteria were more commonly male, between the ages of 2 and 12 years, and African American [12].

3 Challenges in Determining Death by Neurologic Criteria in Children

The fundamental tenants of determination of death by neurologic criteria in children are the same as those in adults—patients must have coma, brainstem areflexia, and apnea after a catastrophic brain injury with a mechanism that is known to lead to death by neurologic criteria [6, 7, 20, 21]. Patients must be observed for a sufficient time to ensure that there has been no recovery of brain function, and reversible

	Adult[6]	Infants and children[7]
General principles	► All physicians making a determination of BD/DNC be intimately familiar with criteria and have demonstrated competence in this complex examination. ► Recommends standardised checklist	► Must be attending physician competent/qualified to perform evaluation. ► BD/DNC should not be determined for patients <36 weeks EGA ► Recommends standardised checklist
Prerequisites/ confounders	► Establish cause of coma ► Establish that brain injury is irreversible ► Exclude mimicking conditions ► Normal physiological parameters ► Observation period: 'Certain period of time has passed to exclude the possibility of recovery (usually several hours)' ► Drug intoxication* (wait five half-lives) ► Temp >36°C	► Establish cause of coma ► Establish that brain injury is irreversible ► Exclude mimicking conditions ► Normal physiological parameters for age ► Observation period: 'Consider deferring BD evaluation for 24–48 hours after resuscitation' ► Drug intoxication* (wait several half-lives—tables for elimination half-life provided) ► Temp >35°C
Neurological examination	► Number of examinations: 1 ► Observation period: none	► Number of examinations: 2 ► Observation period 12 hours (>30 days) ► Observation period 24 hours (37 weeks EGA to 30 days) ► Two different attending evaluators
Apnoea testing	► Number of apnoea tests: 1 ► Criteria: no respiratory effort, $PaCO_2$ ≥60 or ≥20 rise from baseline	► Number of apnoea tests: 2 ► Criteria: no respiratory effort, $PaCO_2$ ≥60 or ≥20 rise from baseline
Ancillary testing	► Acceptable reasons to use ancillary testing – Uncertainty exists about the reliability of parts of the neurological examination – Apnoea test cannot be performed ► Acceptable tests: angiography, EEG, radionucleotide cerebral perfusion scan	► Acceptable reasons to use ancillary testing – Components of the examination or apnoea testing cannot be completed safely due to patient's underlying medical condition – Uncertainty about the results of the neurological examination – Medication effect may be present – Reduce inter-examination observation period – May be helpful for social reasons allowing family members to better comprehend the diagnosis of BD/DNC ► Acceptable tests: angiography, EEG, radionucleotide cerebral perfusion scan
Death declaration	► Time of death: time of blood gas with appropriately elevated CO_2 or time ancillary test results	► Time of death: not specified

*Drug intoxication includes any medications that can suppress the central nervous system. Both adult and paediatric guidelines state that clearance of drugs should be based on total amount of medication administered, elimination half-life of the drug and any active metabolites, presence of organ (eg, hepatic or renal) dysfunction, and if applicable, the age of the child. Additionally, they caution that hypothermia may alter drug metabolism and clearance.

Fig. 1 Adult and pediatric standards for determination of death by neurologic criteria in the United States. Reproduced with permission from: Kirschen MP, Lewis A, Rubin M, Kurtz P, Greer DM. New perspectives on brain death. J Neurol Neurosurg Psychiatry. 2021;92(3):255–62

mimics of death by neurologic criteria and confounders (e.g., hypotension, hypothermia, severe metabolic derangements) have been excluded. The accepted pediatric standards reflect a conservative approach to confirming death by neurologic criteria in children, recognizing that the pediatric clinical exam may be more easily confounded than the adult exam (Fig. 1). Standards for determination of death by neurologic criteria in children vary throughout the world including whether separate pediatric standards exist, qualifications of examiners, and aspects of the evaluation that differ based on age [22, 23].

3.1 Qualifications of Physicians to Evaluate Children for Death by Neurologic Criteria

One of the main challenges to pediatric determination of death by neurologic criteria is the infrequent nature with which evaluations for death by neurologic criteria are performed. There is a linear association between PICU size and number of patients declared dead by neurologic criteria each year [12]. Larger PICUs, those with 2000–4000 patients per year, declare a median of 10 patients dead by neurologic criteria per year, whereas PICUs with fewer than 500 patients per year only declare a median of 1 patient dead by neurologic criteria per year. Thus, it is possible that a given physician may only perform an evaluation for death by neurologic criteria once every few years. Similarly, there are limited learning opportunities for trainees to gain experience and expertise in the evaluation for death by neurologic criteria. While evaluations for death by neurologic criteria are most commonly performed by pediatric intensivists and neurologists, per the accepted standards, potentially qualified clinicians also include neonatologists, pediatric neurosurgeons, and pediatric trauma surgeons if they have sufficient education about death by neurologic criteria and experience in their training or practice [6].

The rarity of the procedure underscores the need for residency and fellowship programs, as well as continuing education programs for practicing physicians, to incorporate education on death by neurologic criteria into their curricula, and for institutions to create comprehensive protocols and checklists. Some online educational platforms include sections on pediatric declaration of death by neurologic criteria like OPENPediatrics (www.openpediatrics.org) and the Neurocritical Care Society's Brain Death Determination Course (www.neurocriticalcare.org/education/braindeath) [24].

3.2 Challenges to Performing the Neurologic Examination for Determination of Death by Neurologic Criteria in Children

While the essence of the evaluation for death by neurologic criteria is similar in adults and children, several special considerations must be taken into account for children [6, 20, 25–27]. First and foremost, in order to determine that the brain injury is irreversible without the possibility of recovery, the child must be observed for a period of time with serial examinations to evaluate for evidence of brain function. Based on pathophysiology that occurs after brain injury, and after hypoxic-ischemic brain injury in particular, many experts feel that longer periods of observation after brain injury are prudent in children. In infants and young children with open fontanelles and unfused sutures, the distensibility of the skull and dura may reduce intracranial hypertension and prevent herniation syndromes. Additionally, the brainstem in infants can be more resistant to hypoxic-ischemic brain injury and may show some residual function after cerebral edema subsides [28–30]. Older children and adolescents can experience similar physiology to adults after catastrophic brain injury, where the intracranial pressure exceeds the mean arterial blood pressure causing brain circulatory arrest.

Thus, the guidelines recommend waiting at least 24 h after birth (for infants born after 37 weeks gestational age), following cardiopulmonary resuscitation, or after other severe brain injury prior to initiating the evaluation for determination of death by neurologic criteria. Erring on the conservative side, many physicians will observe children for longer than 24 h. This additional time also allows for recovery of other potentially injured organs (e.g., heart, lungs, kidneys, liver), clearance of medications which can confound the examination, and discussions with families about the nature, severity, and implications of the brain injury. The accepted standards recommend that two physicians each perform a neurologic examination separated by an additional age-dependent observation period (24 h for newborns 37 weeks gestational age to 30 days old, and 12 h for patients 31 days to 18 years old), again to monitor for recovery of neurologic function [6]. Evaluations for death by neurologic criteria are not performed in newborns of less than 37 weeks corrected gestational age because some brainstem reflexes may not be completely developed so the neurologic examination may be less reliable.

It is essential to meticulously ensure that the evaluation for death by neurologic criteria is not confounded by factors that suppress brain function. These include hypotension, hypothermia, and severe metabolic disturbances; recommended thresholds are provided in the accepted standards [6, 7, 25, 31]. Clearance of medications that can suppress brain function or confound the evaluation should be confirmed by either checking levels or waiting at least five half-lives. Age-dependent metabolism of certain drugs, as well as renal and hepatic dysfunction, should be considered when determining the appropriate observation period after medication administration or stopping sedative infusions. Additionally, hypothermia, either from exposure or induced as part of targeted temperature management, can impact clearance of medications. Pediatric pharmacists can provide essential guidance in this regard.

The evaluation should be performed or directly supervised by an attending physician with training and competency in the pediatric evaluation for death by neurologic criteria. Most components of the evaluation are identical between adults and children. Additional components include determining the absence of sucking or rooting reflexes in children less than 6 months of age. Automated pupillometers can be used to augment the examination, but these devices are not validated specifically for the evaluation for death by neurologic criteria in children [32]. The pupillary border may not be sufficiently formed to obtain an accurate measurement in children younger than 6 months of age. Spinally mediated reflexes of the head/face, neck, trunk, and extremities can be seen in children and can often be challenging to distinguish from volitional or brain-mediated movements [7, 33, 34]. If it is unclear if a movement is spinal-mediated or not, providers can consult physicians with additional expertise or perform an ancillary test, presuming all other components of the evaluation for death by neurologic criteria including the apnea test are consistent with death by neurologic criteria.

3.3 Challenges to Conducting an Apnea Test for Determination of Death by Neurologic Criteria in Children

The apnea test is an essential part of the evaluation for death by neurologic criteria since it interrogates the respiratory control centers in the medulla oblongata of the lower brainstem [6, 7, 35]. It involves removal of the patient from intermittent mandatory mechanical ventilation and observation for spontaneous respirations while the patient becomes more hypercarbic and acidotic, both triggers for respiration. With disconnection from the ventilator, there is a drop in airway pressure which can result in atelectasis and desaturation [36]. Other possible adverse events during the apnea test include hypotension, arrhythmia, pneumothorax, and cardiac arrest [35]. Several measures can be taken to avoid these complications. First, a pediatric intensivist should determine the patient's risk level for cardiopulmonary decompensation during the apnea test. If the risk level is too high, the apnea test can be deferred until it can be performed with less risk, or the evaluation for death by neurologic criteria can proceed with the addition of an ancillary test. In one single center study, patients who had an apnea testing deferred given concern for cardiopulmonary instability had higher mean arterial pressure, positive end expiratory pressure (PEEP), fraction of inspired oxygen (FiO2), and oxygenation index than patients in whom apnea testing was completed [37].

Second, apneic oxygenation should be used to reduce the risk of desaturation and hypoxemia. In children, apneic oxygenation should include the application of continuous positive airway pressure (CPAP) using 100% FiO_2 at the same PEEP the patient required on the ventilator prior to the apnea test [6, 37, 38]. CPAP can be provided with a flow inflating resuscitation bag with functioning PEEP valve, T-piece with functioning PEEP valve, or mechanical ventilator in CPAP mode. CPAP is preferred to tracheal insufflation in children since maintaining PEEP can prevent atelectasis and hypoxemia. Tracheal insufflation can lead to pneumothoraces from air trapping if the oxygen cannula is too large relative to diameter of the endotracheal tube or from trauma. Tracheal insufflation can also lead to CO_2 washout, which could prolong the apnea test or prevent an adequate rise in $PaCO_2$.

Third, physicians conducting the apnea test should follow a protocol that includes pre-oxygenation and be prepared to manage any potential complications including hypoxemia and hypotension [6, 37, 38]. If these steps are followed, the apnea test complication rate is low. In one study of more than 120 apnea tests in children, hypotension was reported during 6% of tests, but improved after titration of vasopressors or with resumption of ventilation after apnea test completion in nearly all of these patients. Fewer than 1% of apnea tests were terminated early due to hypotension and no apnea tests were terminated early for hypoxemia [37]. There has been concern that increases in $PaCO_2$ during apnea testing could result in cerebral vasodilation and further increases in intracranial pressure which could theoretically lead to secondary brain injury [39]. A study in adults failed to demonstrate any increases in intracranial pressure during apnea testing [40].

3.4 Ancillary Testing in Children

Ancillary tests are not required to make the determination of death by neurologic criteria in children but can be used in selected circumstances to provide physicians with supporting evidence of the severity and irreversibility of brain injury. According to the accepted standards, ancillary testing can be used to support the determination of death by neurologic criteria when: components of the evaluation or apnea test cannot be completed safely because of the patient's underlying medical condition; if there is uncertainty about the results of the neurologic evaluation; when a medication effect may be present; or to shorten the duration of the observation period between the two neurologic evaluations [6]. The accepted standards additionally state, "ancillary studies may also be helpful for social reasons allowing family members to better comprehend the diagnosis of brain death [6]." In clinical practice, physicians perform ancillary testing for a variety of indications, some that deviate from the recommendations in the accepted standards [41]. A single-center retrospective study from a large children's hospital found that nearly half of the children evaluated for determination of death by neurologic criteria underwent an ancillary test, primarily for the inability to perform or interpret components of the neurologic evaluation due to a confounding condition or injury [42].

Permissible ancillary tests in children include cerebral angiography, electroencephalography (EEG), and radionuclide cerebral blood flow studies [6, 43–45]. The ideal ancillary study does not exist; each test has strengths and limitations that physicians must consider [7, 46, 47]. Particularly for infants, data are limited, and tests may be inaccurate leading to either false positive or false negative results. For this reason, the Royal College of Paediatrics and Child Health recommended that for infants less than 2 months old "in cases where a clinical diagnosis of death by neurologic criteria is not possible, ancillary tests are not sufficiently robust to help confidently diagnose death by neurologic criteria in infants. [48]" Ancillary testing is mandated in about 15% of institutional standards in the United States, primarily in children less than 1-year-old [49].

Among the reasons physicians use ancillary tests beyond the recommendations in the accepted standards are personal preference and to convince families who object to determination of death by neurologic criteria that a patient is truly dead [41]. Some families have difficulty comprehending the concept of death by neurologic criteria and experience cognitive dissonance when trying to understand why their child is dead now when their physical appearance is exactly the same as it was an hour ago, with similar numbers on the cardiopulmonary monitor, skin that is warm to the touch, and bodily functions that work. In these situations, a picture may be worth a thousand words, and showing a family a nuclear medicine scan with lack of perfusion to the brain may help them comprehend the severity and irreversibility of the brain injury, and finality of the determination of death by neurologic criteria. However, given that ancillary tests could show artifact or residual electrical cerebral activity even in patients with no functional neurologic responses, ancillary tests should be used with extreme caution in these situations. Physicians who use ancillary tests outside of the indications recommended in the accepted standards should have a clearly articulated plan for next steps if the test is inconsistent with death by neurologic criteria.

4 Systems Issues in Pediatric Death by Neurologic Criteria Determination

4.1 Variability in Institutional Pediatric Standards on Determination of Death by Neurologic Criteria

After publication of the original 1987 pediatric standards on determination of death by neurologic criteria, studies demonstrated variability in determination of death by neurologic criteria practices [5, 50–52]. The goal of the 2011 update to the pediatric standards on determination of death by neurologic criteria was to achieve better uniformity in practice throughout the United States. Despite this, 10 years after their publication, a 2021 study found that among 130 pediatric hospitals, there was variability in institutional standards on determination of death by neurologic criteria with respect to prerequisites, apnea testing, and ancillary testing [49]. Some variability is understandable if institutions choose to be more conservative than the accepted standards, but it is concerning that 9% of hospital standards had not been updated since the publication of the 2011 standards. Of the 118 institutional standards published after the 2011 standards, most were concordant with the accepted standards in terms of identification of a mechanism of brain injury, prerequisites, and the neurologic examination. Areas of discordance were observation periods, $PaCO_2$ targets during the apnea test, and some aspects of ancillary testing [49].

Some protocols were proscriptive in how the evaluation for death by neurologic criteria should be conducted, while others stated the general components required for determination of death by neurologic criteria and individual physicians were responsible for operationalizing them. Computerized templates and notes within the electronic health record improve adherence to the accepted standards for determination of death by neurologic criteria and documentation practices [53, 54]. Given the low frequency and high stakes of determination of death by neurologic criteria in children, comprehensive institutional standards that adhere to the accepted standards and are explicit about procedural details, may help improve the accuracy and consistency of determination of death by neurologic criteria. These standards can provide reminders in domains where common errors are made and give guidance for management of challenging aspects of the evaluation like spinal reflexes and metabolic disturbances. Future work, guided by the principles of implementation science, is needed to ensure alignment between pediatric institutional standards for determination of death by neurologic criteria and nationally accepted standards.

4.2 Objections to Determination of Death by Neurologic Criteria

Up to 60% of pediatric neurologists and intensivists who evaluate children for death by neurologic criteria reported that they had been asked at least once to continue organ support for a patient declared dead by neurologic criteria outside of organ donation [55, 56]. The most common reason for this request was that families believed their child could regain neurologic function [55]. These requests are likely more frequent in children than adults. A survey of adult neurologists found that approximately 50% of respondents had been involved in situations in which families requested continuation of organ support after determination of death by neurologic criteria [57].

A review of determinations of death by neurologic criteria at 145 children's hospitals over a period of 5 years found that among 1500 children declared dead by neurologic criteria who were not organ donors, the bodies of 2.7% were physically present in the ICU more than 48 h after the declaration of death, and 0.7% were physically present in the ICU more than 5 days after the declaration [12]. Taken together, these studies suggest that while requests for continuation of organ support after declaration of death by neurologic criteria may be common, physicians, working in concert with consultants from ethics, legal, and hospital administration, rarely continue organ support for a prolonged time period. Presumably, many families are supported through grief or denial in these circumstances to a place of better understanding and acceptance.

Rare situations where families contest either the initiation of the evaluation for determination of death by neurologic criteria or removal of organ support after declaration of death can be extraordinarily challenging for the entire medical team and the institution. These cases are occasionally prolonged and can involve legal action and the judicial system. Such objections are discussed more fully in other chapters in this text. We believe that a focus on proactive, supportive communication can frequently avoid intractable disputes, and offer suggestions for communication practices with a particular focus on the pediatric setting (Fig. 2) [55].

In the clinical care of pediatric patients, a shared decision-making approach is typically ideal, in which the goals, values, and preferences of a patient or family are explored and considered prior to making medical recommendations for interventions that have a reasonable chance of achieving those goals [58]. In the case of determination of death by neurologic criteria, a goal of survival or neurologic recovery is unfortunately not achievable. The finality of the determination can therefore justify a more directive approach on the part of the medical team [59]. The steps of the determination and declaration of death by neurologic criteria, and the subsequent removal of organ support, should be presented to a family as the appropriate next steps in the medical care of the patient, rather than being presented as choices that the family has to consider. There is a duty to inform and support the family, but not to seek their explicit consent, particularly if doing so leads them to feel responsible for allowing their child to die [60].

Support of the family requires recognizing that emotional distress is to be expected in the circumstances. Emotion can be expressed as sadness, anger, disbelief, guilt, or many other responses. Recognizing the emotion (e.g., "Any family would be overwhelmed") and using aligning language (e.g., "I wish I had different news for you" or "We've done everything we can to try to save him. You fought for him so hard") can help a family feel heard. When conducting the evaluation for death by neurologic criteria, it may be useful to let the family observe so that they are aware of the protocolized approach and see the lack of response for themselves.

What	When	Who	Where	How
Explain the concept of DNC to the family in terms they can understand. Explain that DNC can result from devastating and nonrecoverable brain injury incompatible with return of consciousness Explain that the goals of the DNC evaluation are to meticulously look for any signs of brain function. Review the components of the evaluation and examination, explaining that multiple exams will occur over time and additional testing may be indicated. Explain that if no signs of brain function are found, the child will be declared dead. Explain that machines that support the organs (i.e., the ventilator) will be removed after the child is declared dead.	Begin to introduce the concept of DNC soon after it is apparent that a child has sustained a severe brain injury. Give families time to accept and understand the gravity and finality of the situation. Inform families when a DNC evaluation will occur so they can be present if they wish Coordinate with families if they request religious rituals be performed for their child during the DNC evaluation process. Avoid introducing the concept or possibility of organ donation until a conversation has been had with a local organ procurement organization representative about the best time and personnel to discuss with the family.	It is often helpful to have a physician with a therapeutic relationship with the family, who also has knowledge and experience regarding the DNC process, introduce the concept of DNC. Consultants who are involved solely for the DNC evaluation should introduce themselves to the family and explain the purpose for their involvement. Ensure that the entire medical team (physicians, nurses, therapists, social workers, case managers, etc) understands the implications of DNC so that consistent information is provided to the family. Consider involving the ethics committee, palliative care team, social workers, psychologists, and chaplaincy to ensure appropriate support for the medical team and the family. These team members can help families come to terms with the meaning of DNC. Ensure that the physicians doing the DNC evaluation are competent to perform the examination and knowledgeable about institutional policies and local and state laws.	Ask families where they prefer to have conversations about prognosis and the DNC process. Some families prefer to stay at the bedside while others may request a private conference room. DNC evaluation and examination should be performed in the ICU. Removal of organ support typically occurs in the ICU, although in some circumstances, can occur in other locations if requested by the family and is in accordance with institutional policies. Organ support is removed in the operating room for children who are organ donors.	Provide education and frequent reeducation about DNC to medical team members and families. Communicate candidly and patiently with families. The DNC examination should be performed without interruption and in accordance with insitutional policy and national guidelines. Allow families to observe the DNC examination and explain findings in accessible terms. Explain the possibility that spinal reflexes may occur in patients who are DNC. Reassure families that the medical team will continue to provide full medical care to their child during the DNC evaluation and will only stop if and when the child is declared dead. Use ther term "death" not "brain death" to avoid confusion between brain death and cardiopulmonary death. Avoid terms like "life-support" when referring to the ventilator after DNC declared.

DNC = death by neurologic criteria.

Fig. 2 Communication with families about death by neurologic criteria. Reproduced with permission from: Lewis A, Adams N, Chopra A, Kirschen MP. Organ Support After Death by Neurologic Criteria in Pediatric Patients. Crit Care Med. 2017;45(9):e916-e24

Phrasing the process as one where you are looking for any signs of life, rather than trying to confirm death, lets them know that it is not an overly hasty or biased evaluation. Listening to doubts and normalizing such responses (e.g., "Many families find it hard to believe") can help build trust. While such an approach may not prevent all objections, it can hopefully help a family feel that they are in partnership with the medical team.

5 Conclusions

Death by neurologic criteria in children is tragic. It typically occurs to previously healthy children after catastrophic brain injury from either hypoxic-ischemic brain injury or trauma. The evaluation for determination of death by neurologic criteria in children can be challenging, but careful planning and adherence to the accepted standards can help avoid pitfalls and false positive determinations. Proactive, supportive communication can help avoid situations where emotion leads to objections to determination of death by neurologic criteria.

References

1. Uniform Determination of Death Act. 12 uniform laws annotated 589 (West 1993 and West suppl 1997).
2. A definition of irreversible coma. Report of the Ad Hoc Committee of the Harvard Medical School to Examine the Definition of Brain Death. Jama 1968;205(6):337–40.
3. Guidelines for the determination of death. Report of the medical consultants on the diagnosis of death to the President's Commission for the Study of Ethical Problems in Medicine and Biomedical and Behavioral Research. Crit Care Med. 1982;10(1):62–4.
4. Banasiak KJ, Lister G. Brain death in children. Curr Opin Pediatr. 2003;15(3):288–93.
5. Report of special Task Force. Guidelines for the determination of brain death in children. American Academy of Pediatrics Task Force on Brain Death in Children. Pediatrics. 1987;80(2):298–300.
6. Nakagawa TA, Ashwal S, Mathur M, et al. Clinical report-Guidelines for the determination of brain death in infants and children: an update of the 1987 task force recommendations. Pediatrics. 2011;128(3):e720–40.
7. Greer DM, Shemie SD, Lewis A, et al. Determination of brain death/death by neurologic criteria: the World Brain Death Project. JAMA. 2020;324(11):1078–97.
8. Lewis A, Bernat JL, Blosser S, et al. An interdisciplinary response to contemporary concerns about brain death determination. Neurology. 2018;90(9):423–6.
9. Burns JP, Sellers DE, Meyer EC, Lewis-Newby M, Truog RD. Epidemiology of death in the PICU at five U.S. teaching hospitals. Crit Care Med. 2014;42(9):2101–8.
10. Meert KL, Keele L, Morrison W, et al. End-of-life practices among tertiary care PICUs in the United States: a multicenter study. Pediatr Crit Care Med. 2015;16(7):e231–8.
11. Sands R, Manning JC, Vyas H, Rashid A. Characteristics of deaths in paediatric intensive care: a 10-year study. Nurs Crit Care. 2009;14(5):235–40.
12. Kirschen MP, Francoeur C, Murphy M, et al. Epidemiology of brain death in pediatric intensive care units in the United States. JAMA Pediatr. 2019;173(5):469–76.
13. Lago PM, Piva J, Garcia PC, et al. End-of-life practices in seven Brazilian pediatric intensive care units. Pediatr Crit Care Med. 2008;9(1):26–31.
14. Agra-Tunas C, Rodriguez-Ruiz E, Rodriguez Merino E. MOdos de Morir en UCI Pediátrica-2 (MOMUCIP-2) study group of the Spanish Society of Paediatric Intensive Care (SECIP). How do children die in PICUs nowadays? A multicenter study from Spain. Pediatr Crit Care Med. 2020;21(9):e610–e6.
15. Yener N, Paksu MS, Koksoy O. Brain death in children: incidence, donation rates, and the occurrence of central diabetes insipidus. J Crit Care Med (Targu Mures). 2018;4(1):12–6.
16. Bonetto G, Taffarel P, Gamerman M, et al. Brain death and organ donation in Argentine pediatric intensive care units. A multicenter study. Arch Argent Pediatr. 2018;116(1):e54–60.
17. Parker BL, Frewen TC, Levin SD, et al. Declaring pediatric brain death: current practice in a Canadian pediatric critical care unit. CMAJ. 1995;153(7):909–16.

18. Hon KL, Tse TT, Au CC, et al. Brain death in children: a retrospective review of patients at a paediatric intensive care unit. Hong Kong Med J. 2020;26(2):120–6.
19. Al-Eyadhy A, Temsah MH, Hasan GM, et al. Causes, timing, and modes of death in a tertiary pediatric intensive care unit: five years' experience. Saudi Med J. 2021;42(11):1186–94.
20. Mathur M, Ashwal S. Pediatric brain death determination. Semin Neurol. 2015;35(2):116–24.
21. Wijdicks EF, Varelas PN, Gronseth GS, Greer DM. Evidence-based guideline update: determining brain death in adults: report of the Quality Standards Subcommittee of the American Academy of Neurology. Neurology 2010;74(23):1911–8.
22. Wahlster S, Wijdicks EF, Patel PV, et al. Brain death declaration: practices and perceptions worldwide. Neurology. 2015;84(18):1870–9.
23. Lewis A, Bakkar A, Kreiger-Benson E, et al. Determination of death by neurologic criteria around the world. Neurology. 2020;95(3):e299–309.
24. Rubin MA, Kirschen MP, Lewis A. The Neurocritical Care Brain Death Determination Course: purpose, design, and early findings. Neurocrit Care. 2021;35(3):913–5.
25. Lewis A, Kirschen MP. Brain death/death by neurologic criteria determination. Continuum (Minneap Minn). 2021;27(5):1444–64.
26. Nakagawa T, Ashwal S, Mathur M, Mysore M. Brain death in children: why does it have to be so complicated? Ann Neurol. 2012;72(2):300.
27. Kirschen MP, Lewis A, Rubin M, Kurtz P, Greer DM. New perspectives on brain death. J Neurol Neurosurg Psychiatry. 2021;92(3):255–62.
28. Sieber FE, Palmon SC, Traystman RJ, Martin LJ. Global incomplete cerebral ischemia produces predominantly cortical neuronal injury. Stroke. 1995;26(11):2091–5. discussion 6
29. Martin LJ, Brambrink A, Koehler RC, Traystman RJ. Primary sensory and forebrain motor systems in the newborn brain are preferentially damaged by hypoxia-ischemia. J Comp Neurol. 1997;377(2):262–85.
30. Smith ML, Auer RN, Siesjo BK. The density and distribution of ischemic brain injury in the rat following 2-10 min of forebrain ischemia. Acta Neuropathol. 1984;64(4):319–32.
31. Lerner DP, Bassil R, Tadevosyan A, et al. Metabolic values precluding clinical death by neurologic criteria/brain death: survey of neurocritical care society physicians. J Clin Neurosci. 2021;88:16–21.
32. Olgun G, Newey CR, Ardelt A. Pupillometry in brain death: differences in pupillary diameter between paediatric and adult subjects. Neurol Res. 2015;37(11):945–50.
33. Saposnik G, Bueri JA, Maurino J, Saizar R, Garretto NS. Spontaneous and reflex movements in brain death. Neurology. 2000;54(1):221–3.
34. Saposnik G, Basile VS, Young GB. Movements in brain death: a systematic review. Can J Neurol Sci. 2009;36(2):154–60.
35. Busl KM, Lewis A, Varelas PN. Apnea testing for the determination of brain death: a systematic scoping review. Neurocrit Care. 2020.
36. Paret G, Barzilay Z. Apnea testing in suspected brain dead children—physiological and mathematical modelling. Intensive Care Med. 1995;21(3):247–52.
37. Puccetti DF, Morrison W, Francoeur C, Mai M, Kirschen MP. Apnea testing using continuous positive airway pressure when determining death by neurologic criteria in children: retrospective analysis of potential adverse events. Pediatr Crit Care Med. 2020.
38. Kirschen MP, McGowan N, Topjian A. Brain death evaluation in children with suspected or confirmed coronavirus disease 2019. Pediatr Crit Care Med. 2021;22(3):318–22.
39. Tibballs J. A critique of the apneic oxygenation test for the diagnosis of "brain death". Pediatr Crit Care Med. 2010;11(4):475–8.
40. Salih F, Hoffmann O, Brandt SA, et al. Safety of apnea testing for the diagnosis of brain death: a comprehensive study on neuromonitoring data and blood gas analysis. Eur J Neurol. 2019;26(6):887–92.
41. Lewis A, Adams N, Chopra A, Kirschen MP. Use of ancillary tests when determining brain death in pediatric patients in the United States. J Child Neurol. 2017;32(12):975–80.
42. Kirschen MP, Puccetti DF, Morrison W. Incidence and indications for ancillary testing in the determination of death by neurological criteria in children. Pediatr Neurol. 2020;106:68–9.

43. Henderson N, McDonald MJ. Ancillary studies in evaluating pediatric brain death. J Pediatr Intensive Care. 2017;6(4):234–9.
44. Heran MK, Heran NS, Shemie SD. A review of ancillary tests in evaluating brain death. Can J Neurol Sci. 2008;35(4):409–19.
45. Zuckier LS. Radionuclide evaluation of brain death in the post-McMath era. J Nucl Med. 2016;57(10):1560–8.
46. Young GB, Lee D. A critique of ancillary tests for brain death. Neurocrit Care. 2004;1(4):499–508.
47. Wijdicks EF. The case against confirmatory tests for determining brain death in adults. Neurology. 2010;75(1):77–83.
48. Royal College of Paediatrics and Child Health. The diagnosis of death by neurological criteria in infants less than two months old. 2015.
49. Francoeur C, Weiss MJ, Macdonald JM, et al. Variability in pediatric brain death determination protocols in the United States. Neurology. 2021.
50. Mathur M, Petersen L, Stadtler M, et al. Variability in pediatric brain death determination and documentation in southern California. Pediatrics. 2008;121(5):988–93.
51. Mejia RE, Pollack MM. Variability in brain death determination practices in children. JAMA. 1995;274(7):550–3.
52. Chang MY, McBride LA, Ferguson MA. Variability in brain death declaration practices in pediatric head trauma patients. Pediatr Neurosurg. 2003;39(1):7–9.
53. Stockwell JA, Pham N, Fortenberry JD. Impact of a computerized note template/checklist on documented adherence to institutional criteria for determination of neurologic death in a pediatric intensive care unit. Pediatr Crit Care Med. 2011;12(3):271–6.
54. Krawiec C, Ceneviva GD, Thomas NJ. Assessing and improving documentation of pediatric brain death determination within an electronic health record. Neuropediatrics. 2019;50(2):80–8.
55. Lewis A, Adams N, Chopra A, Kirschen MP. Organ support after death by neurologic criteria in pediatric patients. Crit Care Med. 2017;45(9):e916–e24.
56. Mataya L, Ross LF, Ghavam A, Paquette ET. Pediatric intensivist and pediatric neurologist perspectives and practices on death by neurologic criteria. J Clin Ethics. 2021;32(3):195–205.
57. Lewis A, Adams N, Varelas P, Greer D, Caplan A. Organ support after death by neurologic criteria: results of a survey of US neurologists. Neurology. 2016;87(8):827–34.
58. Kon AA, Davidson JE, Morrison W, Danis M, White DB, American College of Critical Care M, et al. Shared decision making in ICUs: an American College of Critical Care Medicine and American Thoracic Society Policy Statement. Crit Care Med. 2016;44(1):188–201.
59. Morrison W, Clark JD, Lewis-Newby M, Kon AA. Titrating clinician directiveness in serious pediatric illness. Pediatrics. 2018;142(Suppl 3):S178–S86.
60. Truog RD, Morrison W, Kirschen M. What should we do when families refuse testing for brain death? AMA J Ethics. 2020;22(12):E986–94.

Arguments in Favor of Requiring the Absence of Brain Circulation to Determine Death by Neurologic Criteria

Nathaniel M. Robbins

The Uniform Determination of Death Act (UDDA) defined death by neurologic criteria (i.e. brain death) as "irreversible cessation of all functions of the entire brain, including the brain stem … in accordance with accepted medical standards" [1] The American Academy of Neurology and other organizations have outlined these "accepted medical standards" [2, 3].

Recent scholarly, legal, and public discourse has highlighted controversy around the fact that these accepted medical standards are not sufficient to identify *irreversibility* or interrogate *all functions of the entire brain*, as stipulated in the UDDA. Aware of the mismatch between statutory definitions of death and accepted practices for determination of death, at least two solutions to address this mismatch have been proposed: amend the law, or change accepted practice [4, 5].

In this chapter, I discuss three arguments in favor of changing accepted medical standards, in order to require the absence of brain circulation to determine death by neurologic criteria. While it is possible that patients can have irreversible loss of brain function without loss of brain circulation, technical factors can make it challenging for physicians to determine irreversibility or loss of all functions of the entire brain. Mistakenly determining death is an unacceptable practice. One way to prevent false positives is to change accepted medical standards by mandating confirmation of the absence of brain circulation. This change in practice would align the determination of death by neurologic criteria with the definition of death codified in the UDDA, by demonstrating both *irreversibility* and cessation of *all functions of the entire brain*.

N. M. Robbins (✉)
Geisel School of Medicine at Dartmouth, Hanover, NH, USA

Dartmouth-Hitchcock Medical Center, Lebanon, NH, USA

Department of Neurology, One Medical Center Drive, Lebanon, NH, USA
e-mail: Nathaniel.M.Robbins@dartmouth.edu

A. Lewis, J. L. Bernat (eds.), *Death Determination by Neurologic Criteria*, Advances in Neuroethics, https://doi.org/10.1007/978-3-031-15947-3_12

1 What Are the Accepted Medical Standards for Determining Death by Neurologic Criteria?

In 2010, the American Academy of Neurology published updated guidelines for the determination of death by neurologic criteria in adults. These guidelines have been adopted by a number of other professional organizations and are widely considered to be the "accepted medical standard," as stipulated in the UDDA [6]. Guidelines have also been published for use in newborns, infants, and children [3]. It is worth noting, however, that these "accepted standards" are based on expert opinion and expert consensus, rather than empiric studies or evidence [2].

An international panel of experts recently affirmed these guidelines [7]. These experts list prerequisites for death by neurologic criteria, which include identification of "an established neurologic diagnosis that can lead to the complete and irreversible loss of all brain function" and confirmation that confounding conditions and mimickers are absent. They reaffirm that death by neurologic criteria can be determined when a bedside clinical exam demonstrates coma, brainstem areflexia, and inability to breathe spontaneously. They reinforce that ancillary tests should only be considered when the "clinical exam cannot be completed" [7], and emphasizing the clinical bedside exam for responsiveness, brainstem reflexes, and apnea is the gold standard.

1.1 Do the Accepted Medical Standards for Determination of Death by Neurologic Criteria Assess all Functions of the Entire Brain?

The current accepted medical standards ("standard brain death exam") for determining death by neurologic criteria, when done correctly, are accurate for identifying irreversible loss of brain function in the vast majority of cases—particularly cases in which high intracranial pressure leads to complete loss of brain circulation. However, the standard brain death exam assesses neither all functions of the brain nor irreversibility. This mismatch between accepted practices and legal definitions opens the possibility of false positives, in which a person can be determined to be dead by neurologic criteria according to accepted standards *without* losing all functions of the entire brain, and *without* irreversible loss of such functions. Rare cases of "false positive" brain death determination have been reported: despite the correct and complete use of accepted medical standards to determine death by neurologic criteria, the patient subsequently demonstrates brain functions [8–12]. Jahi McMath, for example, was maintained on organ support for years after a determination of death by neurologic criteria, and she subsequently underwent puberty and menarche [9, 11–13]. She may have also displayed autonomic reactivity and intermittent purposeful motor movements [9, 11, 12, 14].

The details of these exceptional cases have been debated, but the inability of the accepted standards to assess *all functions of the entire brain* is not debatable. The standard brain death exam assesses arousal, motoric brainstem reflexes, motoric limb responses, and respiratory drive. There are a number of other brain functions not contained on this list. For example, "super locked-in patients" who have completely destroyed brainstems and loss of efferent activity, might still have preserved afferent visual and olfactory pathways (which bypass the brainstem on the way to the thalamus and cortex, respectively) [4]. Also, while it is generally assumed that damage to the reticular activating system of the brainstem destroys consciousness, there is no way to empirically verify this [15, 16]. Certain neurohormonal and autonomic functions, controlled by the hypothalamus or higher cortical areas rostral to the brainstem, could also be preserved despite the absence of brainstem function on a bedside exam.

Proponents of the brainstem exam as accepted practice have responded that individuals can be dead despite preservation of discrete brain functions, such as neurohormonal and autonomic functions, because the "brain-as-a-whole" is no longer functioning [17–19]. Over the last few decades the meaning of the phrase "all functions of the entire brain" has been debated, but a general consensus has coalesced around the idea that the phrase should be interpreted as the functioning of the *brain-as-a-whole*, or the core function of the brain, and not the persistence of every single brain function [4, 18]. However, there remains at least two problems with this argument. First, the UDDA clearly states "all functions of the entire brain." The "brain-as-a-whole" is a reasonable interpretation of intent, but it does not follow the letter of the law. Second, it has not been possible to precisely define what functions constitute the "brain-as-a-whole." Which are the "core" functions or "critical" functions [4, 20, 21]? Is it self-preservation [22]? Somatic integration [21, 22]? Why are neurohormonal and autonomic functions excluded as important (i.e. core, critical) brain functions [9, 12]? Until the core functions that define the brain-as-a-whole are empirically verified or achieve universal consensus, debate will continue regarding the question of whether a person with a brain that is severely damaged, but not completely destroyed, is dead.

There remains uncertainty regarding how best to resolve the opposition in the meaning of death by neurologic criteria: does it require irreversible cessation of "all functions of the entire brain" (i.e. whole brain function) or merely the function of the brain-as-a-whole? Defenders of the accepted medical standards highlight the fact that no patients determined to be dead by neurologic criteria, applied properly, have ever regained meaningful neurologic function [6, 7]. However, this statement defends the prognostic accuracy of the accepted standards, and is not a comment on whether the standard exam accurately reflects the binary, biologic state of death. The philosophical conception of death, as conceived by the vast majority of the public, is more in line with the law than accepted medical practice. In other words, most people would agree that death clearly requires both irreversibility and loss of all functions, and neither alone is sufficient. For example, someone who suffers a massive stroke has irreversible loss of *some* brain functions, and someone with a massive sedative or paralytic overdose could have an exam that mimics brain death prior to recovery. However, neither individual is dead, which begs the question: how can we ensure that the accepted medical standards for determining death by neurologic criteria are 100% specific in identifying both the irreversibility and loss of all brain functions that define death?

1.2 Do the Accepted Medical Standards for Determination of Death by Neurologic Criteria Assess Irreversibility?

Irreversibility goes beyond prognosis and reflects the future, or what *will* occur. In contrast to *permanent loss of functions,* which means that "ceased functions will not recover because they will not restart spontaneously and no medical attempts will be made to restart them" [23], irreversibility refers to a biologic state of certainty: once functions have ceased, they "cannot restart spontaneously and cannot be restored by any available technology" [23].

Current practices for the determination of death by neurologic criteria require coma, brainstem areflexia, and loss of respiratory drive. However, it is not technically possible to distinguish irreversible neuronal death from quiescent, nonfunctional tissue, which can mimic irreversible destruction in every way. Inevitably, before brain cells die from hypoxic injury, they transition through a state of stunned hypoperfusion [4, 9]. The "ischemic penumbra" is well-recognized in stroke care, but is not sufficiently considered by proponents of the accepted standards for the determination of death by neurologic criteria [4, 9, 24, 25]. Individuals can suffer a global hypoxic injury to the brain, due to increased intracerebral pressure or another mechanism, but the degree of hypoxia and hypoperfusion may be insufficient to cause widespread neuronal death. In these cases, the individual would be pronounced dead by neurologic criteria according to accepted medical standards, but could later regain some functions if the quiescent brain tissue is supported sufficiently through the time of convalescence.

The accepted standards account for some of these scenarios by recommending that an "irreversible and proximate cause of coma" be established, and that "a certain period of time has passed since the onset of the brain insult to exclude the possibility of recovery (in practice, usually several hours)" [2]. However, the amount of time past which brain recovery is impossible is not currently known, and depends on many factors such as overall health of the patient, age, collateral circulation, degree of global ischemia, state of medical technology and available supportive capabilities, and other patient-level factors—many of which are not yet known. In addition, the requirement of an irreversible cause of coma does not clarify how to assess irreversibility.

In practice, there is no standardized wait time across countries and institutions to ensure lack of recovery, and there is no empirically derived waiting period that can ensure irreversibility. In spite of this, proponents of the accepted standards point out that the current criteria have excellent value for predicting lack of neurologic recovery [6, 7]. In other words, the current criteria are prognostically accurate. However, this line of defense is flawed for at least two reasons.

First, once a death determination is made, organ support is withdrawn; there have not been high quality natural history studies to determine the accuracy of the accepted standards. The relatively few cases with long-term observation periods actually suggest relatively poor prognostic accuracy [8, 9, 26]. Second, death (by circulatory-respiratory or neurologic criteria) requires that life *cannot* be restored (irreversibility), not just that it *will not* be restored (permanence). In other words, a determination of death precludes recovery by definition, so the diagnostic criteria must be formulated in such a way that recovery is impossible—not just improbable.

Currently there is no way to assess some of the core, critical functions that one may include in the functions of the brain-as-a-whole. The accepted practices for determination of death by neurologic criteria only assess responsiveness, brainstem reflexes, and respiratory drive. If those are absent, it is assumed that all functions of the entire brain are lost. However, as was discussed above, this does not assess all functions of the entire brain, and there is currently no definition of the functions that constitute the brain-as-a-whole. Furthermore, there are currently no validated tests to assess brain-as-a-whole functions when the brainstem has been damaged and motor activity is lost. Better tests are required to ensure irreversible cessation of all functions of the entire brain, as required by law. So how can death by neurologic criteria be better determined?

2 Argument 1: Only the Absence of Brain Circulation Can Ensure Loss of All Functions of the Entire Brain

One solution to better align accepted practice with law is to require the absence of brain circulation. Absence of brain circulation is incompatible with life, since all functions of the entire brain inevitably cease without perfusion. Experts in the determination of death by neurologic criteria assume that loss of responsiveness, brainstem reflexes, and respiratory drive reflect loss of brain circulation, but, as discussed above, this assumption can be misguided in cases of global hypoperfusion (i.e. a global ischemic penumbra) and hypofunctional (but not dead) brain, and in cases of direct brainstem injury. An ancillary test demonstrating the absence of brain circulation may better guarantee loss of all functions of the entire brain, if it could ensure the death of all neurons.

There remains a number of concerns with requiring the absence of brain circulation to determine death by neurologic criteria, however. First, tests of brain circulation are not universally available. Newer, more widely available tests such as CT and MR angiography are not as well validated as radionucleotide scanning [27]. Requiring perfusion imaging would preclude determination of death by neurologic criteria in most areas of the world that might not have access to newer technologies.

Second, these tests are not perfect. It is possible that neurons can survive below the threshold for detection of circulation [9, 11]. In other words, currently available tests might still be plagued by insufficient specificity for brain death, if they fail to distinguish a global state of hypoperfusion from true sustained and absent brain circulation. Another concern is that although the death of all neurons would inevitably result from sustained absence of intracerebral circulation, a test such as radionucleotide imaging can confirm absent or low flow only at one moment in time. In other words, perfusion imaging cannot confirm *sustained* absence of flow unless it is repeated several times. Even in cases such as global anoxia causing secondarily increased intracranial pressure with ensuing brain herniation, intracranial pressure would correct according to the Monroe-Kellie doctrine as brain contents herniate out of the intracranial space. Therefore, a test of brain circulation would need to be timed correctly to confirm absent circulation, before subsequent changes in tissue compliance and parenchymal movements (i.e. herniation) reduce intracranial pressure and restore perfusion. Until the thresholds and dynamics of neuronal death resulting from absent perfusion are better understood, a single test showing absent intracranial circulation will be plagued by the same issues that plague the clinical brain death examination, and may not guarantee irreversible loss of all function of the entire brain. At the least, further tests of validation are necessary before the currently available ancillary tests of brain circulation can be universally recommended to improve the determination of death by neurologic criteria.

Finally, it must be noted that requiring the absence of brain circulation could have unfortunate detrimental downstream effects. Patients who are dead by currently accepted medical standards, with no chance of meaningful neurologic recovery, could be considered alive on the basis of some minimal amount of preserved brain circulation, prolonging the uncertainty and suffering of grieving family members. There is also no way to know if these patients, who cannot communicate without efferent motor activity, might be suffering needlessly. Organ donor recipients might also suffer if the dead donor standard, which requires that organ donors be dead before their organs are retrieved, is not simultaneously reconsidered [4, 28–30].

3 Argument 2: Only the Absence of Brain Circulation Can Ensure Irreversible Loss of All Functions of the Entire Brain

Irreversibility is required by the UDDA. At the current time, only the absence of brain circulation can ensure irreversibility. A clinical bedside exam for brain death can confirm the loss of brain function at a given point in time (assuming the accepted standards sufficiently examine all functions of the entire brain). Care must be taken to ensure the absence of toxic-metabolic causes. Serial exams may be required to rule-out ongoing hypoperfusion or "shock" brain, which can mimic loss of all functions of the entire brain [4, 9, 27]. Even with serial exams, irreversibility could only be assured with sufficient wait periods, and the amount of time necessary to ensure that the absence of brain function reflects dead brain, not shocked brain, remains unknown [31]. Without this knowledge, the only way to ensure irreversible loss of function is to ensure that the neural networks are irreversibly interrupted or dead. In addition, as discussed below, judging reversibility depends on available technologies. As a result, absence of brain circulation, especially if sustained past a threshold amount of time, is the only way to ensure that neurons are dead, and that all brain functions are irreversibly lost. Relying on a clinical examination to assess irreversibility, done at a single point of time, without requiring the absence of brain circulation, creates an opportunity for false-positive declaration of death by neurologic criteria.

For example, Shewmon and others have described cases of "chronic brain death," in which systemic collapse did not occur for long periods well after correct determination of death by neurologic criteria [8, 11, 14]. The largest neuropathology studies have concluded similarly, finding large swaths of intact (i.e. nondissolved) brain even long after the determination of death by neurologic criteria [26]. Functional brain tissue does not necessarily mean preservation of functions of the brain-as-a-whole, since it is difficult to determine precisely how much brain, and which parts, demarcate life from death. But, confirming the absence of brain circulation, especially over an extended time, is probably the safest way to guarantee widespread death of neurons, thereby ensuring the irreversibility of death by neurologic criteria and avoiding an erroneous determination of death.

4 Argument 3: Requiring the Absence of Cerebral Circulation Better Aligns Death by Neurologic Criteria with Death by Circulatory-Respiratory Criteria

Currently there is widespread disagreement and confusion about death by neurologic criteria, both in the public and amongst physicians [32, 33]. Part of this confusion stems from the disconnect between the meaning of death, which is a binary biologic state, and the medical practice of death determination, which is nuanced and prone to error.

There is broad consensus among scientists, the public, policymakers, and religious scholars that death is an irreversible state; those who are dead cannot be brought back. This is easy to recognize in retrospect, but difficult to determine in real time. One obvious corollary to this is that there are not two kinds of death, but rather two ways to determine death: circulatory-respiratory and neurologic. Historically, these two methods of death determination have generally been held to different standards. Death by circulatory-respiratory criteria is currently identified by the permanent absence of circulation and respiration, not the irreversible absence [23, 34]. Without resuscitation (either auto-resuscitation or external resuscitation), there is no distinction: permanent cessation transitions to irreversible cessation. In the case of death by circulatory-respiratory criteria, it is not necessary to confirm irreversibility since it is a physiologic fact that the body cannot survive without circulation. For example, when a patient with a "do not resuscitate order" suffers an in-hospital asystolic arrest, an examining physician can pronounce death almost immediately after confirming the absence of heartbeat, respirations, pupillary response, and motor responsiveness. At this point, irreversibility has not been confirmed; circulation could possibly be restored with resuscitation or perhaps extracorporeal membrane oxygenation (ECMO). This technologic innovation, can facilitate circulatory-respiratory function and allows for an individual to be "conscious without a heartbeat or even a heart" [35]. More recently, scientists have restored some pig brain neuronal functions 4 h after decapitation [36]. While restored cellular activity may not meet the threshold of meaningful brain function, one can foresee a time in the future when more advanced technology is applied to humans, further blurring the boundaries of when lost brain function becomes irreversible. Regardless, at some point the threshold of irreversibility is crossed, and an individual is unmistakably dead. This irreversible, binary state of death is independent of whether the determination is done using circulatory-respiratory or neurologic criteria.

One way to align the two methods for determining death is to add a requirement that the determination of death by neurologic criteria require the absence of brain circulation. This would mirror the determination of death by circulatory-respiratory criteria in two ways. First, circulatory-respiratory arrest leads to absence of brain circulation, so the determination of death by circulatory-respiratory criteria already requires the absence of brain circulation. Second, requiring the absence of brain circulation to determine death by neurologic criteria removes the need to confirm irreversibility, since the absence of circulation ensures irreversibility.

Many countries (such as Switzerland) already consider brain death to be the sole criterion for death in organ donation [31]. This concept is increasingly adopted around the world: "the onset of cardiorespiratory arrest is merely a prospective predictor that irreversible loss of brain functions is inevitable unless the circulation and cerebral perfusion are restored" [37]. There is no need to confirm irreversible cessation of all functions of the entire brain after the determination of death by circulatory-respiratory criteria, since brain death inevitably follows loss of circulatory-respiratory function, with accompanying loss of brain circulation [31]. Requiring the absence of brain circulation to determine death by neurologic criteria would mirror the absence of systemic circulation (which includes brain circulation) required during the determination of death by circulatory-respiratory criteria. This alignment may help abate the confusion that results from two distinct methods to determine death, which are held to different standards of permanence and irreversibility, helping to garner support and consensus for the concept and practice of death by neurologic criteria.

5 Conclusions

The UDDA currently requires "irreversible cessation of all functions of the entire brain, including the brain stem … in accordance with accepted medical standards," but the accepted medical standards cannot adequately assess either irreversibility or all functions of the entire brain.

In this chapter, we discussed three arguments for revising the accepted medical standards to require the confirmation of the absence of brain circulation. Required absence of brain circulation would align accepted medical standards with the law and with consensus conceptualization of the binary biologic state of death, by ensuring both irreversibility and loss of all functions of the entire brain. There could be detrimental downstream effect consequences for organ donor recipients and grieving families, however.

A number of other solutions exist, such as amending the UDDA to focus on permanent cessation of brain function, and better aligning the criteria for death determination no matter what the cause [35, 38]. A full discussion of these alternatives is beyond the scope of this chapter, but covered in more detail elsewhere in this book.

References

1. Guidelines for the determination of death. Report of the medical consultants on the diagnosis of death to the President's Commission for the Study of Ethical Problems in Medicine and Biomedical and Behavioral Research. JAMA. 1981;246(19):2184–6.
2. Wijdicks EF, Varelas PN, Gronseth GS, Greer DM. Evidence-based guideline update: determining brain death in adults: report of the Quality Standards Subcommittee of the American Academy of Neurology. Neurology. 2010;74(23):1911–8.
3. Nakagawa TA, Ashwal S, Mathur M, Mysore M. Clinical report—guidelines for the determination of brain death in infants and children: an update of the 1987 Task Force Recommendations. Pediatrics. 2011;128(3):e720–40.
4. Robbins NM, Bernat JL. What should we do about the mismatch between legal criteria for death and how brain death is diagnosed? AMA J Ethics. 2020;22(12):E1038–46.
5. Bernat JL, Dalle Ave AL. Aligning the criterion and tests for brain death. Camb Q Healthc Ethics. 2019;28(4):635–41.
6. Russell JA, Epstein LG, Greer DM, et al. Brain death, the determination of brain death, and member guidance for brain death accommodation requests. AAN position statement Neurology. 2019;92(5):228–32.
7. Greer DM, Shemie SD, Lewis A, et al. Determination of brain death/death by neurologic criteria: the world brain death project. JAMA. 2020;324(11):1078–97.
8. Shewmon DA. Chronic "brain death": meta-analysis and conceptual consequences. Neurology. 1998;51(6):1538–45.
9. Shewmon DA. Truly reconciling the case of Jahi McMath. Neurocrit Care. 2018;29(2):165–70.
10. Shewmon DA. False-positive diagnosis of brain death following the pediatric guidelines: case report and discussion. J Child Neurol. 2017;32(14):1104–17.
11. Shewmon DA, Salamon N. The MRI of Jahi McMath and Its Implications for the Global Ischemic Penumbra Hypothesis. J Child Neurol. 2022;37(1):35–42.
12. Shewmon DA, Salamon N. The extraordinary case of Jahi McMath. Perspect Biol Med. 2021;64(4):457–78.
13. Shewmon DA. The case of Jahi McMath: a neurologist's view. Hastings Cent Rep. 2018;48(S4):S74–S6.
14. Machado C. Jahi McMath: a new state of disorder of consciousness. J Neurosurg Sci. 2021;65(2):211–3.
15. Machado C. Consciousness as a definition of death: its appeal and complexity. Clin Electroencephalogr. 1999;30(4):156–64.
16. Laureys S. Death, unconsciousness and the brain. Nat Rev Neurosci. 2005;6(11):899–909.
17. Bernat JL, Culver CM, Gert B. On the definition and criterion of death. Ann Intern Med. 1981;94(3):389–94.
18. Bernat JL. Refinements in the organism as a whole rationale for brain death. Linacre Q. 2019;86(4):347–58.
19. Russell JA. Author response: brain death, the determination of brain death, and member guidance for brain death accommodation requests: AAN position statement. Neurology. 2019;93(21):948.
20. Shewmon DA. The "critical organ" for the organism as a whole: lessons from the lowly spinal cord. Adv Exp Med Biol. 2004;550:23–42.
21. Shewmon DA. The brain and somatic integration: insights into the standard biological rationale for equating "brain death" with death. J Med Philos. 2001;26(5):457–78.

22. President's Commission for the Study of Ethical Problems in Medicine and Biomedical and Behavioral Research, Controversies in the Determination of Death: A White Paper by the President's Council on Bioethics. Washington, DC; 2008.
23. Dalle Ave AL, Bernat JL. Donation after brain circulation determination of death. BMC Med Ethics. 2017;18(1):15.
24. Coimbra CG. Implications of ischemic penumbra for the diagnosis of brain death. Braz J Med Biol Res. 1999;32(12):1479–87.
25. Astrup J, Siesjo BK, Symon L. Thresholds in cerebral ischemia—the ischemic penumbra. Stroke. 1981;12(6):723–5.
26. Wijdicks EF, Pfeifer EA. Neuropathology of brain death in the modern transplant era. Neurology. 2008;70(15):1234–7.
27. Robbins NM, Bernat JL. Practice current: when do you order ancillary tests to determine brain death? Neurology Clinical Practice. 2018;8(3):266–74.
28. Truog RD, Miller FG. The dead donor rule and organ transplantation. N Engl J Med. 2008;359(7):674–5.
29. Veatch RM. Killing by organ procurement: brain-based death and legal fictions. J Med Philos. 2015;40(3):289–311.
30. Truog RD, Robinson WM. Role of brain death and the dead-donor rule in the ethics of organ transplantation. Crit Care Med. 2003;31(9):2391–6.
31. Dalle Ave AL, Bernat JL. Using the brain criterion in organ donation after the circulatory determination of death. J Crit Care. 2016;33:114–8.
32. Braksick SA, Robinson CP, Gronseth GS, et al. Variability in reported physician practices for brain death determination. Neurology. 2019;92(9):e888–e94.
33. Miller FG, Nair-Collins M, Truog RD. It is time to abandon the dogma that brain death is biological death. Hastings Cent Rep. 2021;51(4):18–21.
34. Bernat JL. On irreversibility as a prerequisite for brain death determination. Adv Exp Med Biol. 2004;550:161–7.
35. Gardiner D, McGee A, Bernat JL. Permanent brain arrest as the sole criterion of death in systemic circulatory arrest. Anaesthesia. 2020;75(9):1223–8.
36. Vrselja Z, Daniele SG, Silbereis J, et al. Restoration of brain circulation and cellular functions hours post-mortem. Nature. 2019;568(7752):336–43.
37. Manara AR. All human death is brain death: the legacy of the Harvard criteria. Resuscitation. 2019;138:210–2.
38. Shemie SD, Gardiner D. Circulatory arrest, brain arrest and death determination. Frontiers in Cardiovascular Medicine 2018;5(15).

Arguments Opposing the Requirement to Demonstrate Absence of Brain Circulation to Determine Death by Neurologic Criteria

Joel Neves Briard and Michaël Chassé

In 1968, the Harvard Medical School Ad Hoc Committee formulated the first formal definition of death by neurologic criteria, the cornerstone of which is the permanent loss of brain function [1]. This work eventually led national scientific societies around the globe to develop standards to guide physicians involved in determination of death by neurologic criteria [2]. Although the construct of death by neurologic criteria is generally accepted worldwide, national standards still provide somewhat heterogeneous definitions and diagnostic criteria [3]. This variability may reflect differences in societal and religious views on the construct of death by neurologic criteria, as well as varying clinical realities, notably systemic differences in access to patient care, among other phenomena.

Recently, the World Brain Death Project conducted seminal work in offering a universal definition of death by neurological criteria, endorsed by international and national professional societies, which states that the condition is the "complete and permanent loss of brain function as defined by an unresponsive coma with loss of capacity for consciousness, brainstem reflexes, and the ability to breath independently" [4]. Furthermore, the consensus statement indicates that "the determination of death by neurologic criteria is a clinical diagnosis," for which a rigorous process must be followed to ensure diagnostic accuracy. Overall, the process should include an appropriate investigation to identify the etiology and severity of the causative brain injury, a scrupulous exclusion of confounding conditions that may alter the validity of the neurologic evaluation, and finally the completion of a rigorous clinical evaluation assessing for the permanent loss of brain function based on strict standards. The consensus statement does not endorse the notion that the absence of

J. Neves Briard
Department of Neuroscience, Université de Montréal, Montréal, QC, Canada
e-mail: joel.neves.briard@umontreal.ca

M. Chassé (✉)
Department of Medicine, Université de Montréal, Montréal, QC, Canada
e-mail: michael.chasse@umontreal.ca

A. Lewis, J. L. Bernat (eds.), *Death Determination by Neurologic Criteria*, Advances in Neuroethics, https://doi.org/10.1007/978-3-031-15947-3_13

brain circulation should be required to support the absence of brain function, and we argue in this chapter that there is currently no scientific rationale to support such a requirement.

There have been significant advances over the past decades in medical science and technology, particularly clinical neuroscience. New diagnostic tools have catalyzed physicians' ability to investigate a variety of paraclinical surrogates for brain function. Neuroimaging and neurophysiology have provided clinicians elegant and rigorous methods to assess brain circulation (e.g. conventional four-vessel cerebral angiography, transcranial Doppler ultrasound, CT-angiography), perfusion (e.g. CT-perfusion scan, nuclear perfusion studies), and electrical function (e.g. electroencephalography, evoked potentials) [5]. Although these tests were not developed specifically for determination of death by neurologic criteria, clinicians often use these paraclinical assessments in the determination of death by neurologic criteria [6]. Despite their contemporary omnipresence in determination of death by neurologic criteria, research has shown that the indications and actual use of these ancillary tests in clinical practice vary worldwide [7]. Ancillary test accessibility and cost, as well as societal preferences and perspectives on the utility of these examinations, likely contribute to this heterogeneity.

From a diagnostic accuracy perspective, the utility of ancillary tests for determination of death by neurologic criteria depends mainly on the population subjected to such exams. First, ancillary tests can be used in comatose patients suspected to be dead by neurologic criteria for whom a valid and complete neurologic evaluation is not feasible or not possible (e.g. in the presence of significant facial trauma, high cervical spinal cord injury, or when an apnea test cannot be performed adequately). In such instances, ancillary tests assess surrogates for brain function, such as circulation and perfusion, aiming to identify patients with paraclinical findings entirely consistent or discordant with the construct of death by neurologic criteria [6]. Second, ancillary tests are sometimes employed to support the determination of death by neurologic criteria in patients who already fulfill all clinical criteria. This application of ancillary testing is frequent worldwide, and in some jurisdictions, it is mandatory for determination of death by neurologic criteria determination [7].

Whether or not a paraclinical examination is required to determine death by neurologic criteria is an important contemporary issue. Some hold the position that ancillary testing should be mandatory for determination of death by neurologic criteria. Specifically, some propose that demonstration of absent brain circulation should be a requirement for the determination. In this chapter, we argue that the determination of death by neurologic criteria does not require the demonstration of absent brain circulation. Our arguments are based on three main points. First, since death by neurologic criteria is based on the clinical permanent loss of brainstem function, it is not relevant to add surrogates of function, such as circulation and perfusion, to the determination process. Further, we argue that brain circulation is not an ideal surrogate for brain function. Finally, brain circulation ancillary tests are of variable diagnostic accuracy, and are not properly validated for determination of death by neurologic criteria.

1 Death by Neurologic Criteria Is the Permanent Loss of Brainstem Function

A clinical neurologic evaluation is the reference standard for determination of death by neurologic criteria, and the demonstration of absent brain circulation should not be mandatory, as it is an unnecessary evaluation of an oversimplified surrogate for brain function.

There is robust international consensus that death by neurologic criteria is the "complete and permanent loss of brain function as defined by an unresponsive coma with loss of capacity for consciousness, brainstem reflexes, and the ability to breathe independently" [4]. Although experts may have different perspectives on the role of ancillary testing in the determination of death by neurologic criteria, there is essentially no debate over the requirement for *permanent loss of brain function*. This is not surprising, as it is widely recognized that death by neurologic criteria is based on the failure of the brain to elicit consciousness (arousal), to mediate any response to the environment (awareness), or to physiologically react to any stimulus. Clinically, this translates into a state of permanent and deep coma, and an absence of brainstem reflexes, including the capacity to initiate breathing (central apnea) [4].

The fact that not a single element of the clinical requirements to determine death by neurologic criteria is specific to functions of the supratentorial brain means that the clinical examination for death by neurological criteria is essentially based on the permanent loss of *brainstem* function, rather than that of the *entire brain* itself [8]. This distinction is important, since it is linked to the long-held opposing perspectives on the anatomical formulation of death by neurologic criteria, discussed in greater detail elsewhere in this book. Briefly, some hold the position that the *entire brain* must permanently cease to function to determine death by neurologic criteria (whole-brain criterion), whereas others are satisfied with the permanent loss of *brainstem* function (brainstem criterion). Although this particular debate has been largely framed as a divide among jurisdictions [9], it is likely that individuals within a same location have different opinions on the matter. For instance, Canadian intensivists appear to be evenly split on the question [10]. This may reflect the paradox that the clinical conditions for death by neurologic criteria focus on permanent loss of *brainstem* function, whereas ancillary tests evaluate the *whole-brain* (e.g. assessment for absence of brain circulation, perfusion, or electrical function).

The debates over (1) the mandatory need for demonstration of absent brain circulation for determination of death by neurologic criteria; and (2) whole-brain versus brainstem criterion are intertwined. Those who favor the requirement of absent brain circulation for determination of death by neurologic criteria are often concerned with the possibility that clinical evaluation is insufficient to prove permanent loss of capacity for consciousness, particularly in individuals with a primary infratentorial lesion [11]. Whereas we know that the ascending reticular activation system, located in the brainstem, is of capital importance for consciousness and vigilance, it is currently uncertain if complete failure of this network is still compatible with consciousness, for instance, through autonomous activity of the hypothalamus, thalamus, or basal forebrain. The consideration that an individual may fulfill all clinical conditions for death by neurologic criteria but still have the capacity for consciousness is a serious one and mandates reflection and scientific investigation. However, to the present day, no such case has been described despite decades of experience with determination of death by neurologic criteria [12, 13]. Thus, this remains a theoretical concern, with no pragmatic implications on the gold standard process of determination of death by neurologic criteria, that is, the demonstration of permanent loss of brainstem function.

Despite incredible technological progress over the past 50 years, the contemporary determination of death by neurologic criteria remains fundamentally based on a clinical evaluation demonstrating the *permanent loss of brainstem function*. There is no evidence that a patient who fulfills all clinical conditions for death by neurologic criteria can recover documentable brain function, regardless of whether absence of brain circulation has been confirmed. There have been some reports of cases where patients initially had a clinical evaluation suggestive of death by neurologic criteria but had persistent brain circulation recovered some brain function (e.g. regained spontaneous breathing); however, the authors of these reports retrospectively recognized that their neurologic evaluations were either incomplete or not reliable due to confounding factors [14, 15]. Conversely, blood circulation is not uncommon among patients who fulfill all clinical conditions for death by neurologic criteria [16, 17]. Therefore, why would we require brain circulation testing if it offers no additional pertinent information on the determination being considered?

2 Brain Circulation Is Not an Ideal Surrogate for Brainstem Function

We have argued that the gold standard evaluation for death by neurological criteria is the clinical demonstration of permanent loss of brainstem function. However, it is not always possible to perform a complete and valid clinical evaluation. For instance, a patient may have a high cervical spine injury limiting the capacity to assess peripheral response to stimuli and the ability to perform a valid apnea test. Ancillary tests that investigate brain circulation are not limited by such factors. Hence, if brain circulation adequately reflects brainstem function, should we not use it as a surrogate measure of brainstem function to determine death by neurologic criteria? We believe not.

Surely, ancillary tests have an important role in the determination of death by neurologic criteria. There are selected situations in which ancillary testing can be useful, as they may provide paraclinical information that support or cast doubt on the determination of death by neurologic criteria. The World Brain Death Project recommends ancillary testing in circumstances where (a) it is not possible to complete all aspects of the minimum clinical examination (including the apnea test), (b) there may be confounding conditions that cannot be resolved, or (c) there is uncertainty regarding interpretation of possible spinally mediated movements [4]. In all these situations, clinicians are conflicted as they cannot complete a comprehensive clinical assessment and must rely on useful, but imperfect, paraclinical data. Albeit the utility of ancillary tests in such circumstances, it is capital to remember that these paraclinical exams assess a surrogate for brain function, namely brain circulation, cerebral perfusion, or electrical function [6]. In other words, no ancillary test directly evaluates the entire brainstem function, which is the basis of the gold standard neurologic clinical examination. Rather, these paraclinical assessments allow inspection of the brain's arterial and/or venous blood flow, the brain's tissular perfusion, or the brain's electrical function, but never the brain's function itself. To illustrate the fundamental differences between circulation, perfusion, and flow, we previously proposed a conceptual model comparing ancillary test surrogates to the process of watering a plant (Fig. 1) [6].

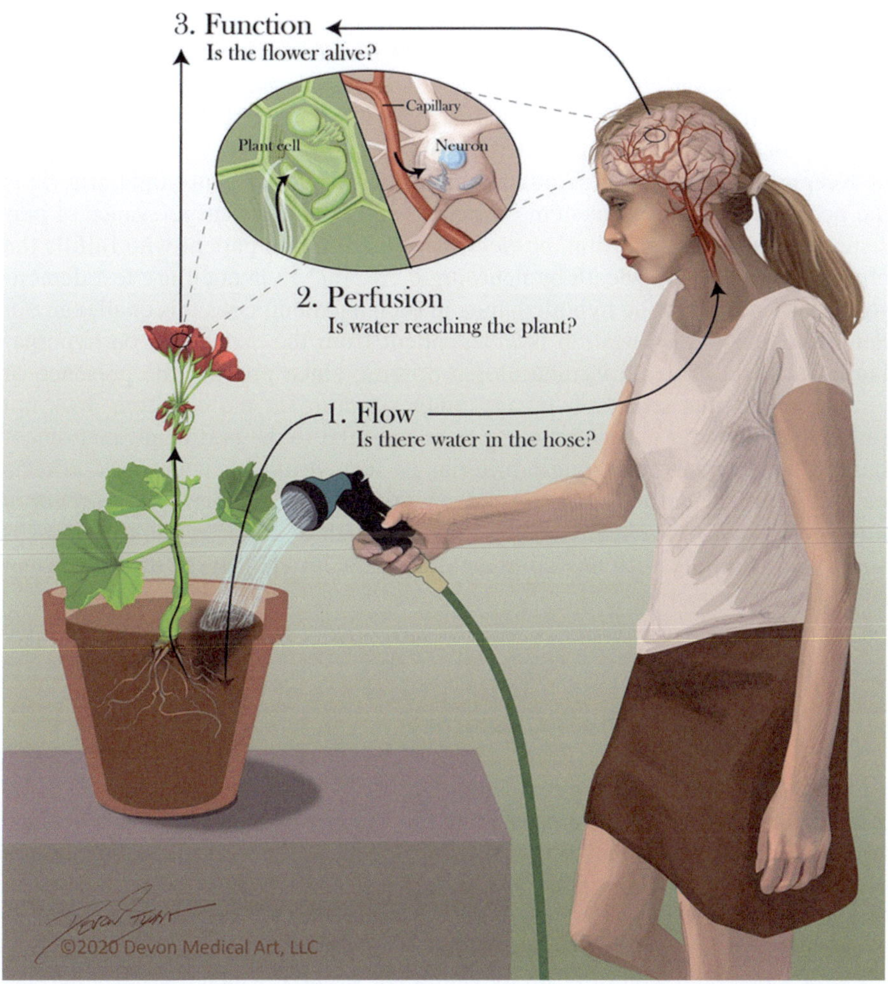

Fig. 1 Conceptual model illustrating the differences between brain circulation, perfusion and function. Reproduced with permission from: Plourde G, Neves Briard J, Shemie SD, Shankar J, Chassé M. Flow is not perfusion, and perfusion is not function: Ancillary testing in brain death. *Can J Anaesth*. 2021;68:953–61

Overall, ancillary test results are considered compatible with death by neurologic criteria when they demonstrate absence of brain circulation, absence of brain perfusion, or absence of brain electrical function. The analogues in the flower model are absence of water in the hose, absence of water reaching the floral cells, and absence of signs of floral life. However, the presence of brain circulation does not guarantee adequate brain perfusion. Furthermore, the presence of brain perfusion is not necessarily sufficient for brain electrical activity. Finally, the latter is an oversimplification of the complexity of brain function. These phenomena have been demonstrated in numerous studies where patients fulfilling all clinical conditions for death by neurologic criteria had persistent brain circulation [16, 17], perfusion [18, 19], or electrical function [20, 21]. For instance, a patient fulfilling clinical conditions for death by neurologic criteria may have preserved brain circulation on conventional four-vessel angiography despite an isoelectric electroencephalogram and absent brainstem auditory evoked potentials [22]. Another patient who meets all clinical conditions for death by neurologic criteria may have brain circulation on radionuclide imaging, but have absent somatosensory evoked potentials and brainstem auditory evoked potentials [23]. In sum, brain circulation is not entirely sensitive for death by neurologic criteria.

Keeping in mind the previous argument that death by neurologic criteria is the permanent loss of brainstem function, what is the clinical relevance of persistent circulation, perfusion, or electrical function in a patient who fulfills the clinical conditions for death by neurologic criteria? If an ancillary test demonstrates circulation to the hypothalamus in such a patient, does this challenge our determination? This question is independent from the discussion on hypothalamic function in death by neurologic criteria, since proving the presence of circulation to this structure, or any other structure in the cranium, does not imply that the latter is functional, or even essential to the pertinent functions of interest in death by neurologic criteria. In fact, death by neurologic criteria being the permanent loss of brainstem function *does not equate to the death of all neurons in the cranium*. Documenting the death of all intracranial neurons is simply not feasible with any contemporary paraclinical test, as they all assess surrogates for function [24].

3 Brain Circulation Ancillary Tests Are of Variable or Unclear Diagnostic Accuracy

Although the presence of brain circulation may not directly infer function in the brainstem, is the absence of brain circulation useful to support the determination of death by neurologic criteria? In other words, is the absence of brain circulation highly specific for death by neurologic criteria? Although this idea is based on strong theoretical reasoning, high-quality epidemiological evidence is currently lacking to support it.

We previously discussed the imperfect sensitivity of brain circulation ancillary tests for a determination of death by neurologic criteria, which translates into the possibility of false negative results. As a determination of death by neurologic criteria must be made with the highest level of specificity, this is not in itself a major issue. The most worrisome error would be in determining a patient to be deceased when they are in fact still alive (false positive). Unfortunately, brain circulation ancillary tests are not free of this error and may indeed cause false positive results. Although it makes physiological sense that absent brain circulation causes permanent death of the brain or is the result of such death (by the absence of cellular metabolic activity), there is no high-quality diagnostic accuracy study that has demonstrated that brain circulation ancillary tests have perfect specificity for death by neurologic criteria. In a recent systematic review and meta-analysis of the accuracy of ancillary tests for determination of death by neurologic criteria that is pending publication (4208 records screened and 133 studies included), we only found one study on brain circulation ancillary testing with low risk of methodological bias as assessed using the QUADAS-2 tool [25]. In this study, Brunser and colleagues assessed the diagnostic accuracy of transcranial Doppler ultrasound in 53 consecutive comatose patients, of which 25 fulfilled the clinical conditions for death by neurologic criteria; there was a false positive result, yielding a specificity of 96%. In a subsequent systematic review and meta-analysis of the diagnostic accuracy of transcranial Doppler ultrasound for death by neurologic criteria, the pooled specificity was found to be 98% (95% confidence interval: 96–99%) [16]. For many medical conditions, such specificity is sufficient, but in the case of death by neurologic criteria, all agree that false positive diagnoses are unacceptable. Unfortunately, the body of evidence for CT-angiography and MR-angiography is also weak, with no currently published studies with low methodological risk of bias. In fact, the World Brain Death Project has warned that these two ancillary tests should not be used for determination of death by neurologic criteria until they are subjected to additional investigation [4]. Finally, although conventional four-vessel cerebral angiography is widely considered a valid brain circulation ancillary test, it is interesting to note that there are in fact no high-quality studies demonstrating that this paraclinical exam is specific.

Thus, the scientific evidence on brain circulation ancillary tests' accuracy for determination of death by neurologic criteria is not robust. Imposing the requirement to demonstrate absent brain circulation to determine death by neurologic criteria carries a risk of making considerable diagnostic errors. This is not, unfortunately, the only limitation in the utility of brain circulation ancillary testing. Conventional four-vessel cerebral angiography is a specialized test and is not easily accessible, even in developed systems of care [3, 7, 26]. Moreover, the reliability of transcranial Doppler ultrasound is operator-dependent, and this test cannot be used in the significant proportion of patients with an inadequate bone window [16]. Overall, even if we believed that absent brain circulation was necessary to determine death by neurologic criteria, tests assessing brain circulation in this context have not yet reached a level of accuracy, reliability, and accessibility that justify their routine inclusion in determination of death by neurologic criteria.

4 Conclusion

In this chapter, we argued that the demonstration of absent brain circulation is not required to determine death by neurologic criteria. Despite decades of advances in technology and science, death by neurologic criteria remains a clinical determination based on the permanent loss of brainstem function. No single case of clinically determined death by neurologic criteria has ever recovered, even in the presence of initially preserved brain circulation. Brain circulation is not an appropriate surrogate for brainstem function, and it lacks sensitivity for clinically relevant permanent loss of brainstem function. Finally, current evidence does not support the claim that brain circulation ancillary tests have perfect specificity for death by neurologic criteria, and these tests are subject to different technical and logistical caveats. Together, these points provide a strong case against the requirement to demonstrate absent brain circulation to determine death by neurologic criteria. Although brain circulation remains an interesting, although imperfect, surrogate for function in selected patients who require ancillary testing, there is currently no convincing reason to mandate the demonstration of absent brain circulation beyond the currently accepted and robust clinical conditions required for determination of death by neurologic criteria.

References

1. A definition of irreversible coma: report of the ad hoc Committee of the Harvard Medical School to examine the definition of brain death. JAMA. 1968;205(6):337–40.
2. Wijdicks EFM. Brain death worldwide: accepted fact but no global consensus in diagnostic criteria. Neurology. 2002;58:20–5.
3. Lewis A, Bakkar A, Kreiger-Benson E, et al. Determination of death by neurologic criteria around the world. Neurology. 2020;95(3):e299–309.
4. Greer DM, Shemie SD, Lewis A, et al. Determination of brain death/death by neurologic criteria: the world brain death project. JAMA. 2020;324(11):1078–97.
5. Heran MK, Heran NS, Shemie SD. A review of ancillary tests in evaluating brain death. Can J Neurol Sci. 2008;35(4):409–19.
6. Plourde G, Neves Briard J, Shemie SD, Shankar J, Chassé M. Flow is not perfusion, and perfusion is not function: ancillary testing in brain death. Can J Anaesth. 2021;68:953–61.
7. Lewis A, Liebman J, Kreiger-Benson E, et al. Ancillary testing for determination of death by neurologic criteria around the world. Neurocrit Care. 2021;34(2):473–84.
8. Varelas PN, Brady P, Rehman M, et al. Primary posterior fossa lesions and preserved supratentorial cerebral blood flow: implications for brain death determination. Neurocrit Care. 2017;27(3):407–14.
9. Wijdicks EF. The transatlantic divide over brain death determination and the debate. Brain. 2012;135(Pt 4):1321–31.
10. Chassé M, Neves Briard J, Yu M, et al. Clinical evaluation and ancillary testing for the diagnosis of death by neurological criteria: a cross-sectional survey of Canadian intensivists. Can J Anaesth. 2022;69:353–63.
11. Walter U, Fernandez-Torre JL, Kirschstein T, Laureys S. When is "brainstem death" brain death? The case for ancillary testing in primary infratentorial brain lesion. Clin Neurophysiol. 2018;129(11):2451–65.

12. Manara A, Varelas P, Smith M. Neurological determination of death in isolated brainstem lesions: a case report to highlight the issues involved. J Intensive Care Soc. 2020;21(3):269–73.
13. Manara A, Varelas P, Wijdicks EF. Brain death in patients with "isolated" brainstem lesions: a case against controversy. J Neurosurg Anesthesiol. 2019;31(2):171–3.
14. Roberts DJ, MacCulloch KA, Versnick EJ, Hall RI. Should ancillary brain blood flow analyses play a larger role in the neurological determination of death? Can J Anaesth. 2010;57(10):927–35.
15. Webb AC, Samuels OB. Reversible brain death after cardiopulmonary arrest and induced hypothermia. Crit Care Med. 2011;39(6):1538–42.
16. Chang JJ, Tsivgoulis G, Katsanos AH, Malkoff MD, Alexandrov AV. Diagnostic accuracy of transcranial doppler for brain death confirmation: systematic review and meta-analysis. AJNR Am J Neuroradiol. 2016;37(3):408–14.
17. Kramer AH, Roberts DJ. Computed tomography angiography in the diagnosis of brain death: a systematic review and meta-analysis. Neurocrit Care. 2014;21(3):539–50.
18. Facco E, Zucchetta P, Munari M, et al. 99mTc-HMPAO SPECT in the diagnosis of brain death. Intensive Care Med. 1998;24:911–7.
19. MacDonald D, Stewart-Perrin B, Shankar JJS. The role of neuroimaging in the determination of brain death. J Neuroimaging. 2018;28(4):374–9.
20. Su Y, Yang Q, Liu G, et al. Diagnosis of brain death confirmatory tests after clinical test. Chin Med J. 2014;127(7):1272–7.
21. Zhang Y, Sun B, Wu XY. Spectrum analysis of the EEG in patients with brain death. J Clin Neurol. 2008;21(3):213–5.
22. Nau R, Prange HW, Klingelhöfer J, et al. Results of four technical investigations in fifty clinically brain dead patients. Intensive Care Med. 1992;18:82–8.
23. Hansen AVE, Lavin PJM, Moody EB, Sandler MP. False negative cerebral radionuclide flow study, in brain death, caused by ventricular drain. Clin Nucl Med. 1993;18:502–5.
24. Wijdicks EF. The case against confirmatory tests for determining brain death in adults. Neurology. 2010;75:77–83.
25. Chassé M, Glen P, Doyle MA, et al. Ancillary testing for diagnosis of brain death: a protocol for a systematic review and meta-analysis. Syst Rev. 2013;2:100.
26. Greer DM, Wang HH, Robinson JD, et al. Variability of brain death policies in the United States. JAMA Neurol. 2016;73(2):213–8.

Reports of "Recovery" from Death by Neurologic Criteria

Christos Lazaridis and Fernando D. Goldenberg

1 Some cases of "recovery" from death by neurologic criteria

Death is irreversible, so there can be no recovery from death. The title of this chapter thus needs further explication. By "reports of recovery," we mean cases in which further observation and subsequent testing following a determination of death by neurologic criteria demonstrate that the determination that there was irreversible loss of function of the brain was in fact a false-positive (positive in the sense of being consistent with death by neurologic criteria). Several pediatric and adult case reports describing this type of "recovery" have been reported [1].

One unique example is the recent case of Jahi McMath; in her case, both her family and Dr. Alan Shewmon claimed she recovered to a minimally conscious state (MCS) after determination of death by neurologic criteria [2]. To our knowledge, this is the only case of purported recovery of consciousness after appropriate formal determination of death by neurologic criteria.

In this chapter, we offer brief summaries followed by a commentary of 4 cases of "recovery" from death by neurologic criteria, including the McMath case; these are selected examples and do not represent the findings of an exhaustive systematic review of all published reports. Our purpose in summarizing these cases is not to reproduce published details, but rather to provide our own commentary on these cases. It is outside our scope to discuss the accuracy of different clinical and ancillary tests used in these, or any, determinations of death by neurologic criteria.

C. Lazaridis
Neurocritical Care Section, Department of Neurology, University of Chicago, Chicago, IL, USA

MacLean Center for Clinical Medical Ethics, University of Chicago, Chicago, IL, USA
e-mail: lazaridis@uchicago.edu

F. D. Goldenberg (✉)
Neurocritical Care Section, Department of Neurology, University of Chicago, Chicago, IL, USA
e-mail: fgoldenb@uchicago.edu

A. Lewis, J. L. Bernat (eds.), *Death Determination by Neurologic Criteria*, Advances in Neuroethics, https://doi.org/10.1007/978-3-031-15947-3_14

However, it must be recognized that considerations related to the accuracy of different tests should take into account the fact that determination of death by neurologic criteria is generally followed by withdrawal of artificial support inevitably leading to circulatory-respiratory arrest. After discussing these cases, we conclude by reviewing the implications of false-positive determinations of death by neurologic criteria.

1.1 Roberts et al. 2010

Roberts et al. presented two patients who regained spontaneous respiration following determination of death by neurologic criteria [3]. In both cases, the patients did not have absence of brain circulation.

The first patient, a 26-year-old man, was found to be comatose in the setting of a temporal lobe abscess with surrounding vasogenic edema and 1.3 cm midline shift. He was given antibiotics, mannitol and dexamethasone. Seven hours after he became comatose, he was determined to be dead by neurologic criteria based on a clinical evaluation, including apnea testing (in which $PaCO_2$ increased from 42 mm Hg to 69 mm Hg over 10 min). Clinical management became focused on organ preservation for the purposes of transplantation. To further delineate the anatomy of the temporal lobe abscess and to exclude involvement of extracranial vascular structures (which could present a contraindication to donation), brain magnetic resonance imaging (MRI) was performed 2 h after he was declared dead by neurologic criteria. The MRI revealed a flow void in the middle cerebral artery (MCA) and MCA enhancement in the axial T1-weighted post-gadolinium images, indicating the presence of flow. Twenty-eight hours after the declaration of death by neurologic criteria, he began triggering the ventilator and was found to have a spontaneous respiratory rate of 10 breaths/min. The rest of the neurologic examination remained unchanged with absence of brainstem reflexes. Spontaneous respirations persisted for 5 days before determination of death by circulatory-respiratory criteria.

In the second case, a 50-year-old woman suffered a severe traumatic brain injury with a basal skull fracture, subdural hematoma, subarachnoid hemorrhage, generalized cerebral edema, and effacement of the basal cisterns, followed by circulatory-respiratory arrest with return of spontaneous circulation after 5 min. Six hours after admission, she was determined to be dead by neurologic criteria based on her clinical evaluation, including apnea testing (in which $PaCO_2$ increased from 56 mm Hg to 80 mm Hg over 8 min). Nevertheless, a cerebral radionuclide scan was performed because of the authors' experience in the prior case which showed evidence of brain circulation. During subsequent donor management, 11 h after coma onset, she began triggering the ventilator and was found to have spontaneous respirations, though no other brainstem reflexes. Her family decided to proceed with withdrawal of artificial support and she was declared dead by circulatory-respiratory criteria.

The authors concluded that for both patients, several unrecognized confounding factors could have contributed to false-positive determinations. In the first case, ongoing treatment with glucocorticoids and antibiotics may have promoted some resolution of cerebral edema or limited further abscess expansion resulting in restoration of brain circulation. It is also possible that cold caloric testing was not reliable due to otitis media and mastoiditis. In the second case, one could question the short observation period between return of spontaneous circulation after cardiac arrest and the determination of death by neurologic criteria. These cases may suggest that more routine use of studies to evaluate brain circulation should be recommended.

1.2 Webb et al. 2011

Webb et al. described a 55-year-old man with a 20-min period of circulatory-respiratory arrest [4]. He was initially hypothermic (35.2 °C) on arrival to the intensive care unit, but then rapidly became febrile. Therapeutic hypothermia was initiated, and he eventually reached a nadir of 33 °C at 48 h, then rewarming began at 50 h, and his temperature was 36.5 °C at 56 h after presentation. He was treated with propofol and fentanyl from 14 h to 50 h. 72 after return of spontaneous circulation, he had absent brainstem reflexes, then 6 h later, he was determined to be dead by neurologic criteria based on apnea testing and a repeat clinical evaluation. His family consented to organ donation. Twenty-four hours after declaration of death by neurologic criteria (98 h after admission), and on arrival to the operating room for organ procurement, the patient regained corneal reflexes and the cough reflex, and began having spontaneous respirations. 145 hours after admission, his clinical evaluation was again consistent with death by neurologic criteria. A nuclear study showed absence of brain circulation 200 h after admission, after which treatment was withdrawn and death was declared by circulatory-respiratory criteria.

In discussing this case, the authors considered three etiologies that could account for the false-positive determination: (1) fluctuating functional loss of lower brainstem function, which the authors named "brainstem ischemic shock"; (2) the application of therapeutic hypothermia as a confounder; and (3) the administration of glucocorticoids. The case generated controversy. Critics raised concern with the authors' conclusions highlighting the potential confounding effects of hypothermia in conjunction with administration of high dose infusions of fentanyl and propofol [5].

1.3 Latorre et al. 2020

Latorre et al. presented a 59-year-old man with a right basal ganglia and temporal lobe intracerebral hemorrhage causing 1.1 cm midline shift, who lost all brainstem reflexes over 48 h. Apnea testing was not performed due to hemodynamic instability [6]. Instead, a brain SPECT scan was obtained to complete the evaluation for death by neurologic criteria. The results showed absence of Tc-99 m Bicisate uptake in both supra- and infratentorial compartments that was interpreted as consistent with death by neurologic criteria. The family subsequently consented to organ donation. However, the following day, he was noted to have a cough reflex, intermittent spontaneous respirations, and extensor posturing of the right arm and leg to noxious stimulation. Shortly thereafter, he had a generalized seizure and a circulatory-respiratory arrest and was declared dead by circulatory-respiratory criteria.

The authors noted that this was the first report of a false-positive determination of death by neurologic criteria using a brain SPECT scan with diffusible radiopharmaceutical tracers. The authors concluded that death by neurologic criteria should remain a clinical determination, and that an apnea test should be performed unless contraindicated due to severe shock or hypoxemia. They also suggested that if the clinical evaluation cannot be completed, a longer observation period may be necessary prior to ancillary testing unless demonstration of negative or zero cerebral perfusion pressure can be shown for an extended period of time.

1.4 Jahi McMath

The case of Jahi McMath has generated large interest and controversy in the medical literature as well in public media. A PubMed search for "Jahi McMath" in February 2022 yielded over 30 publications; a similar search in Google Scholar found 576 results. This case is particularly contentious and interesting because the patient's family and Dr. Alan Shewmon (the only neurologist permitted by the family to examine Jahi in the post-acute phase) claim that after determination of death by neurologic criteria, Jahi subsequently emerged into an MCS [2].

Jahi McMath was a 13-year-old girl who was declared dead by neurologic criteria on December 12, 2013, after a hemorrhagic complication of oropharyngeal surgery. Despite the issuance of a death certificate in California, Jahi's family objected to the declaration of death, and eventually transferred her to New Jersey where the law allows rejection of death by neurologic criteria on religious grounds.[1] Four and a half years later, she was issued a second death certificate after being declared dead by circulatory-respiratory criteria.

Determination of death by neurologic criteria was made and confirmed by capable examiners including a court-appointed independent child neurologist. She underwent several apnea tests, four electroencephalograms that were isoelectric, and a radionuclide SPECT scan with Tc 99 m bicisate, which showed no brain circulation on the dynamic sequence and no supratentorial or infratentorial parenchymal uptake. Interestingly, her brain MRI 9.5 months after declaration of death by neurologic criteria showed gross integrity of the cortex, basal ganglia, thalamus, and upper brainstem. Extensive demyelinating and cystic changes were noted in the subcortical white matter and lower brainstem.

Shewmon and Salamon, plausibly in our view, argue that despite this devastating degree of brain injury, persistence of gross structural integrity speaks against sustained absence of brain circulation, and suggests the potential for persistence of flow at levels not detectable by SPECT (at least at the time it was performed) [7].

[1] New Jersey is the only state with an exemption clause to determination of death by neurologic criteria.

2 Implications

There are two ways to conceptualize and draw implications from the above cases. The first approach is to consider them to be false-positive determinations of death by neurologic criteria made because of the failure to ensure that prerequisites are met or mimics are excluded before conducting an evaluation, or by inaccurate examination technique [5, 8]. The second approach is to take these reports at face value by accepting that the prerequisites were properly met, no known confounders were present, and that testing and determination were competently performed.

By the first approach, there is not much more to learn beyond unequivocally endorsing the recommendation that determinations of death by neurologic criteria should strictly adhere to published standards and clinicians performing these determinations must have adequate training and experience with determination of death by neurologic criteria.

However, we think that the second approach could generate fruitful discussion, even if one rejected the validity of the aforementioned case reports. In what follows, we provide support for the following two propositions: (1) Absence of brain circulation should be required to determine death by neurologic criteria, and, ideally, this finding should be demonstrated before apnea testing; (2) Death by neurologic criteria is a state, along the spectrum of devastating brain injury, sufficient for a person to be assigned the legal status of death.

2.1 Absence of Brain Circulation

The current neurologic standard in the Uniform Determination of Death Act (UDDA) explicitly calls for "irreversible" cessation of functions of the "entire brain" [9]. It follows that the only pathophysiologic avenue to meet the required burden of proof is via the complete and sustained cessation of brain circulation [10]. This understanding is supported in the recently published World Brain Death Project (WBDP) where it was suggested to ensure neuroimaging evidence of intracranial hypertension or measurement of intracranial pressure equal to or exceeding mean arterial pressure [11]. Nevertheless, it is important to note that the WBDP reiterates the guidance in most national medical standards that death by neurologic criteria is a clinical determination, and any further radiographic or brain circulation testing is merely ancillary, and is warranted if (1) part of the clinical evaluation or apnea testing cannot be completed; (2) uncertainty exists about the interpretation of findings; (3) to reduce the inter-examination observation period (if more than one evaluation is performed); (4) there is that concern medication effect may be present; (5) it is felt that this would be helpful for family members to accept death by neurologic criteria; or (6) there is isolated brainstem pathology (if the whole-brain formulation is being followed). These multiple reasons seem to us to justify requiring testing to assess for brain circulation, rather than considering it optional and ancillary, as discussed in detail elsewhere in this book.

The following modalities are available to assess brain circulation: four-vessel catheter angiography, radionuclide cerebral perfusion scan, transcranial Doppler, computed tomographic angiography, and magnetic resonance angiography. All these tests have pitfalls [12]. Nevertheless, as the aforementioned case reports show, the clinical evaluation is not immune from pitfalls, and does not have perfect sensitivity and specificity. Furthermore, the biases of self-fulfilling and self-reinforcing prophecies, in our opinion, call for epistemic humility in regard to claims about sensitivity and specificity of any of the available tests, clinical or not. We support the current American Academy of Neurology standards for determination of death by neurologic criteria and the WBDP in considering four-vessel catheter angiography, radionuclide cerebral perfusion scan, and transcranial Doppler (in adults) to be acceptable tests to evaluate brain circulation, and we further argue that at least one of these tests ought to be performed as part of all determinations of death by neurologic criteria. Furthermore, we recommend that a study to evaluate brain circulation be performed prior to apnea testing as, at least theoretically, the induction of hypercapnia could contribute to, rather than confirm, the absence of brain circulation [11 p. 1083; supp.4, p. 14].

2.2 Legal Status

A legal status is a category conferring rights and duties on those who fall within it [13, 14]. Just as legal blindness is recognized to be on a spectrum of visual impairment, death by neurologic criteria should be understood to be a threshold state along the spectrum of devastating brain injury, which sufficiently ascertains the permanent loss of consciousness and makes death behaviors appropriate.[2] There are arguments for and against such a proposal, and this concept is discussed in detail elsewhere in this book. The first supporting argument is that it addresses the false notion that death of the brain is sufficient for the irreversible loss of organismal homeostatic integration. Adhering to this position simply ignores contemporary understanding of homeostatic integration as an emergent property of biologic organisms not dependent on any single organ system [15]. By thinking of the medical determination of death as a legal status, one recognizes that by necessity, there must be some degree of "line drawing" in how we identify the transition from living to dead. Line drawing is relevant to the second supporting argument in changing the discussion from irreversibility to permanence. Irreversibility is an implausibly high, and even impossibly high, requirement to meet within the current state of scientific and medical knowledge. Permanence is, in fact, the standard employed in current clinical practice of determination of death by circulatory-respiratory criteria, and there is no justification for why it should be different for death by neurologic criteria [16]. This move is relevant in the case of Jahi McMath; if she indeed transitioned

[2] Death behaviors are behaviors and activities that are appropriate after the declaration of death such as discontinuation of artificial support, initiation of organ donation, burial or cremation, grieving, change of a survivor's marital status, and transfer of property.

from death by neurologic criteria to an MCS, then the current edifice of death by neurologic criteria based on irreversibility is mistaken and would require temporary cessation and full review from its foundation. Death by neurologic criteria as a legal status which requires permanent, not irreversible, loss of the capacity of consciousness is less vulnerable when considering the case of Jahi McMath. It also denotes that since it is a legal stipulation and a medical and social convention, it should be open to revision and update in concordance with increased knowledge or technological enhancements [17].

There are counterarguments to a legal status approach. Although it is a more palatable term than "legal fiction," it still implies that the patient is not really dead but can be treated as such under the law. This argument leads to the issue of transparency or acknowledged vs. unacknowledged fictions [18]. Legal blindness is a transparent legal status, while treating a corporation as a person is an acknowledged one. For death by neurologic criteria to be considered a legitimate legal status, must it be acknowledged as such and publicly deliberated? There are two responses to this question. The first is to grant that indeed transparency is essential and engage in public deliberation. Such deliberation could take the form of allowing personal choice or setting a default and permitting opt-out in the determination of death, as discussed elsewhere in this book. The second response is to reject the requirement for explicit acknowledgement. Despite occasional challenges, death by neurologic criteria has withstood the test of time and is widely considered to be sensible and socially beneficent [19, 20].

Without intending too wide of a digression, this last point can be put in more technical terms, and made stronger, as it pertains to public reason and justification. The recognition of citizens as free and equal moral persons requires that laws applying to them should be justified with reasons that they could recognize as valid [21]. This explains why it would be problematic for death by neurologic criteria to be a non-transparent legal status. Non-transparent reasons, policies and laws may not allow citizens to evaluate them and recognize them as valid by their own lights. However, there are different conceptions on the nature of justification that is in play, and one of them that may be relevant in our discussion about death by neurologic criteria is *evolutionary* justification [22]. According to evolutionists, a law or a policy is justified for members of the public when that law or policy is a stable and evolved equilibrium for the public (even if it is a legal fiction or status). This line of thought may conclude with the (controversial) claim that indeed death by neurologic criteria, by having withstood the test of time, is a stable and evolved equilibrium for societies that have accepted death by neurologic criteria for several decades. It follows that death by neurologic criteria as legal status may not require further justification.

3 Conclusion

There are several published reports of alleged "recovery" from death by neurologic criteria. Recovery from death is not possible, and so the more precise interpretation of these cases is that they are false-positive determinations. A common response to such cases is to explain them away by identifying possible confounders or technical problems with the process of determination. Here, without necessarily committing to the validity of these reports, we discuss the implications of taking them at face value. Two potentially important implications follow. The first is the need to demonstrate absence of brain circulation to make a determination of death by neurologic criteria. Absence of circulation is a necessary pathophysiologic requirement for the cessation of function of any organ, and the brain is no exclusion. Although current knowledge about how long circulation needs to be lost before there is complete, irreversible loss of function of the brain is insufficient, and there are pitfalls of all techniques to evaluate brain circulation, we believe it is necessary to incorporate a study to evaluate brain circulation into all determinations of death by neurologic criteria. The second implication is that the declaration of death by neurologic criteria should be considered a legal status which relies on best available medical technology, with the understanding that the process of death declaration requires both societal acceptance and a focus on beneficence.

References

1. Joffe AR, Khaira G, de Caen AR. The intractable problems with brain death and possible solutions. Philos Ethics Humanit Med. 2021;16(1):11.
2. Shewmon DA, Salamon N. The extraordinary case of Jahi McMath. Perspect Biol Med. 2021;64(4):457–78.
3. Roberts DJ, MacCulloch KA, Versnick EJ, Hall RI. Should ancillary brain blood flow analyses play a larger role in the neurological determination of death? Can J Anaesth. 2010;57(10):927–35.
4. Webb AC, Samuels OB. Reversible brain death after cardiopulmonary arrest and induced hypothermia. Crit Care Med. 2011;39(6):1538–42.
5. Wijdicks EF, Varelas PN, Gronseth GS, Greer DM. There is no reversible brain death. Crit Care Med. 2011;39(9):2204–5. author reply 2206
6. Latorre JGS, Schmidt EB, Greer DM. Another pitfall in brain death diagnosis: Return of cerebral function after determination of brain death by both clinical and radionuclide cerebral perfusion imaging. Neurocrit Care. 2020;32(3):899–905.
7. Shewmon DA, Salamon N. The MRI of Jahi McMath and its implications for the global ischemic penumbra hypothesis. J Child Neurol. 2022;37(1):35–42.
8. Lewis A. Reconciling the case of Jahi McMath. Neurocrit Care. 2018;29(1):20–2.
9. President's Commission for the Study of Ethical Problems in Medicine and Biomedical and Behavioral Research. Defining death: Medical, legal and ethical issues in the determination of death. Washington, DC: US Government Printing Office; 1981.
10. Frank J, Goldenberg F, Ardelt A. Brain death: the contemporary neurological imperative. Crit Care Med. 2011;39(11):2589.
11. Greer DM, Shemie SD, Lewis A, et al. Determination of brain death/death by neurologic criteria: the world brain death project. JAMA. 2020;324(11):1078–97.
12. Wijdicks EFM. The case against confirmatory tests for determining brain death in adults. Neurology. 2010;75(1):77–83.
13. Truog RD, Miller FG. Changing the conversation about brain death. Am J Bioeth. 2014;14(8):9–14.
14. Shah SK. Rethinking brain death as a legal fiction: Is the terminology the problem? Hastings Cent Rep. 2018;48(Suppl 4):S49–52.

15. Miller FG, Nair-Collins M, Truog RD. It is time to abandon the dogma that brain death is biological death. Hastings Cent Rep. 2021;51(4):18–21.
16. Bernat JL. Conceptual issues in DCDD donor death determination. Hastings Cent Rep. 2018;48(Suppl 4):S26–8.
17. Lazaridis C. Defining death: Reasonableness and legitimacy. J Clin Ethics. 2021;32(2):109–13.
18. Shah SK, Miller FG. Can we handle the truth? Legal fictions in the determination of death. Am J Law Med. 2010;36(4):540–85.
19. Magnus DC, Wilfond BS, Caplan AL. Accepting brain death. N Engl J Med. 2014;370(10):891–4.
20. Wahlster S, Wijdicks EF, Patel PV, et al. Brain death declaration: practices and perceptions worldwide. Neurology. 2015;84(18):1870–9.
21. Vallier K. Public justification. The Stanford encyclopedia of philosophy (Spring 2018 Edition), Edward N. Zalta (ed.). https://plato.stanford.edu/archives/spr2018/entries/justification-public. Accessed 2 Jan 2022.
22. Gaus G. The order of public reason. New York: Cambridge University Press; 2011.

Observation Time Prior to Determination of Death by Neurologic Criteria

Jeanne Teitelbaum and Murdoch Leeies

Medical standards on determination of death by neurologic criteria require demonstrating that, prior to commencing an evaluation, a patient has an established neurologic diagnosis, the nature and severity of which is capable of resulting in the irreversible loss of the capacity for consciousness, all brainstem reflexes, and the ability to breathe spontaneously [1–11]. In other words, the patient must have a neurologic diagnosis that could cause death by neurologic criteria and imaging or other data that suggests the injury is severe enough that it may have caused death by neurologic criteria. Additionally, prior to the evaluation, the presence of confounding factors must be excluded.

Establishing that a diagnosis is severe enough to lead to death by neurologic criteria is easiest when there is a structural injury and the damage is evident on imaging, such as after extensive head trauma, hemorrhage, or massive edema. In such cases, the time at which one begins the evaluation for death by neurologic criteria is not of major importance once confounding conditions have been ruled out. The structural damage is evident and clearly severe enough that the patient may be dead by neurologic criteria. Thus, the identification of an irreversible mechanism that has led to a severe injury and the elimination of all confounders are the relevant considerations prior to beginning an evaluation, rather than observation for a specific amount of time [12–29].

Contrastingly, there can be uncertainty about when to begin an evaluation for determination of death by neurologic criteria in the absence of imaging consistent with irreversible damage and edema, as may be the case after circulatory-respiratory arrest. Neurologic assessments may be unreliable in the acute post-resuscitative

J. Teitelbaum (✉)
Montreal Neurological Institute and Hospital, McGill University, Montreal, QC, Canada

M. Leeies
Max Rady College of Medicine, University of Manitoba,
Winnipeg, MB, Canada
e-mail: Murdoch.Leeies@umanitoba.ca

A. Lewis, J. L. Bernat (eds.), *Death Determination by Neurologic Criteria*,
Advances in Neuroethics, https://doi.org/10.1007/978-3-031-15947-3_15

phase after circulatory-respiratory arrest, so many standards recommend that an evaluation for determination of death by neurologic criteria be delayed for at least 24 h post-arrest or an ancillary test be performed [3, 30, 31].

In this chapter, we examine the current standards and evidence for timing of the evaluation for determination of death by neurologic criteria after structural brain injury and hypoxic-ischemic brain injury (with consideration of the impact of targeted temperature management). We then review the advantages and disadvantages of delaying the evaluation for determination of death by neurologic criteria.

1 Current Standards and Evidence for Timing of the Evaluation for Determination of Death by Neurologic Criteria After Structural Brain Injury

1.1 Current Standards

In general, there is very little controversy about the observation time required prior to evaluation for determination of death by neurologic criteria in patients with structural brain injuries [1–11, 30, 31]. Medical standards for determination of death by neurologic criteria indicate that an evaluation for determination of death by neurologic criteria can commence in the setting of coma and absence of motor function and brainstem reflexes if there is a known, proximate cause for the injury that is irreversible. Although there must be definite clinical or neuroimaging evidence of an acute event that has led to the irreversible loss of brain function, only 27% of international medical standards require neuroimaging prior to the evaluation [1]. Additionally, confounders must be ruled out, but there is variation in the factors that are considered across international standards such as: drug clearance (82%), temperature (78%), laboratory values (72%), and blood pressure (44%) [1]. After this step, the question is how much time must elapse in this state to be absolutely sure that the situation is irreversible?

In the United States [2, 5], Canada [8], and many other countries [1, 3, 4, 9–11, 30, 31], once these prerequisites have been met, the evaluation can begin; no minimum observation time is specified. Contrastingly, 24% of international standards require an observation period between 1 and 48 h for all determinations of death by neurologic criteria [1]. Further, a few standards specify the need for an observation period prior to evaluation for death by neurologic criteria after intracerebral hemorrhage (3 standards require a delay of 6 h), major neurosurgical procedures (2 standards require a delay of 4–6 h), secondary brain injury (6 standards require a delay of 12–72 h), or traumatic brain injury (5 standards require a delay of 6 h) [1].

The World Brain Death Project provides some guidance on the observation period prior to performance of an evaluation for determination of death by neurologic criteria that is relevant to patients with structural or hypoxic-ischemic brain injury [30]. The authors note that the period of observation prior to an evaluation for determination of death by neurologic criteria for patients with structural brain injuries should be determined on a case-by-case basis and should be the time thought necessary to confidently exclude reversibility. They emphasize the need for caution. They further advise that prior to commencing an evaluation for determination of death by neurologic criteria, it must be demonstrated that the patient has an established neurologic diagnosis, the nature and severity of which is capable of resulting in the irreversible loss of the capacity for consciousness, all brainstem reflexes, and the ability to breathe spontaneously during a carbon dioxide and acidosis challenge. Additionally, they suggest that there be: (1) neuroimaging evidence of intracranial hypertension (severe cerebral edema and herniation) or (2) intracranial pressure measurements that equal or exceed the mean arterial pressure. In the absence of herniation on neuroimaging, it is suggested that caution be taken when considering performance of an evaluation for determination of death by neurologic criteria. Lastly, they note that confounders and/or reversible conditions that may mimic death by neurologic criteria must be excluded prior to commencing an evaluation for determination of death by neurologic criteria.

1.2 Evidence

There is insufficient evidence to determine the minimally acceptable observation period to ensure that neurologic functions have ceased irreversibly after structural brain injury. However, performance of serial evaluations to assess for evidence of return of function can be considered. The number of evaluations required to determine death by neurologic criteria is discussed elsewhere in this book. The appropriate observation period prior to evaluation for determination of death by neurologic criteria is the focus of a systematic review being conducted by a working group representing Canadian Blood Services, the Canadian Critical Care Society and the Canadian Medical Association A Brain-Based Definition of Death and Criteria for its determination After Arrest of Circulatory ir Neurologic Function in Canada: A Clinical Practice Guideline.

1.3 Recommendation

As there is no firm evidence to dictate the observation time prior to evaluation for determination of death by neurologic criteria after structural brain injury, in accordance with most standards, we believe there is no minimal observation time necessary, as long as (1) there is an established cause for the patient's condition and the severity is sufficient to lead to irreversible loss of brain function and (2) physiologic factors that may confound the evaluation have been screened for and excluded.

2 Current Standards and Evidence for Timing of the Evaluation for Determination of Death by Neurologic Criteria After Hypoxic-Ischemic Brain Injury

During circulatory-respiratory arrest, there is cessation of oxygen and glucose delivery to the entire body, including the brain. If this deprivation is not reversed, cessation of brain circulation will inevitably lead to permanent loss of brain function. When return of spontaneous circulation is achieved, it is not possible to evaluate the extent of recovery without a period of observation. Several studies on prognostication after circulatory-respiratory arrest demonstrate that a patient who is comatose and has absent brainstem reflexes immediately after return of spontaneous circulation can subsequently have return of brain function [32]. The role of targeted temperature management in survivors of cardiac arrest remains controversial in terms of outcome modification, but it is widely accepted that therapeutic hypothermia represents a potential confounder of accurate neurologic prognosis and should be resolved prior to definitive prognostic evaluation or neurologic determination of death [8, 29, 30, 33]. Imaging after hypoxic-ischemic brain injury may not show signs of intracerebral edema, herniation, or brain damage in the first 48 h post-arrest [34, 35].

2.1 Current Standards

Guidance on the length of an observation period after hypoxic-ischemic brain injury vary both around the world and within the United States [6, 36]. In most countries, the observation time is not specified, but 17% of standards note that the evaluation should be delayed 24 h after hypoxic-ischemic brain injury in the absence of targeted temperature management, and 10% indicate there is a need for a delay when targeted temperature management is used [1]. The 2006 standard on determination of death by neurologic criteria in Canada noted that neurologic assessments may be unreliable in the acute post-resuscitation phase after circulatory-respiratory arrest, so evaluation for determination of death by neurologic criteria should be delayed for 24 h or an ancillary test should be performed [8]. The 2010 standard published by the American Academy of Neurology did not differentiate an observation time after hypoxic-ischemic brain injury as compared with structural brain injury; rather, it stated that there is insufficient evidence to determine the minimally acceptable observation period to ensure that neurologic functions have ceased irreversibly [5]. In European standards, timing prior to and between evaluations vary from 6–12 h, and many countries with a stipulated observation period recommend consideration of ancillary testing in lieu of waiting [37].

In addition to the aforementioned guidance provided by the World Brain Death Project about the observation time prior to performance of an evaluation for determination of death by neurologic criteria, the authors provide specific guidance that is relevant to patients with hypoxic-ischemic brain injury [30]. They recommend waiting a minimum of 24 h after hypoxic-ischemic brain injury before performing an evaluation. Further, after treatment with targeted temperature management, they advise waiting a minimum of 24 h after rewarming to ≥36 °C, taking clearance of medications that depress the central nervous system into consideration.

2.2 Evidence

There are no studies of the appropriate observation time after hypoxic-ischemic brain injury prior to evaluation for determination of death by neurologic criteria. The best surrogate for this data is the results of neuroprognostication studies. Sandroni et al. recently (2020) performed a systematic review of studies predicting poor neurologic outcome after circulatory-respiratory arrest [32]. They identified a 0% false-positive rate for prediction of poor outcome with bilateral absence of standard pupillary light reflex at 48 h after return of spontaneous circulation in some studies, which became consistent after 4 days, with sensitivities ranging from 18 to 36%. Similarly, they found a 0% false-positive rate for prediction of poor outcome at 48 h after return of spontaneous circulation or later (range 0–4%). Bilateral absence of the corneal reflex 4 days after return of spontaneous circulation consistently predicted a poor outcome; sensitivity ranged from 23 to 41%. Absence of pupillary light reflex in combination with the absence of corneal reflex on admission or within 72 h predicted poor neurologic outcome with false-positive rate ranging from 0 to 9% and sensitivity ranging from 27 to 28%. The absence of other brainstem reflexes, such as oculocephalic, gag, and cough reflexes, had 0% false-positive rate starting from 48 h after return of spontaneous circulation, but precision was low. An absent or extensor motor response to pain had high sensitivity, but low specificity, for poor outcome.

Although Sandroni's meta-analysis is helpful for neuroprognostication, death by neurologic criteria was not isolated from other poor outcomes, and there was no predictive value for absence of all brainstem reflexes. Nonetheless, we can extrapolate that the presence of coma, the absence of oculocephalic, gag and cough reflexes (false-positive rate = 0% at 48 h) combined with the absence of pupillary light reflex and corneal reflex at 48 h after return of spontaneous circulation is almost universally consistent with a poor neurologic outcome. However, while these findings are consistent with a poor neurologic outcome, Sandroni's study did not specifically address recovery of absent brainstem reflexes or motor responses.

2.3 Recommendation

Data is lacking on the appropriate time to observe a patient after hypoxic-ischemic brain injury, particularly if they are treated with targeted temperature management. However, extrapolating from the literature discussed above, we believe that unless there is clear evidence of severe hypoxic-ischemic damage on imaging after circulatory-respiratory arrest, it is probably best to wait at least 48 h before performing an evaluation for determination of death by neurologic criteria.

3 Advantages and Disadvantages of Using a Longer Observation Period Before Evaluation for Determination of Death by Neurologic Criteria

3.1 Opportunity to Address Diagnostic Uncertainty

In cases where there is any uncertainty about whether it is appropriate to perform an evaluation for death by neurologic criteria, it is *always* appropriate to delay the evaluation for further monitoring and discussion with senior specialized clinicians.

3.2 Resolution of Confounders

It would seem logical that, the greater the time elapsed between neurologic injury and evaluation for determination of death by neurologic criteria, the greater the certainty that modifiable confouders have resolved. For example, the more time that passes after a patient received a medication that could depress the central nervous system, the more certain one can be that the effect of that medication has resolved. Common time-dependent confounders include sedative medications, neuromuscular blocking medications, intoxicants (i.e. alcohol, illicit substance ingestion), hypothermia, electrolyte derangements, and oxygenation or ventilation derangements [26–29, 38].

3.3 Availability of Senior Specialized Clinicians

Many healthcare systems operate using a model in which the most senior clinicians provide coverage and oversight of their patients with a combination of in-person presence and remote supervision of trainees. An accurate evaluation for determination of death by neurologic criteria is needed to avoid a false-positive determination (determination that a patient is dead when they are not) or a false-negative determination (determination that a patient is not dead when in fact, they are). Delaying an evaluation to ensure the most experienced clinician is available to perform an in-person assessment at the bedside facilitates the greatest degree of accuracy. Of course, any clinician performing this evaluation should have adequate training and experience to do so with maximal accuracy. In some cases, it is necessary for a specialist with neurosurgical or neurological expertise to perform the evaluation or for a specialist with neuroradiology or electrophysiology training to interpret an ancillary test, so the evaluation should be delayed until they are available.

3.4 Communication and Education for Families

Injuries that lead to death by neurologic criteria are generally unexpected. Further, as discussed elsewhere in this book, families often have a poor understanding of death by neurologic criteria. Therefore, it is necessary for clinicians to take the time to carefully educate them about the patient's condition before performing an evaluation for determination of death by neurologic criteria. Communication and education about death by neurologic criteria often requires time. A delay between brain injury and evaluation for determination of death can allow families to come to terms with the situation [39].

3.5 Respect for Patients and Their Families

An unnecessary delay prior to an evaluation for determination of death by neurologic criteria can be considered disrespectful to both patients and families. It exposes patients to non-therapeutic interventions related to ongoing critical care and can lead to prolongation of suffering for families who may be anxiously awaiting a definitive answer about the patient's condition. Once it is believed that a patient may be dead by neurologic criteria and the prerequisites are met, it is incumbent on the medical team to perform an evaluation and communicate their findings to the patient's family in a timely fashion.

3.6 Resource Utilization

Another potential harm of delaying the evaluation for determination of death by neurologic criteria is the unnecessary use of critical care resources. In healthcare systems with limited resources, this is particularly problematic. Inappropriate critical care increases cost and has the potential to negatively impact the quality of care for other patients who could benefit from critical care. Lustbader et al. found a substantial increase in intensive care unit costs with an increased time interval between evaluations for determination of death by neurologic criteria, and an increase in observation time undoubtedly also leads to increase in cost [40]. There is no doubt that a longer observation time means a longer ICU stay, and an increase in resource utilization. However, cost and resource utilization should not lead to rushed evaluations.

3.7 Organ Donation

Although the primary purpose of an evaluation for determination of death by neurologic criteria is to distinguish alive from dead, it is important to note that patients who are determined to be dead by neurologic criteria have the potential to be organ donors. Allowing a period of observation prior to the evaluation can facilitate the engagement of organ donation organizations with families. Delaying the evaluation, however, can reduce organ donation. Lustbader et al. reported that performing a second evaluation for determination of death by neurologic criteria negatively affects organ donation [40]. In this series of 1229 adults and 82 pediatric patients declared dead by neurologic criteria, the mean interval between evaluations was 19.9 h in smaller hospitals compared to 16 h in larger centers. The authors noted an increase in rate of refusal, a decrease in consent, and an increase in organ loss due to circulatory-respiratory arrest between evaluations. No patient regained brainstem function between the evaluations. In another small study [41], the authors found that a delay of more than 6 h in the determination of death by neurologic criteria negatively affected consent for organ donation.

4 Conclusion

Given the ramifications of an evaluation for determination of death by neurologic criteria and the risks of a false-positive determination, patients with devastating brain injuries should be admitted to an intensive care unit to allow sufficient time for a specialist to assess them, exclude confounders, and communicate with families before performance of an evaluation for determination of death by neurologic criteria [42]. The absolute duration of observation cannot be specified. We believe that in patients with a structural brain injury leading to edema and herniation, an evaluation for determination of death by neurologic criteria can be performed whenever confounders have been eliminated. In patients with hypoxic-ischemic brain injury, if imaging is done in the first hours after the event (or not at all) and does not demonstrate that the injury is irreversible and catastrophic, we believe the evaluation for determination of death by neurologic criteria should not be performed in the first 24 h post-arrest (or 24 h after rewarming if treated with targeted temperature management), and should probably be delayed for 48 h (noting that drugs are metabolized much more slowly in this situation, especially in the context of hypothermia). Clinicians should take individual circumstances into consideration on a case-by-case basis.

References

1. Lewis A, Bakkar A, Kreiger-Benson E, et al. Determination of death by neurologic criteria around the world. Neurology. 2020;95:e299–309.
2. Greer DM, Wang HH, Robinson JD, et al. Variability of brain death policies in the United States. JAMA Neurol. 2016;73:213.

3. Wahlster S, Wijdicks EFM, Patel PV, et al. Brain death declaration: practices and perceptions worldwide. Neurology. 2015;84:1870–9.
4. Sprung CL, Truog RD, Curtis JR. Seeking worldwide professional consensus on the principles of end-of-life care for the critically ill. The consensus for worldwide end-of-life practice for patients in intensive care units (WELPICUS) study. Am J Respir Crit Care Med. 2014;190:855–66.
5. Wijdicks EFM, Varelas PN, Gronseth GS, Greer DM. Evidence-based guideline update: determining brain death in adults: report of the quality standards Subcommittee of the American Academy of Neurology. Neurology. 2010;74:1911–8.
6. Wijdicks EFM Brain death worldwide: Accepted fact but no global consensus in diagnostic criteria. Neurology. 2002;58:20–5.
7. Nakagawa TA, Ashwal S, Mathur M, Mysore M. The Committee for Determination of guidelines for the determination of brain death in infants and children: an update of the 1987 task force recommendations-executive summary. Ann Neurol. 2012;71:573–85.
8. Shemie SD. Severe brain injury to neurological determination of death: Canadian forum recommendations. Can Med Assoc J. 2006;174:S1–S12.
9. The Australian and New Zealand Intensive Care Society Statement on Death and Organ Donation. 2021. Melbourne.
10. Cronberg T, Brizzi M, Liedholm LJ, et al. Neurological prognostication after cardiac arrest—recommendations from the Swedish resuscitation council. Resuscitation. 2013;84:867–72.
11. NHS Blood and transplant. Diagnosing death using neurological criteria. 2021. https://www.odt.nhs.uk/deceased-donation/best-practice-guidance/donation-after-brainstem-death/diagnosing-death-using-neurological-criteria/.
12. Kung NH, Dhar R, Keyrouz SG. Diffuse leptomeningeal carcinomatosis mimicking brain death. J Neurol Sci. 2015;352:132–4.
13. Rigamonti A, Basso F, Stanzani L, Agostoni E, Lauria G. Guillan-Barre synrome mimicking brain death. J Peripher Nerv Syst. 2009;14:316–9.
14. Vargas F, Hilbert G, Gruson D, et al. Fulminant Guillain-Barré syndrome mimicking cerebral death: case report and literature review. Intensive Care Med. 2000;26:623–7.
15. Moussouttas M, Chandy D, Dyro F. Fulminant acute inflammatory demyelinating polyradiculoneuropathy: case report and literature review. Neurocrit Care. 2004;1:469–74.
16. Young GB. De-efferentation and De-afferentation in fulminant polyneuropathy: lessons from the isolated brain. Can J Neurol Sci. 2003;30:305–6.
17. Liik M, Puksa L, Luus S, Haldre S, Taba P. Fulminant inflammatory neuropathy mimicking cerebral death. Case Rep. 2012;2012:bcr1020114906.
18. Bakshi N, Maselli RA, Gospe SM, et al. Fulminant demyelinating neuropathy mimicking cerebral death. Muscle Nerve. 1997;20:1595–7.
19. Martí-Massó JF, Suárez J, López de Munain A, Carrera N. Clinical signs of brain death simulated by Guillain-Barré syndrome. J Neurol Sci. 1993;120:115–7.
20. Ravikumar S, Poysophon P, Poblete R, Kim-Tenser M. A case of acute motor axonal neuropathy mimicking brain death and review of the literature. Front Neurol. 2016;7:63.
21. Hantson P, Guérit J, De Tourtchaninoff M, et al. Rabies encephalitis mimicking the electrophysiological pattern of brain death. Eur Neurol. 1993;33:212–7.
22. John J, Gane BD, Plakkal N, Aghoram R, Sampath S. Snake bite mimicking brain death. Cases J. 2008;1:16.
23. Dayal M, Prakash S, Verma P, Pawar M. Neurotoxin envenomation mimicking brain death in a child: a case report and review of literature. Indian J Anaesth. 2014;58:458.
24. Freund B, Hayes L, Rivera-Lara L, et al. Adult intestinal colonization botulism mimicking brain death. Muscle Nerve. 2017;56:E27–8.
25. Joffe AR, Anton N, Blackwood J. Brain death and the cervical spinal cord: a confounding factor for the clinical examination. Spinal Cord. 2010;48:2–9.
26. López-Navidad A, Caballero F, Domingo P, et al. Early diagnosis of brain death in patients treated with central nervous system depressant drugs. Transplantation. 2000;70:131–5.
27. Morrow SA, Young GB. Selective abolition of the vestibular-ocular reflex by sedative drugs. Neurocrit Care. 2007;6:45–8.

28. Schmidt JE, Tamburro RF, Hoffman GM. Dilated nonreactive pupils secondary to neuromuscular blockade. Anesthesiology. 2000;92:1476.
29. Webb AC, Samuels OB. Reversible brain death after cardiopulmonary arrest and induced hypothermia. Crit Care Med. 2011;39:1538–42.
30. Greer DM, Shemie SD, Lewis A, et al. Determination of brain death/death by neurologic criteria. JAMA. 2020;324:1078.
31. Shemie SD, Hornby L, Baker A, et al. International guideline development for the determination of death. Intensive Care Med. 2014;40:788–97.
32. Sandroni C, D'Arrigo S, Cacciola S, et al. Prediction of poor neurological outcome in comatose survivors of cardiac arrest: a systematic review. Intensive Care Med. 2020;46:1803–51.
33. Scales DC, Golan E, Pinto R, et al. Improving appropriate neurologic prognostication after cardiac arrest. A stepped wedge cluster randomized controlled trial. Am J Respir Crit Care Med. 2016;194:1083–91.
34. Lopez Soto C, Dragoi L, Heyn CC, et al. Imaging for Neuroprognostication after cardiac arrest: systematic review and metaanalysis. Neurocrit Care. 2020;32:206–16.
35. Schick A, Prekker ME, Kempainen RR, et al. Association of hypoxic ischemic brain injury on early CT after out of hospital cardiac arrest with neurologic outcome. Am J Emerg Med. 2022;54:257–62.
36. Wang MY, Wallace P, Gruen JP. Brain death documentation: analysis and issues. Neurosurgery. 2002;51:731–5. discussion 735-6
37. Citerio G, Murphy P. Brain death: the European perspective. Semin Neurol. 2015;35:139–44.
38. Murphy L, Wolfer H, Hendrickson RG. Toxicologic confounders of brain death determination: a narrative review. Neurocrit Care. 2021;34:1072–89.
39. Sarti AJ, Sutherland S, Healey A, et al. A multicenter qualitative investigation of the experiences and perspectives of substitute decision makers who underwent organ donation decisions. Prog Transplant. 2018;28:343–8.
40. Lustbader D, O'Hara D, Wijdicks EFM, et al. Second brain death examination may negatively affect organ donation. Neurology. 2011;76:119–24.
41. Fernández-Torre JL, Hernández-Hernández MA, Muñoz-Esteban C. Non confirmatory electroencephalography in patients meeting clinical criteria for brain death: scenario and impact on organ donation. Clin Neurophysiol. 2013;124:2362–7.
42. Healey A, Leeies M, Hrymak C, et al. CAEP, CCCS, and CNSF position statement – management of devastating brain injuries in the emergency department: enhancing neuroprognostication and maintaining the opportunity for organ and tissue donation. CJEM. 2020;22:658–60.

Temperature Considerations in the Determination of Death by Neurologic Criteria

Jeffrey M. Singh and Andrew J. Baker

Determination of death by neurologic criteria is predicated on the accurate clinical evaluation for the permanent loss for the capacity of consciousness and brainstem function. One of the prerequisites for the determination of death by neurologic criteria is the elimination of any confounders of death by neurologic criteria. Hypothermia is simultaneously a potential consequence of death by neurologic criteria, due to loss of autoregulation, and a potential direct and indirect confounder to the determination of death by neurologic criteria. Hypothermia reduces brain metabolism and may exacerbate pharmacological confounding by alterations of pharmacokinetics of medications during hypothermia.

These issues require attention by clinicians considering performance of an evaluation for determination of death by neurologic criteria. In this chapter, we review the physiology, pharmacokinetics, and clinical issues relevant to temperature considerations in the determination of death by neurologic criteria, focusing on hypothermia. We also provide recommendations on managing temperature/hypothermia to facilitate determination of death by neurologic criteria.

J. M. Singh
Krembil Brain Institute, University Health Network, Toronto, ON, Canada

Department of Medicine, University of Toronto, Toronto, ON, Canada

Interdepartmental Division of Critical Care Medicine, University of Toronto, Toronto, ON, Canada
e-mail: Jeff.Singh@uhn.ca

A. J. Baker (✉)
Interdepartmental Division of Critical Care Medicine, University of Toronto, Toronto, ON, Canada

Department of Anesthesiology and Pain Medicine, University of Toronto, Toronto, ON, Canada

St. Michael's Hospital, Unity Health Toronto, Toronto, ON, Canada
e-mail: andrew.baker@unityhealth.to

A. Lewis, J. L. Bernat (eds.), *Death Determination by Neurologic Criteria*, Advances in Neuroethics, https://doi.org/10.1007/978-3-031-15947-3_16

1 Alterations in the Regulation of Body Temperature After Death by Neurologic Criteria

The normal range of human core temperature is 36.4–37.5 °C. Heat is produced by metabolic activity, either basal and voluntary activity or though shivering and non-shivering thermogenesis, and is lost primarily through radiation and evaporation from the skin and respiratory tract [1]. Core body temperature is maintained in the physiological range by a variety of heat-generation and heat-conservation mechanisms. These mechanisms, however, can be overwhelmed under extreme environmental conditions or through application of therapeutic cooling (e.g. induced hypothermia) [1].

Thermogenesis may be lost after death by neurologic criteria. Because non-shivering and shivering thermogenesis are governed by the anterior and posterior hypothalamus, respectively, loss of these homeostatic functions after death by neurologic criteria may lead to disturbances in temperature with the onset of poikilothermia. This may be manifest clinically by hypothermia resulting from heat loss into the ambient environment without compensatory thermogenesis. The potential loss of thermoregulatory function after death by neurologic criteria requires clinicians to vigilantly monitor for hypothermia, which may confound the determination of death by neurologic criteria and may impact organ function in patients who will donate organs after death by neurologic criteria.

2 Relationship Between Temperature, Brain Metabolism, and Function

The human brain is metabolically highly active, and although it comprises only a few percent of human body mass, it accounts for one-quarter of the body's total glucose utilization and one-fifth of oxygen utilization at resting state [2]. This high metabolic rate also produces a considerable amount of heat, and activity-related heat generation is sufficient to generate regional variations of brain temperature corresponding to local metabolic activity [2]. The excess heat produced by the brain is cleared in the normal state by perfusing blood, leading to a venous-to-arterial temperature gradient [3]. Excess heat may also be removed through direct conduction to the cerebrospinal fluid and skull.

In addition to the brain producing heat through metabolic activity, almost all cerebral processes are passively affected by temperature. Oxygen consumption and energy expenditure decrease approximately 7% with every centigrade degree decrease in brain temperature, down to approximately 25 °C at which point cerebrovascular autoregulation is thought to fail [1]. The function of individual neurons and global cognitive function are sensitive to temperature fluctuations, as demonstrated in both in vitro and animal experiments [4, 5]. Over a range of physiological temperatures, in vitro experiments show neuron discharge rates decrease as temperature drops, with this decrease becoming more dramatic once temperature drops below physiological temperatures. This phenomenon of temperature-metabolism coupling is used therapeutically in the application of targeted temperature management (TTM), induced hypothermia, and fever avoidance after acute brain injuries to try to limit secondary brain injury. In the context of an ischemic insult, this reduction in metabolism may also be neuroprotective, with multiple published cases of remarkable survival and neurological recovery after accidental severe hypothermia and prolonged circulatory arrest [6].

In the context of death by neurologic criteria, brain temperature falls because of cessation of brain metabolic activity as well and the loss of hemostatic and autoregulatory functions. This drop in temperature has been observed clinically, and the reversal of the normal positive brain temperature-core temperature gradient can occur, reflecting a lack of metabolic activity and cerebral perfusion [7]. One study found that brain and trunk temperature run parallel in time in patients who were unconscious but alive, but dissociated with loss of brain function: core body temperature fell over 6–12 h with a significantly greater decrease in brain temperature—such that the brain was the coldest part of the body, 2–4 °C lower than core body temperature [8].

3 Clinical Implication of Temperature on Determination of Death by Neurologic Criteria

Brain temperature, specifically hypothermia, may influence the determination of death by neurologic criteria in several ways. The key underlying requirement for determination of death by neurologic criteria is that there must be an established neurologic diagnosis with a sufficient severity to lead to the complete and irreversible loss of all brain function, and exclusion of conditions that either confound the clinical evaluation or mimic death by neurologic criteria. Hypothermia may confound determination of death by neurologic criteria either directly (by a reduction in brain metabolism) or indirectly through altered pharmacokinetics of drugs and medications (particularly sedative/hypnotic medications), leading to plasma accumulation and subsequent confounding of the determination.

3.1 Direct Confounding of Determination of Death by Neurologic Criteria by Hypothermia

Although the depression of cerebral metabolism and function by hypothermia is well-established from physiologic experiments and animal studies, there are no data to indicate a threshold temperature that precludes confounding of the clinical determination for death by neurologic criteria. Most of the published literature on the impact of temperature on the neurologic evaluation is in the context of therapeutic hypothermia or TTM after circulatory-respiratory arrest. It is difficult to draw inferences from these data because the application of therapeutic hypothermia is usually associated with concurrent administration of other sedative medications which may also confound the determination of death by neurologic criteria.

There are some informative neurophysiologic data demonstrating that electroencephalographic silence occurs only at very low temperatures (below 20 °C) [9]. One neurophysiology study of 109 patients with hypothermic circulatory-respiratory arrest during surgery found that the mean core temperature when electroencephalographic silence appeared was 20.6 °C, with the highest nasopharyngeal temperature associated with electroencephalographic silence in their cohort being 27.2 °C. Likewise, the mean core temperatures associated with disappearance of the N20-P22 and N13 complexes on somatosensory evoked potentials were 24.7 °C and 20.1 °C respectively [9]. It should be noted that these patients also received induction of anesthesia with midazolam, fentanyl, and isoflurane, which may have further suppressed cortical function and raised the temperature threshold for electrophysiological silence. Another study found that the cortically generated component of somatosensory evoked potentials (N19) was consistently recordable at core temperatures above 26 °C, and disappeared after decreasing the temperature down to 20 °C [10]. These same authors studied brainstem auditory evoked potentials during induced hypothermia for cardiac surgery and found that the components were present in all patients at temperatures above 23 °C and absent below 20 °C [11].

These indirect data would suggest that decreased brain temperature by itself is unlikely to mimic death by neurologic criteria in the healthy brain except at very low temperatures (less than 30 °C). It is possible, however, that mild to moderate hypothermia sufficiently depresses the function of an injured brain below the threshold for clinical detection. Cognitive dysfunction has been observed with temperature exposure and mild fluctuations in core body temperature [12, 13].

The rate of rewarming from hypothermia is also an important consideration because rapid rewarming can exacerbate supply–demand mismatch and induce metabolic crisis and thus confound the clinical exam by inducing transient ischemic brain dysfunction. Animal and pediatric studies have demonstrated reversible impairment of pressure autoregulation after rewarming [14], and a transient mismatch between cerebral metabolic oxygen demand and supply [15]. One pediatric study of children undergoing cardiac procedures with profound hypothermia found that jugular bulb desaturation during rewarming correlated with rate of temperature rise, with lower saturations and higher arterio-venous saturation extraction associated with rapid warming [16]. The threshold rate of rewarming to avoid these phenomena is not known, and likely is dependent on the underlying brain injury, depth of hypothermia, and hemodynamic parameters. Suggested rates for rewarming after hypothermia or TTM after circulatory-respiratory arrest are often conservative, with rates of 0.15–0.5 °C/h [17, 18]. Care should be taken to warm slowly from moderate and severe hypothermia, or an adequate period of observation has passed to ensure transient energy debt or ischemia from rewarming is not present and confounding the evaluation.

3.2 Impact of Temperature on Drug Pharmacokinetics

Hypothermia has an important effect on the pharmacokinetics of medications which may confound the determination of death by neurologic criteria. Hypothermia may impact all or some of the enteral absorption, distribution, and elimination of medications. The rate of absorption of enteral medications, particularly for drugs reliant on active transport, is delayed and decreased with decreasing temperature. The volume of distribution (Vd) of drugs is affected by both the state of physiology and the physicochemical properties of the drug; hypothermia may increase or decrease Vd through complex and interdependent mechanisms including altered tissue perfusion, depressed organ function and disturbances in acid–base status and drug–protein binding [19]. Finally, drug clearance is the most significantly altered; impairment in renal and hepatic clearance can lead to increases in plasma levels of drug or active metabolites [19]. Studies have found that mild to moderate hypothermia has a significant impact on the pharmacokinetics of cytochrome P450-metabolized drugs with a decrease in systemic clearance between 7–22% for every degree below 37 °C, and an increase in serum concentrations that can persist for days post-rewarming [20, 21]. This was also true for common sedative agents, including midazolam [22]. This delayed clearance has been implicated in cases where there have been "reversible" findings consistent with death by neurologic criteria or a false positive determination of death by neurologic criteria related to confounded clinical examinations [23, 24].

4 Determination of Death by Neurologic Criteria After Therapeutic Hypothermia or Targeted Temperature Management

One particularly challenging situation is the determination of death by neurologic criteria in a patient who appears to have lost all brainstem reflexes following circulatory-respiratory arrest but is being treated with TTM. TTM, usually targeting mild hypothermia, became the standard of care in survivors of circulatory-respiratory arrest following the publication of studies demonstrating improved outcomes in patients who were cooled after out-of-hospital arrest with both shockable [25, 26] and unshockable [27] initial cardiac rhythms. Although subsequent studies have demonstrated the equivalency of targeting mild hypothermia and 36 °C [28, 29], cooling post-arrest remains a common practice. Unfortunately, despite best medical care, mortality following circulatory-respiratory arrest is high and many of these deaths are determined by neurologic criteria. In one study comparing treatment with controlled hypothermia and normothermia, over 10% of all deaths were determined by neurologic criteria, and over 40% of all patients who were rewarmed early at the discretion of their treating physician did so because of suspected death by neurologic criteria [27].

Determination of death by neurologic criteria following circulatory-respiratory arrest and TTM is challenging due to the multiple potential confounders of the evaluation. Firstly, if cooling to moderate hypothermia is administered, brainstem reflexes may be temporarily depressed or absent due to the impact of hypothermia on brain metabolism and function. Secondly, the administration of sedatives and opioids, which may confound the determination for death by neurologic criteria, is common during TTM. Most studies implementing a temperature target of 32 °C administered sedative and opioid infusions in all patients for some period of time [25–29], and many studies also administered sedation to all patients treated with targeted normothermia for some period to prevent or treat shivering [27, 28]. The choice of sedative regimen is also important, as some regimens may shorten the time to awakening in comatose survivors of circulatory-respiratory arrest treated with TTM [30]. Since the abolition of brainstem reflexes has been documented with sedatives [31] and neuromuscular blockade agents [32] respectively, clinicians must exercise caution to ensure that these potential confounders have not accumulated during the period of hypothermia as a result of the temperature-related disturbances in pharmacokinetics and pharmacodynamics. One study found that sedative medications were commonly used in proximity of neurological assessment in comatose survivors of arrest, thus potentially confounding accurate neuroprognostication [33].

5 Existing Guidance on Temperature Considerations for the Clinical Evaluation of Death by Neurologic Criteria

Despite these issues, only 78% of international standards on determination of death by neurologic criteria address temperature considerations [34]. There is considerable variability in the recommend minimum core temperature, ranging from 32 °C to 36 °C, and some standards merely specify that the patient be normothermic, or not be hypothermic. Further, standards do not routinely stipulate that clearance of drugs may be prolonged after hypothermia due to increased accumulation when the core temperature was low. There is no standard waiting period or delay from TTM or rewarming before determination of death by neurologic criteria [35–39]. Although eight standards specifically address determination of death by neurologic criteria after hypothermia, only two clearly indicate how long clinicians should delay the determination in this setting; Australia/New Zealand advise waiting 24 h and Poland advises waiting at least 24 h [34].

The variability across standards is problematic from the point of view of construct validity for the determination of death by neurologic criteria. Standards should be consistent across jurisdictions, so determination of death is the same everywhere.

6 Ancillary Testing and Core Body Temperature

While the clinical evaluation for death by neurologic criteria can be impacted by temperature, there is less concern about this with the use of ancillary testing to provide support for the determination. Most ancillary testing modalities evaluate for absence of brain circulation. Given the neuroprotective effects of hypothermia, it is necessary to consider the potential impact of temperature, particularly severe hypothermia, on both the diagnostic accuracy of ancillary testing and the assumptions in a patient who is hypothermic.

There are no human studies evaluating the receiver operating characteristics of any modern ancillary testing modalities for the determination of death by neurologic criteria during hypothermia. Cerebral blood flow may fluctuate during hypothermia and rewarming because of metabolic-flow coupling, but there are no reports of temperature-related reversal of absence of brain circulation demonstrated on ancillary testing. There are reassuring data from animal studies that show cerebral blood flow may decrease with hypothermia, but is preserved even at very low temperatures. One study demonstrated persistent brain circulation in the microvasculature of rats during severe hypothermia, even at temperatures below 18 °C [40]. Another study in pigs found that brain circulation decreased with cooling, but persisted despite suppression of metabolic activity even at temperatures as low as 8 °C [41]. These data are helpful because they support the idea that absence of brain circulation cannot be reversibly induced by hypothermia.

7 Our Recommendations on Considerations Pertaining to Temperature Management in the Determination of Death by Neurologic Criteria

Temperature may have a significant impact on the clinical determination of death by neurologic criteria. The effects may be direct through hypothermia related depression of brain metabolism and function, or indirect through the accumulation of confounding drugs from hypothermia-related changes in pharmacokinetics. These effects may also be additive such that mild hypothermia and a sub-therapeutic level of sedative may together effectively abolish clinical responses, so elimination of all possible confounders and restoration of normothermia is recommended. Given the importance and implications of death determination, we recommend a cautious and conservative approach to determination of death by neurologic criteria in the context of accidental hypothermia or recent application of TTM.

Given the potential for the evaluation for coma and absence of brainstem function to be impacted by hypothermia, patients suspected to be dead by neurologic criteria should not be hypothermic at the time of the determination. Firstly, patients with a clinical evaluation suggestive of death by neurologic criteria and a brain injury consistent in extent and severity to cause death should be warmed to normothermia using external warming, fluid warmers, heated ventilator circuits and automated temperature regulation devices as required. In jurisdictions where multiple independent evaluations are required, physicians should be aware that loss of thermoregulation due to loss of brain function may result in significant changes in body temperature between evaluations as patients become poikilothermic and verify the patient's temperature prior to each evaluation. We recommend a minimum body temperature of 36 °C prior to performing the clinical evaluation for determination of death by neurologic criteria. This reflects the aforementioned theoretical potential for incremental confoundment at lower temperatures, the fact that the risks of warming to normothermia (36 °C) are few, and is consistent with the recent recommendation of the World Brain Death Project [36].

After normothermia has been established, a thorough review of all potentially confounding medications should occur. Caution should be exercised to ensure adequate time for drug clearance prior to clinical evaluation for determination of death by neurologic criteria, given the likelihood of pharmacological confounding from either medication administered during hypothermia or accumulation related to changes in drug pharmacokinetics after cooling. Careful attention should be paid to the timing of administration in relation to the timing and duration of hypothermia. If there is no concern regarding confounding of the clinical exam, determination of death by neurologic criteria can proceed as per accepted standards. If any confounding medications have been administered, the clinician has two options: delay the determination to allow clearance of accumulated drugs, or perform the clinical evaluation and an ancillary study to evaluate for absence of brain circulation. It is extremely difficult to estimate the required delay to ensure elimination of accumulated drug during hypothermia, due to the complex and competing changes in pharmacokinetics related to temperature for different drugs. Indeed, this issue has come into play in high-profile published cases of reversible or false positive determinations of death by neurologic criteria [23, 24]. Accordingly, we recommend extreme caution and a conservative approach in this regard, with prolonged delay of the determination when possible and use of ancillary testing to support the clinical evaluation when necessary.

8 Conclusion

In summary, variations in temperature outside of the normal range, particularly hypothermia, have important effects on brain function, metabolism, and physiology and may confound the determination of death by neurologic criteria. This effect may be exacerbated by loss of thermoregulation when brain function is lost leading to poikilothermia and a decrease in core temperature in response to ambient heat loss. Careful attention to restoration of normal core body temperature and to the accumulation and clearance of potentially confounding medications are required to ensure that determination of death by neurologic criteria is accurate. Special attention is warranted in patients who are treated with therapeutic hypothermia due to the ubiquitous administration of confounding medications and the potential for their delayed clearance. Finally, if there is concern that pharmacological confounders cannot be excluded after hypothermia, ancillary testing should be performed to support the clinical evaluation for determination of death by neurologic criteria.

References

1. Leikin SM, Korley FK, Wang EE, Leikin JB. The spectrum of hypothermia: from environmental exposure to therapeutic uses and medical simulation. Dis Mon. 2012;58(1):6–32.
2. Wang H, Wang B, Normoyle KP, et al. Brain temperature and its fundamental properties: a review for clinical neuroscientists. Front Neurosci. 2014;8:307.

3. Yablonskiy DA, Ackerman JJ, Raichle ME. Coupling between changes in human brain temperature and oxidative metabolism during prolonged visual stimulation. Proc Natl Acad Sci U S A. 2000;97(13):7603–8.
4. Kalmbach AS, Waters J. Brain surface temperature under a craniotomy. J Neurophysiol. 2012;108(11):3138–46.
5. Guatteo E, Chung KK, Bowala TK, et al. Temperature sensitivity of dopaminergic neurons of the substantia nigra pars compacta: involvement of transient receptor potential channels. J Neurophysiol. 2005;94(5):3069–80.
6. Paal P, Gordon L, Strapazzon G, et al. Accidental hypothermia–an update. Scand J Trauma Resusc Emerg Med. 2016;24(1):1–20.
7. Orita T, Izumihara A, Tsurutani T, Kajiwara K. Brain temperature before and after brain death. Neurol Res. 1995;17(6):443–4.
8. Lysoń T, Jadeszko M, Mariak Z, Kochanowicz J, Lewko J. Intracranial temperature measurements in brain death. Neurol Neurochir Pol. 2006;40(4):269–75.
9. Stecker MM, Cheung AT, Pochettino A, et al. Deep hypothermic circulatory arrest: I. Effects of cooling on electroencephalogram and evoked potentials. Ann Thorac Surg. 2001;71(1):14–21.
10. Markand ON, Warren C, Mallik GS, et al. Effects of hypothermia on short latency somatosensory evoked potentials in humans. Electroencephalogr Clin Neurophysiol. 1990;77(6):416–24.
11. Markand ON, Lee BI, Warren C, et al. Effects of hypothermia on brainstem auditory evoked potentials in humans. Ann Neurol. 1987;22(4):507–13.
12. Taylor L, Watkins SL, Marshall H, Dascombe BJ, Foster J. The impact of different environmental conditions on cognitive function: a focused review. Front Physiol. 2016:6–372.
13. Walter EJ, Carraretto M. The neurological and cognitive consequences of hyperthermia. Crit Care. 2016;20(1):199.
14. Joshi B, Brady K, Lee J, et al. Impaired autoregulation of cerebral blood flow during rewarming from hypothermic cardiopulmonary bypass and its potential association with stroke. Anesth Analg. 2010;110(2):321–8.
15. Enomoto S, Hindman Bradley J, et al. Rapid rewarming causes an increase in the cerebral metabolic rate for oxygen that is temporarily unmatched by cerebral blood flow: a study during cardiopulmonary bypass in rabbits. Anesthesiology. 1996;84(6):1392–400.
16. van der Linden J, Ekroth R, Lincoln C, Pugsley W, Scallan M, Tydén H. Is cerebral blood flow/metabolic mismatch during rewarming a risk factor after profound hypothermic procedures in small children? Eur J Cardiothorac Surg. 1989;3(3):209–15.
17. Scirica BM. Therapeutic hypothermia after cardiac arrest. Circulation. 2013;127(2):244–50.
18. Taccone FS, Picetti E, Vincent J-L. High quality targeted temperature management (TTM) after cardiac arrest. Crit Care. 2020;24(1):1–6.
19. van den Broek MP, Groenendaal F, Egberts AC, Rademaker CM. Effects of hypothermia on pharmacokinetics and pharmacodynamics: a systematic review of preclinical and clinical studies. Clin Pharmacokinet. 2010;49(5):277–94.
20. Tortorici MA, Kochanek PM, Poloyac SM. Effects of hypothermia on drug disposition, metabolism, and response: A focus of hypothermia-mediated alterations on the cytochrome P450 enzyme system. Crit Care Med. 2007;35(9):2196–204.
21. Anderson KB, Poloyac SM, Kochanek PM, Empey PE. Effect of hypothermia and targeted temperature management on drug disposition and response following cardiac arrest: A comprehensive review of preclinical and clinical investigations. Ther Hypothermia Temp Manag. 2016;6(4):169–79.
22. Hostler D, Zhou J, Tortorici MA, et al. Mild hypothermia alters midazolam pharmacokinetics in normal healthy volunteers. Drug Metab Dispos. 2010;38(5):781–8.
23. Webb AC, Samuels OB. Reversible brain death after cardiopulmonary arrest and induced hypothermia. Crit Care Med. 2011;39(6):1538–42.
24. Joffe AR, Kolski H, Duff J, deCaen AR. A 10-month-old infant with reversible findings of brain death. Pediatr Neurol. 2009;41(5):378–82.
25. Mild therapeutic hypothermia to improve the neurologic outcome after cardiac arrest. N Engl J Med. 2002;346(8):549–56.

26. Bernard SA, Gray TW, Buist MD, et al. Treatment of comatose survivors of out-of-hospital cardiac arrest with induced hypothermia. N Engl J Med. 2002;346(8):557–63.
27. Lascarrou JB, Merdji H, Le Gouge A, et al. Targeted temperature management for cardiac arrest with nonshockable rhythm. N Engl J Med. 2019;381(24):2327–37.
28. Nielsen N, Wetterslev J, Cronberg T, et al. Targeted temperature management at 33°C versus 36°C after cardiac arrest. N Engl J Med. 2013;369(23):2197–206.
29. Dankiewicz J, Cronberg T, Lilja G, et al. Hypothermia versus Normothermia after Out-of-Hospital Cardiac Arrest. N Engl J Med. 2021;384(24):2283–94.
30. Paul M, Bougouin W, Dumas F, et al. Comparison of two sedation regimens during targeted temperature management after cardiac arrest. Resuscitation. 2018;128:204–10.
31. Morrow SA, Young GB. Selective abolition of the vestibular-ocular reflex by sedative drugs. Neurocrit Care. 2007;6(1):45–8.
32. Schmidt JE, Tamburro RF, Hoffman GM. Dilated nonreactive pupils secondary to neuromuscular blockade. Anesthesiology. 2000;92(5):1476–80.
33. Samaniego EA, Mlynash M, Caulfield AF, Eyngorn I, Wijman CA. Sedation confounds outcome prediction in cardiac arrest survivors treated with hypothermia. Neurocrit Care. 2011;15(1):113–9.
34. Lewis A, Bakkar A, Kreiger-Benson E, et al. Determination of death by neurologic criteria around the world. Neurology. 2020;95(3):e299–309.
35. Greer DM, Wang HH, Robinson JD, et al. Variability of brain death policies in the United States. JAMA Neurol. 2016;73(2):213–8.
36. Greer DM, Shemie SD, Lewis A, et al. Determination of brain death/death by neurologic criteria: The World Brain Death Project. JAMA. 2020;324(11):1078–97.
37. Simpson P, Bates D, Bonner S, et al. A code of practice for the diagnosis and confirmation of death. London: Acadamy of Medical Royal Colleges; 2008.
38. Shemie SD, Doig C, Dickens B, et al. Severe brain injury to neurological determination of death: Canadian forum recommendations. CMAJ. 2006;174(6):S1–13.
39. Cronberg T, Brizzi M, Liedholm LJ, et al. Neurological prognostication after cardiac arrest--recommendations from the Swedish Resuscitation Council. Resuscitation. 2013;84(7):867–72.
40. Mel'nikova NN, Petrova LA. Effect of hypothermia-induced respiratory arrest on cerebral circulation in rats. Bull Exp Biol Med. 2016;160(5):593–5.
41. Ehrlich MP, McCullough JN, Zhang N, et al. Effect of hypothermia on cerebral blood flow and metabolism in the pig. Ann Thorac Surg. 2002;73(1):191–7.

How Many Evaluations Are Required to Determine Death by Neurologic Criteria?

Panayiotis N. Varelas

1 The History Behind the Number of Evaluations and the Observation Period Before an Evaluation

The first mention of the need for more than one evaluation to determine death by neurologic criteria is in the landmark 1968 Journal of the American Medical Association paper entitled "Report of the Ad Hoc Committee of the Harvard Medical School." This Committee, chaired by Dr. Henry K. Beecher, an anesthesiologist, suggested that in order to assure irreversibility of the condition, after the first evaluation, repeat testing is needed after at least 24 h to confirm unreceptivity and unresponsivity, no movements or breathing, no reflexes, and a flat electroencephalogram [1].

In 1981, a report of the medical consultants on the determination of death to the President's Commission for the Study of Ethical Problems in Medicine and Biomedical and Behavioral Research established that the cessation of all brain functions should persist for an appropriate period of observation and/or trial of therapy. Although the paper recognized that the duration of observation periods was a matter of clinical judgment, it also allowed a 6-h period between the evaluations if a "confirmatory" electroencephalogram was utilized. In the absence of confirmatory tests, a period of observation of at least 12 h was recommended when an irreversible condition was well established. For patients with hypoxic-ischemic brain damage, an observation for 24 h was "desirable." However, this observation period could be reduced if a test showed cessation of brain circulation or if an electroencephalogram showed electrocerebral silence in an adult patient without drug intoxication, hypothermia, or shock [2].

In 1995, the Quality Standards Subcommittee of the American Academy of Neurology published a practice parameter for determining death by neurologic criteria in adults. The practice parameter included an option for a repeat neurologic

P. N. Varelas (✉)
Department of Neurology, Albany Medical Center, Albany, NY, USA
e-mail: varelap@amc.edu

A. Lewis, J. L. Bernat (eds.), *Death Determination by Neurologic Criteria*, Advances in Neuroethics, https://doi.org/10.1007/978-3-031-15947-3_17

evaluation and stated that "the interval is arbitrary, but a 6-h period is reasonable" [3, 4].

The most recent update by the Quality Standards Subcommittee of the American Academy of Neurology was published in 2010 [5]. This guideline introduced the concept of a single evaluation to determine death by neurologic criteria. It is noteworthy that the authors were cautious enough to state, "*If a certain period of time has passed since the onset of the brain insult to exclude the possibility of recovery (in practice, usually several hours), one neurologic evaluation should be sufficient to pronounce brain death. However, some US state statutes require two evaluations.*" Interestingly, some states which required two evaluations at the time (like New York) switched to a single evaluation after this update was published [6].

In 1987, the American Academy of Pediatrics Task Force on Brain Death in Children, published guidelines for determination of death by neurologic criteria for this patient population [7]. In this guideline, the recommendations for the number of evaluations required to determine death were age-dependent: for infants 7 days to 2 months, two clinical evaluations and two electroencephalograms separated by at least 48 h; for children 2 months to 1 year, two evaluations and electroencephalograms separated by at least 24 h (if a cerebral radionuclide angiography is performed without visualization of cerebral arteries, then a second evaluation was not required); for children older than 1 year, ancillary testing was not required, but the evaluations were separated by a 12-h period (after hypoxic-ischemic injury, a 24-h period was recommended, but this could be reduced if the electroencephalogram showed electrocerebral silence or the cerebral radionuclide angiography did not show visualization of cerebral arteries).

In 2012, the Pediatric Section of the Society of Critical Care Medicine, the Section on Critical Care of the American Academy of Pediatrics, and the Child Neurology Society revised the Pediatric Guidelines [8]. To be determined dead by neurologic criteria, these guidelines indicated that pediatric patients must have two evaluations performed by two different physicians, with an observation period between the clinical evaluations and apnea tests (although these may be performed by the same physician). Moreover, age-specific recommendations address the challenge of determining death by neurologic criteria in the pediatric population. For pre-term infants, no recommendation was included due to limited literature. For term neonates (37 weeks and up to 30 days of age), the observation period between evaluations was recommended to be at least 24 h, but only 12 h was required for infants and children from >30 days to 18 years of age. The guidelines indicated that the first evaluation may be performed 24 h after birth or 24 h following circulatory-respiratory resuscitation or other severe brain injury. The guidelines required ancillary testing (1) when any components of the evaluation or apnea testing cannot be completed; (2) if there is uncertainty about the results of the neurologic evaluation; or (3) if a medication effect may be present. Interestingly, ancillary testing was also recommended as an option to reduce the inter-evaluation period. Even if this period was shortened, though, a second neurologic evaluation was still required. Ancillary studies in neonates may have limited sensitivity, so it may be necessary to rely on repeated evaluations [9].

The inconsistency between the adult and pediatric guidelines raises questions about the observation period prior to determination of death by neurologic criteria and the number of evaluations required to determine death by neurologic criteria. Further, there is notable variability in the observation period and the number of evaluations required across national guidelines, state laws, hospital policies, and bedside practice [10–13]. The observation time prior to determination of death by neurologic criteria is discussed elsewhere in this book. Here, I review considerations that could impact the number of evaluations performed, advantages and disadvantages of performing multiple evaluations, variability in the number of required evaluations, and, finally, provide my personal recommendations for the suggested number of evaluations for determination of death by neurologic criteria in various clinical circumstances.

2 Considerations That Could Impact the Number of Evaluations Performed

The two major concerns of physicians and families when determining death by neurologic criteria are whether the patient's condition may, in fact, be reversible and the possibility of faulty determination. To mitigate these risks, more than one evaluation has been utilized to reduce the risk of mistakes via repetition and allowance for more time to pass following the patient's injury.

2.1 Evaluation for Irreversibility

Irreversibility of the determination is key; the need for this dates back to the Declaration of Sydney in 1968 and its amendment in 1983 [14]. If an observation period is allowed and the evaluation is repeated, the performance of an additional evaluation may theoretically increase the chances of detecting some residual brain function that emerged between the first and second evaluation. However, the counterargument to this point is that if it is believed that there is a chance of reversibility, the determination should not be initiated in the first place. So, what is this period of no-return? Many variables could play a role in defining this. Pathology studies may shed light on the period of time required to have severe, irreversible brain damage either through necrosis or apoptosis.

Post-global hypoxic-ischemic brain injury (for example after circulatory-respiratory arrest), the timeframe to determine a condition is irreversible may be different than it is in patients with other types of brain injuries (for example, after a large ischemic stroke). In both situations, this period should be long enough to allow for the cells to be irreversibly damaged and/or their connections (reflecting the functions they serve) to be irreversibly lost. In an autopsy study after hypoxic-ischemic brain injury, a median (interquartile range) period of 25 (15–40) min to return of spontaneous circulation was associated with severe selective eosinophilic neuronal death scores [15] indicating severe irreversible injury in >30% neuronal cells in the cortex and/or brainstem. However, in this study only 2 patients were determined to be dead by neurologic criteria (49% died following withdrawal of organ support) and in one of them the entire determination process was not even completed (*Christian Endisch, MD, personal communication*). In another neuropathology study of 23 patients, the total pathological damage score had a modest correlation with the time to return of spontaneous circulation (Spearman correlation coefficient of 0.66), indicating that other factors play a role beyond no-flow time, such as the duration and quality of cardiopulmonary resuscitation [16]. Because hypoxic-ischemic injury frequently spares the brainstem, the overall incidence of death by neurologic criteria in this patient population is relatively low (10%) [17]. Patients who eventually are determined to be dead by neurologic criteria invariably have diffuse brain edema on neuroimaging, but this finding may be missed if the neuroimaging is performed very early after the event. There is no evidence that, in this small subgroup of patients who have diffuse brain edema and herniation, brainstem reflexes can return over time, but in other patients without these features, brainstem reflexes may emerge with time.

One large single center study reported the results of dual evaluations for determination of death by neurologic criteria in 71 patients with hypoxic-ischemic brain injury performed by two different physicians (separated by at least 24 h and followed by an apnea test after the second evaluation) [18]. The median time between circulatory-respiratory arrest and the first evaluation ("observation period") was 51 h, the median time between the first and second evaluation was 25 h and the median time between the second evaluation and organ donation (for those who became donors) was an additional 49 h. None of these patients regained any brain function during any of these time intervals. Half (36) of these patients had ancillary testing that revealed no brain circulation, indicating massively elevated intracranial pressure and no tissue perfusion. These data suggest that if a significant no-flow or low-flow period has elapsed and if there is diffuse bilateral cerebral edema with signs of central herniation, loss of brain function is irreversible and a single evaluation may suffice.

In patients with other catastrophic brain injuries (large ischemic strokes, large intracranial or subarachnoid hemorrhages, severe head trauma with significant shifts of tissue, etc.) who have rostrocaudal evolution of brain damage with effacement of the basilar cisterns, transtentorial and then tonsillar herniation seen on imaging or intracranial pressure equal or greater than the mean arterial pressure, reversibility is easier to establish. Contrary to global ischemia, where the cerebral perfusion is nil during pulselessness, in most of these focal processes, the tissue cerebral perfusion may be decreased within an interval spanning minutes to several hours, but eventually reaches zero or negative values [19, 20]. Accordingly, in this patient population, there is no evidence that more than one evaluation is required if all the criteria are met (based on these observational data supporting lack of reversibility [18]). However, despite the fact that cerebral perfusion pressure can be measured, there is no data on the time between reaching zero after such a gradual process and the point of no return after which there could not be any reversible brain function if tissue perfusion somehow resumed.

A different pathophysiologic process occurs in patients with posterior fossa lesions (brainstem ischemic strokes or hemorrhages). In those patients, the clinical evaluation (including an apnea test) may be consistent with death by neurologic criteria, but there may be residual brain circulation supratentorially and even functional circuits (electroencephalographic activity [21, 22], or visual-evoked potentials [23]). In patients with these lesions who initially have residual brain circulation or electrical activity observed on ancillary testing, a small case series suggests that this brain circulation ceases on serial ancillary testing after a period that extends from 2 h to 6 days [24]. The cause of supratentorial blood flow and function loss may be explained by the development of obstructive hydrocephalus at the level of the cerebral aqueduct of Sylvius, leading to elevated intracranial pressure, inability of blood pressure to compensate due to primary cardiomedullary center collapse (absent Cushing's triad), and additional cerebral ischemia, with further cytotoxic cerebral edema development and ultimately complete cessation of brain circulation. An alternative, or contributing, mechanism could be collapse of the vein of Galen, its tributaries, and the basal venous plexuses due to high intracranial pressure from developing hydrocephalus or upward herniation, leading to additional edema and a vicious cycle of intracranial pressure surge. Interestingly, in a 2008 white paper on controversies in the determination of death by the President's Council on Bioethics, *primary brainstem lesions* were excluded from the group considered to have "total brain failure" emphasizing that the condition of the brainstem was not by itself considered a reliable indicator of the condition of the higher brain centers [25]. More recently, the World Brain Death Project suggests that if an isolated brainstem injury is the cause of coma with absence of brainstem reflexes and a positive apnea test, ancillary testing should be performed and supratentorial and infratentorial blood flow loss should be confirmed before determination of death by neurologic criteria [26].

2.2 Prevention of False-Positive Determinations of Death by Neurologic Criteria

The second concern associated with determination of death by neurologic criteria is that the determination could be incorrect such that a patient who is determined to be dead by neurologic criteria is not (*false-positive death*) due to a mistake made during the evaluation (for example, omission of parts of the evaluation or misinterpretation of the findings). In theory, a second evaluation by a different or more competent and experienced physician at a different time may correct these problems. However, how many evaluations would need to be performed to facilitate confidence in the determination is unknown. In some cases, more than two evaluations are presently required [10–12]. Repetition of the evaluation does not definitively prevent false-positive determination, though. In fact, there are no data to date supporting the hypothesis that more than one evaluation for death by neurologic criteria by different or the same examiner decreases the risk of a false-positive determination.

Similarly, there is no proof that escalating the number of evaluations further, such as two versus three versus more, inversely correlates with errors.

To prevent false-positive determinations, it is important to recognize that another factor is perhaps even more important than the number of evaluations— the level of experience of the person performing the evaluation. The 2010 American Academy of Neurology Guidelines state that, "*Legally, all physicians are allowed to determine brain death in most US states. Neurologists, neurosurgeons, and intensive care specialists may have specialized expertise. It seems reasonable to require that all physicians making a determination of brain death be intimately familiar with brain death criteria and have demonstrated competence in this complex examination. Some US state or hospital guidelines require the examiner to have certain expertise*" [5]. In a study evaluating US hospital policies on determination of death by neurologic criteria, 33% of policies required expertise in neurology or neurosurgery to perform the evaluation [11]. Parts of the evaluation may be missed or misinterpreted by inexperienced physicians, especially in smaller hospitals where determinations of death by neurologic criteria may be infrequent. In addition to area of specialization, it is also worth considering whether the amount of experience or type of credentials a healthcare provider has could increase the risk of a false-positive determination. Should residents or fellows or advanced practice professionals, even if trained in neurology or neurosurgery, be allowed to independently perform an evaluation for determination of death by neurologic criteria or should they need to be closely supervised by an attending physician? In the US, 62% of 342 US hospital policies stipulated that an attending physician determines death by neurologic criteria, while eight policies allowed advanced practice professionals to make the determination [11].

3 Advantages and Disadvantages of Performing Multiple Evaluations

The decision to perform only one evaluation to determine death by neurologic criteria offers many theoretical advantages: it shortens the time for the patient, simplifies the process of identifying available and skilled examiners, and curtails the pain and uncertainty that many families experience. Additionally, organ donation consent rate decreases as the time to determination increases [27]. Moreover, performance of a single evaluation could decrease the incidence of organ dysfunction, thereby increasing the number and quality of organs procured per donor [18]. Another argument against performing a second evaluation is that up to 12% of patients may experience circulatory-respiratory arrest during the waiting period between evaluations [27]. A single evaluation may also decrease the cost of hospitalization, especially since patients being evaluated for death by neurologic criteria require management in the expensive intensive care unit environment: in an older single-center study, a single evaluation shortened the time to determination by an average of 14.4 h, and this decreased the cost of hospitalization by an estimated $1200/patient [28].

On the other hand, there may also be disadvantages to performing a single evaluation: theoretically, some of these patients might have regained some detectable neurologic function by the time of a second evaluation, if one had been performed (*vide supra* re: the concern about reversibility of the patient's condition). This is reinforced by the fact that an adequate period to wait between injury onset and performance of the first (or only) evaluation is not established. As mentioned, the 2010 American Academy of Neurology practice parameter for determination of death by neurologic criteria notes that it is necessary to ensure "*a certain period of time has passed since the onset of the brain insult to exclude the possibility of recovery [in practice, usually several hours*]" prior to performing a single evaluation, introducing a potential risk of inadequate time antedating a single evaluation [5].

However, there is no proof that recovery can occur between repeated evaluations. In a large study from New York with 1229 adult and 82 pediatric patients who underwent two separate evaluations mandated by New York State statute, nobody regained brain function after the first evaluation [27]. More recently, in a large single center study from Detroit, none of the 266 patients determined to be dead by neurologic criteria after 1 or 2 evaluations (the second trailing the first by an average of 20.9 h) regained any signs of brain function after the first evaluation [18].

Another disadvantage to the performance of only one evaluation is that the grieving period for families is shortened. This could affect the consent for organ donation if families do not have enough time to understand cognitively and accept emotionally the concept of death by neurologic criteria. This may be one of the reasons, in addition to mistrust of the medical establishment, why Black families consistently demonstrate lower consent rates for organ donation than other racial groups. However, there was no association between the number of clinical evaluations and consent rate in a recent study where Black families consented to organ donation 3.7 times less than White families [29].

4 Variability in the Number of Clinical Evaluations

The single evaluation has not been espoused by neonatologists or pediatricians and is not widely accepted in other countries [8, 11, 30, 31]. A recent assessment of death by neurologic criteria protocols from around the world reported that 83/136 contacted countries had protocols and the rest did not. Three countries used protocols by other countries, reducing the number of unique protocols to 78. Although a third of protocols (25; 32%) did not indicate the number of evaluations required to determine death by neurologic criteria, 44 (83%) of the 53 protocols that did indicate the number of evaluations required ≥2 evaluations. 38 of those (86%) stated that the evaluations needed to be conducted consecutively, separated by a range of 1 to 72 h. An observation period was required before conducting the evaluation for 47% of protocols; this ranged from 1 to 48 h (all protocols that mentioned hypoxic-ischemic brain injury without hypothermic treatment mandated an observation period of at least 24 h). Although 45% of protocols did not mention the number of required apnea tests, 23 (53%) required one, 19 (44%) required two, and 1 (2%) required three [13].

Guidance on the examiners for multiple evaluations also varies. In Canada, when a postmortem transplant is planned, death by neurologic criteria should be determined by at least two physicians. The evaluations can be done concurrently, but if done at different times, they should both be full evaluations including apnea tests. There is only a fixed interval of time between evaluations for neonates and children [32, 33]. In the United Kingdom, two full consecutive evaluations (including two apnea tests) by two physicians observing each other are required. No inter-observer period is needed. A short period of time is needed between apnea tests to allow the patient's arterial blood gases and baseline parameters to return to the pre-test state [34]. In Australia and New Zealand, two full evaluations (including apnea testing with each) by two different physicians who have sufficient qualifications and experience (as defined by jurisdiction) are required. The tests may be done consecutively, but not simultaneously. There is no requirement for one physician to be present during the test performed by the other physician, but such presence is acceptable. There is a minimum 4-h observation period prior to determination of death by neurologic criteria based on a clinical evaluation and apnea test without ancillary testing which extends to 24 h for hypoxic-ischemic injury in patients cooled >6 h. No fixed interval between the two clinical tests is required, except where age-related criteria apply (neonates and children) [35].

5 Guidance by the World Brain Death Project on the Number of Evaluations Required to Determine Death by Neurologic Criteria

Recognizing these worldwide differences, the World Brain Death Project authors suggested that a single evaluation, including apnea testing, is the minimum standard for determination of death by neurologic criteria for adults. However, if two evaluations were to be performed, they suggested that: a) an intervening period is unnecessary and a second observation period is redundant, if the prerequisite of irreversibility (which includes an observation period prior to initiating testing) has been satisfied; b) the evaluations be performed by two separate examiners; and c) only one apnea test be performed in adults [26].

6 Conclusion

There is clearly some subjectivity to the number of evaluations required to determine death by neurologic criteria. Below, and in Table 1, I provide my personal recommendations for the suggested number of evaluations for determination of death by neurologic criteria in various clinical circumstances. These recommendations are formulated based on my personal experience and review of the literature, but they should not be used in place of national guidelines or institutional protocols. I advocate for routine performance of a study to assess brain circulation as part of

Table 1 Suggested best practice for determination of death by neurologic criteria in adults: number of evaluations

	Catastrophic brain injury with rostrocaudal herniation	Anoxic-ischemic injury with edema and central herniation	Anoxic-ischemic injury without edema nor central herniation	Primary posterior fossa catastrophic injuries
Observation period	Not defined, initiate first exam after all brain function lost	≥24 h post event if no hypothermia 72 h post rewarming if hypothermia used	≥24 h post event if no hypothermia 72 h post rewarming if hypothermia used	Not defined, initiate first evaluation after all brain function lost
Single evaluation	Sufficient	Probably sufficient, but few data exist to support it	Not sufficient based on currently available data	Not sufficient if ancillary test shows any brain circulation
Dual consecutive evaluations (no waiting period)	Not indicated	Probably redundant, but few data exist to support it	Probably not sufficient based on currently available data	Not sufficient if ancillary test shows any brain circulation
Dual evaluations separated by hours	Not indicated	Not indicated	Probably sufficient	Sufficient if ancillary test shows no brain circulation in close proximity to the last clinical evaluation
Multiple evaluations separated by hours	Not indicated	Not indicated	Probably redundant	Sufficient if ancillary test shows no brain circulation in close proximity to the last clinical evaluation
Apnea test	Single apnea test after the clinical evaluation	Single apnea test after the single or second clinical evaluation	Single apnea test after the second clinical evaluation	Following each clinical evaluation
Brain circulation test	Following the single clinical evaluation and apnea test	Following the single or second clinical evaluation and apnea test	Following the second clinical evaluation and apnea test	Following each clinical evaluation and apnea test until there is cessation of brain circulation

all evaluations until data demonstrate its redundancy. Arguments in favor of and opposing this are discussed in detail elsewhere in this book.

In the majority of patients with catastrophic brain injury, the process begins supratentorially and progresses rostrocaudally with elevated intracranial pressure, decreased cerebral perfusion pressure, cytotoxic or venous edema, transtentorial herniation, and progressive brainstem dysfunction and ends with the collapse of all medullary functions leading to loss of spontaneous breathing. In these patients, irreversible loss of brain function can be established with a single evaluation including an apnea test and a confirmatory test to evaluate brain circulation. In comparison to performance of multiple evaluations, this does not negatively affect consent for donation and may lead to better organ procurement and transplantation rates.

For patients who have hypoxic-ischemic brain injury, an adequate observation period should be allowed **before** an evaluation for death by neurologic criteria is initiated. This is usually at least 24 h from the index arrest, but may be extended to 72 h post rewarming if hypothermia was induced. In patients who develop bilateral massive cerebral edema and central transtentorial herniation, a single evaluation for death by neurologic criteria including an apnea test and a confirmatory test to evaluate brain circulation may suffice. However, there are no adequate data to support or refute this approach. Alternatively, two consecutive evaluations by two examiners without a waiting period, followed by a single apnea test after the second evaluation and a confirmatory test to evaluate brain circulation may be considered until more data are published. For those rare patients who exhibit no clinical brain function after hypoxic-ischemic injury, but who do not show massive cerebral edema or signs of herniation, a more conservative approach should be espoused with an observation period (as aforementioned) and two or more clinical evaluations by different examiners separated by several hours followed by apnea tests, repeat neuroimaging, and a confirmatory test to evaluate brain circulation.

For catastrophic brainstem or posterior fossa injuries, single or multiple clinical evaluations including an apnea test are not enough to exclude a false-positive determination of death by neurologic criteria, if performed without a test to evaluate brain circulation. An ancillary test that shows no supratentorial or infratentorial circulation is key for the determination. Multiple tests to evaluate brain circulation may be required over a period of hours or days and brain death should not be declared until cessation of circulation is identified in close chronological proximity to a clinical exam and an apnea test that both demonstrate absence of brain function.

References

1. A Definition of Irreversible Coma. Report of the Ad Hoc Committee of the Harvard Medical School to Examine the Definition of Brain Death. JAMA. 1968;205(6):337–40.
2. Guidelines for the Determination of Death. Report of the medical consultants on the diagnosis of death to the President's Commission for the Study of Ethical Problems in Medicine and Biomedical and Behavioral Research. JAMA. 1981;246(19):2184–6.
3. Practice parameters for determining brain death in adults (summary statement). The Quality Standards Subcommittee of the American Academy of Neurology. Neurology. 1995;45(5):1012–4.
4. Wijdicks EF. Determining brain death in adults. Neurology. 1995;45(5):1003–11.
5. Wijdicks EF, Varelas PN, Gronseth GS, Greer DM. Evidence-based guideline update: determining brain death in adults: report of the quality standards Subcommittee of the American Academy of neurology. Neurology. 2010;74(23):1911–8.
6. New York State Department of Health. Guidelines for Determining Brain Death [Internet]. 2011. https://www.health.ny.gov/professionals/hospital_administrator/letters/2011/brain_death_guidelines.htm. Accessed 1 July 2022.

7. Report of Special Task Force. Guidelines for the determination of brain death in children. American Academy of Pediatrics Task Force on Brain Death in Children. Pediatrics. 1987;80(2):298–300.
8. Nakagawa TA, Ashwal S, Mathur M, Mysore M. Guidelines for the determination of brain death in infants and children: an update of the 1987 task force recommendations-executive summary. Ann Neurol. 2012;71(4):573–85.
9. Fainberg N, Mataya L, Kirschen M, Morrison W. Pediatric brain death certification: a narrative review. Transl Pediatr. 2021;10(10):2738–48.
10. Greer DM, Varelas PN, Haque S, Wijdicks EFM. Variability of brain death determination guidelines in leading us neurologic institutions. Neurology. 2008;71(22):1839–40.
11. Greer DM, Wang HH, Robinson JD, et al. Variability of brain death policies in the United States. JAMA Neurol. 2016;73(2):213–8.
12. Shappell CN, Frank JI, Husari K, et al. Practice variability in brain death determination: a call to action. Neurology. 2013;81(23):2009–14.
13. Lewis A, Bakkar A, Kreiger-Benson E, et al. Determination of death by neurologic criteria around the world. Neurology. 2020;95(3):e299–309.
14. Machado C, Korein J, Ferrer Y, et al. The declaration of Sydney on human death. J Med Ethics. 2007;33(12):699–703.
15. Endisch C, Westhall E, Kenda M, et al. Hypoxic-ischemic encephalopathy evaluated by brain autopsy and neuroprognostication after cardiac arrest. JAMA Neurol. 2020;77(11):1430–9.
16. Bjorklund E, Lindberg E, Rundgren M, et al. Ischaemic brain damage after cardiac arrest and induced hypothermia—a systematic description of selective eosinophilic neuronal death. A neuropathologic study of 23 patients. Resuscitation. 2014;85(4):527–32.
17. Mulder M, Gibbs HG, Smith SW, et al. Awakening and withdrawal of life-sustaining treatment in cardiac arrest survivors treated with therapeutic hypothermia. Crit Care Med. 2014;42(12):2493–9.
18. Varelas PN, Rehman M, Mehta C, et al. Comparison of 1 vs 2 brain death examinations on time to death pronouncement and organ donation: a 12-year single center experience. Neurology. 2021;96(10):e1453–e61.
19. Salih F, Hoffmann O, Brandt SA, et al. Safety of apnea testing for the diagnosis of brain death: a comprehensive study on neuromonitoring data and blood gas analysis. Eur J Neurol. 2019;26(6):887–92.
20. Smith ML, Counelis GJ, Maloney-Wilensky E, et al. Brain tissue oxygen tension in clinical brain death: a case series. Neurol Res. 2007;29(7):755–9.
21. Ferbert A, Buchner H, Ringelstein EB, Hacke W. Brain death from infratentorial lesions: clinical neurophysiological and transcranial Doppler ultrasound findings. Neurosurg Rev. 1989;12(Suppl 1):340–7.
22. Ogata J, Imakita M, Yutani C, Miyamoto S, Kikuchi H. Primary brainstem death: a clinico-pathological study. J Neurol Neurosurg Psychiatry. 1988;51(5):646–50.
23. Ferbert A, Buchner H, Ringelstein EB, Hacke W. Isolated brain-stem death. Case report with demonstration of preserved visual evoked potentials (VEPs). Electroencephalogr Clin Neurophysiol. 1986;65(2):157–60.
24. Varelas P, Bardy P, Rehman M, et al. Primary posterior fossa lesions and preserved supratentorial cerebral blood flow: implications for brain death determination. Neurocrit Care. 2017;27(3):407–14.
25. Controversies in the Determination of Death. A White Paper of the President's Council on Bioethics [Internet]. 2008. https://bioethicsarchive.georgetown.edu/pcbe/reports/death. Accessed 1 June 2022.
26. Greer DM, Shemie SD, Lewis A, et al. Determination of brain death/death by neurologic criteria: the world brain death project. JAMA. 2020;324(11):1078–97.
27. Lustbader D, O'Hara D, Wijdicks EF, et al. Second brain death examination may negatively affect organ donation. Neurology. 2011;76(2):119–24.
28. Varelas P, Rao CV, Rehman M, et al. Single brain death examination does not improve organ transplantation. Neurocrit Care. 2010;13:S39.

29. Kananeh MF, Brady PD, Mehta CB, et al. Factors that affect consent rate for organ donation after brain death: a 12-year registry. J Neurol Sci. 2020;416:117036.
30. Wahlster S, Wijdicks EF, Patel PV, et al. Brain death declaration: practices and perceptions worldwide. Neurology. 2015;84(18):1870–9.
31. Wijdicks EF. Brain death worldwide: accepted fact but no global consensus in diagnostic criteria. Neurology. 2002;58(1):20–5.
32. Shemie SD, Doig C, Dickens B, et al. Severe brain injury to neurological determination of death: Canadian forum recommendations. Can Med Assoc J. 2006;174(6):S1–13.
33. Teitelbaum J, Shemie SD. Neurologic determination of death. Neurol Clin. 2011;29(4):787–99.
34. Academy of Medical Royal Colleges. A code of practice for the diagnosis and confirmation of death [Internet]. 2008. https://www.aomrc.org.uk/reports-guidance/ukdec-reports-and-guidance/code-practice-diagnosis-confirmation-death/. Accessed 1 June 2022.
35. ANZICS. The ANZICS statement on death and organ donation [Internet]. 2021. https://www.anzics.com.au/wp-content/uploads/2021/06/ANZICS-Statement-on-Death-and-Organ-Donation-4.1.pdf. Accessed 1 July 2022.

Part III

Scientific Issues

Research Questions to Inform the Determination of Death by Neurologic Criteria

Guillaume Maitre and Sam D. Shemie

More than 50 years after the Ad Hoc Committee of Harvard Medical School's 1968 publication on determination of death by neurologic criteria [1], it is time to build a structured international research agenda about death by neurologic criteria. As scientists, we must be transparent about what we know and what we do not know in order to develop research questions and use evidence to answer them. While critique is essential to motivate inquiry, criticism without action serves little purpose. As previously demonstrated, international collaboration including countries with different cultural perspectives should be encouraged to continue making progress in this issue [2].

In the wake of the work done as part of the World Brain Death Project [3], this chapter establishes a list of research questions about death by neurologic criteria and addresses the feasibility of answering these questions. We acknowledge that defining death is not solely a matter of science. It also has legal and ethical implications and that its determination necessitates important metaphysical, cultural, religious, and spiritual considerations [4]. For the purposes of this discussion, our scope will be confined to science and practice. We support the view of Lazaridis that the definition of human death should conform to the best available and pragmatic

G. Maitre
Pediatric Critical Care Unit, Montreal Children's Hospital, McGill University Health Center, Montréal, QC, Canada

Pediatric Intensive Care Unit, Woman-Mother-Child Department, Centre Hospitalier Universitaire Vaudois, Biology and Medicine Faculty, University of Lausanne, Lausanne, Switzerland
e-mail: guillaume.maitre@mail.mcgill.ca

S. D. Shemie (✉)
Division of Critical Care, Montreal Children's Hospital, McGill University Health Centre and Research Institute, McGill University, Montréal, QC, Canada

System Development, Canadian Blood Services, Ottawa, ON, Canada
e-mail: sam.shemie@mcgill.ca

A. Lewis, J. L. Bernat (eds.), *Death Determination by Neurologic Criteria*, Advances in Neuroethics, https://doi.org/10.1007/978-3-031-15947-3_18

medical science and practice [5]. Current practice is largely based on large volume historical clinical experience, clinical reports, and expert-derived consensus due to insufficient amounts of available prospective evidence.

Historically, humanity's understanding of death has been situated within existing physiological knowledge, diagnostic techniques, and technological support. Regardless of the primary or secondary cause, the physiological pathway to death occurs in a consistent and predictable pattern. Occurring either after primary neurologic failure or secondary to cessation of circulatory-respiratory function, all human death evolves toward permanent cessation of brain functions. The circulatory-respiratory criteria for death are surrogates for determining permanent loss of brain functions after a certain duration in the absence of restoration of oxygenated circulation [6]. Indeed, some countries, including the United Kingdom, stipulate death determination criteria in this manner [7–9]. Switzerland and the United Kingdom include assessment of some brain functions in the determination of circulatory-respiratory death [7]. Canada is currently evolving towards a single brain-based definition of death after the arrest of circulation and respiration or neurologic function [10]. Regardless, continued adherence to dual criteria for death, circulatory-respiratory versus neurologic, poses ongoing conceptual challenges to the acceptance of death by neurologic criteria. A fundamental health policy research question arises: if one accepts the premise that science is evolving towards all death being brain-based, what are the strategies and obstacles to be addressed (societal, legal, ethical, clinical) in moving beyond the existing dual definitions of death?

1 Feasibility of Research on the Determination of Death During and After the Determination

While some could argue that prospective research about end-of-life is challenging to perform and obtain consent for, a number of publications have shown that research projects about dying and dead patients in intensive care units are feasible and that the consent rate is high. The DePPaRt study by Dhanani et al. was successfully performed in 20 centers in three countries. Its goal was to document the physiology of the process of death by circulatory criteria and develop a tool to predict how long patients take to die after removal of life-sustaining therapies. The overall family consent rate was more than 90%. [11]. A related study called NeuPaRT aims to establish the temporal relationship between circulatory arrest and the cessation of brain electrical activity [12]. The INDEX study which is prospectively evaluating the use of CT perfusion in and around the time of death by neurologic criteria has completed enrollment [13]. These projects, among others, demonstrate the feasibility and the acceptance by patients' families of prospective research about determination of death and the ability to generate high-quality evidence in this setting.

2 Research Questions/Topics About the Clinical Evaluation for Death by Neurologic Criteria

There are several aspects of the clinical evaluation for death by neurologic criteria that merit research. These include demographics, the clinical assessment, apnea testing, the number of evaluations, neuroendocrine secretion, specific considerations based on the patient's age and the mechanism and location of injury, the use of interventions to decrease intracranial pressure and recovery of function after determination of death by neurologic criteria (Table 1).

Table 1 Research questions/topics about the clinical evaluation for death by neurologic criteria

1. Demographics
1a. What is the national/international incidence and point prevalence of death by neurologic criteria?
1b. Report death by neurologic criteria as a distinct and separate outcome after devastating brain injury in all observational, interventional, or prognostic outcome studies
2. Clinical assessment
2a. Establishment of a standardized serial neuromonitoring registry of brainstem reflexes
2b. Should there be a hierarchy of the components of the clinical evaluation whose absence portend a higher risk of death by neurologic criteria?
2c. What is the pattern and role of heart rate variability from brain injury to death by neurologic criteria?
3. Apnea testing
3a. How often does the apnea test show evidence of the ability to breathe in a patient whose clinical evaluation is otherwise consistent with death by neurologic criteria?
3b. What method of apnea testing has the lowest rate of complications?
3c. What is the incidence of retained brain circulation prior to performance of apnea testing?
3d. Does apnea testing increase brain circulation and intracranial pressure?
3e. What level of P_aCO_2 is adequate to generate respiratory drive when considering age, altitude, and baseline level?
3f. Routine reporting of apnea test results and those where the P_aCO_2 far exceeds 60 mmHg
4. Number of evaluations
4a. How often does a second evaluation contradict the first one?
4b. What are the circumstances in which a second evaluation contradicts the first one?
5. Neuroendocrine secretion
5a. Does pathophysiology (imaging, brain circulation) differ between death by neurologic criteria with diabetes insipidus versus ongoing antidiuretic hormone secretion?
5b. Does the persistence of antidiuretic hormone secretion have any influence on the reversibility of cessation of consciousness, brainstem areflexia, and inability to breathe spontaneously?
6. Special considerations related to hypoxic-ischemic brain injury
6a. How long after return of spontaneous circulation can clinical findings be considered reliable means to determine death by neurologic criteria?
6b. Does treatment with targeted temperature management change the natural history (incidence, determinants, temporal evolution) of death by neurologic criteria?
6c. How much time after rewarming is required to remove the potential for temperature to be a confounder to the evaluation?
6d. Should naloxone be administered routinely before performing a clinical evaluation for death by neurologic criteria in case of exposure to opiates? What dose and frequency of naloxone administration could ensure elimination of opiates as a confounder?

(continued)

Table 1 (continued)

7. Special considerations related to infratentorial lesions
7a. What is the incidence in which primary infratentorial lesions evolve to death by neurologic criteria?
7b. Do all patients with primary infratentorial lesions who fulfill clinical criteria for the brainstem criterion for death evolve to meet the whole-brain criterion? What are predictors of evolution of infratentorial lesions to whole-brain injury?
7c. What imaging technique can confirm complete and permanent destruction of the brainstem?
7d. What is the expected evolution of electroencephalography and brain blood flow in patients with infratentorial injury who meet the clinical criteria for the brainstem criterion for death?
7e. Is there potential for recovery of brainstem function in a patient with a primary infratentorial injury who fulfills the clinical criteria for the brainstem criterion for death, but not the whole-brain criterion?
7f. Does the destruction of the reticular activating system assure the absence of consciousness? What technique might confirm the absence of covert consciousness?
8. Special consideration related to the pediatric population
8a. Does the pathophysiology of death by neurologic criteria differ between adults and children after complete ossification of the skull?
8b. Does the compliant skull in infants increase the risk of preserving brain circulation, perfusion, and function? Are there specific issues in newborns?
8c. Are there valid ancillary tests in newborns and infants?
8d. What is the scientific rationale for repeated evaluations in children?
8e. Establishment of a standardized reporting registry of all components of the evaluation for death by neurologic criteria in children
9. Surgical interventions to decrease intracranial pressure
9a. Can decompressive craniectomy potentially reverse death by neurologic criteria?
9b. Is the evolution to death after decompressive craniectomy influenced by the type of primary brain injury?
9c. How long after decompressive craniectomy is a determination of death by neurologic criteria reliable? Should ancillary testing be routinely required?
9d. Should patients presenting with brainstem compression who meet the clinical criteria for neurologic death undergo a surgical decompression or a ventriculostomy to guarantee irreversibility?
10. Recovery of brain function after death by neurologic criteria
10a. Can brain function resume after death by neurologic criteria and if so, under what conditions? What are the best methods to determine if patients recover any brain functions after meeting neurologic criteria for death? Are there any neuroimaging or neurodiagnostic features that might predict potential brain function recovery?
10b. Can the spinal cord develop patterned reproducible responses to stimuli over time in a patient who is dead by neurologic criteria?
10c. Might functional MRI or brain electrical stimulation differentiate spinal and brain-mediated movements?

2.1 Demographics

While the causes of devastating brain injury leading to death have been well described, the international incidence of death by neurologic criteria and its point prevalence remain unknown. Death by neurologic criteria is not a disease, but a determination that requires recognition and evaluation to establish. The incidence of death by neurologic criteria in the United States was reported as 47 per million population in 2016 [14]. Inferences about the international incidence of death by neurologic criteria have been extrapolated by the annual worldwide incidence of deceased organ donation to be approximately 30,000 per year, but this estimate may be inaccurate [15]. Any reported incidence of death by neurologic criteria should be indexed by per million population. While exact data is lacking, it is well-known that the clinical experience with death by neurologic criteria is substantial, while the research base lags behind.

2.2 Clinical Assessment

The fundamental conditions for determination of death by neurologic criteria are the permanent absence of consciousness, brainstem areflexia, and the inability to breathe spontaneously in the setting of a clear etiology without confounding or reversible conditions. There is no established hierarchy of the components of the clinical evaluation, and this order becomes relevant in conditions where a complete evaluation is not possible, e.g., following limb amputation or trauma to the ears or eyes. We advocate the creation of a standardized registry of the neurologic evaluations, which includes serial testing of components of the clinical assessment for death by neurologic criteria in patients with devastating brain injury from admission until determination of death. This registry could improve prognostication about progression to death by neurologic criteria. Interventional or prognostic studies should also report death by neurologic criteria as a distinct outcome, rather than combining it with death by circulatory-respiratory criteria or grouping it together with unfavorable neurological outcomes.

In addition to the above components of the clinical evaluation, a research question to consider is whether there are additional conditions that should be fulfilled to determine death by neurologic criteria. For example, a decrease in heart rate variability was identified in a small cohort as a significant predictive factor for progression to death by neurologic criteria [16]. This finding raises questions that could be answered by empirical studies. What is the pattern of heart rate variability during the path from brain injury to death by neurologic criteria? Would close monitoring of heart rate variability help predict or determine death?

2.3 Apnea Testing

The apnea test generates many controversies [17, 18]. The risks of cardiopulmonary instability when performing an apnea test can be minimized by various techniques, including maintaining CPAP during the test, gradual hypoventilation during testing prior to ventilator disconnection, or use of inhaled carbon dioxide [19–23]. How often does the apnea test show evidence of the ability to breathe in a patient whose clinical evaluation is otherwise consistent with death by neurologic criteria? What method of apnea testing is related to the lowest rate of complications? A registry of evaluations for death by neurologic criteria would help answer these questions.

Apnea testing is arguably regarded as the most crucial test to evaluate the brainstem; it is widely considered essential and generally reserved as the final component of the evaluation to ensure the highest pre-test probability for the determination of death by neurologic criteria. Theoretical concerns have been raised regarding the potential for hypercarbia-induced cerebral vasodilation and increased intracranial pressure [17]. Brain circulation prior to, during, and after performance of apnea testing has not been studied. This type of study could be beneficial to address theoretical criticisms and questions regarding the necessity for consent prior to apnea testing [24–26]. Hypercarbia cannot increase brain circulation and intracranial pressure if there is no circulation to begin with. However, if there is no brain function but retained brain circulation, it is theoretically conceivable that performing an apnea test might enhance flow and thereby increase intracranial pressure. How might a clinician distinguish between these states? Does making this distinction require direct intracranial pressure measurements or transcranial Doppler prior to and during an apnea test? Studies should be performed when intracranial pressure is being measured already to evaluate if apnea testing actually increases intracranial pressure and decreases cerebral perfusion pressure.

While the P_aCO_2 target for apnea testing is generally ≥60 mmHg, there are two case reports of children who began to breathe at a higher P_aCO_2 [27–29]. What level of P_aCO_2 is adequate to ensure respiratory drive would be initiated (taking age, altitude, and baseline P_aCO_2 levels into consideration)? A standardized registry of apnea test results would be informative as the final P_aCO_2 achieved may vary greatly between tests. Prospective studies comparing different P_aCO_2 targets can also provide additional guidance.

2.4 Number of Evaluations

Previous pediatric and adult studies demonstrated that the second sequential evaluation for death by neurologic criteria does not contradict the first one if performed in accordance with national standards [30, 31]. However, many countries require repeated evaluations over time. What is the purpose of this second exam, and how often does it contradict the first exam if the latter is done in accordance with medical standards and includes all components of the clinical evaluation performed in the absence of any confounding or reversible conditions? Is the purpose ensuring irreversibility or merely obtaining confirmation of proper testing by another physician? For any system that insists on repeated examinations over time, data collection and routine reporting of any inconsistency, and corresponding details, are needed.

2.5 Neuroendocrine Secretion

One area of criticism about the whole-brain criterion for death is that some brain functions, like neuroendocrine secretion, are not tested by the standards for determination of death by neurologic criteria, but that the Uniform Determination of Death Act requires absence of neuroendocrine secretion. Nair-Collins and Joffe argue that over half of the determinations of death by neurologic criteria are misdiagnosed because of the persistence of neuroendocrine functions. They report a 50% (95% CI 47.2–52.3%) incidence of central diabetes insipidus in 2546 patients determined to be dead by neurologic criteria [32]. Arguments continue about the relevance of neuroendocrine secretion in the context of determination of death by neurologic criteria, as discussed elsewhere in this book. Does pathophysiology (imaging, perfusion) differ between cases of death by neurologic criteria with diabetes insipidus versus those with maintained antidiuretic hormone secretion? Does the persistence of antidiuretic hormone secretion have any influence on the irreversibility of cessation of consciousness and brainstem areflexia?

2.6 Special Considerations Related to Hypoxic-Ischemic Brain Injury

Hypoxic-ischemic brain injury after resuscitated circulatory-respiratory arrest has emerged as the most common etiology for death by neurologic criteria in both children [33] and adults [34]. This finding is in part due to the opiate overdose epidemic [35]. Should naloxone be administered routinely before performance of a clinical evaluation for determination of death by neurologic criteria in case the patient has taken opiates?

It is generally recommended to wait 24 h prior to the first clinical evaluation for determination of death by neurologic criteria after hypoxic-ischemic brain injury as there is no clear data on the potential for reversibility of coma or brainstem function during the period of ischemia-reperfusion injury. Outcome studies of prognostic clinical signs after hypoxic-ischemic brain injury often do not distinguish poor neurologic outcome in survivors from death after withdrawal of life-sustaining treatment and death by neurologic criteria [36]. How long after return of spontaneous circulation can clinical findings be considered reliable means to determine death by neurologic criteria? Notably, this question is distinct from the question about how long after return of spontaneous circulation clinical neuroprognostication is reliable.

Targeted temperature management is used after hypoxic-ischemic (and other types of) brain injury, but there is variability in target temperature and duration.

Recent research showed no beneficial outcome after out-of-hospital circulatory-respiratory arrest for patients treated with hypothermia compared to normothermia [37]. Does targeted temperature management change the natural history of death by neurologic criteria? What are the incidence and the determinants that predict evolution to death by neurologic criteria after targeted temperature management? While the therapeutic impact of hypothermia or targeted temperature management continues to be debated [38], how much time is required after rewarming to remove the confounding potential of temperature and the indirect effect of temperature on drug elimination?

2.7 Special Considerations Related to Infratentorial Lesions

Clinical assessment often cannot distinguish the whole-brain and brainstem formulations of death. The source of this controversy is cases with primary infratentorial lesions where there is absence of brainstem function but there is potential for persistent cerebral electrical activity or brain blood flow. There is international variability in the criterion used for determination of death by neurologic criteria [39] and the requirements for ancillary testing in this patient population [40]. Many patients with primary infratentorial lesions who meet the brainstem criterion for death will meet the whole-brain criterion with time, but this may be dependent on the type of lesion and the performance of ventricular drainage or posterior fossa decompressive craniectomy [41].

There are unanswered questions related to the incidence of death by neurologic criteria in patients with infratentorial injuries, the utility of brainstem evoked potentials in this population and the predictors of evolution to whole-brain injury. Is there an imaging technique that can confirm complete and permanent destruction of the brainstem? Do all patients with primary infratentorial injury evolve to whole-brain injury with time? What is the expected evolution of electroencephalography and brain circulation evaluations in patients with infratentorial injuries? Is there potential for reversibility of brainstem function? Does the destruction of the reticular activating system assure the absence of consciousness? What does the persistence of cerebral electrical activity or brain circulation mean in terms of actual potential for consciousness? What techniques might confirm the absence of covert consciousness in this patient group [42]? Is there a role for functional MRI in this patient population?

2.8 Special Considerations in the Pediatric Population

In some parts of the world, there are different standards for determination of death by neurologic criteria in pediatric patients than adults, as discussed elsewhere in this book. However, the justification for different standards in the pediatric population is unclear. Does the pathophysiology differ between adults and children after complete ossification of the skull? Does the compliant skull in infants increase the risk of preserving blood flow, perfusion, and thus function? Are there specific issues in newborns? What is the scientific rationale for repeated evaluations in children [43]? Are there valid ancillary tests that can be used in newborns and infants?

2.9 Surgical Interventions to Decrease Intracranial Pressure: Ventricular Drainage and Decompressive Craniectomy

Unpublished sporadic cases have been reported of resumption of brainstem function after death by neurologic criteria following decompressive craniectomy. The United Kingdom recently issued "red flag" concerns about this phenomenon [44]. While decompressive craniectomy results in an immediate decrease in intracranial pressure, the indications, therapeutic effectiveness, and extent of skull decompression vary. Regardless, both ventricular drainage and decompressive craniectomy can be deployed during life-saving neuroprotective phases of care intended to improve survival and quality-of-life. Is decompressive craniectomy a potential means to reverse death by neurologic criteria, and how is this impacted by the surgical approach (e.g., extent of decompression, the type of dural substitute being used)? Is the evolution to death after decompressive craniectomy influenced by the type of primary brain injury-- ischemic, hemorrhagic, or traumatic? How long after decompressive craniectomy is it reliable and safe to determine death, and should ancillary testing be routinely required?

Salih et al. reported 7.3% of adult patients are declared dead by neurologic criteria following decompressive craniectomy. However, these patients all demonstrated a secondary increase of intracranial pressure, and death followed a severe decrease of cerebral perfusion pressure despite the craniectomy [45]. Can the adverse impact of intracranial hypertension-related brain ischemia be separated out from the impact of direct cellular injury?

Death by neurologic criteria has traditionally been rooted in "irreversibility," meaning "cannot be reversed." However, intracranial pressure-reducing surgical interventions are often available, but not employed, such that death by neurologic criteria needs to be considered permanent, meaning "will not be reversed," predicated on a clinical decision that these interventions would not be therapeutically effective. The idea of confirming the irreversibility of death by neurologic criteria by the nontherapeutic use of ventricular drains or decompressive craniectomy seems extreme, but it is worth asking if this should be done. Should patients presenting with brainstem compression who meet the clinical criteria for neurologic death undergo a surgical decompression or a ventriculostomy to guarantee irreversibility? If not, death by neurologic criteria should be reconceptualized as a permanent state, similar to death by circulatory-respiratory criteria after withdrawal of life-sustaining treatments whereby cardiopulmonary resuscitation and extra-corporeal membrane oxygenation are not indicated and therefore not provided.

2.10 Recovery of Brain Function After Death by Neurologic Criteria

There are some case reports of alleged retained or recovered brain function in patients maintained on organ support after determination of death by neurologic criteria [46, 47]. The accuracy of these determinations is questionable, and objective methodological review is lacking. The distinction between complex spinal versus centrally mediated movements can be unclear. Can the spinal cord develop patterned reproducible responses to stimuli over time in a patient who is dead by neurologic criteria? What are the best methods to determine if patients recover any brain functions after meeting neurologic criteria for death? Are there neuroimaging features that might predict potential recovery? Might functional MRI or brain electrical stimulation differentiate spinal and brain-mediated movements?

3 Research Questions About Neuroimaging and Ancillary Tests for Determination of Death by Neurologic Criteria

Research questions about neuroimaging and ancillary tests are summarized in Table 2.

3.1 Neuroimaging

The World Brain Death Project suggested that prior to performing an evaluation for determination of death by neurologic criteria, evidence of increased intracranial pressure should be documented on neuroimaging [3]. While the absence of neuroimaging evidence of devastating brain injury warrants caution, recent studies suggest that the presence of brain herniation or other signs of cerebral swelling are poor predictors of whether a patient will meet neurologic criteria for death [48]. Further study is required to assess the correlation between signs of herniation on neuroimaging, findings on the clinical assessment, a positive apnea test, and absence of blood flow on an ancillary test. Are there findings on neuroimaging, like the severity of markers of intracranial pressure and brain edema or the degree of pontomedullary compression, that are predictive of coma, brainstem areflexia, inability to breathe spontaneously, and absence of brain blood flow? Can brain functions be irreversibly lost in the absence of herniation, and if so, how often and under what conditions may this occur?

3.2 Ancillary Testing

Classical ancillary tests (catheter or CT angiography, radionuclide imaging) measure brain circulation. A challenge for all comparative research on ancillary testing is the question of what should be considered the gold standard for comparison. Is it

Table 2 Research questions about neuroimaging and ancillary tests for determination of death by neurologic criteria

1. Neuroimaging
1a. What is the correlation between various findings on neuroimaging and fulfillment of the clinical conditions for death by neurologic criteria, a positive apnea test, and absence of blood flow on ancillary test?
1b. Can brain function be irreversibly lost in the absence of herniation on neuroimaging? Under what conditions?
2. Ancillary testing
2a. What is the validity of the various ancillary testing modalities?
2b. How often does an ancillary test contradict the clinical determination of death by neurologic criteria?
2c. What should be considered the gold standard for comparison?
2d. What is the time-based correlation between continuous transcranial doppler findings and cessation of clinical brain function?
2e. Does EEG still have a role? What is the impact of combining EEG with brainstem evoked potentials?
2f. How might functional MRI be applied to determination of death by neurologic criteria?
2g. What is the value of various experimental ancillary testing methods in neurologic death determination?
3. Brain circulation
3a. How much blood flow is required to generate any organized brain function?
3b. What is the potential for heterogenous flow depending on regional compartmental pressures?
3c. What are the lower limits of brain circulation detection?
3d. What are the lower limits of brain circulation, perfusion, and duration of ischemia associated with cessation of brain function?
3e. What are the critical cerebral perfusion pressure and ischemic thresholds under which an already damaged brain tissue is not perfused and function is irreversibly lost?
3f. Is there any differential risk for resumption of brain function in cases of no-flow versus low-flow states?
3g. Should absent brain circulation on an ancillary test be sufficient to determine death? Do all patients with absent brain blood flow on an ancillary test have no brain function?
3h. What are the fundamental differences between death by neurologic criteria with preserved versus absent brain circulation?
3i. What is the distinction between intracranial-pressure-related hypoxia-ischemia versus cytotoxic brain nonfunction in the presence of preserved blood flow?

an unconfounded clinical evaluation or comparison to another ancillary test? Although Dalle Ave et al. reported cases of inconsistency between the clinical evaluation and ancillary test results [46], we do not have precise data about the validity (specificity/sensitivity) of all modalities of ancillary testing. A systematic review and meta-analysis is pending [49]. To better understand the performance of ancillary tests, we need to know how often they contradict a clinical determination. How many patients who meet neurologic criteria for death on clinical evaluation would have absent brain circulation? Do all patients with absent brain circulation have no brain function? What are the fundamental differences between death by neurologic criteria with persistent vs. absent brain circulation? What is the natural history of patients who clinically appear to be dead by neurologic criteria, but have preserved brain circulation?

The distinction between brain circulation, perfusion, and function has been well-described [50], but it is unknown if there should be a distinction between relevant and irrelevant brain circulation? If so, what is the definition of irrelevant brain circulation? Previous studies showed the minimum amount of flow required for neuronal function [51]. However, we would like to know how much circulation is required to resume any organized brain function. Further, what is the potential for heterogeneous flow depending on regional compartment pressures? Fundamentally, is there any differential risk for resumption of brain function in cases of no flow versus low flow states? What are the lower limits of detection of brain circulation for each ancillary test? Is this lower limit compatible with the critical cerebral perfusion under which brain function is irreversibly lost?

We also need to have a better understanding of brain perfusion physiology. What are the lower limits of brain circulation, perfusion, and duration of ischemia associated with cessation of brain function? While stroke and ischemic penumbra studies are helpful, they are generally performed in patients with normal baseline brain function, unlike brain perfusion physiology after devastating brain injury. What are the critical cerebral perfusion pressure and ischemic thresholds under which already damaged brain tissue is not perfused and function is irreversibly lost? Can the concept of ischemic penumbra be applied to the whole brain [32, 52]?

3.3 Types of Ancillary Tests

While the electroencephalogram has the longest historical use for determination of death by neurologic criteria and has widespread availability and experiential use, many countries have abandoned it for this purpose due to its inability to evaluate the deep cerebral hemispheres and the brainstem. Does it still have a role in the determination of death by neurologic criteria, perhaps in conjunction with brainstem-evoked potentials?

Transcranial Doppler imaging has the advantage of being performed at the bedside repetitively or continuously. It allows documentation of the stepwise evolution of the flow pattern preceding death from decreased diastolic to reverse flow. The time-based correlation between arrest of flow and cessation of clinical brain function is unknown. CT-perfusion is another ancillary testing currently under investigation via a prospective multicenter study in Canada [13].

Another test which has been used to distinguish covert consciousness in a subset of patients with vegetative states is the functional MRI [53]. Could functional MRI be applied to determination of death by neurologic criteria?

Lastly, ultrasound of the retina, electrical impedance measurement, transcranial magnetic stimulation, craniovascular flowmetry, and brain tissue oximetry are experimental approaches that could have a potential role in the determination of death by neurologic criteria [54–57]. More investigations are needed in this domain.

4 Research Questions About the Aftermath of Determination of Death by Neurologic Criteria

Formerly, circulatory-respiratory arrest quickly followed death by neurologic criteria [58–60]. With tremendous advances in critical care medicine, in the absence of somatic injuries that are incompatible with circulatory-respiratory stability, it is possible to maintain systemic functions after death by neurologic criteria with attentive intensive medical and nursing care [61, 62]. In countries where death by neurologic criteria is not commonly accepted, and withdrawal of support is refused, the time between death by neurologic criteria and circulatory-respiratory arrest can be several months [63–65]. Organ support can also be continued after death by neurologic criteria in the setting of organ donation or pregnancy. How often is organ support continued for organ donation, pregnancy, and family objections to the determination of death by neurologic criteria? Is the ability to maintain systemic organ functionality impacted by the presence or absence of brain circulation on ancillary testing? Can the necrotic brain revascularize with time, and if so, does this have any potential for resumption of brain function or consciousness? How does prolonged organ support influence spinal reflexes? Is it possible for the spinal cord to "learn" over time, inferring patterned responses to peripheral sensory stimuli? What is the best technique to ensure these responses are spinal in origin?

A Canadian survey among critical care physicians discusses requests for continuation of organ support after determination of death. 55% (128/231) of responding physicians reported having been asked to continue organ support after death determination. Half of these requests were accompanied by a threat of legal action [66]. Similar proportions of requests to continue organ support after death by neurologic criteria were reported in surveys among American neurologists [67] and pediatricians [68]. There are a number of reasons for families to reject determination of death by neurologic criteria [69]. Multidisciplinary research is required to evaluate the incidence of these objections and develop effective strategies to anticipate, prevent, or mitigate conflicts during the emotionally challenging process of death determination. This research should include spiritual care providers, ethicists, and experts in end-of-life care communication.

Research questions about the aftermath of determination of death by neurologic criteria are listed in Table 3.

Table 3 Research questions/topics about the aftermath of determination of death by neurologic criteria

a. How often is organ support continued in the setting of organ donation, pregnancy, and family requests due to legal objection?
b. Is the ability to maintain systemic organ functionality impacted by the presence or absence of brain circulation on ancillary testing?
c. Can the necrotic brain revascularize with time? Is it associated with any brain function resumption?
d. How does prolonged organ support influence the presence of spinal reflexes? What is the best technique to ensure that movements are spinal in origin?
e. Is it possible for the spinal cord to learn over time?
f. Development of effective strategies to anticipate, prevent, or mitigate conflicts during the process of death determination
g. Comparative measurement of the incidence of legal objections to determination of death by neurologic criteria

5 Research Questions About Worldwide Variance in Determination of Death by Neurologic Criteria

Eliminating any potential for diagnostic error in the determination of death by neurologic criteria is paramount. The variability in standards for determination of death by neurologic criteria among and within countries has been well-studied [70–72]. While persistently demonstrated and disconcerting, it is reassuring that the fundamental conditions for death by neurologic criteria (establishment of etiology, the absence of reversible conditions, absence of consciousness, brainstem reflexes and ability to breathe spontaneously) remain remarkably consistent. The World Brain Death Project attempted to increase harmonization to address the variability in this regard, but the impact of that attempt remains unstudied [3].

Further, adherence to standards is not well-studied. What are the differences between the standards and what clinicians actually do at the bedside? How might these actual practices affect the reliability of the determination and prevention of diagnostic error, confirm consistency and accuracy of medical record documentation, neuroimaging features, and ancillary testing? We advocate for the establishment of a standardized registry of all components of the evaluation for death by neurologic criteria to inform research and quality assurance in this domain. Questions about worldwide variance in determination of death by neurologic criteria are summarized in Table 4.

Table 4 Research questions/topics about worldwide variance in determination of death by neurologic criteria

a. What is the impact of the World Brain Death Project?
b. What is the adherence of actual clinical practices to local standards? How do the actual practices affect the reliability of death determination and the accuracy of medical record documentation?
c. Establishment of a standardized registry of all the components of the evaluation for death by neurologic criteria used by clinicians

6 Conclusion

Death by neurologic criteria remains a highly debated question in both the scientific and philosophic fields. As researchers, we can make significant progress in the first domain with high-quality methodology and respond effectively to the oft repetitive criticism. This kind of research is feasible and essential to bring scientific knowledge, clarity, and uniformity among definitions and practices. International collaboration is critical to bring together experienced researchers with different perspectives. The research questions listed above in various categories give an overview of the main unanswered and debated issues and establishes an agenda whose purpose is to inform the determination of death by neurologic criteria. With answers to these questions, healthcare practitioners and society in general will make definitive progress in understanding the process of dying.

References

1. A definition of irreversible coma: Report of the Ad Hoc Committee of the Harvard Medical School to Examine the Definition of Brain Death. JAMA. 1968;205(6):337–40.
2. Dogan G, Kayir S. Global scientific outputs of brain death publications and evaluation according to the religions of countries. J Relig Health. 2020;59(1):96–112.
3. Greer DM, Shemie SD, Lewis A, et al. Determination of brain death/death by neurologic criteria. JAMA. 2020;324(11):1078–97.
4. Baker A, Shemie SD. Biophilosophical basis for identifying the death of a person. J Crit Care. 2014;29(4):687–9.
5. Lazaridis C. Defining death: reasonableness and legitimacy. J Clin Ethics. 2021;32(2):109–13.
6. Shemie SD, Gardiner D. Circulatory arrest, brain arrest and death determination. Front Cardiovasc Med. 2018;5:15.
7. Swiss Academy of Medical Sciences. Determination of Death with Regard to Organ Transplantation and Preparations for Organ Removal. 2017:31.
8. Academy of Medical Royal Colleges: A Code of Practice for the Diagnosis and Confirmation of Death [Internet]. https://aomrc.org.uk; 2008. https://aomrc.org.uk/wp-content/uploads/2016/04/Code_Practice_Confirmation_Diagnosis_Death_1008-4.pdf. Accessed 24 Feb 2022.

9. Oram J, Murphy P. Diagnosis of death. Contin Educ Anaesthes Crit Care Pain. 2011;11(3):77–81.
10. Organ Donation and Transplantation Collaboration: Developing a Brain-based Definition of Death and Evidence- based Criteria for its Determination after Arrest of Circulation or Neurologic Function in Canada [Internet]. https://profedu.blood.ca. https://profedu.blood.ca/sites/default/files/odtc_death_determination_guideline_project_snapshot.pdf. Accessed 24 Feb 2022.
11. Dhanani S, Hornby L, van Beinum A, et al. Resumption of cardiac activity after withdrawal of life-sustaining measures. N Engl J Med. 2021;384(4):345–52.
12. Norton L, Gibson RM, Gofton T, et al. Electroencephalographic recordings during withdrawal of life-sustaining therapy until 30 minutes after declaration of death. Can J Neurol Sci. 2017;44(2):139–45.
13. ClinicalTrials.gov: CT-Perfusion for Neurological Diagnostic Evaluation (INDex-CTP) [Internet]. www.clinicaltrial.gov.; 2017. https://clinicaltrials.gov/ct2/show/NCT03098511. Accessed 24 Feb 2022.
14. Seifi A, Lacci JV, Godoy DA. Incidence of brain death in the United States. Clin Neurol Neurosurg. 2020;195:105885.
15. Global Observatory on Donation and Transplantation [Internet]. www.transplant-observatory.org. http://www.transplant-observatory.org/data-charts-and-tables/. Accessed 24 Feb 2022.
16. Piantino JA, Lin A, Crowder D, et al. Early heart rate variability and electroencephalographic abnormalities in acutely braininjured children who progress to brain death. Pediatr Crit Care Med. 2019;20(1):38–46.
17. Joffe AR, Hansen G, Tibballs J. The world brain death project: the more you say it does not make it true. J Clin Ethics. 2021;32(2):97–108.
18. Tibballs J. It is time to abandon apneic-oxygenation testing for brain death. Archiv Organ Transplant. 2020;5(1):006–10.
19. Giani M, Scaravilli V, Colombo SM, et al. Apnea test during brain death assessment in mechanically ventilated and ECMO patients. Intensive Care Med. 2016;42(1):72–81.
20. Sharpe MD, Young GB, Harris C. The apnea test for brain death determination: an alternative approach. Neurocrit Care. 2004;1(3):363–6.
21. Scott JB, Gentile MA, Bennett SN, Couture M, MacIntyre NR. Apnea testing during brain death assessment: a review of clinical practice and published literature. Respir Care. 2013;58(3):532–8.
22. Lang CJ. Apnea testing by artificial CO2 augmentation. Neurology. 1995;45(5):966–9.
23. Busl KM, Lewis A, Varelas PN. Apnea testing for the determination of brain death: a systematic scoping review. Neurocrit Care. 2021;34(2):608–20.
24. Berkowitz I, Garrett JR. Legal and ethical considerations for requiring consent for apnea testing in brain death determination. Am J Bioeth. 2020;20(6):4–16.
25. Lewis A, Greer D. POINT: should informed consent be required for apnea testing in patients with suspected brain death? No Chest. 2017;152(4):700–2.
26. Truog RD, Tasker RC. COUNTERPOINT: should informed consent be required for apnea testing in patients with suspected brain death? Yes Chest. 2017;152(4):702–4.
27. Vardis R, Pollack MM. Increased apnea threshold in a pediatric patient with suspected brain death. Crit Care Med. 1998;26(11):1917–9.
28. Brilli RJ, Bigos D. Apnea threshold and pediatric brain death. Crit Care Med. 2000;28(4):1257.
29. Brilli RJ, Bigos D. Altered apnea threshold in a child with suspected brain death. J Child Neurol. 1995;10(3):245–6.
30. Joffe AR, Shemie SD, Farrell C, Hutchison J, McCarthy-Tamblyn L. Brain death in Canadian PICUs: demographics, timing, and irreversibility. Pediatr Crit Care Med. 2013;14(1):1–9.
31. Lustbader D, O'Hara D, Wijdicks EF, et al. Second brain death examination may negatively affect organ donation. Neurology. 2011;76(2):119–24.
32. Nair-Collins M, Joffe AR. Frequent preservation of neurologic function in brain death and brainstem death entails false-positive misdiagnosis and cerebral perfusion. AJOB Neurosci. 2021:1–14.

33. Kirschen MP, Francoeur C, Murphy M, et al. Epidemiology of brain death in pediatric intensive care units in the United States. JAMA Pediatr. 2019;173(5):469–76.
34. Kramer AH, Baht R, Doig CJ. Time trends in organ donation after neurologic determination of death: a cohort study. CMAJ Open. 2017;5(1):E19–27.
35. Keenan S, Kramer A, Healey A, Weiss MJ, Dhanani S, Beed S, et al. The variable impact of the overdose crisis on organ donation among five Canadian provinces: a retrospective study. Can J Anaesth. 2021;68(6):846–54.
36. Sandroni C, Geocadin RG. Neurological prognostication after cardiac arrest. Curr Opin Crit Care. 2015;21(3):209–14.
37. Dankiewicz J, Cronberg T, Lilja G, et al. Hypothermia versus normothermia after out-of-hospital cardiac arrest. N Engl J Med. 2021;384(24):2283–94.
38. Nolan JP, Sandroni C, Bottiger BW, et al. European Resuscitation Council and European Society of Intensive Care Medicine guidelines 2021: post-resuscitation care. Intensive Care Med. 2021;47(4):369–421.
39. Wijdicks EF. The transatlantic divide over brain death determination and the debate. Brain. 2012;135(Pt 4):1321–31.
40. Hoffmann O, Masuhr F. Use of observational periods or ancillary tests in the determination of brain death in Germany. Eur Neurol. 2015;74(1–2):11–7.
41. Varelas PN, Brady P, Rehman M, et al. Primary posterior fossa lesions and preserved supratentorial cerebral blood flow: implications for brain death determination. Neurocrit Care. 2017;27(3):407–14.
42. Walter U, Fernandez-Torre JL, Kirschstein T, Laureys S. When is "brainstem death" brain death? The case for ancillary testing in primary infratentorial brain lesion. Clin Neurophysiol. 2018;129(11):2451–65.
43. Nakagawa TA, Ashwal S, Mathur M, et al. Clinical report-guidelines for the determination of brain death in infants and children: an update of the 1987 task force recommendations. Pediatrics. 2011;128(3):e720–40.
44. The Faculty of Intensive Care Medicine: Diagnosing Death using Neurological Criteria [Internet]. www.ficm.ac.uk; 2021. https://www.ficm.ac.uk/diagnosing-death-using-neurological-criteria. Accessed 24 Feb 2022.
45. Salih F, Finger T, Vajkoczy P, Wolf S. Brain death after decompressive craniectomy: incidence and pathophysiological mechanisms. J Crit Care. 2017;39:205–8.
46. Dalle Ave AL, Bernat JL. Inconsistencies between the criterion and tests for brain death. J Intensive Care Med. 2020;35(8):772–80.
47. Bernat JL, Dalle Ave AL. Aligning the criterion and tests for brain death. Camb Q Healthc Ethics. 2019;28(4):635–41.
48. Ray A, Manara AR, Mortimer AM, Thomas I. Brain herniation on computed tomography is a poor predictor of whether patients with a devastating brain injury can be confirmed dead using neurological criteria. J Inten Care Soc. 2021.
49. Chasse M, Glen P, Doyle MA, et al. Ancillary testing for diagnosis of brain death: a protocol for a systematic review and meta-analysis. Syst Rev. 2013;2:100.
50. Plourde G, Briard JN, Shemie SD, Shankar JJS, Chasse M. Flow is not perfusion, and perfusion is not function: ancillary testing for the diagnosis of brain death. Can J Anaesth. 2021;68(7):953–61.
51. Hossmann KA. Viability thresholds and the penumbra of focal ischemia. Ann Neurol. 1994;36(4):557–65.
52. Coimbra CG. Implications of ischemic penumbra for the diagnosis of brain death. Braz J Med Biol Res. 1999;32(12):1479–87.
53. Owen AM. The search for consciousness. Neuron. 2019;102(3):526–8.
54. Basano L, Ottonello P, Nobili F, et al. Pulsatile electrical impedance response from cerebrally dead adult patients is not a reliable tool for detecting cerebral perfusion changes. Physiol Meas. 2001;22(2):341–9.
55. Lapitska N, Gosseries O, Delvaux V, et al. Transcranial magnetic stimulation in disorders of consciousness. Rev Neurosci. 2009;20(3–4):235–50.

56. Palmer S, Bader MK. Brain tissue oxygenation in brain death. Neurocrit Care. 2005;2(1):017–22.
57. Riggs BJ, Cohen JS, Shivakumar B, et al. Doppler ultrasonography of the central retinal vessels in children with brain death*. Pediatr Crit Care Med. 2017;18(3):258–64.
58. Pallis C. Prognostic value of brainstem lesion. Lancet. 1981;317:8216.
59. Hung TP, Chen ST. Prognosis of deeply comatose patients on ventilators. J Neurol Neurosurg Psychiatry. 1995;58(1):75–80.
60. Jennett B, Gleave J, Wilson P. Brain death in three neurosurgical units. Br Med J (Clin Res Ed). 1981;282(6263):533–9.
61. Shewmon DA. Chronic "brain death": meta-analysis and conceptual consequences. Neurology. 1998;51(6):1538–45.
62. Mallampalli A, Powner DJ, Gardner MO. Cardiopulmonary resuscitation and somatic support of the pregnant patient. Crit Care Clin. 2004;20(4):747–61.
63. Takeuchi K. Report on the criteria for the determination of brain death in children. JMAJ. 2002;45(7):291–307.
64. Al-Shammri S, Nelson RF, Madavan R, Subramaniam TA, Swaminathan TR. Survival of cardiac function after brain death in patients in Kuwait. Eur Neurol. 2003;49(2):90–3.
65. George S, Thomas M, Ibrahim WH, et al. Somatic survival and organ donation among brain-dead patients in the state of Qatar. BMC Neurol. 2016;16(1):207.
66. van Beinum A, Healey A, Chandler J, et al. Requests for somatic support after neurologic death determination: Canadian physician experiences. Can J Anaesth. 2021;68(3):293–314.
67. Lewis A, Adams N, Varelas P, Greer D, Caplan A. Organ support after death by neurologic criteria: results of a survey of US neurologists. Neurology. 2016;87(8):827–34.
68. Lewis A, Adams N, Chopra A, Kirschen MP. Organ support after death by neurologic criteria in pediatric patients. Crit Care Med. 2017;45(9):e916–e24.
69. Pope TM. Brain death forsaken: growing conflict and new legal challenges. J Leg Med. 2017;37(3–4):265–324.
70. Braksick SA, Robinson CP, Gronseth GS, et al. Variability in reported physician practices for brain death determination. Neurology. 2019;92(9):e888–e94.
71. Greer DM, Varelas PN, Haque S, Wijdicks EF. Variability of brain death determination guidelines in leading US neurologic institutions. Neurology. 2008;70(4):284–9.
72. Wahlster S, Wijdicks EF, Patel PV, et al. Brain death declaration: practices and perceptions worldwide. Neurology. 2015;84(18):1870–9.

Research on the Newly Deceased Following Declaration of Death by Neurologic Criteria

Tamar Schiff and Brendan Parent

Nearly half a century ago, Willard Gaylin considered the promise of "neomorts" or "new cadavers," bodies with sustained circulatory-respiratory function following declaration of death by neurologic criteria. In his 1974 article "Harvesting the Dead," Gaylin described the possibility of a "bioemporium" full of such bodies, maintained for a period of years for uses ranging from clinical training and experimentation to serving as continuous sources of transfusable blood and immunoglobulins [1]. Sporadic reports of the testing of medications and devices in bodies of patients declared dead by neurologic criteria were documented in the medical literature in the following decade [2–4]. After a period of quieted interest, new calls to regulate and standardize methods for research after death by neurologic criteria arose from the experiences of investigators at MD Anderson Cancer Center and the University of Pittsburgh [5–7].

The potential of research after death by neurologic criteria is immense, allowing for testing of medical products in a human physiologic environment without subjecting living research participants or laboratory animals to risk of harm. As a prominent recent example of this research model, in late 2021 investigators at the NYU Langone Transplant Institute attached kidneys from genetically altered pigs to bodies donated by the families of two people recently declared dead by neurologic criteria [8]. These time-limited trials allowed researchers to assess the function of the novel kidney transplants and monitor for potential pathologic sequelae in human bodies. While the deceased cannot suffer bodily harm after death, regulatory and ethical safeguards are still necessary to ensure conduct respectful of human bodies and to prevent the macabre future Gaylin envisioned.

T. Schiff · B. Parent (✉)
Division of Medical Ethics, Department of Population Health, NYU Grossman School of Medicine, New York, NY, USA
e-mail: tamar.schiff2@nyulangone.org; brendan.parent@nyulangone.org

A. Lewis, J. L. Bernat (eds.), *Death Determination by Neurologic Criteria*, Advances in Neuroethics, https://doi.org/10.1007/978-3-031-15947-3_19

There are legal requirements to follow a person's documented directives for how their body may be used postmortem for purposes of therapy, research, or education [9]. More broadly, assurance that wishes regarding the disposition of one's body are respected even after death can serve as a general comfort to the living, contemplating their legacies and reputations and thinking of the eventualities of their own bodies [2]. Recognition of and adherence to autonomous choices made before death can also have implications on the treatment of living patients who have lost the ability to make decisions for themselves, such as those temporarily incapacitated, in a persistent vegetative state/unaware-wakeful state, or with advanced dementia [10]. Finally, appropriate treatment of bodies is an important measure in providing the respect owed to the loved ones of the deceased, who often identify the body as an extension of the individual prior to their death [2, 10, 11].

Without oversight and uniform standards, research using bodies after death by neurologic criteria could cause erosion of trust in the medical and scientific communities, creating harm greater than the benefit of advances yielded by this research. Conducted inappropriately, research after death by neurologic criteria could breed fear that end-of-life care may suffer or death by neurologic criteria may be declared prematurely in order to recruit bodies for this type of research. It could likewise cause public hesitation to register for organ donation for fear bodies could be used or misused for these purposes [5]. Limited studies of stakeholder perspectives indicate support for research after death by neurologic criteria [3, 12]. However, notable reservations underscore the requirement for guidelines attentive to key factors highlighted in this chapter. This discussion focuses on practice standards and relevant law in the United States.

1 Determination of Death by Neurologic Criteria by Accepted Standards

For research after death by neurologic criteria to proceed, there must be confidence that death has occurred. Each state legally recognizes death by neurologic criteria, but with variations [13]. Most states have adopted the Uniform Law Commission's (ULC) Uniform Determination of Death Act (UDDA), which states, in relevant part, that a person is dead who has sustained irreversible cessation of all functions of the entire brain, including the brainstem [14]. The UDDA also requires that death be declared in accordance with accepted medical standards, but there is not universal agreement on the standards for death by neurologic criteria across institutions or between clinicians (as discussed elsewhere in this book). As long as there is professional disagreement about when and how death by neurologic criteria can be determined, the enterprise of research after death by neurologic criteria is subject to ethical misgivings, demonstrating good reason to unify such standards.

2 Institutional Research Oversight Body

While there are significant federal regulations in place to protect human subjects involved in research, they do not apply to research on the deceased. The Federal Policy for the Protection of Human Subjects, or "Common Rule," and parallel regulations of the Food and Drug Administration set standards to minimize harm through Institutional Review Board (IRB) oversight and respect the autonomy of research subjects through informed consent requirements [15, 16]. But such subjects are defined as living humans in the Common Rule, and as a healthy human or patient in FDA regulations. Those declared dead by neurologic criteria are not living, healthy, or patients. Although these are not living subjects, research on bodies after death by neurologic criteria implicates the rights and interests of surviving family members, the legacy of the deceased individual, and the well-being of other patients who could benefit from organ donation. Accordingly, standards for oversight tailored to this kind of research are ethically necessary [6].

Although research after death by neurologic criteria is not under the purview of IRBs, there should be equivalent committees trained to identify and understand issues specific to research on those declared dead by neurologic criteria (Table 1). In addition to ensuring that the authorization process and provision of information to a family is adequate, these committees should also ensure that the research is appropriate and could not or should not be carried out in nonhuman animals or living humans; nothing unduly influential has been promised to authorizers and bodies are not commodified; that the team responsible for managing the deceased is appropriate and the use of resources does not strain the institution's acute care capacity; the recently deceased is managed with dignity and with significant privacy

Table 1 Sample guidance for oversight committee for research on newly deceased

Composition
Group of diverse, multidisciplinary representatives including:
• Clinician(s)
• Clinical/translational researcher(s)
• Ethics committee member(s)
• Clergy representative(s)
• Community member(s)
• Representative(s) from office overseeing institutional review board (IRB)
Responsibilities
Tasked with ensuring:
• Proper authorization for use in research obtained from deceased or surrogate decision-maker
• Use of newly deceased is warranted and advantageous in place of nonhuman animals or living humans
• Adequate information about study is provided to authorizing party
• Bodies are not commodified
• Team conducting study is appropriately trained in respectful treatment of bodies
• Resources do not compromise clinical care of living patients
• Conflicts of interests between end-of-life clinicians and investigators are minimized
• Proper recordkeeping of research protocols and procedures

protections; and that only those deceased with particular distinguishing features that make them advantageous for this research are involved [7, 17, 18]. Policies should be in place to guarantee proper recordkeeping of proposals and their findings, as well as a procedure for internal audit of the committee's review process [17].

Examples of such committees existing in parallel to IRBs are the Committee for Oversight of Research Involving the Deceased (CORID) at the University of Pittsburgh [17] and the Research on the Deceased Oversight Committee at New York University Grossman School of Medicine. Membership of oversight committees may differ across institutions, dependent on specific needs or circumstances, but should include multidisciplinary and diverse representatives with relevant expertise. As an example, CORID membership includes physicians, nurses, ethics committee members, clergy, community representatives, and representation from their institutional office with jurisdiction over the IRB; the committee is chaired by the institution's Associate Vice Chancellor for Clinical Research [17].

3 Authorization of Deceased Donation for Research Purposes

It is critical that valid authorization has been obtained for research after death by neurologic criteria. The concept of informed *consent* is tied to the rights of living individuals to decide how they are treated in medicine and in research. Choosing what happens to one's body after death, or choosing what happens to a loved one's body, is thus distinguished as *authorization*, which should also be informed enough for autonomous decision-making. Each state has adopted a legal framework for gifting one's body after death. Most of these frameworks are variations of the ULC's Uniform Anatomical Gift Act, which allows individuals to specify their body gifts after they die for organ transplant and research, and for others to make this gift if the deceased did not specify their wishes [9, 19]. In many states, when people register for donation via a registry or at the DMV, the language includes authorization for transplantation and research. But when most people register their donation wishes, they are often only aware of gifting for purposes of organ donation.

It is unclear when organ donation authorization should be a valid proxy for authorization for research after death by neurologic criteria. In accordance with respect for persons, and to align practice with the theory behind donation as an anatomical gift, adequate education should accompany the authorization decision, and individuals should have clear means to opt in to different uses. The option to select different purposes of an anatomical gift exists only in some registries, but few if any explain what research donation, such as research after death by neurologic criteria, might entail. States such as Delaware, New Jersey, and Pennsylvania legally require education around organ donation, but none require education on postmortem research [20–22]. Improving education and awareness regarding research after death by neurologic criteria would mean more informed and potentially explicit authorizations for this purpose and could prevent distrust manifesting as refusals to authorize any kind of donation, including organs for transplantation.

As research after death by neurologic criteria grows more common, so too will the consideration of donating one's body for this purpose. Since most people do not currently consider donating their bodies for this use, explicit first-person authorization is unlikely and the authorization decision therefore falls to family members or other surrogate decision-makers. While a family's decision to depart from common social and cultural practices of death may cause distinctive anxieties [5], families value the opportunity to make something positive out of their loved one's death and to honor their altruistic desires to contribute to science [12]. Postmortem authorization for organ donation often provides an opportunity to find a sense of meaning in loss, rather than heightened distress [23]. Similar positive sentiments have been reported to extend to a family's ability to contribute to research on behalf of the decedent [7].

Whether authorization is given by the decedent before their death or by a surrogate decision-maker, it must be determined what information is morally and logistically relevant based on the parameters of the study. This determination should account for the ways in which respectful treatment of the dead differs between religions, cultures, regions, families, and individuals. For example, in trials of novel organs either manufactured or derived from nonhuman animals, authorizers should at least know the nature of some tissues or materials that are used (human stem cells, nonhuman animal parts); how long the body will be used; whether and when the body will be returned; how the body will look upon return (impacting whether an open-casket funeral will be possible); and what costs or reimbursements will be involved. Authorizers should also have the opportunity to ask questions and gather additional information they find important.

4 Appropriate Identification of Research Candidates

There are a number of factors to consider to appropriately identify candidates for research after death by neurologic criteria (Table 2). Patients declared dead by neurologic criteria who require an autopsy should be excluded from consideration for research after death by neurologic criteria [7]. Infants born before 37 weeks of gestation should likewise be excluded as death by neurologic criteria cannot be determined prior to this age [24]. Owing to emotional, religious, cultural, or other factors, there will be individuals who will not agree with a determination of death by neurologic criteria of a loved one. Despite legal protections in most states for proceeding with withdrawal of organ support after declaration of death by neurologic criteria, this disagreement should preclude pursuit of research authorization to avoid unnecessary conflict and fostering distrust in medicine and research. For similar reasons, investigators should exclude the uniquely vulnerable population of unrepresented patients who lack decisional capacity, advanced care directives, and a known surrogate decision-maker [25].

There is also current consensus that organ donation should be prioritized over use of a body for research purposes after death by neurologic criteria as the clinical impact of transplantation is immediate compared to the possible, and often remote,

Table 2 Suggested minimum criteria for selection of research candidates

Eligibility criteria for inclusion in study on recently deceased
• Death by neurologic criteria in accordance with AAN 2010/AAP-CNS-SCCM 2011 standards (or any update to these standards)
• Death by neurologic criteria accepted by family or surrogate decision-makers
• Authorization can be obtained from deceased or surrogate decision-maker or authorization indicated in advanced care directive
• Ineligible for organ donation[a]
• No prior commitment of body donation to other organization or purpose[a]
• No autopsy required[a]
• Not pregnant
• >37 gestational weeks in age
• Additional factors as warranted by specific study

AAN American Academy of Neurology, *AAP* American Academy of Pediatrics, *CNS* Child Neurology Society, *SCCM* Society of Critical Care Medicine

[a]Relative exclusions: Bodies should be excluded if participation in study precludes organ donation, body donation to other organization with prior commitment, or autopsy. If participation does not preclude these options, coordination of both (e.g., participation in study followed by organ donation) may be pursued

benefit of research outcomes [6, 7, 18]. State frameworks accordingly give legal precedence to organ donation for transplant over donation of a body or body parts for research [26]. Ensuring this prioritization requires coordination with organ procurement organizations, responsible for facilitating recovery of organs and tissues for transplant therapy.

Strict prioritization of organ recovery for transplantation has potential to complicate or limit recruitment for research after death by neurologic criteria. Families are first approached for authorization for organ donation and the deceased is only considered for research after death by neurologic criteria when organs are determined ineligible for transplant. After such delay, families may prefer to withdraw support and proceed with funeral, cremation, or burial arrangements. Moreover, when organ donation is authorized, even marginal organs – those of questionable quality for transplant success – are often recovered without any accepting transplant centers. This means that transplantation does not occur and research after death by neurologic criteria is no longer possible. Considering the potential of research after death by neurologic criteria in advancing translation of basic and nonhuman animal research to living human subjects, it might be time to reconsider the legal prioritization of transplant donation over research donation in certain cases.

5 Confidentiality

The American Medical Association Code of Medical Ethics states that "information contained within a deceased patient's medical record, including information entered postmortem, should be kept confidential to the greatest possible degree" [27]. Still, postmortem research presents particularly unique opportunity, by intention or incident, to obtain information able to impact living relatives directly and pose perceived harm to decedents' reputations or legacies [2, 11]. Examples include information pertaining to genetic data or disease, HIV status, or paternity [6]. Accordingly, the authorization process should explicitly review possibly discoverable information and the planned scope of its disclosure [10]. Distinct from medical pertinence or emotional impact, this information could impact insurance status or employment, requiring protections against its discovery by such agencies [6]. Lastly, as research in the newly deceased has potential to draw significant public attention, possible media coverage and permissible release of information need also be considered in the planning of the research and the authorization process [2, 6].

6 Concluding Considerations in Study Design

Research after death by neurologic criteria can provide unique insights while eliminating risks incurred by living study participants. To achieve these ends ethically, protocols must be designed to ensure respectful treatment of both human bodies and surviving loved ones. Oversight committees should review proposals for these considerations. For example, duration of the research period should be limited to the shortest amount of time necessary, and efforts must be made to minimize invasive interventions to the body [18].

Respect also requires attention to appropriate research location. A site suitable for research after death by neurologic criteria will require systems, staffing, and technology able to accommodate circulatory and ventilatory support, surgical procedures, clinical monitoring, and data collection. This will most often be consistent with intensive care units, operating rooms, or another designated area in a hospital outfitted with comparable capabilities [5]. Research after death by neurologic criteria must not interfere with care of living patients in its utilization of resources and personnel. Attention should also be given to avoiding the potential of patients or families witnessing ongoing research after death by neurologic criteria, as this may be unsettling in a care environment.

Every effort should also be made to eliminate conflicts of interest among clinicians, investigators, and research staff. Those involved in obtaining authorization for research after death by neurologic criteria or conducting the study should have as minimal involvement as possible in the clinical care of patients prior to death and the determination of death by neurologic criteria [17]. Finally, there should be an emphasis on transparency in report of research practices and results to promote public trust and disseminate knowledge about research after death by neurologic criteria.

References

1. Gaylin W. Harvesting the dead. Harper's. 1974;249(1492):23–8.
2. Wicclair MR. Ethics and research with deceased patients. Camb Q Healthc Ethics. 2008;17(1):87–97.

3. Morris MC, Sachdeva T, Hardart GE. Enrolling brain-dead humans in medical research: stakeholder opinions. AJOB Empirical Bioethics. 2014;5(4):22–9.
4. Coller BS. Inhibition of human platelet function in vivo with a monoclonal antibody. Ann Intern Med. 1988;109(8):635.
5. DeVita MA, Wicclair M, Swanson D, Valenta C, Schold C. Research involving the newly dead: an institutional response. Crit Care Med. 2003;31(Supplement):S385–90.
6. Wicclair MR, DeVita M. Oversight of research involving the dead. Kennedy Inst Ethics J. 2004;14(2):143–64.
7. Pentz RD, Cohen CB, Wicclair M, et al. Ethics guidelines for research with the recently dead. Nat Med. 2005;11(11):1145–9.
8. Montgomery RA, Stern JM, Lonze BE, et al. Results of two cases of pig-to-human kidney xenotransplantation. N Engl J Med. 2022;386(20):1889–98.
9. National Conference of Commissioners on Uniform State Laws. Uniform Anatomical Gift Act §4 2006.
10. Wicclair MR. Informed consent and research involving the newly dead. Kennedy Inst Ethics J. 2002;12(4):351–72.
11. Tomasini F. Research on the recently dead: an historical and ethical examination. Br Med Bull. 2008;85(1):7–16.
12. Pentz RD, Flamm AL, Pasqualini R, Logothetis CJ, Arap W. Revisiting ethical guidelines for research with terminal wean and brain-dead participants. Hast Cent Rep. 2003;33(1):20.
13. Nikas NT, Bordlee DC, Moreira M. Determination of death and the dead donor rule: a survey of the current law on brain death. J Med Philos. 2016;41(3):237–56.
14. National Conference of Commissioners on Uniform State Laws. Uniform Determination of Death Act 1981.
15. FDA Policy for the Protection of Human Subjects [Internet]. [cited 2022 Jan 10]. https://www.fda.gov/science-research/clinical-trials-and-human-subject-protection/fda-policy-protection-human-subjects
16. HHS Regulations for the Protection of Human Subjects in Research [Internet]. [cited 2022 Jan 10]. https://www.hhs.gov/ohrp/regulations-and-policy/regulations/45-cfr-46/index.html
17. Yasko LL, Wicclair M, DeVita MA. Committee for oversight of research involving the dead (CORID): insights from the first year. Camb Q Healthc Ethics. 2004;13(04)
18. Parent B, Gelb B, Latham S, et al. The ethics of testing and research of manufactured organs on brain-dead/recently deceased subjects. J Med Ethics. 2020;46(3):199–204.
19. National Conference of Commissioners on Uniform State Laws. Uniform Anatomical Gift Act §9 2006.
20. 14 DE § 851.2.1.3.4.
21. 18A NJ §7F-4.3.
22. 20 Pa. C.S. § 8627.1.
23. Batten HL, Prottas JM. Kind strangers: the families of organ donors. Health Aff. 1987;6(2):35–47.
24. Nakagawa TA, Ashwal S, Mathur M, Mysore M. Guidelines for the determination of brain death in infants and children: an update of the 1987 task force recommendations-executive summary. Ann Neurol. 2012;71(4):573–85.
25. Kim H, Song M-K. Medical decision-making for adults who lack decision-making capacity and a surrogate: state of the science. Am J Hosp Palliat Med. 2018;35(9):1227–34.
26. National Conference of Commissioners on Uniform State Laws. Uniform Anatomical Gift Act §11(d) 2006.
27. AMA Council on Ethical and Judicial Affairs. AMA code of medical ethics' opinions on confidentiality of patient information. Virtual Mentor. 2012;14(9):705–7. https://journalofethics.ama-assn.org/article/ama-code-medical-ethics-opinions-confidentiality-patient-information/2012-09

Part IV

Legal Issues

U.S. State Laws on the Determination of Death by Neurologic Criteria

Ben Sarbey, Samuel A. Thumma, and Nita A. Farahany

The declaration of death of an individual has significant importance for various reasons, including for both law and medicine. In the vast majority of cases, the determination of death is based on the cessation of circulatory and respiratory functions. In less than 2% of cases, however, death is determined based on neurologic criteria, sometimes referred to as "brain death." The Uniform Determination of Death Act (UDDA), promulgated in the early 1980s, has had a significant role in standardizing the legislative criteria for determining death in the United States, although there remains variation between states in the established laws in this area. In this chapter, we provide an overview of state laws in the United States concerning the determination[1] of death and explore some legal controversies that have arisen since the adoption of the UDDA. We begin with a brief history of the inclusion of neurologic criteria in law for the determination of death, then proceed to a

[1] In this chapter, we use "determination" to describe state laws that address how death is diagnosed. A declaration of death specifies the moment in time that death is declared, which does not always follow immediately from a determination of death.

The views expressed are solely those of the authors.

B. Sarbey
Department of Philosophy, Duke University, Durham, NC, USA
e-mail: ben.sarbey@lawnet.duke.edu

S. A. Thumma
Arizona Court of Appeals, Division One, ULC Drafting Committee to Revise the Uniform Determination of Death Act (UDDA), Phoenix, AZ, USA
e-mail: sthumma@appeals.az.gov

N. A. Farahany (✉)
Duke University, Durham, NC, USA

ULC Drafting Committee to Revise the Uniform Determination of Death Act (UDDA), Durham, NC, USA
e-mail: farahany@duke.edu

A. Lewis, J. L. Bernat (eds.), *Death Determination by Neurologic Criteria*, Advances in Neuroethics, https://doi.org/10.1007/978-3-031-15947-3_20

discussion of some of the variations of state and international laws. We conclude by examining selected U.S. cases that have centered on ambiguity concerning determination of death based on neurologic criteria.

1 Historical U.S. State Laws on the Determination of Death

Historically, death was determined based on the loss of circulatory (cardiac) and respiratory (pulmonary) functions in an individual. To determine death, physicians typically would "feel for the pulse, listen for breathing, hold a mirror before the nose to test for condensation, and look to see if the pupils were fixed" [1]. This worked well before means of artificial support were developed; circulatory-respiratory criteria were clear and unproblematic. When an individual stopped breathing or their heart stopped pumping blood for a sufficient amount of time, medicine was powerless to sustain the patient's life and death and decay soon began. Beginning in the 1950s, however, medicine began using artificial technology including ventilators. These artificial support advances allowed for circulatory and respiratory functions to be technologically sustained, even in cases of severe brain injury [2].

In 1968, given the disconnect that artificial support introduced for the determination of death, an ad hoc committee of the Harvard Medical School convened to consider whether standards under which death was determined should be developed. The committee sought to develop standards to identify "irreversible coma" as the definition of death [3]. It recognized that provision of care for individuals who had irreversible loss of brain functions placed a significant burden on families and hospitals warranting consideration of a new way to determine death. As they saw it, the traditional method of determining death was "obsolete" and they were concerned about the externalities this created, including for organ donation after death to save the lives of patients waiting on transplant lists [3]. Focusing on the narrow situations in which individuals with catastrophic brain injuries were sustained by artificial support, the committee recommended unreceptivity and unresponsiveness, lack of movements or breathing, lack of reflexes, and a flat electroencephalogram as tests for death [3].

This new medical standard for death prompted some of the earliest state laws on the determination of death. It was evident that there were novel legal challenges associated with the determination of death for patients who were on artificial support. These included variability within the practice of medicine in handling such cases, the legal propriety of removing artificial support in these circumstances, questions about when the medical duty to treat ended, the requisite consent needed to remove artificial support, and the impact on possible organ donation [4]. With respect to organ donation, the "dead donor rule" prohibits the taking of organs from patients in a manner that causes death [5]. Although at times described as not being explicitly a legal doctrine, the dead donor rule is commonly accepted hospital policy and a commonly understood ethical requirement [6]. Indirectly, it reflects state laws against homicide. The dead donor rule has been codified into law in only three states (Hawaii, Maryland, and New Mexico) but is a significant factor for organ transplantation. The issue lies in a double bind: organs which can save the lives of others quickly deteriorate after death, but donation of vital organs (the heart, lungs, and other organs essential for life) could cause death. Those early ad hoc enactments were inconsistent, and only enacted in some form by about half of the states [7].

In 1980, the President's Commission for the Study of Ethical Problems in Medicine and Biomedical and Behavioral Research was asked to consider the question of how to define death. The Commission, composed of bioethical, medical, and legal experts, aimed to provide recommendations that could form the basis for a uniform definition of death [1]. As it noted, "[c]riminal prosecution, inheritance, taxation, treatment of the cadaver, and mourning are all affected by the way society draws the dividing line between life and death" [1]. After an extended process of receiving information from a variety of sources and consensus building, the Commission agreed upon a "whole brain death" standard. It found "the vital functions of the entire brain—and not merely portions thereof, such as those responsible for cognitive functions—as the only proper neurologic basis for declaring death"[1]. This early rigorous characterization of neurologic criteria for a determination of death set the stage for a uniform recommendation.

One year before the Commission's report, the Uniform Law Commission (ULC) in 1980 undertook the task of recommending a single standard for the determination of death. The ULC is an organization composed of legal experts organized to draft uniform or model acts for purposes of clarity and uniformity among states [8]. The ULC is not a legislative body, but instead formulates and promulgates statutory laws for consideration and adoption by the states. The purpose of this ULC effort to craft a statute on the determination of death has been described as creating a clear, shared standard among the states and to allow for organ donation in a manner consistent with the dead donor rule [9]. The ULC proposed a Uniform Determination of Death Act (UDDA) for consideration by the states, which included a provision for determining death by neurologic criteria, in addition to existing circulatory-respiratory criteria. The UDDA provides in relevant part:

> An individual who has sustained either (1) irreversible cessation of circulatory and respiratory functions, or (2) irreversible cessation of all functions of the entire brain, including the brainstem, is dead. A determination of death must be made in accordance with accepted medical standards [10].

The UDDA was soon approved by the American Bar Association and the American Medical Association and subsequently adopted by a large majority of states [11]. It has had significant influence, both in the United States and internationally, and it maintains broad (although not universal) support [12].

The UDDA does not specify which diagnostic tests should be used to determine death. States generally have followed the UDDA in omitting such specification, and instead set general criteria for determining death, leaving the details to the medical establishment "in accordance with accepted medical standards." Perhaps because of the lengthy history and nature of declaring death by cessation of circulatory and respiratory functions, there have not been efforts to generate nationally accepted medical standards for the determination of death by circulatory-respiratory criteria [13]. By contrast, in 1995 the American Academy of Neurology issued standards for determination of death by neurologic criteria for adults [14]. These standards specify methods for determining death as well as observation periods and applicable tests [14]. These standards were revised in 2010, reaffirmed in 2017, and are being revised again currently. As of 2010, the standards were described as follows:

> To determine "cessation of all functions of the entire brain, including the brainstem," physicians must determine the presence of unresponsive coma, the absence of brainstem reflexes, and the absence of respiratory drive after a CO2 challenge. To ensure that the cessation of brain function is "irreversible," physicians must determine the cause of coma, exclude mimicking medical conditions, and observe the patient for a period of time to exclude the possibility of recovery [15].

These standards, and the pediatric standards issued by the Society of Critical Care Medicine, American Academy of Pediatrics, and Child Neurology Society in 2011 [16], have proven influential in establishing the tests to determine death by neurologic criteria. These standards are commonly, but not universally, recognized as the "accepted medical standards" that are called for in many state statutes.

2 Contemporary State Laws

In the United States, the practice of medicine is regulated primarily at the state level, and all states have passed legislation concerning the determination of death, including enactments that permit the use of neurologic criteria to declare death. But there exist variations in those laws that have, at times, created confusion and discord between law and medicine (see Fig. 1) Thirty-nine states, the District of Columbia, and the U.S. Virgin Islands have adopted the UDDA in its entirety or with very similar wording [17–56]. Of these, Washington state did so by case law reflected in a decision from 1980 by its state supreme court [57]. All others have done so through statutory enactments.

Two states have adopted aspects of the UDDA, but solely in the context of patients whose respiratory and circulatory functions are maintained by artificial means. Connecticut requires that death be declared by circulatory-respiratory or neurologic criteria in the context of making a determination concerning continuation or removal of artificial support [58]. Florida mandates that neurologic criteria be used when respiratory and circulatory functions are maintained artificially [59].

Other jurisdictions differ in their description of requirements for determination of death by circulatory-respiratory or neurologic criteria. Some jurisdictions (Iowa, Louisiana, Texas, and Guam) appear to set circulatory-respiratory criteria as the primary standard for declaring death, with neurologic criteria as a secondary standard [60–63]. Three states (Hawaii, Kentucky, and Missouri) set circulatory-respiratory criteria as the general standard, with neurologic criteria as the standard when artificial support for breathing/circulation is used [64–66]. Hawaii's statute is typical of this approach: death occurs when "the person has experienced irreversible cessation of spontaneous respiratory and circulatory functions," but "[i]n the event that artificial means of support preclude a determination that respiratory and circulatory functions have ceased, a person shall be considered dead if, based on ordinary

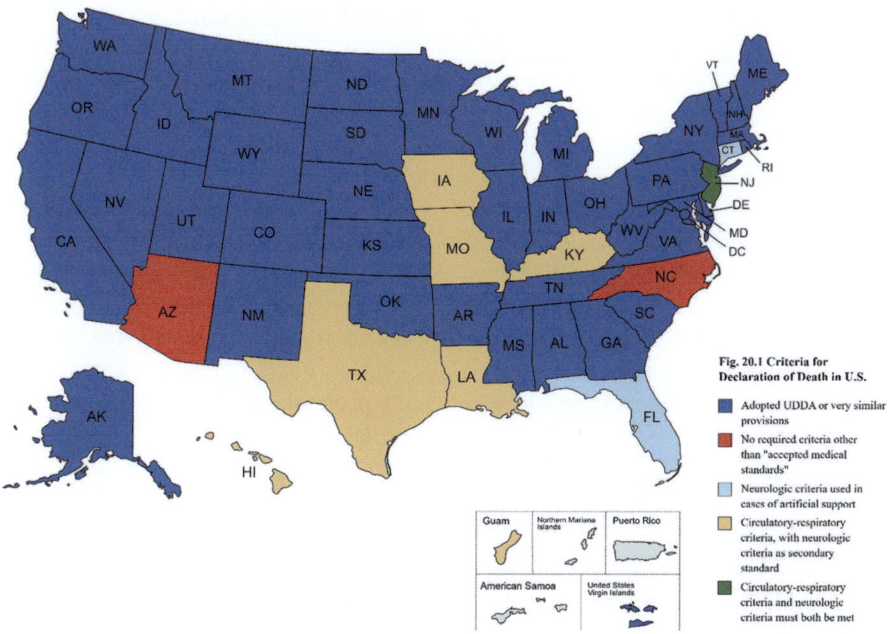

Fig. 1 Criteria for declaration of death in the U.S.

standards of current medical practice, the person has experienced irreversible cessation of all functions of the entire brain, including the brainstem" [65]. New Jersey requires both conditions: an individual is dead when circulation and breathing can only be supported artificially *and* there is "irreversible cessation of all functions of the entire brain" [67].

Two states (North Carolina, Arizona) do not directly specify criteria for determination of death. North Carolina states "ordinary and accepted standards of medical practice" are to be used for a determination of death [25]. However, North Carolina specifies that neurologic criteria *may* be used, along with other unspecified criteria (presumably circulatory-respiratory). Arizona states that "[a] determination of death must be made in accordance with accepted medical standards" [68, 69].

There is also some variability in the language describing death by neurologic criteria. Most states specify the UDDA requirement of "irreversible cessation of all functions of the entire brain, including the brainstem" as the criteria for death. Louisiana slightly changes this wording to "irreversible total cessation of brain function" [61]. Iowa and Texas have significantly different wording, requiring the cessation of "spontaneous brain function(s)" but removing the "irreversibility" requirement [60, 62].

Several states have diverged from the UDDA requirement that "[a] determination of death must be made in accordance with accepted medical standards." For example, Hawaii, Iowa, Louisiana, Maryland, Texas, and Virginia require that the determination of death be "based on ordinary standards of medical practice" [31, 51,

60–62, 65]. North Carolina requires that "accepted" medical standards apply [68]. Florida adds that the accepted medical standards also be "reasonable," without further defining that word [59]. Kentucky and Missouri reference "usual and customary standards" [64, 66]. Georgia and New Jersey, by contrast, omit language concerning medical standards and instead state which medical professionals are qualified to make the determination of death [24, 67].

Finally, state laws vary in whether they specify the number and qualifications of clinicians needed to make a death determination and, if so, how many and what is required. Georgia lists "a qualified physician," or a registered professional nurse or a physician assistant "authorized to make a pronouncement of death" [24]. Some states require one physician make the determination of death, others two physicians; some require particular types of physicians such as neurologists or specialists in critical care. Others allow nurses or other health professionals to make the determination of death.

3 Neurologic Criteria Across the World

Many countries do not have legal standards in place for determination of death by neurologic criteria [70]. Of those countries that do have a legal standard for the determination of death based on neurologic criteria, 87% require cessation of function of the entire brain, and the remaining 13% require brainstem death [71]. While several Asian countries have not established legal criteria for determinations of death by neurologic criteria, Japan, South Korea, and Pakistan have done so, but solely in the context of organ donation [72, 73].

The United Kingdom is a unique jurisdiction in several ways—it historically relied on the cessation of function of the brainstem as the criterion for death [74] but now has a standard based on loss of consciousness as well as loss of brainstem function. The current standard in the U.K. is that death requires "the irreversible loss of the capacity for consciousness, combined with irreversible loss of the capacity to breathe... and therefore irreversible cessation of the integrative function of the brainstem" [75].

Canadian law varies by province, but most accept some version of neurologic criteria. Manitoba is the only Canadian province to require loss of all brain functioning [76]. The provinces of Prince Edward Island as well as Northwest Territories both recognize "brain death as determined by generally accepted medical criteria" [77, 78]. Remaining provinces have a mix of neurologic criteria or no legal standard concerning the determination of death.

4 Recent Legal Controversies in the United States

Several court cases have tested the application and interpretation of law on the determination of death by neurologic criteria. The case of Jahi McMath had a particularly high profile among such cases. Jahi was a 13-year-old girl in California who had a circulatory-respiratory arrest in 2013 after tonsil surgery that resulted in irreversible brain damage [79]. A coroner in California issued a death certificate after the hospital completed an evaluation and found her to meet the standards for death by neurologic criteria, but the family insisted that Jahi was still alive, and filed and won a suit to continue artificial support [79]. Later, the hospital obtained a court ruling that it could discontinue artificial support due to medical futility and a court declaration that Jahi was dead [80]. However, before support was discontinued, the McMath family transferred Jahi to a New Jersey hospital and then an apartment in New Jersey, a state that statutorily permits religious-based exemption from the application of neurologic criteria for determining death, which the McMath family invoked. This created a paradoxical scenario where Jahi was legally dead in one state but alive in another. Jahi was declared dead in New Jersey after circulatory-respiratory arrest, approximately 4 years after California issued her death certificate [79]. Jahi's father filed a medical malpractice suit, seeking damages for medical negligence and emotional distress [81]. In the complaint, he argued that Jahi had not met the legal requirement for death in California because Jahi still had hypothalamic function and therefore did not meet the legal requirement of "irreversible cessation of all functions of the entire brain" [82].

The case of Aden Hailu concerned a related issue: whether statutory criteria were fulfilled by the medical tests performed on the patient. Hailu was a college student in Nevada who in 2015 had surgery to remove her appendix. During the surgery, an insufficient supply of oxygen to her brain resulted in severe hypoxic-ischemic brain damage [83]. The hospital performed a series of tests to determine whether Hailu should be pronounced dead. These included three electroencephalogram (EEG) tests, which showed some residual electrical activity in the brain. Later, an apnea test showed that Hailu could not breathe on her own and the hospital declared death consistent with the American Academy of Neurology (AAN) standard for determination of death by neurologic criteria. Hailu's father sought an injunction to prevent the hospital from discontinuing artificial support [84]. The case reached the Nevada Supreme Court, which focused on whether the AAN standard was the "accepted medical standard" required by the Nevada determination of death statute. The court relied on a New Jersey report on the status of the New Jersey Declaration of Death Act, which found significant disagreement concerning the AAN standard and noted that state statutes did not explicitly incorporate that standard [85]. The court held that this lack of consensus prohibited the AAN standard from being properly considered the "accepted" medical standard [84]. In response, the Nevada legislature amended its determination of death statute in response to the Hailu case to specify the AAN and Society of Critical Care Medicine standards as the "accepted medical standards" [37].

Other legal challenges have arisen from issues unrelated to medical standards. Several cases have centered on whether consent is required to conduct testing pursuant to a determination of death by neurologic criteria [4]. Addressing different issues, a Kentucky case from 2014 involving 2-month-old Isaac Lopez centered on parental rights. Isaac was declared dead by neurologic criteria after sustaining severe blunt injuries. His mother sought an injunction to prevent the hospital from removing artificial support, arguing that such action would violate her constitutional rights to make decisions for Isaac. The court ruled that "with death, no parental decision making survives, (save decisions regarding burial)" [86]. Consequently, the court found that the hospital no longer had an obligation to continue treatment, as there was no longer a living patient and artificial support was discontinued [87].

These cases raise a variety of issues concerning laws on the determination of death. First, lack of uniformity among state laws concerning determinations of death have created or exacerbated conflict related to determination of the end-of-life. Although not extraterritorial, state laws are not islands in our modern interconnected world, as the transport of Jahi McMath to New Jersey after a declaration of death in California aptly illustrates. Some states allow for an individual or others to seek an exemption from a determination of death on neurologic criteria based on the individual's personal religious beliefs. (New Jersey, and, indirectly, Illinois, California, and New York). States that do so may result in disagreement between physicians and families over determinations of death.

Second, legal challenges have and will likely continue to occur regarding whether medical tests for death by neurologic criteria meet existing statutory requirements [4]. The Aden Hailu case addressed whether the AAN standard is the "accepted medical standard" required in that jurisdiction. In other jurisdictions, there is no specificity, and may be little consistency, in what constitutes "accepted medical standards."

Third, defining death by neurologic criteria as "irreversible cessation of all functions of the entire brain, including the brainstem" has resulted in legal challenges when some functions of the brain—specifically hypothalamic functions— persist in patients who meet the AAN's standard for death by neurologic criteria. Accepted medical standards do not require "all functions of the entire brain" to satisfy a determination of death by neurologic criteria, creating a conflict between language in statutory enactments and accepted medical practice [15]. Ambiguity concerning the regions of the brain specified in state statutes also may result in litigation. Confusion and conflict may also arise from ambiguity concerning the meaning of "medically accepted standards," "entire brain," and whether cessation of brain functions is "irreversible."[2]

Finally, ambiguities resulting in legal challenges may have a snowball effect. The Jahi McMath and Aden Hailu cases, in particular, have drawn significant public attention. They also have been reported as undermining public understanding and trust in both the law and medical practice concerning determination and declaration of death, perhaps leading to an increase in legal challenges [88].

[2] The UDDA and most state laws specify "irreversible" cessation of circulatory or brain function for determinations of death. Some commentators and scholars have advocated for using the word "permanent" instead of irreversible. They argue "permanent" implies that the cessation of brain function will not be restored, whereas "irreversible" implies that functioning *cannot* be restored [90–92].

5 Conclusion

A declaration of death has significant legal, medical, social, religious, and financial implications including marking the point at which medical duties to the individual may end, organ donation is ethically permissible, and the transfer of an estate [89]. Determining death is typically uncontroversial in both law and medicine because it is based on the cessation circulatory and respiratory functions. However, the introduction of technological circulatory-respiratory support has led healthcare facilities to be able to maintain these functions in individuals with catastrophic brain injury that would lead to permanent loss of circulatory and respiratory functions in the absence of such support.

To address these cases, states legally permit a determination of death by neurologic criteria, advisory bodies weighed in with suggested approaches, and the ULC promulgated the UDDA. The UDDA has had a significant role in standardizing the legislative criteria for declaring death. But despite this move toward greater standardization of state laws on this issue, there remain variations in state statutory approaches to the determination of death.

States have largely, but not universally, adopted neurologic criteria for the determination of death. Those that have done so have variations in the wording and requirements of their laws. State statutes differ over whether circulatory-respiratory or neurologic criteria are the primary way to determine the death of an individual, and whether criteria apply generally or specifically in the context of considering discontinuation of artificial support for respiration and circulation. At times, what constitutes "accepted medical standards" and the "entire brain" have been approached differently in different jurisdictions. Judicial interpretation of similar or identical statutory text has also varied between jurisdictions.

References

1. President's Commission for the Study of Defining Death: A Report on the Medical, Legal and Ethical Issues in the Determination of Death 1981. https://repository.library.georgetown.edu/bitstream/handle/10822/559345/defining_death.pdf. Accessed 28 Mar 2022.
2. Belkin GS. Death before dying. Oxford: Oxford University Press; 2014.
3. A Definition of Irreversible Coma: Report of the Ad Hoc Committee of the Harvard Medical School to Examine the Definition of Brain Death. JAMA. 1968;205:337–40.
4. Pope TM. Brain death forsaken: growing conflict and new legal challenges. J Leg Med. 2017;37:265–324.
5. Robertson JA. The dead donor rule. Hast Cent Rep. 1999;29:6–14.
6. Omelianchuk A. How (not) to think of the 'dead-donor' rule. Theor Med Bioeth. 2018;39:1–25.
7. Nikas NT, Bordlee DC, Moreira M. Determination of death and the dead donor rule: a survey of the current law on brain death. J Med Philos. 2016;41:237–56.
8. Uniform Law Commission. 2022. https://www.uniformlaws.org/aboutulc/overview. Accessed 5 Apr 2022.
9. Veatch RM. Abandon the dead donor rule or change the definition of death? Kennedy Inst Ethics. 2004;14:261–76.
10. National Conference of Commissioners on Uniform State Laws. Uniform Determination of Death Act 1980. https://www.uniformlaws.org/HigherLogic/System/DownloadDocumentFile.

ashx?DocumentFileKey=4d19d096-be64-3c0f-ae71-a514b64c06a6&forceDialog=0. Accessed 28 Mar 2022.
11. Sarbey B. Definitions of death: brain death and what matters in a person. J Law Biosci. 2016;3:743–52.
12. Bernat JL. The whole-brain concept of death remains optimum public policy. J Law Med Ethics. 2006;34:35–43.
13. Shemie SD, Hornby L, Baker A, et al. International guideline development for the determination of death. Intensive Care Med. 2014;40:788–97.
14. The Quality Standards Subcommittee of the American Academy of Neurology. Practice parameters for determining brain death in adults (summary statement). Neurology. 1995;45:1012–4.
15. Wijdicks EFM, Varelas PN, Gronseth GS, Greer DM. Evidence-based guideline update: determining brain death in adults: report of the quality standards subcommittee of the american academy of neurology. Neurology. 2010;74:1911–8.
16. Nakagawa TA, Ashwal S, Mathur M, Mysore M. Clinical report—guidelines for the determination of brain death in infants and children: an update of the 1987 task force recommendations. Pediatrics. 2011;128:e720–40.
17. Ala. Code § 22-31-1.
18. Alaska Stat. § 09.68.120.
19. Ark. Code Ann. § 20-17-101.
20. Cal Health & Safety Code § 7180.
21. Colo. Rev. Stat. Ann. § 12-36-136.
22. Del. Code Ann. Tit. 24, § 1760.
23. D.C. Code § 7—601.
24. Ga. Code Ann. § 31-10-16.
25. Idaho Code Ann. §54-1819.
26. IL § 755 ILCS 50/1-10.
27. Ind. Code Ann. § 1-1-4-3.
28. Kan. Stat. Ann. § 77—205.
29. Ma. Gen. Laws 106 § 1-107.
30. Me. Rev. Stat. Ann. Tit. 22, § 2811.
31. Md. Code Ann., Health-Gen. § 5-202.
32. Mich Comp Laws Ann §333.1033.
33. Minn. Stat. Ann. § 145.135.
34. Miss. Code Ann. § 41-36-3.
35. Mont. Code Ann. § 50-22-101.
36. Neb. Rev. Stat. § 71-7202.
37. Nev. Rev. Stat. Ann. § 451.007.
38. N.H. Rev. Stat. Ann. § 141-D:2.
39. N.M. Stat. Ann. § 12-2-4.
40. N.Y. 10 CRR-NY 400.16.
41. N.D. Cent. Code § 23-06.3-01.
42. Oh. Rev. Code § 2105.35.
43. Okla. Stat. Ann. Tit.63, § 3122.
44. Or. Rev. Stat. Ann. § 432.300.
45. 35 Pa. Stat. Ann. § 10203.
46. R.I. Gen. Laws § 23-4-16.
47. S.C. Code Ann. § 44-43-460.
48. S.D. Codified Laws § 34-25-18.1.
49. Tenn. Code Ann. § 68-3-501.
50. Utah Code Ann. § 26-34-2.
51. Va. Code § 54.1-2972.
52. V.I. Code Tit. 19, § 869.
53. Vt. Stat. Ann. Tit. 18, § 5218.
54. W.Va. Code Ann. § 16-10-1.

55. Wis. Stat. Ann. § 146.71.
56. Wyo. Stat. Ann. §35-19-101.
57. Supreme Court of Washington. 94 Wn.2d 407. 1980. https://law.justia.com/cases/washington/supreme-court/1980/46582-1.html.
58. Conn. Gen. Stat. Ann. § 19A-504A.
59. Fla. Stat. Ann. § 382.009.
60. Iowa Code Ann. § 702.8.
61. La. Rev. Stat. Ann. § 9:111.
62. Tex. Health & Safety Code Ann. § 671.001.
63. Guam §§ 83B101 - 83B103.
64. Ky. Rev. Stat. Ann. § 446.400.
65. Haw. Rev. Stat. § 327C-1.
66. Mo. Ann. Stat. § 194.005.
67. N.J. Rev. Stat. § 26:6A.
68. N.C. Gen. Stat. § 90--323.
69. Az. Rev. Stat. § 14-1107.
70. Wahlster S, Wijdicks EFM, Patel PV, et al. Brain Death Declaration: Practices and Perceptions Worldwide. Neurology. 2015;84:1870–9.
71. Lewis A, Bakkar A, Kreiger-Benson E, et al. Determination of death by neurologic criteria around the world. Neurology. 2020;95:e299–309.
72. Terunuma Y, Mathis BJ. Cultural sensitivity in brain death determination: a necessity in end-of-life decisions in Japan. BMC Med Ethics. 2021;22:58.
73. Wasay M, Mujtaba SB, Sheerani M, Barech S, Jooma R. A review of brain death protocols across the globe and need for brain death guideline for Pakistan. Pak J Neurol Sci. 2020;15:38–47.
74. Airedale NHS Trust v. Bland. 1 All ER 831. 1993. https://www.globalhealthrights.org/wp-content/uploads/2013/01/HL-1993-Airedale-NHS-Trust-v.-Bland.pdf.
75. Academy of Royal Medical Colleges. A Code of Practice for the Diagnosis and Confirmation of Death. 2008. https://www.aomrc.org.uk/wp-content/uploads/2016/04/Code_Practice_Confirmation_Diagnosis_Death_1008-4.pdf. Accessed 17 Apr 2022.
76. Manitoba, The Vital Statistics Act C.C.S.M. c. V60.
77. Prince Edward Island 2011, Human Tissue Donation Act, Ch H-12.1.
78. Northwest Territories, Human Tissue Donation Act, SNWT 2014, c 30.
79. Chow K. Jahi McMath, Teen At Center Of Medical And Religious Debate On Brain Death, Has Died. NPR. 2018. https://www.npr.org/2018/06/29/624641317/jahi-mcmath-teen-at-center-of-medical-and-religious-debate-on-brain-death-has-di. Accessed 4 Apr 2022.
80. Simon D, Shoichet CE. Judge: California teen is brain dead after tonsil surgery. CNN. 2013. https://edition.cnn.com/2013/12/24/health/jahi-mcmath-girl-brain-dead/. Accessed 4 Apr 2022.
81. Rocha V. Family of Jahi McMath sues doctor, Oakland hospital over brain damage. LA Times. 2015. https://www.latimes.com/local/lanow/la-me-ln-family-of-jahi-mcmath-sues-doctor-oakland-hospital-20150303-story.html. Accessed 4 Apr 2022.
82. Complaint, McMath v. Rosen. (Cal. Super Ct. Dec. 09, 2015) (No. RG15796121).
83. Associated Press. Nevada court to decide fate of woman on life-support. CBS News. 2015. https://www.cbsnews.com/news/nevada-court-to-decide-end-of-life-case/. Accessed 4 Apr 2022.
84. Supreme Court of Nevada. Hailu v. Prime Healthcare. 2015. https://cases.justia.com/nevada/supreme-court/2015-68531.pdf?ts=1447722170.
85. New Jersey Law Revision Commission. Final Report Relating to New Jersey Declaration of Death Act. 2013. https://static1.squarespace.com/static/596f60f4ebbd1a322db09e45/t/5cf7f3615c724d000144d259/1559753569691/njddaFR011813.pdf. Accessed 4 Apr 2022.
86. In re Isaac Lopez, a Minor No. 14-CI-354 (Jefferson Circuit Court, Kentucky 22 July 2014) (Order).
87. Pope TM. Legal Briefing: Brain Death and Total Brain Failure. J Clin Ethics. 2014;25:245–57.

88. Debolt D. More families now challenging doctors' brain-death diagnoses. Seattle Times. 2016. https://www.seattletimes.com/nation-world/more-families-now-challenging-doctors-brain-death-diagnoses/. Accessed 4 Apr 2022.
89. Gardiner D, Shemie S, Manara A, Opdam H. International Perspective on the Diagnosis of Death. Br J Anaesth. 2012;108:i14–28.
90. Bernat JL. How the distinction between "irreversible" and "permanent" illuminates circulatory–respiratory death determination. J Med Philos. 2010;35:242–55.
91. Bernat JL. On noncongruence between the concept and determination of death. Hast Cent Rep. 2013;43:25–33.
92. Rodríguez-Arias D, Smith MJ, Lazar NM. Donation after circulatory death: burying the dead donor rule. Am J Bioeth. 2011;11:36–43.

Is Consent Required for Clinicians to Make a Determination of Death by Neurologic Criteria?

Thaddeus Mason Pope

The overwhelming weight of authority in the United States, Canada, and the United Kingdom holds that clinicians are not legally or ethically required to obtain family consent before making a determination of death by neurologic criteria. There is a consensus among professional guidelines, a consensus in statutes, and a near consensus among court decisions. Moreover, prevailing practice does not require consent. Accordingly, increasingly vocal proponents of a consent requirement bear a heavy burden to overcome the presumptive legitimacy of the status quo. While their arguments have some validity, proponents cannot surmount the weightier considerations against imposing a consent requirement. Nevertheless, even though clinicians and hospitals are not legally required to obtain consent, they should still notify families about the intent to make a determination of death by neurologic criteria and offer temporary reasonable accommodations when feasible.

1 The Uniform Determination of Death Act Does Not Require Consent for Determination of Death by Neurologic Criteria

The general rules for medical consent are widely understood. Clinicians must normally obtain patient or surrogate consent before administering any test or procedure. Otherwise, the clinician commits not just malpractice but tortious battery [1, 2]. However, United States law carves out several specific exceptions to this general consent rule. The most salient is the emergency exception. Determination of death by neurologic criteria is another exception.

For four decades, in almost every jurisdiction in the United States, the determination of death has been governed by the Uniform Determination of Death Act

T. M. Pope (✉)
Mitchell Hamline School of Law, Saint Paul, MN, USA
e-mail: Thaddeus.Pope@mitchellhamline.edu

A. Lewis, J. L. Bernat (eds.), *Death Determination by Neurologic Criteria*, Advances in Neuroethics, https://doi.org/10.1007/978-3-031-15947-3_21

(UDDA) [3]. The UDDA provides that "an individual who has sustained ... irreversible cessation of all functions of the entire brain, including the brainstem, is dead" [3]. The UDDA is deliberately silent on exactly how clinicians should measure irreversible cessation. Instead, the UDDA delegates and defers to the medical profession. The UDDA provides that "determination of death must be made in accordance with accepted medical standards."

In other words, drafters deliberately designed the UDDA to leave space for the medical profession to innovate. Accordingly, the UDDA provides only a general standard that delegates responsibility for establishing the precise method and manner for assessing satisfaction of that standard [4]. The President's Commission explained: "legislation should be formulated at the second level, that of general standards" because "operational criteria" with a greater "level of specificity," are "better left to medical bodies to establish" [4].

In short, the UDDA requires that clinicians make determinations of death in accordance with "accepted medical standards." Because those standards themselves do not require consent, the UDDA does not require consent. Evidence on this point is significant and compelling. Among other substantiation, we can look both to (1) professional organization position statements and (2) custom and practice.

1.1 Accepted Medical Standards: Professional Society Statements

Every relevant professional society maintains that consent for determination of death by neurologic criteria is not required. The most consequential and most recent is from the World Brain Death Project [5]. Its 2020 report is supported by 33 medical societies and 5 world federations. The World Brain Death Project concludes that consent is not needed prior to determination of death by neurologic criteria. Indeed, the World Brain Death Project states this in strong and unambiguous terms, addressing consent in two separate recommendations:

> It is recommended that there is ***no need for consent*** for performance of the clinical evaluation, apnea testing, or ancillary testing for determination of brain death/death by neurologic criteria [5].
>
> It is suggested that legislation, regulations, judicial formulations, executive orders, decrees, or legal guidelines about brain death/death by neurologic criteria indicate that there is ***no need for consent*** for performance of the clinical evaluation, apnea testing, or ancillary testing for determination of brain death/death by neurologic criteria [5].

The American Academy of Neurology (AAN) takes a similar position. The AAN is the world's largest association of neurologists. Its position statement says that "its members have ... the moral authority ... to perform a brain death evaluation including apnea testing ... ***without obligation to obtain informed consent***" [6]. Furthermore, the AAN expands on this view and even more strongly states that clinicians have not only the authority but also the "professional responsibility" to make a determination of death by neurologic criteria without the obligation to obtain consent [6]. The AAN maintains that not only "may" clinicians perform an evaluation for determination of death by neurologic criteria, they "should" perform it.

Other professional society guidelines are in accord. For example, the Los Angeles County Medical Association guidelines state that while a patient's surrogate may request confirmation of determination of death by neurologic criteria by a physician of their choosing, "the determination of death remains a medical decision" [7].

1.2 Accepted Medical Standards: Custom and Practice

Looking to professional society position statements is not the only way to ascertain "accepted medical standards." We can also look to custom and practice. Multiple studies of clinicians and hospitals show that they make determinations of death by neurologic criteria without requiring consent.

Most clinicians do not think consent is needed to determine death by neurologic criteria. Surveys of individual clinicians show that 78% of neurologists and 72% of pediatric neurologists strongly disagreed or somewhat disagreed that physicians should obtain consent from a patient's family before making a determination of death by neurologic criteria [8, 9]. Even more cogent is a survey of standards on determination of death by neurologic criteria for pediatric patients from more than 100 hospitals. Not one requires consent. Furthermore, more than one-third of the standards explicitly state that consent is ***not*** required [10].

1.3 Accepted Medical Standards: Concession by Consent Proponents

Notably, even proponents of a legal consent requirement concede that the *status quo* does not require consent. They concede that determination of death by neurologic criteria "has been obtained ***without informed consent*** for over a half century" [11]. Other proponents concede that "consent for brain death evaluations ultimately ***may not be required*** under current legal standards" and "obtaining consent for the death by neurologic criteria evaluation ***would be a change*** from what we know about current practice" [12].

1.4 Conclusion

In sum, the UDDA provides that death should be determined according to "accepted medical standards." Those accepted medical standards do not require consent. Therefore, the UDDA does not require consent.

2 No Other Statute or Regulation Requires Consent

Unlike the UDDA itself, other state statutes or regulations more directly and explicitly address consent for determination of death by neurologic criteria. As with hospital standards, when the law addresses consent, it always addresses it the same way. It always provides that consent is ***not*** required.

For example, the Nevada statute provides:

> A determination of the death of a person ... is a clinical decision that does ***not require*** the consent of the person's authorized representative or the family member with the authority to consent or withhold consent [13].

Similarly, New York law provides that the hospital must "make diligent efforts to notify the patient's Surrogate Decision-maker that the process for determining brain death is underway. Consent ***need not*** be obtained" [14]. And like Nevada and New York, New Jersey guidelines provide that "there is ***no need*** for consent to be obtained for the clinical assessment" [15].

3 Accommodation Laws Do Not Require Consent

Laws like those in Nevada, New York, and New Jersey that explicitly exclude the need for consent are not the only relevant laws. Similarly revealing are statutes and regulations that require "reasonable accommodation" for families when they assert a religious or moral objection to determination of death by neurologic criteria [16]. Despite their enfolding objective, these laws typically address only organ-sustaining treatment ***after*** determination of death by neurologic criteria. There are three key types of accommodation.

3.1 Post-Determination, Post-Declaration Accommodation

California, Illinois, and New York have laws that require hospitals to afford families a reasonable accommodation [16]. These laws essentially mean that hospitals continue mechanical ventilation for a short period of time after determination and declaration of death. Since legal duties under these laws are triggered only after determination of death, this accommodation affects neither determination nor declaration of death.

3.2 Post-determination, Pre-declaration Accommodation

New Jersey law is even stronger, requiring that individuals with religious objections cannot be declared dead on neurological criteria [17, 18]. New Jersey offers a complete exemption from declaration of death on neurologic criteria. But even this broad accommodation law pertains only to "declaration" or "pronouncement" of death, not to "determination" of death. The drafters specifically anticipated that the determination will be performed concurrent with ascertainment of information about religious beliefs. The law does not permit objections to determination. It permits objections only to the post-determination declaration of death on neurologic criteria [19].

3.3 Pre-determination, Pre-declaration Accommodation

New York law is unique in that it requires hospitals to have a "procedure for reasonable accommodation of the individual's religious of moral objections to the ***determination***" [20, 21]. But even this law does not require that clinicians obtain consent, and it leaves hospitals significant discretion in how they will accommodate objections.

4 Courts Confirm That Consent Is Not Required

The UDDA and other statutes and regulations are not the only authorities holding that consent is not required for determination of death by neurologic criteria. Over the past few years, the consent question has repeatedly reached courts in the United States (Table 1) [16].

It might appear, on cursory examination, that the courts are evenly split [16]. But a closer look shows more consistency. Every judgment that was carefully briefed and argued holds that consent is not required. The conclusion of these courts accords with courts in other common law jurisdictions like Canada and the United Kingdom [22–26]. The small handful of court judgments requiring consent were issued under exigent circumstances when the family sought a temporary restraining order.

The sole exception from the line of cases in the United States is the Allen Callaway case [27]. On July 22, 2016, 6-year-old Allen Callaway drowned. Clinicians at Saint Vincent Healthcare in Billings, Montana, performed most of the evaluation for death by neurologic criteria. Based on these results, clinicians

Table 1 Court cases addressing consent for determination of death by neurologic criteria

Year	Court Case	State	Posture	Consent required
2020	Sharon Lucy Frederick	New York	Merits	No
2020	Nick Torres	Texas	TRO	No
2016	Mirranda Lawson	Virginia	Merits	No
2016	Allen Callaway	Montana	Merits	Yes
2016	Alex Pierce	California	TRO	Yes
2015	Aden Hailu	Nevada	Merits	No
2008	Motl Brody	Washington, DC	Merits	No
2007	Tara Hawkins	Georgia	Merits	No
2006	Brett Shively	Kansas	TRO	Yes

strongly suspected that Callaway was dead. However, Callaway's mother objected to performance of the apnea test.

The hospital filed a petition seeking a judicial declaration permitting it to conduct "brain activity tests" on Allen over his mother's objections. But in September 2016, the Pondera County Court denied the hospital's request. The court held that Allen's mother has the "sole authority to make medical decisions on his behalf, including the decision as to whether any future brain functionality examinations should be administered" [27].

The court concluded: "This Court is not willing to create in the medical profession sole and exclusive authority to make a decision whether to conduct a brain death examination. If such an important public policy is to be made, it is the role of the legislature, and not this Court, to do so" [27]. But in fact, the Montana legislature had already made the important public policy decision about authority to conduct an evaluation for death by neurologic criteria. As explained above, the 1983 Montana UDDA "grants" physicians the right to perform an evaluation for death by neurologic criteria. It states that "A determination of death must be made in accordance with ***accepted medical standards***" [28]. Since accepted medical standards do not require consent, neither does the Montana UDDA.

5 Arguments for Requiring Consent

Even if the above legal analysis shows that consent is not legally required for determination of death by neurologic criteria, it is still appropriate to ask whether that should be the case. Even if we already know what the rules are, we should be prepared to offer explanations and justifications for what the rule ***ought*** to be. Proponents for requiring consent make five types of arguments.

5.1 Unreliability and Self-Fulfilling Prophecy

Some advocates for requiring consent charge that the apnea test does not measure what it purports to measure. It is an unreliable test [29, 30]. For example, Rodriguez-Arias questions the very validity of the apnea test [31]. An even stronger version of this argument charges that the apnea test not only fails to determine death by neurologic criteria, but even ***causes*** it [32]. Joffe argues that clinicians should not perform the apnea test even with consent because the test is contraindicated, serves no diagnostic purpose, and produces a self-fulfilling prophecy [33].

5.2 Significant Risks from Apnea Testing

Other advocates for requiring consent are not ready to abandon the apnea test altogether. Nonetheless, they are concerned about potential iatrogenic harm. One commentator observes that "Most discussions on informed consent for [death by neurologic criteria] focus on the safety issue of the apnea test" [11] Unlike ancillary testing, the apnea test imposes significant risks with no countervailing benefit to the patient [31, 34–37]. Indeed, these risks may be enhanced by variability in how the apnea test is conducted [38]. In short, proponents for requiring consent charge that the apnea test is not in the patient's best interest, or at least it is a value-laden judgment balancing risks and benefits that is theirs to make.

5.3 Religion and Conscience-Based Objections

Some advocates for requiring consent do not focus on the diagnostic accuracy or risks of the apnea test. Instead, they defend a consent requirement to respect religion. For example, in a 2016 Virginia case, the parents of 2-year-old Mirranda Grace Lawson physically blocked clinicians from performing the apnea test and gave them a handwritten note: "We are Christians, and it is against our religious beliefs to remove the ventilator" [39]. Unlike arguments concerning unreliability and risks, this argument addresses not only the apnea test but also ancillary testing and the clinical exam.

5.4 Respect for Persons

Some advocates for requiring consent argue that it shows respect for the person [12]. Paquette et al. argue that the justification for seeking consent lies in respect for the moral status and agency of the person. Therefore, they argue that informed consent is required not only for the apnea test but also for all other aspects of the evaluation for death by neurologic criteria [40].

5.5 Race and Trust

Conflicts over determination of death by neurologic criteria occurred disproportionately with families of certain races. A lack of trust between Black and Hispanic communities and the medical community is often cited as the reason [41]. When clinicians proceed to determine death without consent, it further damages the already fragile trust [42, 43]. Johnson contends that consent is important because it preserves trust with historically marginalized and exploited communities [35]. Paquette et al. argue that seeking consent lessens complicated grief and preserves trust [40].

6 Arguments Against Requiring Consent

Those arguing against requiring consent make a broader range of arguments than those arguing for consent [38]. First, they contend the prima facie duty to obtain consent is not even triggered. Second, they argue that even if it is, determination of death by neurologic criteria is so important and fundamental, it must be excepted from the general duty. Third, those arguing against consent contend that an exception is warranted because of the need to protect the integrity of the medical profession and steward scarce resources. Finally, those arguing against consent make arguments from symmetry and harmony [44].

6.1 Prima Facie Duty Not Even Triggered

Opponents to requiring consent contend that the prima facie duty to obtain consent is never triggered in the first place because the apnea test is not "treatment." For example, in court cases, hospitals have successfully argued that the apnea test is not "health care" requiring consent [25, 45]. Instead, the apnea test is just an assessment or evaluation. Therefore, neither the right to consent nor the right to refuse apply [46, 47]. Hester similarly argues that the apnea test is not itself healthcare, but rather a means to determine whether healthcare is appropriate [48]. Bertino and Potter compare the evaluation for death by neurologic criteria with assessing a patient's decision-making capacity [49].

6.2 Need to Answer Fundamental Questions

Even if there is a prima facie duty to get consent, opponents to requiring consent contend that presumption is outweighed by the unique and special importance of determination of death by neurologic criteria. Ascertaining whether a patient is alive, or dead, is the most fundamental aspect of providing medical care [50]. Indeed, some professional societies like the AAN say that clinicians not only "may" test without consent but also have a "responsibility" to do so [6, 11, 51].

Clinicians are obliged to provide appropriate care based on an accurate diagnosis [9]. Accordingly, Vercler and Laventhal argue that physicians may unilaterally perform the apnea test, because they have a fundamental responsibility to determine the suitability of technological interventions that they administer [52]. Hester similarly argues that physicians must determine where their professional obligations point [48]. Furthermore, clinicians must confirm that an individual is eligible for healthcare services. Otherwise, they commit fraud by billing for services that are not "medically necessary" [53].

6.3 Integrity of the Medical Profession

Once clinicians have determined that a patient is dead, they generally have no ongoing duty to "treat" that patient [16, 54]. Accordingly, clinicians will stop physiological support like mechanical ventilation either immediately or after a brief period of reasonable accommodation. But, some families seek to avoid this result by refusing to allow clinicians to perform a determination of death by neurologic criteria [55, 56]. Without a determination of death by neurologic criteria, there can be no declaration of death.

Courts have weighed the integrity of the medical profession against patient rights for decades [57]. Many scholars and professional societies have published accounts of the goods internal to the practice of medicine, including: (1) prevention and/or treatment of disease and illness, (2) restoration of health, (3) relief of pain and/or suffering, (4) healing through the therapeutic encounter, and (5) sustainment, preservation, protection, or prolongation of life. None of these is furthered by not determining death by neurologic criteria [58, 59].

Furthermore, forcing clinicians to treat contrary to professional standards causes moral distress [60–63]. This is especially difficult for nurses and ancillary staff who spend more time with these patients. Moral distress is linked to problems with retention, absenteeism, and even care quality [44].

6.4 Stewardship of Scarce Resources

The need to answer fundamental questions and the need to preserve the integrity of the medical profession are not the only reasons to determine death. Hospitals also need to know whether patients are alive or dead to facilitate appropriate triage of clinician time and material resources [64, 65]. Clinicians must be good stewards of scarce resources like ICU beds [11]. For example, the AMA Code of Ethics "requires physicians to be prudent stewards of the shared societal resources with which they are entrusted" [66].

A consent requirement would permit families to indefinitely prevent the determination, and therefore, the declaration of death. This would cause a misallocation of scarce resources. Hospitals are "not places to maintain the dead" [67]. Some commentators reference the COVID-19 pandemic to illustrate distributive justice arguments about wasting scarce ICU resources on the dead when others with significant prospects for benefit are denied those same resources [47]. Yet, distributive justice concerns weigh heavy even without a pandemic surge.

Admittedly, the inability to declare death is not always an obstacle to withdrawing ICU treatment like mechanical ventilation. For example, hospitals in California and Texas regularly withdraw life-sustaining treatment from living patients over the objections of the patient's legally authorized decision-maker [68]. This is not permitted in many other jurisdictions like Ontario [69].

6.5 Symmetry and Harmony with Death by Circulatory-Respiratory Criteria

In addition to the forgoing arguments, opponents to requiring consent argue that omitting consent fits with adjacent and analogous rules. First, since consent for determining death is not required on one prong of the UDDA, it should not be required for determining death on the other prong [67]. In other words, consent is not required for determining "irreversible cessation of circulatory and respiratory functions." Therefore, it should similarly not be required for determining "irreversible cessation of all functions of the entire brain."

Among others, the AAN argues that the rule for determining death by neurologic criteria should mirror that for determining death by circulatory-respiratory criteria. The AAN supports its position that there is "no obligation" to obtain consent for determination of death by neurologic criteria by demonstrating that position "is ***analogous*** to the authority and responsibility historically granted to the medical profession to determine circulatory death without the requirement for additional informed consent" [6].

The argument from symmetry can be pushed even further. When performing cardiopulmonary resuscitation, chest compressions are critical to maintaining blood flow and positive pressure ventilation until spontaneous circulation is restored. Compressions are regularly interrupted or paused to check for pulse and rhythms [70–72]. While guidelines recommend minimizing these interruptions, they are not only permitted, but recommended. Interrupting chest compressions to check pulse is associated with poorer outcomes. Yet, clinicians do not seek consent for that non-risk-free diagnostic test. Analogously, clinicians should not be required to obtain consent for tests determining death by neurologic criteria.

6.6 Symmetry and Harmony with Accommodations

A second argument from symmetry focuses on accommodations. Since we now disallow families to circumvent declaration of death in other ways, we should disallow it here too. With a consent requirement, families could prevent the declaration of death not because of any specific right to opt out (as in New Jersey), but simply because they can prevent the prerequisite. In other words, while they have no right to object to the "declaration" of death, they could achieve a practically identical result by objecting to the "determination" of death. In short, if we disallow objections to declaration/pronouncement of death, then we should similarly disallow objections to determination of death.

6.7 Rebuttal Points

In addition to their own arguments against consent, opponents to requiring consent also rebut the arguments of proponents. For example, several authors argue that the risks from apnea testing are not as common or serious as consent proponents suggest [11, 46, 47]. When properly conducted, the risks are minimal [12].

6.8 Weighing the Pro/Con Arguments

There is a strong tradition of deference to the status quo. While clinicians must normally obtain informed consent before administering tests and procedures, determination of death by neurologic criteria is a well-established exception to this requirement. This sizable and stable legal and ethical consensus weighs heavily toward not requiring consent. Opponents bear the burden of presenting compelling reasons why the status quo is inadequate and should be replaced. Such justifications do not clearly outweigh the compelling policy reasons not to require consent. Especially compelling are distributive justice concerns.

7 Notification and Reasonable Accommodation

Just because clinicians may legally proceed with determination of death by neurologic criteria without consent (or even over family objections) does not mean that they should immediately do so. Clinicians and hospitals should consider family wishes. This typically entails two duties: (1) notifying family of the intent to perform an evaluation for death by neurologic criteria and (2) making reasonable accommodations to delay testing.

7.1 Family Notification

Since the earliest days of determination of death by neurologic criteria, courts have held that clinicians should apprise the family [73, 74]. Even if there is no "decision" for families to make, clinicians should still consult with families [75]. This is required in law, in professional guidelines, and as a matter of custom and practice.

Laws in some states specifically require family notification of the intent to perform an evaluation for determination of death by neurologic criteria. New York law provides that the facility must "make diligent efforts to ***notify*** the patient's Surrogate Decision-maker that the process for determining brain death is underway" [21]. New Jersey guidelines similarly provide that "the exam should commence ***following notification*** of surrogate decision makers" [15]. Florida requites that the "next of kin of the patient shall be ***notified*** as soon as practicable of the procedures to determine death" [76].

Beyond laws, broadly endorsed standards make this recommendation [77]. The World Brain Death Project provides: "It is recommended that reasonable efforts should be made to ***notify*** a person's next-of-kin before performing a brain death/death by neurologic criteria determination" [5]. Similarly, the AAN advises that clinicians "perform a brain death evaluation including apnea testing only "***after informing*** a patient's loved ones or lawful surrogates of that intention" [6].

The authors of the 2011 Society of Critical Care Medicine/American Academy of Pediatrics/Child Neurology Society pediatric standards make similar recommendations: "Physicians are obligated to provide support and guidance for families ... permitting families to ***be present*** during the evaluation can help them understand that their child has died" [78]. Obviously, permitting the family to observe determination requires informing them [79–81]. Moreover, even when not specifically required, prevailing custom and practice involve notifying the family that the evaluation will be conducted. Even if their consent is not required, respect and compassion dictate that the family be informed.

7.2 Reasonable Accommodation in Delaying Testing

To some degree, notifying families of the intent to perform an evaluation for determination of death by neurologic criteria entails some reasonable accommodation and delay. After all, a key point of notification is to permit the family to gather and say goodbye. For example, hospitals may also delay testing to permit the family time to process the situation, to explore transfer options, or even to observe the evaluation [9]. But reasonable accommodation often goes beyond this. Clinicians regularly ***delay*** testing to be compassionate and respectful. This is evidenced both by professional society standards and by custom and practice.

While the World Brain Death Project clearly recommends against requiring consent, it recommends reasonable accommodation to objections. The World Brain Death Project makes two separate recommendations on this point.

> It is suggested that, in the setting of a request to either forgo a brain death/death by neurologic criteria examination ... a family should be provided with a ***finite period*** of time to seek to arrange transfer to another facility (should they wish to do so) and the health care team should speak to a potential accepting institution if requested to do so [5].

> It is recommended that attempts should be made to ***handle requests*** to either forgo a brain death/death by neurologic criteria examination ... within a given hospital system before turning to the legal system [5].

Other professional societies also recommend reasonable accommodation. A widely respected set of guidelines from the Hastings Center similarly supports accommodating requests to delay determination [82]. Notably, these delays are finite and short-term, typically lasting just hours or days.

Few laws elevate these recommendations into legal duties. Indeed, New York law is unique in requiring that hospitals have a "procedure for reasonable accommodation of the individual's religious or moral objections to the ***determination***" [20]. A court recently interpreted this law in the case of Yechezkel Nakar [83].

New York-Presbyterian Hospital/Columbia University Irving Medical Center conducted the clinician evaluation for determination of death by neurologic criteria of Mr. Nakar and wanted to perform an apnea test. Mr. Nakar's family objected on religious grounds and wanted to consult with their rabbi. Despite these objections, the hospital proceeded to complete the determination. The court held that since there "was no immediate need to declare Mr. Nakar brain dead, it was not reasonable for respondents to take such action." Instead, "it would have been reasonable for [the hospital] to accommodate [the family's] objections by delaying." The court ordered the hospital to vacate the death certificate [83].

Beyond professional society guidelines and legal requirements, most hospitals already regularly afford accommodations to families. While most discussions of accommodation focus on time ***after*** determination and declaration (so families have an opportunity to gather and say goodbye), hospitals also offer accommodations ***before*** determination [84]. Among other evidence, court cases in the United States indicate that clinicians regularly offer brief accommodations of 24–72 h [44, 85].

7.3 Symmetry with Other Accommodations

While only four jurisdictions in the United States legally mandate reasonable accommodation ***after*** determination of death, this is commonly afforded in other jurisdictions. It is the standard of care. If hospitals accommodate families after death, then they should also accommodate families ***before*** death. Still, these accommodations are almost always definite, not indefinite. Accordingly, hospitals should clarify both the reasons for and duration of accommodation.

7.4 Other Reasonable Accommodations

As discussed above, the typical form of accommodation is a short-term delay before performing the evaluation for determination of death by neurologic criteria. Reasonable accommodation can also take other forms. Another accommodation might be changing the components of the evaluation. Lazaridis argues that while clinicians need not obtain consent, they must respect "legally protected conscientious objections," by doing ancillary testing instead of apnea testing [86].

8 Conclusion

The overwhelming weight of legal authority and ethical argument supports not requiring consent for determination of death by neurologic criteria. Therefore, clinicians do not and should not have a legal duty to obtain informed consent or any other consent before making a determination of death by neurologic criteria. There is no "decision" or "choice" for families to make. However, hospitals should notify families that they will perform the determination. Family acquiescence or silence to this announcement constitutes assent. When families do not assent, hospitals should offer short-term accommodation when appropriate. They should also have policies for managing and accommodating objections.

References

1. Pope TM. Certified patient decision aids: solving persistent problems with informed consent law. J Law Med Ethics. 2017;45(1):12–40.
2. Pope TM. Clinicians may not administer life-sustaining treatment without consent: civil, criminal, and disciplinary sanctions. J Health Biomed Law. 2013;9:213–96.
3. Uniform Determination of Death Act, Uniform Law Acts 1980;12:589.
4. President's Commission for The Study of Ethical Problems in Medicine and Biomedical and Behavioral Research, Defining Death: Medical, Legal, and Ethical Issues in the Determination of Death. 1981.
5. Greer DM, Shemie SD, Lewis A, et al. Determination of brain death/death by neurologic criteria: the world brain death project. JAMA. 2020;324(11):1078–97.
6. Russell JA, Epstein LG, Greer DM, et al. Brain death, the determination of brain death, and member guidance for brain death accommodation requests. Neurology. 2019;92(5):228–32.
7. Joint Committee of Bioethics of the LACBA and LACMA, Guidelines for Physicians Forgoing Life-Sustaining Treatment for Adult Patients. 2006.
8. Lewis A, Adams N, Varelas P, et al. Organ support after death by neurologic criteria: results of a survey of US neurologists. Neurology. 2016;87:827–34.
9. Lewis A, Adams N, Chopra A, et al. Organ support after death by neurologic criteria in pediatric patients. Crit Care Med. 2017;45(9):e916–24.
10. Francoeur C, Weiss MJ, Macdonald JM, et al. Variability in pediatric brain death determination protocols in the United States. Neurology. 2021;97(3):e310–9.
11. Muramoto O. Is informed consent required for the diagnosis of brain death regardless of consent for organ donation? J Med Ethics. 2021;47:e5.
12. Leemputte M, Paquette E. Consent for conducting evaluations to determine death by neurologic criteria: a legally permissible and ethically required approach to addressing current controversies. Curr Pediatr Rep. 2019;7(4):152–62.
13. Nev. Rev. Stat. § 451.008.
14. New York State Department of Health, Guidelines for Determining Brain Death. 2011.
15. New Jersey Ad Hoc Committee on Declaration of Death by Neurological Criteria, Guidelines for Determining Death Based on Neurological Criteria. 2014.
16. Pope TM. Brain death forsaken: growing conflict and new legal challenges. J Leg Med. 2017;37(3–4):265–324.
17. N.J. Reg. § 13:35-6A.6.
18. N.J. Stat. Ann. § 26:6A-5.
19. New Jersey Commission on Legal and Ethical Problems in the Delivery of Health Care, New Jersey Advance Directives for Health Care and Declaration of Death Acts: Status, Commentary, and Analysis. 1991.
20. N.Y.C.R.R. § 400.16(c)(2).
21. New York State Department of Health, Guidelines for Determining Brain Death; 2005.
22. Manchester University NHS Trust v. Midrar Namiq (2020) EWHC 180 (Fam).
23. Manchester University NHS Trust v. Midrar Namiq (2020) EWCA Civ 164.
24. Barts Health NHS Trust v. Dance, [2022] EWHC 1165 (Fam) & (2022) EWHC 1435 (Fam).
25. McKitty v. Hayani, No. CV-17-4125 (Ontario Superior Court, Sept. 28, 2017) (Endorsement 2).
26. A.D. v M.O.M., 2018 CanLII 86640 (Ontario HPARB).
27. In re A.C., Order No. DG-16–08, at 8 (9th Dist. Pondera County, Mont., Sept. 26, 2016).
28. Mont. Code Ann. § 50–22-101.

29. Shewmon DA. Point: Informed consent should be obtained for apnea testing in the determination of death by neurological criteria. Yes Chest. 2022;161(5):1143–5.
30. Tibballs J. It is time to abandon apneic-oxygenation testing for brain death. Arch Organ Transplant. 2020;5(1):006–0010.
31. Rodríguez-Arias D, Molina-Pérez A, Díaz-Cobacho G. Death determination and clinicians' epistemic authority. Am J Bioeth. 2020;20(6):44–7.
32. Tibballs J, Bhati N. New challenges to the legal definition and medical determination of brain death: a multi-jurisdictional approach – cases from the United States, the United Kingdom, Canada and Australia. J Law Med. 2022;28(3):831–54.
33. Joffe AR. The apnea test: requiring consent for a test that is a self-fulfilling prophecy, not fit for purpose, and always confounded? Am J Bioeth. 2020;20(6):42–4.
34. Jackson BM. Informed consent for apnea testing: meeting the standard of care. Am J Bioeth. 2020;20(6):49–51.
35. Johnson LSM. Restoring trust and requiring consent in death by neurological criteria. Am J Bioeth. 2020;20(6):33–5.
36. Yanke G, Rady MY, Verheijde J, McGregor J. Apnea testing is medical treatment requiring informed consent. Am J Bioeth. 2020;20(6):22–4.
37. Joffe AR, Anton NR, Duff JP. The apnea test: rationale, confounders, and criticism. J Child Neurol. 2010;25(11):1435–143.
38. Busl KM, Lewis A, Varelas PN. Apnea testing for the determination of brain death: a systematic scoping review. Neurocrit Care. 2021;34(2):608–20.
39. In re Lawson, No. CL16–2358 (City of Richmond Cir. Ct., Va., June 10, 2016) (Order).
40. Paquette E, Frader J, Shah S, Tasker RC, Truog R. Beyond the apnea test: an argument to broaden the requirement for consent to the entire brain death evaluation. Am J Bioeth. 2020;20(6):17–9.
41. Kananeha MF, et al. Factors that affect consent rate for organ donation after brain death: a 12-year registry. J Neurol Sci. 2020;416:117036.
42. Goodwin M. Revisiting death: implicit bias and the case of Jahi McMath. Hastings Center Rep. 2018;48(6):S77–80.
43. Wijdicks EFM. Clinical problems in brain death determination and organ donation, in brain death. 3rd ed. Oxford: Oxford University Press; 2017.
44. Lee BM, Trowbridge A, McEvoy M, et al. Can a parent refuse the brain death examination? Pediatrics. 2020;145(4):e20192340.
45. Morlani et al. v. Haddara, 2021 ONSC 7288.
46. Antommaria AHM, Sveen W, Stalets EL. Informed consent should not be required for apnea testing and arguing it should misses the point. Am J Bioethics. 2020;20(6):25–7.
47. Bhagat D, Lewis A. The case against solicitation of consent for apnea testing. Am J Bioeth. 2020;20(6):20–2.
48. Hester DM. Determining death and the scope of medical obligations. Am J Bioeth. 2020;20(6):37–9.
49. Bertino J, Potter J. Requiring consent for brain-death testing: a perilous proposal. Am J Bioeth. 2020;20(6):28–30.
50. Pope TM. Brain death testing: time for national uniformity. Am J Bioeth. 2020;20(6):1–3.
51. Centers for Medicare and Medicaid Services. State operations manual appendix A - survey protocol, regulations and interpretive guidelines for Hospitals. 2020;482.45(a)(5). https://www.cms.gov/regulations-and-guidance/guidance/manuals.
52. Vercler CJ, Laventhal NT. Schrödinger's cat and the ethically untenable act of not looking. Am J Bioeth. 2020;20(6):40–2.
53. Bosek MS, Anderson JA, Vernaglia LW, Morrigan SP, Bard TR. Refusal of brain death diagnosis. JONAS Healthc Law Ethics Regul. 2007;9(3):87–94.
54. Pope TM. Legal briefing: brain death and total brain failure. J Clin Ethics. 2014;25(3):245–57.
55. Fonseca v. Kaiser Permanente Medical Center Roseville, 222 F. Supp. 3d 850 (E.D. Cal. 2017).
56. Life Guardian Foundation. Wallet Card to Protect and Preserve Life. 2022. https://lifeguardianfoundation.org.

57. Cruzan v. Director, Missouri Department of Health, 497 U.S. 261; 1990.
58. Cassell EJ. The nature of suffering and the goals of medicine. Oxford: Oxford University Press; 2004.
59. Pellegrino ED. What the philosophy of medicine is. Theor Med. 1998;19(4):315–36.
60. Macauley RC. Ethics in palliative care: a complete guide. Oxford: Oxford University Press; 2014.
61. Mobley MJ, et al. The relationship between moral distress and perception of futile care in the critical care unit. Intensive Crit Care Nurs. 2007;23(5):256–63.
62. Bell J, Breslin JM. Healthcare provider moral distress as a leadership challenge. JONAS Healthc Law Ethics Regul. 2008;10(4):94–7.
63. Ulrich CM, Hamric AB, Grady C. Moral distress: a growing problem in the health professions? Hastings Center Rep. 2010;40(1):20–2.
64. Lewis A, Pope TM. Physician power to declare death by neurologic criteria threatened. Neurocrit Care. 2017;26(3):446–9.
65. Wilkinson D, Savulescu J. Ethics, conflict and medical treatment for children: from disagreement to dissensus. Edinburgh: Elsevier; 2019.
66. American Medical Association, Code of Medical Ethics 11.1.1. https://www.ama-assn.org/delivering-care/ethics/code-medical-ethics-financing-and-delivery-health-care.
67. Lewis A, Greer D. Point: should informed consent be required for apnea testing in patients with suspected brain death? No Chest. 2017;152(4):700–2.
68. Pope TM. Dispute resolution mechanisms for intractable medical futility disputes. NYL Sch L Rev. 2014;58:347–68.
69. Cuthbertson v. Rasouli, 2013 SCC 53.
70. Nichol G, et al. Trial of continuous or interrupted chest compressions during CPR. New Eng J Med. 2015;373(23):2204–14.
71. Hanisch JR, Counts CR, Latimer AJ, Rea TD, Yin L, Sayre MR. Causes of chest compression interruptions during out-of- hospital cardiac arrest resuscitation. J Am Heart Assoc. 2020;9:e015599.
72. Panchal AR. Part 3: adult basic and advanced life support: 2020 American Heart Association guidelines for cardiopulmonary resuscitation and emergency cardiovascular care. Circulation. 2020;142(16-suppl-2):S366–468.
73. Dority v. Superior Court, 145 Cal. App. 3d 273 (1983).
74. Alvarado v. New York City Health & Hospital Corporation, 547 N.Y.S.2d 190 (N.Y. Sup. Ct. 1989).
75. R (Tracey) v Cambridge University Hospitals NHS Foundation Trust & Ors (2014) EWCA Civ 822.
76. Fla. Stat. §382.009(3).
77. Lewis A, Varelas P, Bernat JL, et al. Religion and brain death: managing requests to forego a brain death evaluation or continue somatic support after brain death. Supplement 12 to Greer DM et al. Determination of brain death/death by neurologic criteria: the world brain death project. JAMA. 2020;324(11):1078.
78. Nakagawa TA, Ashwal S, Mathur M, Mysore M. Clinical report: guidelines for the determination of brain death in infants and children: an update of the 1987 task force recommendations. Pediatrics. 2011;128(3):e720–40.
79. Carroll E, Lewis A. Brain death discussions, in shared decision making in adult critical care. Cambridge: Cambridge University Press; 2021.
80. Australian and New Zealand Intensive Care Society. The Statement on Death and Organ Donation; 2021.
81. Ettinger NA, Coleman RD, Loftis LL. Pediatric intensivists reply to "can a parent refuse the brain death examination?". Pediatrics. 2020;146(3):e2020008144.
82. Berlinger N, Jennings B, Wolf SM. The Hastings center guidelines for decisions on life-sustaining treatment and care near the end of life. 2nd ed. Oxford: Oxford University Press; 2013.

83. Nakar v. New York Presbyterian Hospital, No. 1780/2017 (Kings County Supreme Court, NY 2019).
84. Flamm AL, Smith ML, Mayer PA. Family members' requests to extend physiologic support after declaration of brain death: a case series analysis and proposed guidelines for clinical management. J Clin Ethics. 2014;25(3):222–37.
85. Summons v. Cook Children's Medical Center, No. 017–303367-1 (17th Judicial District Court of Tarrant County, Texas 2018).
86. Lazaridis C. Accommodating apnea testing not death determination refusal. Am J Bioeth. 2020;20(6):47–9.

Legal Responses to Religious and Other Objections to Declaration of Death by Neurologic Criteria

Jennifer A. Chandler

Twentieth-century advances in resuscitation and mechanical ventilation gave us a new human state—"a 'living' body with a 'dead' brain" [1]. These technological changes gave us a metaphysical question with important ethical and legal implications: is a human in this state alive or dead? The concept of death by neurologic criteria has achieved widespread, but not universal, acceptance around the world since the first published clinical standard [2]. It has been adopted legally in many jurisdictions around the world through legislation or other legal instruments like regulations, decrees, or executive orders, as well as through judge-made law. A survey of the various legal definitions of death around the world was conducted for the World Brain Death Project [3] and is available in a supplement to the main project publication [4]. However, the idea that the irreversible loss of all functions of the brain constitutes death continues to be debated, and this debate periodically surfaces in the legal setting.

In this chapter, I describe briefly how legal challenges have been framed and received by courts, with a focus on religious objection. Other legal challenges, such as the claim that consent is required prior to testing to determine death by neurologic criteria are covered in detail elsewhere in this book. The cases addressed in this chapter come from Canada and the United States; both are jurisdictions in which there have been recent, highly publicized court cases. This chapter also considers how legislators have responded to religious or cultural objections to death by neurologic criteria in various ways around the world followed by offering as a case study a description of *McKitty v Hayani*—a recent Canadian case raising a constitutional right to the accommodation of religious objections to death by neurologic criteria [5, 6]. Although the question remains unsettled in Canada—as it is in the United States—one thing can be confidently asserted: there will be other legal cases challenging the determination of death by neurologic criteria on religious grounds.

J. A. Chandler (✉)
Faculty of Law, University of Ottawa, Ottawa, ON, Canada
e-mail: chandler@uottawa.ca

A. Lewis, J. L. Bernat (eds.), *Death Determination by Neurologic Criteria*, Advances in Neuroethics, https://doi.org/10.1007/978-3-031-15947-3_22

1 The Underlying Motivation for Legal Challenges to the Determination of Death by Neurologic Criteria

It is useful to keep in mind that the underlying motivation for an objection to death by neurologic criteria may differ from the legal framing of that objection. A family that feels that their relative has received poor care, feels rushed, or distrusts the determination may emphasize a religious objection in part because a society has made this kind of claim more easily available. Indeed, many societies put a high value on respect for religious freedoms. Some racial minorities may be less likely to trust health care providers because of historical medical abuses, current systemic disparities in outcomes, and personal experience of social discrimination [7, 8]. Some Eastern philosophical and cultural traditions such as Shintoism, Taoism, and Buddhism differ from Western cultures in their thinking about the relationship between the soul, mind, brain, and body, and their relative importance to personhood and metaphysical status, as discussed elsewhere in this book [9, 10]. Thus, objections may reflect diverse cultural and philosophical positions as well.

Often the flashpoint for legal disputes is the implication of the determination of death for the discontinuation of ventilation. If a person is dead, surrogate consent is not usually needed to withdraw support. But if a person is alive, the rules on the termination of ventilation and other supports vary, with many either requiring surrogate consent, or at least leaving it unclear, in which case risk-averse clinicians often seek consent [11]. Thus, the approach to determining death (i.e. whether neurological or cardiac criteria are applied) affects which end of life legal decision-making regime is applicable and so has important consequences for the patient and family.

2 The Types of Legal Objections to Death by Neurologic Criteria

Four main categories of legal objections to death by neurologic criteria have been raised: (1) objections to evaluation for the determination of death by neurologic criteria without explicit consent, (2) objections to the accuracy of the determination, (3) objections to the legal adequacy of the testing procedures, and (4) objections to the concept of death by neurologic criteria itself.

Opposition to the evaluation for determination of death by neurologic criteria is a legal strategy that can be used by those who doubt the accuracy of the determination or dispute the concept of death by neurologic criteria, even if they are unlikely to win legal arguments on other grounds [11]. An example of this approach is the argument that consent is required to perform the apnea test. The case law on whether consent is required is inconsistent in the United States, as discussed in more detail elsewhere in this book. Some statutes address the point directly. Nevada amended its law in 2017 to indicate that the determination of death by neurologic criteria is a clinical decision that does not require surrogate consent [11]. Litigation is ongoing in Ontario to determine whether the applicable statute does or does not require consent for performance of the apnea test [12].

Sometimes, litigants argue that the determination of death by neurologic criteria is incorrect. For example, in the recent Ontario case of Taquisha McKitty, her family argued both that she should be exempt from the application of the neurologic criteria of death due to her religious beliefs and that the determination of death by neurologic criteria was factually incorrect [5]. She had been determined to be dead by neurologic criteria based on multiple independent evaluations using the accepted medical standards. When her family questioned her movements, further ancillary tests (i.e. a nuclear brain blood flow study and somatosensory evoked potentials) confirmed the determination. McKitty's family argued that the nature and duration of her movements were not consistent with death by neurologic criteria and that this raised a doubt about the accuracy of the determination, although this argument was ultimately rejected by the court [5, 6].

A third form of challenge relates to whether testing procedures are able to meet the legal requirements for determination of death by neurologic criteria. Pope notes several variants of this type of challenge [11]. These are whether the testing procedures applied reflect "medically accepted standards," and, further, whether the most commonly used tests actually measure what the law requires them to measure. In particular, the Uniform Determination of Death Act (UDDA)—a model state law created in 1980 and subsequently adopted by most American states—articulates standards for the determination of death by circulatory/respiratory criteria and by neurologic criteria [13]. The neurologic standard indicates that a person is dead on neurologic criteria when he or she has sustained irreversible cessation of all functions of the entire brain, including the brainstem. The UDDA also states that whether this has happened is to be determined according to "accepted medical standards." The question then is whether the medically accepted tests actually show that *all* functions of the *entire* brain have ceased *irreversibly*.

In the 2015 case of Aden Hailu, the Supreme Court of Nevada questioned whether the American Academy of Neurology's (AAN) standards, which were used for the determination of death, were the "accepted medical standards," and whether they measured what the law required [14]. This was not ultimately settled because Hailu's heart stopped and the issue of death by neurologic criteria became moot during subsequent stages of the litigation [11]. The Nevada legislature amended its state law in response to this case to specify authoritative medical standards by name (the AAN's standards for determination of death by neurologic criteria in adults and the Society of Critical Care Medicine, American Academy of Pediatrics, and Child Neurology Society's standards for determination of death by neurologic criteria in pediatric patients). However, as Pope notes, the new statute does not necessarily answer the question of whether the testing procedures set out in these standards actually measure the irreversible cessation of all functions of the entire brain [11]. In fact, there is an active question as to whether the AAN standards, which acknowledge and allow the persistence of neuroendocrine function after determination of death by neurologic criteria, match the statutory UDDA requirement that refers to loss of all functions of the *entire* brain [15, 16]. This issue may be rectified as the

UDDA is being revised, but it is unknown at present what the revisions will be and how they will be incorporated into state laws on determination of death [17].

Finally, litigants also challenge the concept of death by neurologic criteria. These challenges are based on the position that death by neurologic criteria (or the irreversible loss of all brain functions) is not obviously equivalent to death. The exact mixture of physiological functions necessary to constitute life is a metaphysical rather than a biological question in an age when technology has decoupled vital functions from one another and introduced new states of partial function that may be artificially sustained. Indeed, the slow or limited legal uptake of death by neurologic criteria in some parts of the world reflects the presence of different cultural, philosophical, and religious traditions, as discussed elsewhere in this book. Some Orthodox Jews, Indigenous Americans, Buddhists and Muslims reject death by neurologic criteria and instead maintain that death occurs only when circulatory and respiratory functions cease [11]. On this view, it is immaterial if circulatory and respiratory functions are artificially sustained, and a person is alive while these functions continue.

It is important to note that the legal definition of death is a legal question, even if it is clearly informed and influenced by biomedical knowledge and concepts. The Ontario Court of Appeal recently stated [6]:

> The determination of legal death is not simply, or even primarily, a medical or biological question. The question of who the law recognizes as a human being—entitled to all of the benefits and protections of the law—cannot be answered by medical knowledge alone. Facts about the physiology of the brain-dead patient are needed to determine what obligations are owed to the brain-dead patient, but the enquiry is not ultimately technical or scientific: it is evaluative. Who the common law ought to regard as a human being—a bearer of legal rights—is inescapably a question of justice, informed but not ultimately determined by current medical practice, bioethics, moral philosophy, and other disciplines.

Even if the law adopts a brain-based definition of death, this is not necessarily the end of the inquiry. In many jurisdictions, a constitutional right to freedom of religion and conscience is enshrined in the law, and this constrains the ability of governments to legislate or act in ways that infringe upon that constitutional right. Families disputing a determination of death by neurologic criteria have occasionally framed the dispute as a violation of the constitutional right to freedom of religion. The families of Taquisha McKitty, Jahi McMath, and Israel Stinson all raised religious freedom in their challenges to determination of death by neurologic criteria, although the issues related to religious freedom and death by neurologic criteria remain unsettled, as discussed below.

Another potential challenge to the concept of death by neurologic criteria comes from a disability discrimination perspective. Protections against discrimination on the basis of disability exist in the constitutions and domestic laws of many jurisdictions. A range of views exist on which mixture of functions is necessary for the status of being alive. Although not consistent with the current medical and legal approaches, some suggest that permanent loss of consciousness is enough to constitute death. Still others argue that continued bodily functions like absorbing nutrients, excreting waste, healing wounds, gestating a fetus, and proceeding through puberty are sufficient criteria for life even if all brain functions are lost and other functions persist only with medical support [18]. Essentially the disability discrimination argument is that the law has arbitrarily chosen to categorize a certain class of people as dead on the basis of their particular functional disabilities and to deprive them of legal personhood. The Canadian *Charter of Rights and Freedoms* offers constitutional protection against discrimination on the basis of disability (s.15 of [19]). An argument of this type could satisfy the first step of the analysis, which asks whether the law makes a distinction on the basis of disability that "reinforces, perpetuates or exacerbates disadvantage." This would not be the end of the inquiry as the Canadian *Charter* allows for governments to argue that discrimination is justified in the circumstances. The kinds of justifications proposed for the concept of death by neurologic criteria articulated in the original Harvard Ad Hoc Committee report of 1968 remain present today [2]. The burden on patients, families, hospitals, and society of people who are irreversibly comatose is an important consideration, as are the benefits of the organ donation and transplantation system for donors and recipients.

Disability rights-based arguments have been attempted in several cases in the United States where families have argued that their relative is not dead (i.e. the determination of death by neurologic criteria was in error), or that the use of neurologic criteria to declare death violated their constitutional rights. In those cases, families have also advanced arguments that because their loved ones are not dead, it would be legally impermissible discrimination under legislation like the *Americans with Disabilities Act* to withhold or remove treatment (see e.g. [20, 21]).

It is possible that further disability discrimination-based arguments will arise in the future. Recently there have been calls for greater attention to the human rights of persons with severe brain injury, to ensure better quality of life and access to rehabilitation and assistive technologies [22, 23]. Evidence of consciousness in people who were thought to be in a vegetative/unaware-wakeful state has led to calls for improved education, more rigorous neurological examinations, and multimodal evaluations including functional imaging or electrophysiological techniques [23]. Of course, this pertains to people who are brain-injured, not those who are dead by neurologic criteria. However, concerns about misdiagnosis and treatment pessimism in severe brain injury may spill over into perceptions about misdiagnosis of death by neurologic criteria as well as questions about the metaphysical correctness of the definition of death by neurologic criteria and its moral implications for personhood.

3 Legislative Responses to Religious and Cultural Objections to Death by Neurologic Criteria

While acceptance of death by neurologic criteria was rapid in some places and fairly smooth in many, cultural and religious factors led to a slower and more conservative approach in others [10]. In some places, the law has accepted death by neurologic

Table 1 Legal exemptions and accommodations for cultural or religious objection to death by neurologic criteria

Legal approach[a]	Jurisdiction	Details
Exemption		
Only applied with prior consent	Japan	Death by neurologic criteria is determined only where patient or family consent to organ donation
Not applied if refused	New Jersey	Death by neurologic criteria may not be declared if there are reasons to believe patient would have objected
In practice—Not applied if refused	Israel	Must take views of patient "into consideration" before evaluating the patient to determine death by neurologic criteria. In practice, clinicians do not usually proceed in the face of opposition
At discretion of health care personnel	Illinois	Hospitals must adopt policies that allow health care personnel to take patient's religious views into account concerning time of death
Accommodation		
Ventilation may not be removed	Israel	Ventilation and associated treatment cannot be discontinued until the heart stops if the patient objects to death by neurologic criteria
Delay discontinuation of ventilation	California	Hospitals must adopt policies allowing a reasonably brief period of delay before discontinuing cardiopulmonary support, and must make reasonable efforts to accommodate any special religious or cultural practices
May include delay in discontinuation of ventilation	New York	Hospitals must adopt policies that may include continuation of ventilation for a limited period in the context of moral or religious objection to death by neurologic criteria
Unspecified	Trinidad and Tobago	Religious and cultural requests of family must be met as far as possible before and after removal of ventilation. This requirement is contained in an organ-donation-specific regulation

[a] Here, the use of the term "exemption" refers to a law that prevents the legal declaration of death by neurologic criteria. The use of the term "accommodation" refers to a law that does not prevent the legal determination of death by neurologic criteria, but addresses whether and when ventilation and associated interventions may be discontinued after that determination

criteria only for the purpose of organ donation, or has accepted it while providing exemptions or accommodations for religious dissent. Table 1 briefly describes the types of legal exemptions and accommodations around the world, with more detailed discussion following.

3.1 Japan

Japan adopted its *Organ Transplant Law* in 1997 as a compromise between those who approved of organ donation from people determined to be dead using neurologic criteria and those opposed to the idea that a person whose circulation and respiration persist may be determined dead by neurologic criteria [24, 25]. The 1997 law's acceptance of death determination on neurologic criteria was an exception to the general rule that legal death was defined in cardiopulmonary terms, and this exception was made only for the purposes of organ donation [26]. Akabayashi suggests that this reflects the cultural and religious syncretism of Japan, which can accommodate different philosophies in one legal policy [24].

The 1997 law stated that donation was only possible under the law if the donor had given prior written consent to organ donation and to a declaration of death by neurologic criteria, and the family members did not object [25]. The law was revised in 2010 to remove the requirement for the patient's prior written consent, but family consent is still required and may be given unless the patient had indicated opposition to donation [27]. Some have suggested that the 2010 law recognized that death determined neurologically is legal death generally (i.e. not just in the context of organ donation) [26]. However, the interpretation articulated by the Japan Organ Transplant Network is that the determination of death on neurologic criteria is accepted only in the context of organ donation [28]. The Japan Organ Transplant Network states [28]:

> In Japan, while the Organ Transplant Law is enacted, brain death is acknowledged as human death only when a transplant is to be performed. [...] In Japan, brain death is still not widely accepted as human death. Whether one considers heart arrest or brain death to be the benchmark for human death is a matter of personal judgement-- a decision based on personal views regarding life and death. Therefore, it is important for families to have serious discussions and express their intentions regarding this issue.

3.2 New Jersey, United States

In 1991, New Jersey enacted the *New Jersey Declaration of Death Act*, which legally recognized death determined by both circulatory-respiratory and neurologic criteria [29]. In part due to the large Jewish population in the state, it included an exemption for religious objection [11]. The Act states that a person may not be declared dead by neurologic criteria if a physician has reason to believe that this would violate the personal religious beliefs of the individual [11, 30]. Evidence of this belief may come from information in the medical records, information provided by a family member or from any other person who knows the individual's religious beliefs. The Act also provides for continued coverage under insurance and benefits programs for those with religious objections to death by neurologic criteria [11, 31].

The law is not demanding in terms of what qualifies as a legitimate religious belief, and Pope suggests that moral objections and nontheistic beliefs that are sincerely held would probably qualify [11].

As a result, in New Jersey, death cannot be declared until circulatory-respiratory function ceases if there are reasons to believe this would violate the patient's personal religious beliefs. As a result, legal status is determined according to the values of each individual and that where there is a religious objection to death by neurologic criteria, medical support may continue for a protracted period [11].

3.3 Israel

Diverse views are held among the Jewish Israeli population on the concept of death by neurologic criteria (as discussed elsewhere in this book); part of the ultra-Orthodox community rejects this means to declare death [32]. Although the concept of death by neurologic criteria was accepted by the Chief Rabbinate in 1986, and regulated under a clinical practice guideline since 1996, religious objections continue [33]. Following extensive discussion between medical and religious authorities, Israel passed the *Brain-Respiratory Death Act* in 2008 [34]. This Act validates the concept of death by neurologic criteria, and incorporated practice requirements intended to respond to the views of the ultra-Orthodox communities [32, 33]. In response to the religious conviction that death may only be determined when spontaneous breathing has irreversibly ceased, the Act indicates that the apnea test is mandatory, and notes that when it cannot be performed, death by neurologic criteria cannot be declared [32, 33].

The Act requires clinicians to inquire with the next-of-kin about the patient's views on the determination of death by neurologic criteria, and to take those views "into consideration" before performing an evaluation to determine death by neurologic criteria. Cohen et al. note that this was introduced in response to the strong opposition of the ultra-Orthodox community to death by neurologic criteria and it was understood to offer an exemption from determination of death by these criteria even though the legislation only requires that objection be taken into consideration [33]. Indeed, Cohen et al. note that at least as of the time of writing (2012), most clinicians did not proceed with evaluations for death by neurologic criteria when any opposition—religious or otherwise—was expressed by the next-of-kin [33].

Another key accommodation under the Act is that even if death by neurologic criteria has been determined, a patient may not be disconnected from a ventilator and associated support cannot be discontinued until circulatory-respiratory function ceases if the declaration of death by neurologic criteria is incompatible with the patient's views [34].

3.4 Illinois, United States

Illinois accepted death by neurologic criteria for the purposes of organ donation in its *Uniform Anatomical Gift Act* of 1981 [35]. In the subsequent 1983 case of *Re Haymer,* the Appeals Court broadened the common law acceptance of death by neurologic criteria for general application—i.e., beyond the context of organ donation [36, 37]. In 2007, it amended its law on hospital licensing to include an accommodation requirement [11, 38]:

> Every hospital must adopt policies and procedures to allow health care professionals, in documenting a patient's time of death at the hospital, to take into account the patient's religious beliefs concerning the patient's time of death.

As Pope points out, this is a relatively weak exemption in that the hospital need only allow a health care professional to take religious beliefs into account; it does not require that they be taken into account [11]. On the other hand, in allowing physicians to deviate from "documenting a patient's time of death," it goes further than states like New York or California, which are discussed below. In those two states, time of death is declared on the basis of the determination of death by neurologic criteria, although accommodation in the form of a delay in the removal of ventilation is permitted [11]. The Illinois law appears to leave considerable discretion in the hands of health care professionals on whether and how to take religious beliefs into consideration.

3.5 California, United States

California recognized death by neurologic criteria when it adopted the *Uniform Determination of Death Act* in 1982 [11]. In 2009, it passed a statute requiring hospitals to adopt a policy for providing family or next-of-kin with "a reasonably brief period of accommodation" before discontinuation of cardiopulmonary support [11, 39]. During this period, previously ordered cardiopulmonary support must be continued, but no other medical intervention is required. This delay is not limited to those who have religious or cultural objections to death by neurologic criteria. The statute does not limit the reasons why a family may seek a delay, although the law appears to contemplate the desire to gather family and next-of-kin at the bedside because the statute defines a "reasonably brief period" as the time afforded for that purpose. Pope notes that most hospitals provide 24 h of delay although some permit up to 36 h [11].

As for religious accommodation, the 2009 law requires that if any special religious or cultural practices and concerns of the patient or family are raised, the hospital must also make reasonable efforts to accommodate them.

In addition, the law requires that in determining what is reasonable for delays in discontinuing ventilation or accommodating religious or cultural matters, hospitals must consider "the needs of other patients and prospective patients in urgent need of care." This direction suggests that what is reasonable will vary according to the degree of demand for critical care resources from time to time.

3.6 New York, United States

New York judicially accepted death by neurologic criteria in 1984 [11]. Subsequently, the New York Department of Health (NYDOH) recognized death by neurologic criteria in administrative regulations adopted in 1987. The regulations required hospitals to establish and implement a written policy that sets out "a procedure for the reasonable accommodation of the individual's religious or moral objection to the determination as expressed by the individual, or by the next of kin or other person closest to the individual" [11, 40].

In 2011, NYDOH and the New York State Task Force on Life & the Law published Guidelines for Determining Brain Death, which provided further clarification on the reasonable accommodation requirement [11, 41]. The Guidelines indicate that reasonable accommodations policies could include [41]:

> specific accommodations, such as the continuation of artificial respiration under certain circumstances, as well as guidance on limits to the duration of the accommodation. Policies may also provide guidance on the use of other resources, such as clergy members, ethics committees, palliative care clinicians, bereavement counselors, and conflict mediators to address objections or concerns.

The 2011 Guidelines go on to say that accommodation is not required where objections are not moral or religious in nature, such as objections based solely on psychological denial or the alleged inadequacy of the determination of death by neurologic criteria. However, in these cases, hospital staff should be sensitive to these concerns and consider using similar resources to help family members accept the death [41].

3.7 Trinidad and Tobago

Trinidad and Tobago's *Human Tissue Transplant Act* states that for the purposes of deceased organ donation, death may be determined on circulatory-respiratory or neurological criteria [42]. Part E of the regulations passed under this Act indicates the following [42]:

> The nearest relative shall be given an opportunity to be present on the occasion of the final discontinuance of a system of life support and his religious and cultural requests shall be met as far as possible, before and after the discontinuance.

This legislative acceptance of death by neurologic criteria is limited to the context of organ donation, which requires either consent to posthumous donation or surrogate consent in the case of persons not known to have objected to donation. Legal acceptance of death by neurologic criteria more generally outside of the organ donation context would need to be contained in judicial decisions or other legislation. Yet this donation-focused legislation requires unspecified reasonable accommodation of religious or cultural factors, again presumably in the context of donation.

4 A Legal Case Study in Religious Freedom and Death by Neurologic Criteria: *McKitty v Hayani*

The freedom of religion is an important human right that has been recognized in international treaties and in the domestic laws of numerous legal jurisdictions. The manner in which that right is given legal protection varies considerably among jurisdictions and so it is impossible here to draw generalizable conclusions about how religious freedom claims may affect the determination of death by neurologic criteria [43]. Instead, I offer an overview of how the Canadian *Charter of Rights and Freedoms* was recently invoked by a family challenging the determination of death by neurologic criteria, and how this argument fared before the courts. This case—*McKitty v Hayani*—was ultimately moot because death by circulatory-respiratory criteria occurred during the litigation, and the evidentiary record was incomplete [5, 6]. However, the Court of Appeal proceeded to clarify the legal analysis in anticipation of a future dispute. The question therefore remains open under Canadian law, and will need to be resolved in future litigation.

In 2017, 27-year-old Taquisha McKitty was found unconscious and pulseless on a Brampton sidewalk [5, 6]. She was resuscitated and taken to the hospital where she was ventilated. Initially, she showed respiratory effort, but suffered an additional hypoxic-ischemic injury to her brain in the hospital and ceased breathing spontaneously. Six days after being taken to the hospital, she was determined to be dead by neurologic criteria by two critical care physicians. A physician informed Taquisha's family that she was dead by neurologic criteria and that organ support would be discontinued. The McKitty family requested additional independent tests. These tests, which included ancillary procedures (nuclear brain blood flow study and somatosensory evoked potentials) confirmed the determination. Here, the focus will be on the religious freedom-based claims, rather than certain other legal or factual disputes in the case.

The McKitty family argued that Taquisha was alive "according to the laws and precepts of her Christian faith" and that in order to comply with the Canadian *Charter of Rights and Freedoms*, the law must accommodate religious beliefs by providing an exemption for those who believe that death occurs only with the loss of circulatory-respiratory function.

The key legal questions raised in the case were the following:

1. Can someone who is determined to be dead by neurologic criteria raise a claim under the *Charter?*
2. Does the *Charter* apply to this situation?
3. Does the determination of death by neurologic criteria infringe McKitty's *Charter* right to religious freedom?
4. Is the infringement of her *Charter* right to religious freedom nonetheless a reasonable limit on that right?

A key initial question was whether someone who is dead by neurologic criteria has any remaining *Charter* rights that could be asserted on her behalf. The trial judge found that Taquisha McKitty was foreclosed from raising *Charter* rights because she had been declared dead before the legal process was commenced and no *Charter* breaches had contributed to her death [3]. This reasoning was obviously unsatisfactory, as it presumed the very fact in dispute—namely the constitutional validity of the legal position that death by neurologic criteria is death. The Ontario

Court of Appeal overruled the trial judge, stating that for the purposes of the litigation, one cannot conclude a person is legally dead and has no constitutional rights before asking the question of whether the application of standards for death by neurologic criteria was constitutional ([6] at para. 39, 47).

The second question is whether the *Charter* applies to a situation in which a doctor determines death by neurologic criteria. The complexity here is that the *Charter* applies to government action, and not to purely private action. Therefore, *the Charter* would uncontroversially apply to any statutory definition of death, since legislation falls within government action. It is less clear if and when a doctor determining death would be considered to be performing a government action. The Court of Appeal in *McKitty v Hayani* felt that a doctor determining death in the course of providing medical treatment would not be engaged in "government action," although the discussion of this point was very brief. The McKitty family argued that the completion of a medical certificate of death under the *Vital Statistics Act* was a "government action," but this point was not addressed by the trial judge or appellate court. Thus, we are left with a situation that in some Canadian provinces where there is a legislated definition of death by neurologic criteria, the *Charter* could be invoked to challenge the relevant legislation. Elsewhere, a weaker and more difficult argument for challengers would be that judges are required to develop the common law definition of death in line with "*Charter* values" or the deeper principles underlying constitutional human rights.

The next two questions—whether the determination of death by neurologic criteria infringes the *Charter* right to freedom of religion, and whether that infringement is nonetheless reasonable and justified—are the heart of the dispute. The Court of Appeal did not offer a final ruling on these points, given that the dispute was moot and the evidentiary record incomplete. However, the Court noted that the issue would likely arise again in urgent circumstances, and so it was advisable to clarify the legal analysis and correct errors made at the trial level in anticipation of future litigation ([6] at para 9).

The right to freedom of religion can be invoked where there is (1) a sincere belief having a nexus with religion and (2) state conduct that interferes with the ability to act in accordance with these religious beliefs in a manner that is more than trivial or insubstantial ([6] at para 59). It is not too difficult to establish a prima facie infringement of religious freedom based on these two criteria [44, 45]. It seems quite possible that, with adequate evidence about the sincerity of a person's religious belief that life endures until death by circulatory-respiratory criteria, a court would find that the imposition of the neurologic criteria for death would prima facie infringe a person's *Charter* right to religious freedom.

This is not the end of the discussion, however, since the *Charter* includes a justification provision. The government may argue that the infringement of a *Charter* right is justified by a pressing governmental objective and that the means chosen to achieve that objective are reasonable and proportionate. When it comes to religious freedom, Canadian courts are demanding at this stage because of the ease of establishing a prima facie infringement, along with the fact that "[m]uch of the regulation of a modern state could be claimed by various individuals to have a more than trivial impact on a sincerely held religious belief" ([46] at para 36). The reasons for the adoption of death by neurologic criteria are thus essential to this legal reasoning, as is the question of whether it is possible to offer some form of religious accommodation. The Court of Appeal in *McKitty v Hayani* noted that at this stage, a future court would need to consider whether accommodations, such as those found in some American states, would be possible and reasonable ([6] at para 82). Evidence that these accommodations are not unduly burdensome will most likely be important in any future Canadian disputes, although the difference between publicly funded and private insurance-based health care systems is an important point of distinction.

In conclusion, Canada remains without a clear legal decision on how the *Charter* right to religious freedom affects death by neurologic criteria and, particularly, whether there is a constitutional obligation to provide some form of accommodation of religious dissent. Given that the provinces vary in how they have legally recognized brain death (i.e., in a statute or the common law), the analysis is likely to differ according to the location in which the next dispute arises.

5 Conclusion

The technological changes that brought about a new human state—"a 'living' body with a 'dead' brain" [1]—are still fairly recent in the course of human cultural evolution. Although the concept of death by neurologic criteria was quickly and widely accepted in many locations, this has not been universal. The concept continues to sporadically be the subject of legal challenges whose underlying motivations may be diverse, even if the legal objections are framed according to the legally available arguments. Recent legal challenges in the United States and Canada have taken four main forms: (1) objections to evaluation for the determination of death by neurologic criteria without explicit consent, (2) objections to the accuracy of the determination, (3) objections to the legal adequacy of the testing procedures, and (4) objections to the concept of death by neurologic criteria itself.

This last category includes objections based on religious freedom. So far, the question of whether the constitutional right to religious freedom requires that religious objection to neurologic criteria for death determination be accommodated is unsettled in Canada and the United States. Some jurisdictions around the world have included an explicit exemption or accommodation requirement in laws governing the determination of death based on neurologic criteria. It will be relevant in future litigation whether this can be done without undue hardship in those places where accommodation is offered, taking into consideration differences that may alter the challenges associated with accommodation such as different health care resources and funding systems.

References

1. Gross SE, Lavi S, Boas H. Medicine, technology, and religion reconsidered: the case of brain death definition in Israel. Sci Technol Hum Values. 2019;44(2):186–208.
2. Ad Hoc Committee Harvard Medical School. A definition of irreversible coma: report of the Ad Hoc committee of the Harvard medical school to examine the definition of brain death. JAMA. 1968;205(6):337.
3. Greer DM, Shemie SD, Lewis A, et al. Determination of brain death/death by neurologic criteria: the world brain death project. JAMA. 2020;324(11):1078.
4. Silvester W, Dawson R, Quayyum S, et al. Brain death—death by neurologic criteria and the law. Supplement 13 to Greer DM et al. Determination of brain death/death by neurologic criteria: the world brain death project. JAMA. 2020;324(11):1078.
5. McKitty v Hayani 2018 ONSC 4105 (Ontario Superior Court of Justice). https://canlii.ca/t/hsqbh. Accessed 14 Jan 2022.
6. McKitty v Hayani 2019 ONCA 805 (Ontario Court of Appeal). https://canlii.ca/t/j2rv5. Accessed 14 Jan 2022.
7. Chandler JA. Cultural neuroethics in practice – Human rights law and brain death. In: Farisco M, editor. Neuroethics and cultural diversity. London: ISTE-Wiley; 2022.
8. Brown TR. Medical futility and religious free exercise. First Amendment Law Rev. 2017;15:43–95.
9. Terunuma Y, Mathis BJ. Cultural sensitivity in brain death determination: a necessity in end-of-life decisions in Japan. BMC Med Ethics. 2021;22(1):58.
10. Yang Q, Miller G. East–West differences in perception of brain death: review of history, current understandings, and directions for future research. J Bioeth Inq. 2015;12(2):211–25.
11. Pope TM. Brain death forsaken: growing conflict and new legal challenges. J Legal Med. 2017;37(3–4):265–324.
12. Morlani et al. v. Hadara 2021 ONSC 7288 (Ontario Superior Court of Justice). https://canlii.ca/t/jk476. Accessed 14 Jan 2022.
13. National Conference of Commissioners on Uniform State Laws (NCCUSL). Uniform Determination of Death Act 1980. https://www.uniformlaws.org/viewdocument/final-act-49?CommunityKey=155faf5d-03c2-4027-99ba-ee4c99019d6c&tab=librarydocuments. Accessed 14 Jan 2022.
14. In re guardianship of Hailu, 361 P.3d 524 (Nev. 2015).
15. Lewis A, Bonnie RJ, Pope T. It's time to revise the uniform determination of death act. Ann Intern Med. 2019;172(2):143–5.
16. Miller FG, Nair-Collins M. An incoherent proposal to revise the Uniform Determination of Death Act. Hastings Center Bioethics Forum, 23 January 2020. https://www.thehastingscenter.org/an-incoherent-proposal-to-revise-the-uniform-determination-of-death-act/. Accessed 14 Jan 2022.
17. Lewis A. The uniform determination of death act is being revised. Neurocrit Care. 2022; https://doi.org/10.1007/s12028-021-01439-2.
18. Veatch RM. Controversies in defining death: a case for choice. Theor Med Bioeth. 2019;40:381–401.
19. Canadian Charter of Rights and Freedoms, Part 1 of the Constitution Act, 1982, being Schedule B to the Canada Act 1982 (UK), 1982, c 11.
20. McMath et al. v. State of California et al., Complaint for Declaratory and Injunctive Relief, Petition for Injunctive and Declaratory Relief. Dated 23 December 2015. https://www.thaddeuspope.com/images/Jahi-McMath-Federal-Complaint-20151223.pdf. Accessed 14 Jan 2022.
21. Fonseca v. Kaiser Permanente et al. Complaint for Declaratory Relief and Request for Temporary Restraining Order and Injunctive Relief. Dated 28 April 2016. https://www.thaddeuspope.com/images/Fonseca_v_Kaiser_ED_Cal_04-28-16_complaint.pdf. Accessed 14 Jan 2022.

22. Ezer T, Wright MS, Fins JJ. The neglect of persons with severe brain injury in the United States: an international human rights analysis. Health Hum Rights. 2020;22(1):265–78.
23. Young MJ, Bodien YG, Giacino JT, et al. The neuroethics of disorders of consciousness: a brief history of evolving ideas. Brain. 2021;144(11):3291–310.
24. Akabayashi A. Bioethics across the globe. Singapore: Springer; 2020. p. 13–26.
25. Japan Organ Transplant Network (JOTN). The enactment of the organ transplantation law, and the revised organ transplant act [Internet]. https://www.jotnw.or.jp/en/04/. Accessed 14 Jan 2022.
26. Asai A, Kadooka S, Aizawa K. Arguments against promoting organ transplants from brain-dead donors, and views of contemporary Japanese on life and death. Bioethics. 2012;26(4):215–23.
27. Natori Y. Legal determination of brain death. JMAJ. 2011;54(6):363–7.
28. Japan Organ Transplant Network (JOTN). Views on brain death [Internet]. https://www.jotnw.or.jp/en/05/. Accessed 14 Jan 22.
29. New Jersey Law Revision Commission. Final Report Relating to the New Jersey Declaration of Death Act. 18 January 2013. https://www.thaddeuspope.com/images/NJ_Law_Rev_Comm_-_NJ_Decl_Death_Report_2013.pdf. Accessed 14 Jan 2022.
30. N.J. Rev. Stat. § 26:6A-5; N.J. Admin. Code § 13:35-6A.6.
31. N.J. Rev. Stat. § 26:6A-7.
32. Ashkenazi T, Steinberg A, Cohen J. A national survey of attitudes of the Zionist ultra-orthodox community in Israel to organ donation. Prog Transplant. 2019;29(1):43–7.
33. Cohen J, Ashkenazi T, Katvan E, Singer P. Brain death determination in Israel: the first two years experience following changes to the brain death law-opportunities and challenges: brain death determination. Am J Transplant. 2012;12(9):2514–8.
34. Israel, Cerebro-Respiratory Death Act, 2008. Translation available from the Halachic Organ Donation Society, hods.org. https://www.hods.org/pdf/law51%20Braindead.pdf; Accessed 14 Jan 2022.
35. Ill Rev Stat Ch 110 § 302(b); 1981.
36. Clausing M. The acceptance of brain death as a legal definition of death in Illinois: In Re Haymer. DePaul Law Rev. 1983;33:207–23.
37. In re Haymer 115 Ill. App. 3d 349, 450 N.E.2d 940 (Ill. App. Ct 1983).
38. 210 Ill. Comp. Stat 85/6.23; 2012.
39. Cal Health & Safety Code § 1254.4(a).
40. 10 N.Y.C.R.R. § 400.16; 1987.
41. NY State Department of Health and NY State Task Force on Life & the Law. Guidelines for Determining Brain Death November 2011. https://www.health.ny.gov/professionals/hospital_administrator/letters/2011/brain_death_guidelines.pdf. Accessed 14 Jan 2022.
42. Trinidad and Tobago. Human Tissue Transplant Act c. 28:07, Act 13 of 2000, and Human Tissue Transplant Regulations LN 314/2004. https://rgd.legalaffairs.gov.tt/laws2/Alphabetical_List/lawspdfs/28.07.pdf. Accessed 14 Jan 2022.
43. Durham WC, Evans C. Freedom of religion and religion-state relations. Chapter 19. In: Tushnet M, Fleiner T, Saunder C, editors. Routledge handbook of constitutional law. London/New York: Routledge; 2013.
44. Berger BL. Section 1, constitutional reasoning and cultural difference: assessing the impacts of Alberta v Hutterian Brethren of Wilson colony. Supreme Court Law Rev (2d). 2010;51:25–46.
45. Gilbert D. Faith and/in medicine: religious and conscientious objections to MAiD. Dalhousie Law J. 2020;32:657.
46. Alberta v. Hutterian Brethren of Wilson Colony 2009 SCC 37 (Supreme Court of Canada). https://scc-csc.lexum.com/scc-csc/scc-csc/en/item/7808/index.do. Accessed 14 Jan 2022

Is Death by Neurologic Criteria a Legal Fiction or Status?

Seema K. Shah

Almost since its inception in 1968 [1], scholars have debated the concept of determining death by neurologic criteria (also called "brain death") [2, 3]. Despite over 50 years of determination of death by neurologic criteria, the public's understanding remains murky [4]. There have also been several high-profile court cases of conflict between families and hospitals over patients who were determined to be dead by neurologic criteria [5], and more than a third of pediatric neurologists and intensivists note that they have encountered families who refuse evaluation for determination of death by neurologic criteria [6]. As of this writing, the Uniform Law Commission is considering whether or how to revise the legal conditions for the neurologic determination of death [7] set forth in the Uniform Determination of Death Act—a highly influential model law that sets out standard criteria for death in the United States [8].

One key question for the Uniform Law Commission is whether and to what extent the law governing the neurologic determination of death should reflect a biological truth about death. The standard view from bioethics commissions and prominent scholars (even if they do not necessarily agree on the criterion of the neurologic determination of death) is that the law governing the neurologic determination of death should reflect biological reality—death should not be a legal fiction [9, 10, 11]. Gilbert Meilander felt compelled to issue a personal statement appended to the 2008 President's Council on Bioethics report addressing controversies in the determination of death to make this very point. Meilander argued that: "[w]e should not create 'legal fictions' or 'social agreements' whose aim is less an accurate determination of death than a ready supply of organs. Whatever else human beings may

S. K. Shah (✉)
Department of Pediatrics, Northwestern Feinberg School of Medicine, Chicago, IL, USA

Bioethics Program, Ann & Robert H. Lurie Children's Hospital, Chicago, IL, USA

Northwestern Pritzker School of Law, Chicago, IL, USA
e-mail: Seema.Shah@northwestern.edu

A. Lewis, J. L. Bernat (eds.), *Death Determination by Neurologic Criteria*, Advances in Neuroethics, https://doi.org/10.1007/978-3-031-15947-3_23

be, they are living bodies, and their death is a biological reality that we need to mark as accurately as we are able" [12]. Furthermore, there is discomfort with the idea of doctors practicing "a form of deceit" [13] when they determine whether patients are dead.

In this chapter, I argue somewhat counterintuitively that a legal fictions approach need not be deceptive and can be helpful for understanding the concept of the neurologic determination of death. Conceiving of the neurologic determination of death as a legal fiction provides a way to reconcile seemingly intractable debates about how to determine death, at least as a legal or policy matter.

I first provide a brief overview of the concept of the neurologic determination of death and the debates surrounding this concept that have persisted. Next, I explain legal fictions and how they relate to legal statuses. I briefly attempt to demonstrate that the neurologic determination of death is easier to understand as a legal concept than as a biological one. While in previous writings I have called the neurologic determination of death a legal fiction [14, 15], I explain why it is equally acceptable to call the neurologic determination of death a legal status [16]. I conclude by considering how a legal fictions approach can inform recent proposals to revise the Uniform Determination of Death Act. For example, the responsible use of legal fictions or statuses to laws determining death by neurologic criteria should build in reminders of the gaps in the law, including by providing exemptions for those who do not accept the neurologic determination of death. My goal is to provide practical guidance regarding the dangers to avoid that may be helpful for ongoing and future efforts to revise the neurologic determination of death.

1 Background

For centuries, people believed death occurred when the heart stopped beating, blood stopped flowing, and breathing ceased, and these functions would never start again. In the 1950s and 60s, the invention of ventilators that could maintain patients in what French scientists referred to as *coma dépassé* raised questions about how to conceive of the status of these patients [17]. In a letter inviting people to join a committee to "give further consideration to the definition of brain death," the Dean of Harvard Medical School noted that Henry Beecher, a prominent physician and early medical ethicist, had raised "ethical problems relating to the hopelessly unconscious man" [18]. In less than 4 months, this so-called Ad Hoc Committee coalesced around a new definition of death, which they published in the August of 1968 [1].

This committee was chaired by Henry Beecher. Among the members of the committee were three neurologists, a physiologist, a professor of law, a professor of public health, a biochemist, a historian, a transplant nephrologist, a neurosurgeon, and a professor of social ethics. Not all members attended all meetings or contributed significantly, however. The historian referred to himself as "more of a silent contributor." The professor of social ethics, Ralph Potter, stated that he felt "[i]t was not a deliberative body" [18]. No women served on this committee, and it is not evident that there were any members of minoritized groups.

In correspondence between the committee members, one neurologist noted that he was aware of objections to the neurologic determination of death: "I would suppose that brain death was now almost universally accepted as death of that individual. I know of one lay person, however who still believes that the heart has personal attributes…" [18]. The neurologist concluded that transplantation of corneas from patients determined to be dead by neurologic criteria was acceptable, but heart transplantation may face objections.

Ultimately, the committee explicitly gave two reasons for developing a way for people to be determined to be dead by neurologic criteria: (1) patients with a "permanent loss of intellect" were being kept on ventilators in ways that burdened them, their families, hospitals, and other patients who needed hospital beds; and (2) "Obsolete criteria for the definition of death can lead to controversy in obtaining organs for transplantation" [1]. Yet the Harvard Ad Hoc Committee never offered a scientific or philosophical explanation for why someone in an irreversible coma should be considered dead.

Reservations emerged about whether this was really a new form of death, in part because pregnant people were able to gestate fetuses for months after being determined to be dead by neurologic criteria [19]. Scholars questioned whether being dead was compatible with being able to support fetal development and give birth to a viable infant. Over time, evidence accumulated that people determined to be dead by neurologic criteria could maintain a range of capacities that might seem inconsistent with being dead. First, patients determined to be dead by neurologic criteria still had documented brain functions, typically involving the hypothalamus and pituitary gland [20]. The commonplace nature of some brain functioning continuing after a determination of death by neurologic criteria suggests that there is, at a minimum, a disconnect between what the language of the law requires and what the standard clinical evaluation is using to determine whether someone is dead under the law. Additionally, patients who were determined to be dead can persist on ventilators for years [21], heal wounds, maintain a body temperature in the range of normal, and mount stress responses to incisions [22–24].

Subsequent bioethics commissions and scholars tried to fill in these gaps and explain why the neurologic determination of death could be considered on a par with traditional circulatory-respiratory criteria for death [25, 26]. Recognizing the weight of this evidence and the persistent controversy, in 2008, the President's Council on Bioethics published a report providing a foundation for the neurologic determination of death [11]. The Council acknowledged the evidence and engaged with philosophical debates about the neurologic determination of death in a remarkable way. For instance, some scholars have argued that there could be two kinds of death—the death of the person and the subsequent death of the body. These scholars contend that a person is dead once the so-called higher brain cannot function [27]. The Council countered that defining death as when the person is gone "expands the concept of death beyond the core meaning it has had throughout human history" [11]. Recognizing the death of a person would require determining as dead anencephalic infants and patients in a persistent vegetative state who can still breathe on their own—all individuals typically recognized as alive. Furthermore, a law that defined the death of a person as equivalent to the traditional notion of death would clearly depart from what many people believe.

Ultimately, the President's Council argued that an organism must do "vital work" and "need-driven commerce" with the world, and that this vital work stops once the brain has ceased functioning, so such patients are dead [11]. This conclusion was met critically. Critics noted that because individuals who meet the neurologic criteria for death are able to heal wounds, gestate fetuses to viability, and mount stress responses, they are performing "need-driven commerce" with the world [28, 29]—thus suggesting that the Council's rationale did not succeed. As the Uniform Law Commission works to revise the Uniform Determination of Death Act, controversy over the determination of death by neurologic criteria continues [30, 31].

2 Legal Fictions and Statuses

Before examining the current state of debates on death by neurologic criteria, it is helpful to explain the definitions of legal fictions and statuses. There are several types of legal fictions. Core to all of them is the concept of a gap between what the law states and reality. Henry Sumner Maine provided an early definition of legal fictions, defining them as instances when a judge makes a ruling in which "the law has been wholly changed; the fiction is that it remains what it always was" [32]. Maine was focused on the need to rein in judges who strayed from their duty to interpret the law. He therefore objected to the idea that the letter of the law could remain the same yet the way the law actually functioned was extended beyond what the words would seem to permit. Lon Fuller expanded and enriched the concept of a legal fiction. Fuller proposed the following definition: "A fiction is either (1) a statement propounded with a complete or partial consciousness of its falsity, or (2) a false statement recognized as having utility" [33]. Fuller also recognized that the power or motivation to create legal fictions was not limited to judges; legislatures could use legal fictions if, for instance, the statutes they wrote did not map onto reality.

The fiction that corporations are persons under the eyes of the law is one of the most widely known legal fictions. While everyone knows that a corporation is different from a person, and most corporations are in fact run by several persons, the law *treats* corporations as persons. This means that laws governing what persons can do can be applied to corporations, without having to draft new laws to cover corporate actions. Similarly, the "attractive nuisance" doctrine is another example of a status fiction. Traditionally, property owners could not be held responsible when someone trespassed on their property and became injured; they only faced potential liability if they had *invited* someone onto their property. If a child happened upon someone's property in order to play there and was injured, however, strictly applying the law to deny that child any relief seemed unfair. As a result, the "attractive nuisance" fiction was borne, which turned a trespassing child into an "invitee" by imagining that a property owner with anything on their land that might attract a child has effectively invited children to play on it. This legal fiction served to make property owners liable when children are injured while on their property. These types of legal fictions can be thought of as a *status* fictions, which gives one entity the same status as another under the law [14].

There are other types of fictions. In an anticipatory fiction, the law treats something as if it has already happened even though it has not yet occurred. An example of this is anticipatory breach in contract law, where someone in a contract can breach the contract to mitigate harm if they know the other party will not be able to meet their end of the bargain. Bright line fictions are used when the law draws a somewhat arbitrary line to designate a point in a process or along a spectrum and treats people differently depending on where they fall on that line. For example, the idea that children reach the age of majority and are then presumed to have capacity to consent to a wide range of activities clearly does not capture actual decision-making capacity for everyone. Some people may develop capacity earlier than age 18, while some people may have questionable decision-making capacity throughout their lives. In fact, brain development is based on a variety of factors and extends well beyond adolescence for many people [34].

Fuller argued that legal fictions can be dangerous, as they lack transparency, and make it more difficult for individuals to understand how their government is acting. Additionally, because not all uses of a legal fiction will have equal justification, there is a danger that a legal fiction developed for one purpose will be used for another unthinkingly [33]. Fuller suggested that acknowledging legal fictions is a way to address this danger. Returning to the matter at hand for a pertinent example, if a patient has been determined to be dead by neurologic criteria, can research be conducted on them without oversight or permission [35]? This topic is addressed in further detail elsewhere in this book.

Chunlin Leonhard has explained that one way to acknowledge a legal fiction and prevent its unthinking application to new areas is to build in reminders of its falsity [36]. Leonhard proposes that this could be done by providing language that clues people into the falsity behind the fiction (e.g., "implied," "quasi," "constructive") or by resting a fiction on something that is so patently untrue it will not be confusing. It is important to recognize that this strategy is not bulletproof; the fiction of corporate personhood is a good example to the contrary. While everyone knows corporations are not persons, corporate personhood has now been used to justify allowing corporations freedom of speech rights, which many legal commentators have criticized as a problematic extension of the First Amendment. Political commentators have worried that this extension of free speech rights also allows corporations to drown out the speech of individuals, given the greater resources they have to make political speech, pay for advertisements, or make campaign donations.

Legal fictions, however, are not always that different from legal statuses. Legal statuses are categories that confer rights for and obligations toward those who belong to the category [16]. Legal blindness is one example, where a point is marked on the spectrum of diminished vision, and anyone whose vision falls below that point counts as legally blind. Legal statuses are distinct from legal fictions. Unlike legal fictions, statuses are simply categories that are designated by law—not examples of the law relying on falsity. However, a legal status, like a legal fiction, does not have to track objective reality perfectly or at all.

3 Applying Legal Fictions and Statuses to the Neurologic Determination of Death

Legal fictions are helpful for understanding the neurologic determination of death in two ways. First, as argued above, the fact that some brain functions continue in some patients after a determination of death by neurologic criteria suggests there is, at a minimum, a disconnect between what the language of the law requires and what can be established with the standard clinical evaluation to determine death. The law has stringent language requiring that all functions of the brain have stopped; yet how that law is applied in practice through clinical evaluation ends up counting some people as dead even though they retain some functions of the brain (as discussed elsewhere in this book).

The second type of legal fiction in the neurologic determination of death is the lack of a clear justification for why death can be established by neurologic criteria. The Ad Hoc Committee referred to patients who met certain neurologic criteria as dead without a scientific or philosophical justification. Scholars and bioethics commissions have not developed an uncontested justification for determining death in line with neurologic criteria. While neurologic criteria for death may be consistent with identifying the death of a person, many people and religions would simply not accept the death of a person as a form of biological death.

4 Conclusion: Guidance for Proposals to Revise the UDDA as Fictions or Statuses

There have been several recent proposals to revise the UDDA [30, 31, 37–40]. Proposals that eliminate neurologic criteria for death altogether or abandon the dead donor rule [40], as well as to allow patients to choose how their death is determined [38, 39], are ways to reform the law that does not rely upon falsity. While these proposals involve different tradeoffs, for our purposes, it is sufficient to note that they do not rely on legal fictions.

Contrastingly, some other proposals rely on legal fictions by seeking to retain the use of neurologic criteria for determination of death without requiring the cessation of all functions of the entire brain or justifying why neurologic criteria can be used to establish death for human organisms [30, 37]. For these proposals, it will be important to avoid the dangers associated with unacknowledged legal fictions by building in reminders of their falsity. For example, these approaches could either use labels to clarify the fictive status of this type of death, or—somewhat paradoxically—describe something that is patently false [30]. In taking the former approach, labels such as "implied," "quasi," or "constructive" are terms that can demonstrate something that is legally established may not otherwise be true.

As an example, the Uniform Determination of Death Act could be revised to read as follows: "An individual who has sustained either (1) irreversible cessation of circulatory and respiratory functions, or (2) irreversible cessation of all functions of the entire brain, including the brain stem, is *legally* dead." Alternatively, the act could describe individuals determined to be dead by neurologic criteria as "dead for legal purposes" [41]. Whether such changes will increase transparency about the gap between the legal criteria for death and a biological definition of death is ultimately an empirical question; it is possible that saying someone is "constructively dead for legal purposes" would be clearer, though much clunkier.

Second, to the extent the *fiction* is apparent, the legal fiction is more transparent. For example, indicating that, for the purposes of determining death by neurologic criteria, some parts of the brain simply do not count could be one way to make the legal nature of death clearer and more evident. Finally, allowing exemptions from the neurologic determination of death (but not the traditional determination of death by circulatory-respiratory criteria) would be another way to recognize that there is reasonable disagreement on the use of neurologic criteria to declare death, rather than a biological reality that some wrongly reject. There are also important moral reasons to provide exemptions from or choice regarding the neurologic determination of death [38, 39], as discussed elsewhere in this book.

References

1. Ad Hoc Committee of the Harvard Medical School to examine the definition of brain death. A definition of irreversible coma. JAMA. 1968;205:337–40.
2. Veatch RM. Abandon the dead donor rule or change the definition of death? Kennedy Inst Ethics J. 2004;14(3):261–7.
3. Beecher HK. Ethical problems created by the hopelessly unconscious patient. New Engl J Med. 1968;278:1425–30.
4. Shah SK, Kasper K, Miller FG. A narrative review of the empirical evidence on public attitudes on brain death and vital organ transplantation: the need for better data to inform policy. J Med Ethics. 2015;41(4):291–4.
5. Lewis A, Pope TM. Physician power to declare death by neurologic criteria threatened. Neurocrit Care. 2017;26(3):446–9.
6. Mataya L, Ross LF, Ghavam A, Paquette ET. Pediatric intensivist and pediatric neurologist perspectives and practices on death by neurologic criteria. J Clin Ethics. 2021;32(3):195–205.
7. Uniform Law Commission, New Drafting and Study Committees to be Appointed [Internet]. Uniformlaws.org. https://www.uniformlaws.org/committees/community-home/digestviewer/viewthread?MessageKey=a71e3d9d-4cf9-4529-9169-a08acf5edde6&CommunityKey=d4b8f588-4c2f-4db1-90e9-48b1184ca39a&tab=digestviewer. Accessed 17 Mar 2022.
8. National Conference of Commissioners on Uniform State Laws, Uniform Determination of Death Act (Chicago: National Conference of Commissioners on Uniform State Laws, 1981).
9. Rich B. Structuring conversations on the fact and fiction of brain death. Am J Bioeth. 2014;8:31–3.
10. Marquis D. Death as a legal fiction. Am J Bioethics. 2014;148:28–9.
11. President's Council on Bioethics, Controversies in the Determination of Death (Washington, DC: December 2008).
12. Meilander G. Personal statement of Gilbert Meilaender, Ph.D. President's Council on Bioethics, Controversies in the Determination of Death (Washington, DC: December 2008).

13. Delimiting death. Editorial. Nature. 2009;461:570.
14. Shah SK, Miller FG. Can we handle the truth? Legal fictions in the determination of death. Am J Law Med. 2010;36(4):540–85.
15. Shah SK, Truog RD, Miller FG. Death and legal fictions. J Med Ethics. 2011;37(12):719–22.
16. Shah SK. Rethinking brain death as a legal fiction: is the terminology the problem? Hast Cent Rep. 2018;48(Suppl 4):S49–52.
17. Mollaret P, Goulon M. Le coma dépassé. Rev Neurol. 1959;101:3–15.
18. Wijdicks EFM. The neurologist and Harvard criteria for brain death. Neurology. 2003;61(7):970–6.
19. Siegler M, Wikler D. Brain death and live birth. JAMA. 1982;248:1101–2.
20. Nair-Collins M, Joffe AR. Hypothalamic function in patients diagnosed as brain dead and its practical consequences. Handb Clin Neurol. 2021;182:433–46.
21. Shewmon DA. The brain and somatic integration. J Med Philos. 2001;26:457–78, 467-69
22. Wetzel RC, Setzer N, Stiff JL, Rogers MC. Hemodynamic responses in brain dead organ donor patients. Anesth Analg. 1985;64(2):125–8.
23. Truog RD, Fackler JC. Rethinking brain death. Crit Care Med. 1992;20(12):1705–13.
24. Johnson LSM. A legal fiction with real consequences. Am J Bioeth. 2014;14(8):34–6.
25. President's Commission for the Study of Ethical Problems in Medicine and Medical and Behavioral Research. Defining death: medical, ethical, and legal issues in the determination of death. Washington, DC: U.S. Government Printing Office; 1981.
26. Huang AP, Bernat JL. The organism as a whole in an analysis of death. J Med Philos. 2019;44(6):712–31.
27. Lizza P. The conceptual basis for brain death revisited: loss of organic integration or loss of consciousness? Adv Exp Med Biol. 2004;550:52.
28. Shewmon DA. Brain death: can it be resuscitated? Issues Law Med. 2009;25(1):3–14.
29. Miller FG, Truog RD. The incoherence of determining death by neurological criteria: a commentary on "Controversies in the determination of death", a White Paper by the President's Council on Bioethics. Kennedy Inst Ethics J. 2009;19(2):185–93.
30. Lewis A, Bonnie RJ, Pope T. It's time to revise the uniform determination of death act. Ann Intern Med. 2020;172(2):143–4.
31. Shewmon DA. Statement in support of revising the uniform determination of death act and in opposition to a proposed revision. J Med Philos. 2021:jhab014. Online ahead of print
32. Maine HS. Ancient law: its connection with the early history of society, and its relation to modern ideas. London: Murray; 1861.
33. Fuller LL. Legal fictions. Palo Alto, CA: Stanford University Press; 1967.
34. Giedd J, Blumenthal J, Jeffries NO, et al. Brain development during childhood and adolescence: A longitudinal MRI study. Nature Neurosci. 1999;2:861–3.
35. Shah SK. Piercing the veil: the limits of brain death as a legal fiction. U Mich J Legal Reform. 2015;48:301.
36. Leonhard C. Dangerous or benign legal fictions, cognitive biases, and consent in contract law. St John's L Rev. 2017;91:385.
37. Omelianchuk A, Bernat J, Caplan A, et al. Revise the UDDA to align the law with practice through neuro-respiratory criteria. Neurology. 2022. Online ahead of print.
38. Ross LF. Respecting choice in definitions of death. Hast Cent Rep. 2018;48(Suppl 4):S53–5.
39. Veatch RM. Controversies in defining death: a case for choice. Theor Med Bioeth. 2019;40(5):381–401.
40. Miller FG, Nair-Collins M, Truog RD. It is time to abandon the dogma that brain death is biological death. Hast Cent Rep. 2021;51(4):18–21.
41. Written communication with Erin Paquette and Thaddeus Pope. Meeting of the Uniform Law Commission Drafting Committee on Updating the Uniform Determination of Death Act, March 10, 2022.

Legal Considerations on the Declaration of Death by Neurologic Criteria in the Pregnant Patient

Kimberly Mutcherson

The boundaries which divide Life from Death are at best shadowy and vague. Who shall say where the one ends, and where the other begins?—Edgar Allan Poe

Death at the beginning of life, for newborns or pregnant persons,[1] is difficult to contemplate regardless of whether it is declared using circulatory-respiratory criteria or neurologic criteria. When there is controversy about the declaration of death or when the declaration of death conflicts with someone's desire to maintain the pregnant person in a state of limbo to benefit the fetus, the complex nature of death during pregnancy, childbirth, or soon after is compounded. When a healthcare provider declares that a pregnant person is dead using circulatory-respiratory criteria, the pre-viability fetus typically also dies and a viable fetus will experience the same fate if there is no opportunity for emergency surgery to remove it in a timely fashion. These cases are, of course, tragedies, but they are not controversial. However, when a healthcare provider declares that a pregnant person is dead by neurologic criteria, as described in this chapter, ethical and legal questions about the continuation of medical care can lead to tense clashes between those who want to maintain the pregnant person's body for as long as possible for the sake of the fetus and those

[1] I use the terms pregnant people/person, pregnant woman/women, and people capable of pregnancy in this chapter. While it is the case that people who are not women, including those who are non-binary and trans men, can and do choose to become pregnant, the use of the capacity for pregnancy as a justification for myriad laws, policies, and practices that limit the rights of women because of sexism and hostility to female equality cannot be erased or overstated. Thus, understanding the roots of practices that deny women and other pregnant people the rights given to people who are not pregnant is best understood within the context of centuries of misogyny and ongoing battles for sex equality.

K. Mutcherson (✉)
Rutgers Law School, Camden, NJ, USA
e-mail: Kim.mutcherson@rutgers.edu

A. Lewis, J. L. Bernat (eds.), *Death Determination by Neurologic Criteria*, Advances in Neuroethics, https://doi.org/10.1007/978-3-031-15947-3_24

who see using a dead person's body in this way as desecration of a corpse and possibly an unethical form of experimentation on a fetus trapped inside a dead body.

This chapter places the horror of death by neurologic criteria during pregnancy squarely within broader legal and ethical discussions about whether the interests of a fetus, assuming that a fetus even has interests, should ever supersede the autonomy, bodily integrity, or dignity interests of the pregnant person.

1 Marlise Muñoz and the Texas Advance Directives Act

Marlise Muñoz was not expecting to die. She was healthy and was enjoying her second pregnancy while living with her husband, Erik, and her first child, a son, in their home in Texas. She was 33 years old and 14-weeks pregnant when, just 2 days before Thanksgiving 2013, she collapsed on the floor of her kitchen due to a pulmonary embolism. Because this tragic incident happened in the middle of the night, Marlise's husband, Erik, did not immediately find her, and by the time he did discover her face down on the kitchen floor and not breathing, she, and by extension her fetus, had been deprived of oxygen for more than an hour. She was transported to John Peter Smith Hospital in Fort Worth, Texas, but the prolonged period without oxygenation of the brain had caused such severe injury that the healthcare providers declared her dead by neurologic criteria.

This was a devastating outcome for the loved ones Marlise left behind, but the hospital compounded the family's grief with the decision that followed the determination of death by neurological criteria. Marlise and Erik were paramedics and the couple had previously engaged in serious discussions about their preferences for heroic or life-saving measures if either of them experienced a catastrophic injury. Marlise was clear that she did not wish to be sustained by technology under these conditions. Given this knowledge of what Marlise would have wanted had she been lucid and competent to make choices for herself, Erik, now a widow with a young child to raise as a single father, and Marlise's parents requested that the hospital discontinue support which would lead to the cessation of respiration and circulation. They knew that doing so meant that they would lose both Marlise and the fetus that she was carrying, which was far too early in gestation to be capable of successfully surviving outside of the womb.[2] To their surprise and horror, the hospital refused to discontinue support and declared that they would keep her on the ventilator for purposes of continuing the pregnancy hopefully until a point of fetal viability.

The hospital administrators justified their position based on the Texas Advance Directives Act ("the Act"), which explicitly provides that "a person may not

[2] While viability of a fetus during a pregnancy is not a hard and fast rule, it generally is described to occur at or after 23-24 weeks gestation [1]. There is no set of medical criteria by which a fetus at 14-weeks gestation would be considered medically viable.

withdraw or withhold life-sustaining treatment... from a pregnant patient [2]." In other words, no matter the requests made by a pregnant person when competent or by the person who is the substitute decisionmaker if the pregnant patient is incompetent, the hospital will not honor those wishes if the choice is to withdraw or withhold treatment that is "life sustaining." Thus, the hospital's interpretation of the statute required them to keep Marlise on life support in a ghoulish attempt to continue gestation of the fetus until the point of viability even though they had already declared Marlise to be legally dead. The statute does not categorically exempt any other group of people from the protections of the Act meaning that only in the case of pregnant people could a hospital deny a competent person's desire to withdraw or refuse initiation of life-saving treatment. It is only for pregnant people, the vast majority of whom are cis women, that laws in Texas and elsewhere abridge the right to make healthcare decisions related to life-saving.

When informed that the hospital would not fulfill their request to remove Marlise from support, the family was, of course, appalled and took the only step available to them for relief—they went to court and sought an order to require the hospital to discontinue support. During their court battle with the hospital to act based on their specific knowledge of what Marlise would have wanted for herself and the fetus she carried, a media firestorm raged around them. On one side were anti-choice forces who rallied to the cause of "saving" the future child, even though the pregnancy was weeks from viability when Marlise died. And, frankly, there were many reasons to think that even if the pregnancy continued to the point of medical viability, the fetus would still die or would have experienced irreparable harm either from the detriment of oxygen when Marlise first experienced her catastrophic injury or during the weeks when it was being gestated in a dead body. On the other side of the controversy were patients' rights advocates, women's rights advocates, pro-choice advocates, and many bioethicists who were appalled by the hospital's reliance on the Texas statute to justify continuation of support for a patient they had already decided was dead [3]. For many on this side of the controversy, the use of Marlise's body as a vessel for housing a fetus fed into long-standing and ever-present concerns about the dehumanization of women and the balancing of a woman's interests in bodily autonomy against any fetal interests, an issue discussed in more depth later in this chapter.

Only after Marlise spent 62 days on support in the hospital did the family receive the requested court order that forced the hospital to take the step that they had sought all along. The hospital allowed Marlise and her fetus to finally rest. However, the trauma that the family experienced while the hospital refused to act can likely not be overestimated, especially for Erik Muñoz, who was caring for a young child while fighting for and mourning the death of his wife and the child she carried. In a documentary about this period in his life, Erik discussed the horror of Marlise's body beginning to smell as it broke down over time and his deep sense of helplessness at not being able to effectuate what he knew she would want [4].

2 An Attempt to Amend the Texas Advance Directives Act

Marlise's family used their tragic experience with the Texas Advance Directives Act as a catalyst for action and began a campaign to pass Marlise's Law. If passed, this new law would countermand the directive to ignore the medical decisions of pregnant patients or their lawful decisionmakers if they chose to forego life-sustaining treatment [5]. In March 2015, Representative Elliot Naishtat of the Texas Legislature filed a bill that would repeal the portion of the Texas law that spoke to ignoring the wishes of pregnant people who did not want to be supported artificially. Representative Naishat explained the impetus for the bill as follows: "Being pregnant should not prohibit a woman from having her personal decision respected. The law should reflect the consideration a woman puts into planning the treatment she wishes to receive, or not receive, when she is no longer able to express herself. Planning for end-of-life care is a deeply personal decisionmaking process for all persons, including those who may be pregnant [6]."

The bill from Representative Naishat sharply contrasted with a bill introduced by Republican representative Matt Krause. That bill sought to change the law to make clear that its strictures against withholding treatment applied to those declared dead by neurologic criteria during pregnancy "if the life-sustaining treatment is enabling the unborn child to mature [7]." Further, the bill called for the state's Attorney General to appoint a Guardian ad Litem to "represent the unborn child's interests... in any litigation or other matter regarding the health care decisions made for the pregnant patient [8]." Neither of these bills successfully passed in the State legislature leaving the pregnancy exclusion in the Texas Advance Directives Act intact and as originally passed.

3 Advance Directives During Pregnancy in Other States

Texas is not the only state with an advance directives law that exempts pregnancy from its general standard. In fact, in 2013 most states had some sort of carveout for pregnancy in their laws on advance directives and medical decisionmaking. As described in one opinion piece, "Five states create a presumption in favor of treatment that can be rebutted with a specific advance directive. Fourteen require continued treatment if the fetus can develop to live birth or viability. Twelve, including Texas, categorically require continued treatment for all pregnant women, regardless of the wishes of the patient or her family or the viability of the embryo or fetus [8]."

Whatever the limitations, these carveouts for pregnant people are deeply problematic for a range of reasons beginning with the assumption that bodily integrity and the concomitant constitutionally protected right to make decisions about one's own medical care is diminished or wholly demolished during pregnancy, even when that pregnancy is not viable, and even when the pregnant person is still within a window that would allow them to legally terminate the pregnancy if they so choose.[3]

[3] The U.S. Supreme Court's June 2022 opinion in *Dobbs v. Jackson Women's Health*, has led to abortion bans and other significant pre-viability restrictions on abortion that only compound the issue of bodily autonomy and choice for pregnant women and others as weighed against any state interest in pre-viable fetal life.

4 Pregnancy, Bodily Autonomy, and Dignity

The issue of how and when, if ever, healthcare providers can legally and ethically deny pregnant women and other people capable of pregnancy the rights to bodily integrity and autonomy granted to those who are not pregnant is not unique to the extreme circumstance of a declaration of death by neurologic criteria while pregnant. States have subjected pregnant women to penalties, including arrest and prosecution, for ingesting illicit drugs while pregnant, even without proof of harm to the fetus [9, 10]; women have faced decades of pregnancy discrimination, sometimes justified by concerns about harm to a fetus, that has not been wholly remedied by the Pregnancy Discrimination Act [11];and women in labor have endured forced c-sections when their clinicians sought court orders to override the woman's decision to decline the surgery [12].

In two of the most shocking cases involving forcing a pregnant woman to have a c-section against her will in the interest of the health and well-being of her fetus, judges came to very different conclusions. In the first, *In Re Madyun*, involving a 19-year-old pregnant Black Muslim woman and her husband, a judge ordered a c-section after an obstetrician testified to the risk of infection because of the length of time that Ms. Madyun was in labor after her water initially broke. In its opinion, the Court wrote, "All that stood between the Madyun fetus and its independent existence, separate from its mother, was put simply, a doctor's scalpel. In these circumstances, the life of the infant inside its mother's womb was entitled to be protected [13]." The judge's articulation that all that stood between the Madyun fetus and its independent existence was a scalpel conveniently ignored that, in fact, that scalpel had to go through Ms. Madyun's body in order to do its work of freeing the child from its mother. The erasure of Ms. Madyun, the dismissal of her choices about her body and about what was best for her future child, and the court's decision to force her to undergo a significant surgical intervention all evidenced a profound lack of respect for her as a competent adult.

By contrast, in *In Re AC*, a lower court ordered a c-section for a 27-year-old white woman dying of cancer while 26.5 weeks pregnant who had refused the procedure while competent. In that case, both the woman and her child died soon after the surgery, which, according to the facts of the case, was likely to hasten the pregnant woman's death according to her healthcare providers [14]. On appeal, the Court held that the lower court was wrong to order the c-section because a pregnant woman, like any competent adult, has a right to have her care decisions respected and clinicians and courts may not decide to balance her interests against those of the fetus she carries [14].

Most authorities have sided with the appeals court in *In Re AC* which refused to treat pregnant women as lesser simply because they are pregnant. This is not to say that no courts have made questionable and, arguably, unconstitutional decisions about forced care for pregnant women for the supposed benefit of a fetus. However, the default position in U.S. law is that competent adults, pregnant or not, are the ultimate arbiters of what happens to their bodies. Given the long history of the abuse of women's bodies by the medical profession ranging from unconsented experimentation on enslaved women [15], forced sterilizations of poor women and women deemed "feeble-minded [16]," and the use of the threat of prison or a report to child welfare authorities to force cooperation with a provider's treatment plan, the default position that courts and healthcare providers should give pregnant women the same right to decide their own fates as other competent adults is of paramount importance.

5 Declaring Death by Neurologic Criteria During Pregnancy

In addition to these disputes about how the law treats pregnant people, there is a great deal of controversy surrounding declaration of death by neurologic criteria, as discussed throughout this book. Many lay people simply do not accept the idea that death is the proper category for someone who is still breathing and whose heart is still beating, even if these functions are only continuing through the magic of technology.

However, while many controversies between clinicians and families of patients who are determined to be dead by neurologic criteria develop due to objections to the determination (based on religious beliefs, fear about misdiagnosis, concerns about exploitation to free up organs for donation etc.), the disagreement between the hospital and the Munoz family was not about whether she was alive or dead. The hospital had made its determination of death, and the family accepted that determination. There was no space between the two entities on that issue. Thus, in a typical case, if Marlise had not been pregnant, the hospital would have removed her from the ventilator, and her family could have begun their mourning process. Because both the hospital and family agreed Marlise was dead, this chapter will not address the decisionmaking process for a pregnant woman in a vegetative state/unaware wakeful state, or any other circumstances where the relevant question is about sustaining a pregnant person's life.

As there was no disagreement about whether Marlise was dead, there was no legitimacy to the hospital's argument that, by law, it could not discontinue life-sustaining treatment. Simply referencing the actual text of the statute should have put to rest any argument that the law required the hospital's action because keeping Marlise Muñoz on life-support after the declaration of death by neurologic criteria was, by definition, not contemplated by the Act. The statutory language reads:

> "Life-sustaining treatment" means treatment that, based on reasonable medical judgment, *sustains the life of a patient and without which the patient will die.* The term includes both life-sustaining medications and artificial life support, such as mechanical breathing machines, kidney dialysis treatment, and artificially administered nutrition and hydration.... [2] (emphasis added).

A plain language reading of that definition excludes a patient who is already dead because no treatment is life-sustaining in those circumstances. As such, agreement on why the statute was not applicable to the Muñoz case should be straightforward. This is not to say that it should have been a foregone conclusion to withdraw support, but rather that the statute was not a sufficient basis upon which to justify continuation of support for a dead person.

Some might argue that once a pregnant person is dead, the only true patient is the fetus, assuming there has not yet been fetal demise, but even that argument is not wholly persuasive. First, for a fetus to be a patient, the fetus would need to be a person—a position that the U.S. Supreme Court rejected in *Roe v. Wade* in which the majority of the Court held that a fetus is not a constitutional person [17]. Importantly, however, the lack of constitutional personhood does not mean that states have no interest in a fetus; in fact, the Court has said that states have an interest in potential life "from the outset of the pregnancy," while also holding that states could not ban abortion prior to the point of viability and could not ban abortions post-viability without an exception for the life and health of the pregnant woman [18]. With this in mind, in a situation involving a pregnant woman who could legally decide to end a pregnancy if she was competent, as was true of Marlise Munoz at the time that she died, requiring that her body be maintained after death by neurologic criteria to gestate a pregnancy against her wishes and those of their family seems anathema.

To illustrate how this scenario fits within the expectation of current law, imagine a circumstance where a child needs a kidney transplant in order to survive, and the child's mother, who is a perfect match, refuses to offer up her kidney because she has a religious objection to organ donation or because she simply does not want to undergo major surgery and live the remainder of her life with only one kidney. No court could compel the mother to subject herself to surgery and life with one kidney to benefit her child even if many, if not most, people would think parting with her kidney was the right thing to do. The law does not require parents to make bodily sacrifices of this kind on behalf of their children. Now imagine that this parent, whose refusal was no doubt widely condemned in life, experiences a pulmonary embolism, and her healthcare provider declares death by neurologic criteria. Even in death, the remaining parent and the healthcare team could not simply decide to now take the kidney that the dead patient did not offer in life, if the dead parent had not only refused consent to the kidney donation but also had actively opposed it on religious grounds. Even in death, the interests and beliefs that we had in life can still prevail.

Based on the law as described above, if Erik believed that Marlise would have wanted support continued for the sake of the fetus, there would have been no conflict between the family and the hospital and her wishes would have been honored. Unfortunately, because he indicated she would have wanted support discontinued, the hospital's unsupportable understanding of the Texas Advance Directives Act resulted in an excruciating form of limbo for 2 months before the hospital followed her wishes.

Arguing for what should happen if a pregnant woman with a pre-viability fetus dies by any criteria is easier than arguing about what should happen if a pregnant woman dies as determined by neurologic criteria or otherwise and her fetus is viable. However, even in a situation involving a viable fetus, a dead pregnant person, and a determination of personhood for the fetus, which is not the case under prevailing constitutional law, the fetus would be a minor.[4] For a minor, the law presumes that a living, fit, and competent parent is the proper medical decisionmaker unless a court finds otherwise. Thus, in the case of Marlise Muñoz, Erik Muñoz would be the proper decisionmaker for his dead wife and the viable fetus unless the law changed to require appointment of a Guardian ad Litem to represent the interests of the fetus. And, even in that case, the Guardian would need to make the case that it was in the best interest of the fetus to continue gestating in the body of a dead woman.

Finally, there is a strong argument that pregnant or not, all people are entitled to dignity in death. We have laws against desecrating or abusing dead bodies. We have rituals, secular and religious, about how to respect the dead. We allow people to assign healthcare proxies and write advance directives and living wills to express their wishes about what should happen to them when they are incapacitated upon their death, and we expect those wishes to be carried out. There is no dignity in a woman having her dead body hooked up to machines in a hospital against her wishes simply because she is pregnant while her family postpones their mourning to fight a battle in court to effectuate the preferences of their dead loved one.

6 Conclusion

In life and in death, pregnant people deserve dignity and respect and should have the same rights as people who are not pregnant; concern for the potential life of a fetus should not vitiate these rights. This is especially true because women's bodies have long been subject to abuse by the law and in medicine. Decisionmaking for a pregnant person should mirror decisionmaking for people who are not pregnant, and courts, lawmakers, and healthcare providers should feel bound to honor decisions that a pregnant person made while competent if that person loses capacity to make their own decisions and even in the extreme circumstance of death. If a pregnant person has no advance directive, the opinions of those who knew them best and who that person or the law has legally designated as surrogate decisionmakers must be valued. The status of being pregnant does not divest a pregnant person of their worth. The presence of a fetus in anyone's body, especially a fetus that is not viable, does not make that person incompetent. Life at what cost is a decision that individuals make, not legislators, physicians, or healthcare providers. Consequently, even in death, the body of a dead pregnant person should be treated with the respect that we give to everyone in death, which includes not treating that body as a receptacle available for gestation, experimentation, or desecration.

[4] Though the Supreme Court overruled *Roe v Wade in Dobbs*, the Court did not go so far as to declare a fetus to be a constitutional person.

References

1. Eunjung Cha A, Roubein R. Fetal viability is at the center of Mississippi abortion case. Here's why. Wash. Post 1 Dec. 021. https://www.washingtonpost.com/health/2021/12/01/what-is-viability/#S3YM3REI6VF5RKO3RG4WPOZYDM-2.
2. Texas Advance Directives Act, Title 2, Subtitle H, §166.049.
3. Fernandez M and Eckholm E. Pregnant, and forced to stay on life support. N.Y. Times, 10 Jan. 2014.
4. Haimowitz R. 62 Days. 2018.
5. Salazar A. Bill seeks to give women autonomy regarding end-of-life decisions. San Antonio Current, 12 Mar 2015. https://www.sacurrent.com/news/bill-seeks-to-give-women-autonomy-regarding-end-of-life-decisions-2407380.
6. Goodwin M. "Marlise's Law": Protecting the Autonomy and Dignity of Brain-Dead Pregnant Women, Harv. L.: Bill of Health 2015. https://blog.petrieflom.law.harvard.edu/2015/03/16/marlises-law-protecting-the-autonomy-and-dignity-of-brain-dead-pregnant-women/#:~:text=Marlise's%20Law%20recognizes%20a%20woman's,having%20her%20personal%20decision%20respected. Accessed 5 May 2022.
7. H.B. 1901, 84th Gen. Assemb., Reg. Sess. (Tx. 2015).
8. Caplan A, Pope TM. Pregnant and Dead in Texas: A Bad law, Badly Interpreted. L.A. Times. Jan. 16, 2014. https://www.latimes.com/opinion/la-xpm-2014-jan-16-la-oe-caplan-pope-texas-pregnancy-life-support-20140116-story.html.
9. Guttmacher Inst. Substance Use During Pregnancy. May 1, 2022. https://www.guttmacher.org/state-policy/explore/substance-use-during-pregnancy. Accessed 16 May 2022.
10. Faherty L, Kranz A, Russell-Fritch J, et al. Association of punitive and reporting state policies related to substance use in pregnancy with rates of neonatal abstinence syndrome. JAMA Netw Open. 2019;2(11):e1914078.
11. Kitroeff N, Silver-Greenberg J. Pregnancy Discrimination is Rampant Inside America's Biggest Companies. N.Y. Times, 2-8-2019. https://www.nytimes.com/interactive/2018/06/15/business/pregnancy-discrimination.html
12. ACLU. Coercive and Punitive Governmental Responses to Women's Conduct During Pregnancy. https://www.aclu.org/other/coercive-and-punitive-governmental-responses-womens-conduct-during-pregnancy. Accessed 16 May 2022.
13. In re Madyun, 114 Daily Wash. L. Rptr. 2233 (D.C. Super. Ct. 1986), reprinted in In re A.C., 573 A.2d 1234, 1262 (D.C. 1990).
14. In re A.C., 573 A.2d 1234, 1240 (D.C. 1990).
15. Khabele D, Holcomb K, Connors N, Bradley L. A perspective on James Marion Sims, Md, and antiblack racism in obstetrics and gynecology. J Minimally Invasive Gyn. 2021;28(2):153–5.
16. Roberts D. Killing the Black Body Random House; 1997.
17. Roe v. Wade, 410 U.S. 113; 1973.
18. Planned Parenthood of Southeastern Pennsylvania v. Casey, 505 US 833, 846; 1992.

Part V

Religious Issues

Christian Perspectives on Death by Neurologic Criteria

Courtney S. Campbell

Christian theologians, ethicists, and ecclesiastical leaders have engaged in debates about the criteria for determining death well before the Harvard Ad Hoc Committee report and in the policy processes that endorsed it, culminating in the Uniform Determination of Death Act [1, 2]. In its 1981 landmark report, *Defining Death*, the President's Commission observed that Christian theological beliefs that "the human essence or soul departs at the moment of death is not inconsistent with the establishment, through neurologic examination, of the time when death occurs." The Commission further maintained that "a statute incorporating a brain-based standard is accepted by theologians of all backgrounds" [3, 4].[1] However, theological writing has recently mirrored contemporary scientific, philosophical, and policy controversy about the validity of death determination by neurologic criteria exemplified by sharp critiques especially within the Roman Catholic tradition, calling into question the theological consensus cited by the Commission. This chapter begins by identifying five theological and moral interests that Christian religious traditions (Roman Catholic, Protestant, Orthodox) have in debates regarding the criteria for the determination of death.

[1] For a critique of this coherence of theology and neurologic criteria, see Byrne PA, O'Reilly S, Quay PM. Brain Death—An Opposing Viewpoint," *JAMA* 1979; 242: 1985–1990; Byrne PA, O'Reilly S, Quay PM, Salsich Jr. PW. Brain Death – The Patient, the Physician, and Society. *Gonzaga Law Review* 1982; 18:429–516.

C. S. Campbell (✉)
Program in Medical Humanities, Oregon State University, Corvallis, OR, USA
e-mail: ccampbell@oregonstate.edu

A. Lewis, J. L. Bernat (eds.), *Death Determination by Neurologic Criteria*, Advances in Neuroethics, https://doi.org/10.1007/978-3-031-15947-3_25

1 The Christian Stake in the Determination of Death Controversy

Christians have a stake in the determination of death debate because of its implications for understanding the *nature of the self*, or what Pope John Paul II referred to as a "sound anthropology" [5]. In general, theological teaching in Orthodox, Protestant, and Roman Catholic traditions interprets the self as an ensouled body or embodied soul, with "death" signifying the separation of the soul from the body [6]. This metaphysical and moral anthropology means Christian perspectives primarily engage issues in determining death at what James Bernat designates as the "concept" of death [7]. Moreover, Christian concepts of death contain implications for discussion of "criteria" for death since the cessation of specific bodily functions, such as respiration, circulation, somatic integration, or consciousness, is viewed as an indicator of dis-ensoulment.

Insofar as all Christian traditions acknowledge that judgments that the soul has left the body cannot be empirically substantiated, the determination of death debate requires Christian articulation of the *relationship of religious convictions and medical science.* Christian perspectives regarding death have a significant stake in the integrity of the medical profession and the comprehensiveness and reliability of the empirical methods and clinical tests used by physicians to assess the biological signifiers of death. Christian teaching beginning with Pope Pius XII in 1957 expresses respect for the autonomy and independent integrity of the medical sciences insofar as determining that the criteria of death have been met in any particular person is a medical judgment [2, 6]. However, Christian scholars acknowledge the limitations inherent in relying on finite and fallible physicians to make such determinations with invariable accuracy and likewise resist a scientific reductionism that identifies the person with a particular function or organ such as the brain.

Christian discussion of determining death manifests an important stake in the moral value of *organ donation* and thereby in the integrity of the *dead donor rule.* Organ donation is morally evaluated by most Christian denominations as a praiseworthy "gift" of life to other persons that displays core virtues such as neighbor-love, altruism, sacrifice, and solidarity with strangers [7]. Christian ethicists have nonetheless stressed that the question of determining death must be separated from the question of organ donation; revisions to the criteria for determining death should be deliberated upon and decided upon independent of the impact of such revisions on retrieving organs from the deceased. The "ethical warning" articulated by the pioneering Christian bioethicist Paul Ramsey in the late 1960s against revising the criteria for determining death for the purpose of facilitating transplantation continues to inform contemporary Christian perspectives [8]. The social need for transplantable organs should not compromise the integrity of the criteria, procedures, or judgments for determining death. The current critiques of the scientific basis and legitimacy of neurologic criteria for determining death have led some scholars, especially in the Roman Catholic tradition, to call for ecclesiastical prohibitions on organ donation [9].

Christian traditions distinguish criteria for determining death not only from organ donation but also from the moral perplexities and the moral integrity of *end-of-life care and treatment decisions.* The permissibility of refusing or discontinuing medically futile interventions in most Christian traditions (controversy persists about discontinuing feeding tubes) presumes the patient is alive; a judgment that a person is imminently dying is different than a judgment that they are dead. Revisions to the criteria for determining death should not usurp ethical choices and values of patients (or proxies) in discontinuing life-prolonging treatments. Should these issues be conflated, some Christians contend that applying criteria for determining death becomes a guise for hastening death and sanctioning euthanasia.

Christian engagement with the concept and criteria of death also reflects decisions about *moral authorities* for Christian communities. *Scripture* provides guidance both on theological anthropology, such as the symbol of the human person as created "in the image of God" (*imago Dei*), and on essential somatic functions, such as respiratory activity (see Table 1). Scriptural texts and analogies are supplemented by *moral tradition* that incorporates reflection on the historical experience of religious communities over time. This may encompass hagiographies of the saints, ecclesiastical councils, papal encyclicals, and prophetic teachings of leaders. Catholic scholars divided on the status of neurologic criteria for determining death acknowledge the authority of the 1312 ecclesiastical Council of Vienne for understanding the human soul [10, 11]. Christian traditions also recognize the *moral purposes of law* in guiding decisions. The Catholic commitment to "natural" law necessitates and sanctions empirical inquiry in understanding nature and the human body, while Protestant theology emphasizes the role of law in preserving social order (a purpose embedded in arguments for "uniform" medical criteria and methods in determining death) and the pedagogical purpose of law in teaching social values, such as solidarity with strangers. Finally, religious narratives and personal moral experience disclose the importance of *practical moral wisdom,* a deliberative process of equilibrium between theological and ethical principles and concrete

Table 1 Scriptural passages in Christian scholarship relevant to the concept of death and determination of death by neurologic criteria

Biblical passage	Content
Genesis 1:26	Human beings are created in the image of god
Genesis 2:7	Bestowal of breath makes persons living beings
Genesis 25:8	Abraham dies following his last breath
Genesis 35:18	Rachel names her child with her last breath
Genesis 35:28	Jacob breaths his last breath and dies
1 kings 17:17	A child dies as there is no breath left in him
Job 33:4	The breath of god gives me life
Ecclesiastes 12:7	The spirit and body separate upon death
Ezekiel 37:5	God causes breath to enter dry bones and they live
Luke 23:46, John 19:30	Jesus ceased to breathe and died
John 11:25–16	Jesus promises a resurrection and eternal life
John 19:34	Blood and water issue from the dead body of Jesus
James 2:26	The body without the spirit is dead

situations of moral choice. Practical moral wisdom aims to provide sufficient moral clarity and moral certainty even in the context of scientific uncertainty. Circumstances of contested declarations of death, such as those presented in the Jahi McMath narrative [12], will necessarily inform Christian views on neurologic criteria for determining death.

Having identified vital Christian theological, anthropological, and ethical interests in determining death, I now discuss debates in Catholic, Protestant, and Orthodox traditions prompted by current disputes in scientific, medical, and policy communities regarding the legitimacy of neurologic criteria for determining death.

2 Contemporary Controversies in Roman Catholicism

Roman Catholicism has engaged in the most extensive inquiry into the determination of death of any Christian tradition. The Vatican Pontifical Academy of Sciences (PAS) convened four study groups—1985, 1989, 2005, and 2006—at the behest of Pope John Paul II and Pope Benedict XVI to examine neurologic criteria of death. These forums disclosed deep substantive differences among Catholic physicians, scientists, scholars, and theologians [11, 13].[2] More recently, the Catholic University of America convened a symposium to articulate a "consistent Roman Catholic position" on the validity of neurologic criteria for determining death [14]. A presumed stable consensus has given way to a "crisis" in Catholicism over the status of neurologic criteria for death [15].

Pope John Paul II presented the most authoritative magisterium statement on determining death by neurologic criteria in a 2000 address to the 18th International Congress of the Transplantation Society [5]. This address is considered a "watershed" moment in Catholic teaching on the criteria for determining death and provides an inescapable conceptual framework for both affirming and criticizing the validity of neurologic criteria within subsequent Roman Catholic scholarship [16–18]. It offers a fascinating illustration of the intertwining of theological, moral, anthropological, ecclesiastical, and empirical claims in Catholic understandings of the concept and criteria for determining death.

The context for John Paul's comments is the *ethical* legitimacy of organ donation, and he begins with an ethical affirmation of the dead donor rule: "vital organs which occur singly in the body can be removed only after death, that is, from the body of someone who is certainly dead." The dead donor rule presupposes the resolution of an *empirical* issue that John Paul II presents as his framing question in the dispute, "when can a person be considered dead with complete certainty?" [5]. His analysis of this question initially invokes a *philosophic-theological concept* of death and the limits of empirical observation: death is constituted by the "total disintegration of that unitary and integrated whole that is the personal self. It results from the

[2] It is beyond the scope of this overview to discuss the details of the Pontifical Academy of Sciences (PAS) working groups, but their magisterial authority and public transparency are of considerable dispute among Catholic scholars [12].

separation of the life-principle (or soul) from the corporal reality of the person." The timing of dis-ensoulment is not, however, verifiable by any empirical scientific method, meaning that finite fallible physicians must discern in the body "biological signs" that death has occurred [5].

John Paul II then turned from the concept to the criteria of death, observing that the criteria should provide a "scientifically secure means" for determining that the "biological sign" of total disintegration of the holistic self has occurred. Neurologic criteria verify this disintegration through "the complete and irreversible cessation of all brain activity (in the cerebrum, cerebellum, and brain stem)." The Pope acknowledges that this is a *medical empirical* judgment about which the church has no expertise. However, informed by an ethical commitment to respect human dignity, the church assumes a "Gospel duty to compare the data offered by medical science with the Christian understanding of the unity of the person" [5]. This comparative analysis permits a *conditional ecclesiastical* judgment that neurologic criteria, "if rigorously applied, does not seem to conflict with the essential elements of a sound anthropology" [5, 17, 19]. The Pope contends that this determination can meet, if not the demanding epistemic standard of "complete certainty" that initially framed his analysis, an epistemic standard of *practical moral wisdom* of "moral certainty," that is, the highest level of certainty for moral choice, a level of assurance beyond a reasonable doubt. This returns John Paul II to his initial normative issue: "This moral certainty is considered the necessary and sufficient basis for an ethically correct course of action," that is, organ donation from a cadaveric donor is ethically permissible [5].

While John Paul II's comments have been interpreted as an "unambiguous endorsement" of neurologic criteria for determining death [16], subsequent Catholic commentary has highlighted several equivocal assertions to maintain that the question of criteria for determining death is far from settled theologically. Several theological and ethical flashpoints can be identified.

1. *Moral Authority of Papal Teaching.* While fidelity to papal teaching is advanced by both advocates and critics of neurologic criteria [10, 11], Catholic scholars argue about its scope and bindingness. John M. Haas of the National Catholic Bioethics Center contends that the Catholic magisterium has articulated a consistent and definitive voice over the past three decades on the acceptability of neurologic criteria and of organ transplantation [17]. However, other scholars contend that John Paul II presented only a "conditional approval [of neurologic criteria] pending the fulfillment of several specific presuppositions or conditions" that are not satisfied with the requisite certainty [18, 20]. Others understand the papal teaching to provide pastoral guidance to Catholic medical practitioners and patients rather than expressing incontrovertible principles for policy issues. These views invite a broad-ranging scientific and theological debate even as they limit the scope of papal teaching.
2. *Scientific Controversy Erodes Moral Certainty.* John Paul II specified that determining death by neurologic criteria requires "clearly determined parameters commonly held by the international scientific community" [5], an ecclesiastical

invitation to ongoing research and the generation of empirical knowledge that could support scientific critiques and disagreements on death by neurologic criteria. The analyses of Dr. D. Alan Shewmon, a participant in the working groups for both the Pontifical Academy of Sciences and the Pontifical Academy of Life, that forms of integrative somatic functioning persist subsequent to a determination of death by neurologic criteria, have been a notable catalyst for Catholic (and bioethical) debate [21, 22]. Shewmon's claims of the persistence of integrating activity are cited as "persuasive and irrefutable evidence" by virtually every Catholic critic of neurologic criteria to the effect that death by neurologic criteria does not constitute the death of a person [18]. Shewmon defends his position as entirely consistent with the position of John Paul II, given papal considerations regarding the necessity of scientific consensus, the open question of whether death by neurologic criteria is compatible with embodied anthropology, and the standard of certitude [11].

Consequently, in contrast to the papal specification that "clear" and "common" scientific parameters warrant epistemic and moral certainty in determining death by neurologic criteria, as articulated by one Catholic scholar, "doubt has become an international consensus" [23, 24]. This has created a parallel dispute among Catholic scholars over interpretations of medical evidence [19]. Catholic scholars (as well as policy commissions) defending death by neurologic criteria necessarily must account for Shewmon's empirical claims and their ethical implications in their arguments [25]. These have included philosophical critiques of Shewmon's metaphysics of the soul, epistemic interpretations of integrative functioning, and the practical ethics of organ transplantation (developed more fully below) [17, 26, 27]. Ultimately, debate and resolution of the meaning of empirical evidence about persistent integrative functioning reside within the competency of the medical profession, not ecclesiastical leadership or scholarship. However, in the absence of papal stipulations for common and international "parameters" for determining death, Catholic physicians may no longer possess moral certainty for their decisions.

3. *The Concept of Death and (Dis)integration.* John Paul II's double equivocation on the relationship between the concept of death and the criteria for death has likewise seemed to critics of death by neurologic criteria to reopen rather than settle the issue within Catholicism. The papal claim that the criteria "do not seem to conflict" with anthropology is stated in a negative form and is not a ringing affirmation of coherence, and the language of "seem" opens the possibility of a conflict. As the papal concept of death is comprised of the "total" disintegration of the integrated, unified self, these hesitations have been a catalyst for subsequent Catholic commentary on the meanings of integration. Shewmon distinguishes between "levels" and "types" of integration, the latter of which encompass "life-constituting" and "life-sustaining" integration. He argues that brain-based somatic integration is "indispensable" for life-sustaining but not for life-constituting integrative activities; indeed, "the constitutive integration minimally needed for the existence of a rationally ensouled human organism is *entirely* non-brain-mediated" [11]. Other scholars, while agreeing with Shewmon

that somatic and cellular functions persist following a determination of death by neurologic criteria, argue against his conclusion by interpreting integration as an "intercommunicative" activity coordinated by the brain that maintains the unitary wholeness of the body [26]. Still others distinguish between "coordination" and "integration," holding that even "complex coordination" between localized cells, tissues, organs, and organ systems is insufficient to establish somatic integration of the organism as a whole. This requires "global, autonomous integration of vital functions" mediated by the brain [27, 28].

4. *Organ Donation and the Dead Donor Rule.* Pope Francis has affirmed that organ donation is a supernal "gift of life" and a "witness of love" and a core feature of a "culture of donation" as part of the Catholic witness to life [29]. Given the scientific dispute and its implications for a practice based on moral certainty, critics of death by neurologic criteria have argued that the moral safeguards of transparency and informed consent are insufficient ethical justification for organ donation. Some critics support organ transplantation in principle but contend the gift ethic and donative culture have been supplanted by utilitarian commitments to maximize social welfare, or alternatively have echoed Ramsey's prescient moral warning against conflating determining death with societal needs for transplantable organs [11, 18]. Other scholars have proposed a moratorium on donation, maintaining that until the empirical preconditions for moral certainty are satisfied, "the Catholic Church should call for a halt on vital organ transplantation" [9]. Indeed, the prevalence of moral uncertainty leads some scholars to invoke an earlier encyclical of John Paul II that raised the specter of a connection between donation and euthanasia when "organs are removed without respecting objective and adequate criteria which verify the death of the donor" [30].

   While Catholic scholars in general affirm the moral necessity of the dead donor rule [14], empirical disputes and moral uncertainty have culminated in divergent applications. Catholic physician-bioethicist Edmund Pellegrino, chair of the President's Council on Bioethics in 2005–2009 when it proposed a different scientific rationale for death by neurologic criteria, argued that using circulatory-respiratory criteria provided greater moral assurance of death. Pellegrino acknowledged that relying on circulatory-respiratory criteria to definitively meet the standards of the dead donor rule could diminish organ availability but maintained fidelity to patient welfare and the intrinsic value of every person warranted this moral trade-off [31]. Shewmon has gone the farthest in suggesting the abandonment of the dead donor rule, contending the ontological question of the timing of patient death presumed by the rule is mistaken and the issue of donation is instead a moral matter. Rather than asking whether the patient is dead in order to begin with transplant procedures, Shewmon proposes a non-maleficence reframing of the issue: "When can organs X, Y, Z . . . be removed without causing or hastening death or harming the patient in any way?" Organ donation thereby does not require the death of the prospective donor, but rather refrainment from harm. However, Shewmon expresses reservations about the practical implementation of this pre-death permission of donation [32, 33].

5. *The Clear Conscience.* The concept of moral certainty gives ethical permission for Catholic physicians participating in determining death or in organ retrieval, as well as prospective Catholic organ donors and recipients, to act with the integrity of conscience. In 2008, Pope Benedict XVI emphasized that in questions pertaining to determining death, "there cannot be the slightest suspicion of arbitration [arbitrariness] and where certainty has not been attained, the principle of precaution must prevail" [34]. In circumstances of uncertainty, the moral presumption is to err on the side of continued life since to make a mistake and presume death is an irrevocable and irreversible determination. In this context of scientific and epistemic uncertainty, some Catholic scholars argue there is an "obligation" to treat persons declared brain dead "*as if* they were living human beings" [19, 35]. If the empirical conditions presupposed by moral certainty are not satisfied, the presumption in favor of life would seem to require using circulatory-respiratory criteria to determine death.

   Furthermore, in the absence of definitive empirical evidence and correlative moral certainty, some arguments maintain that Catholic physicians or patients "are not morally obligated" to participate in determinations of death by neurologic criteria or organ transplantation [20]. Notably, some lay Catholics have expressed reticence about either donating or receiving a transplanted organ, lest they be complicit in hastening death [17]. The crisis in Catholic teaching about the status of death determination by neurologic criteria is not simply a scholarly dispute but bears substantive practical and pastoral implications.

Catholic diversity regarding the validity of death by neurologic criteria is conditioned by the tradition's reliance on the professional resolution of scientific uncertainties. A Catholic moral infrastructure regarding organ donation, end-of-life care, and conscience is challenged by contemporary scientific and Catholic disputes about death by neurological criteria. Emerging Catholic opposition to death by neurologic criteria may necessitate a more extensive religious exemption or conscientious objection clause than recognized in current policy. Given the prevalence of Catholics in the general population, the religious exemption statute adopted by New Jersey primarily in recognition of Orthodox Jewish objections to neurologic criteria for death may need to be embraced in other states. If some Catholics are unable to accept in clear conscience an account of death by neurologic criteria, this objection should receive a religious accommodation in the law [36, 37].

3 Contemporary Protestant Perspectives

Protestant theologians such as Paul Ramsey and Stanley Hauerwas vitally informed bioethical discussion regarding death by neurologic criteria prior to the 1981 President's Commission report. Ramsey developed several arguments for, and moral warnings about, "updating" the criteria for determining death to encompass neurologic criteria, including a prescient observation of the prospect of "widespread misunderstanding" among professionals and the public [6]. Framing his analysis through the "biblical norm of fidelity to covenant" [6, 38], Ramsey vigorously argued for conceptually partitioning the issues of caring for dying persons, criteria for determining death, and organ donation processes. He also invoked metaphors from constitutional political philosophy, including "separation of powers" and "checks and balances," to specify that the central professional covenant of loyalty to the primary patient must not be compromised for the needs of other persons [6, 39]. Stanley Hauerwas agreed with Ramsey that Christian thought did not affirm a theological preference regarding circulatory-respiratory or neurologic criteria for determining death, but he maintained the Christian concept of death possessed a prophetic influence, formulating "the kind of questions that *should* be asked" about any definition or criteria of death [40].

While Protestant faith traditions commonly follow Hauerwas's observation in articulating a concept of death situated within the Christian salvation narrative, recent scholarship has not engaged the controversy over death by neurologic criteria in the sustained manner found in the Catholic tradition. In general, Protestant emphases on personal freedom and liberty of conscience are institutionalized through denominational autonomy on many bioethical questions. Notably, a compilation of bioethical teachings of 54 Christian denominations or professional organizations indicates only seven have articulated positions on the criteria for determining death ([41]; see Table 2). Fourteen denominations have formulated position statements or developed congregational resources on organ donation that presume but do not address the acceptability of determining death by neurologic criteria [42]. Moreover, Protestant ecclesiology does not institutionalize a definitive teaching authority [43]; hence, in the context of ecclesiastical bioethical silence, principles of Christian liberty and self-determination prevail. It is possible however to identify central issues embedded in Protestant denominational statements and congregational resources addressing neurologic criteria.

Scripture and Soul. A foundation of historical Protestantism is *sola scriptura*, that is, the word of God revealed in Scripture is essential to both salvation and moral life. Protestant discussions of neurologic criteria for death make extensive use of Scripture as the framing teaching authority for the concept of death. The scriptural references frequently invoked identify "breath" or respiration as the biological sign of the presence of an immortal soul in the human person (see Table 1). The biblical symbol of "breath," or its absence, contains a bivalent meaning. Some Protestant traditions combine the symbol with a stewardship theology that renders the absence of spontaneous respiration consistent with death by neurologic criteria. However, the same passages lead biblicist or fundamentalist denominations and scholars to support circulatory-respiratory criteria as the definitive marker for the presence or absence of the soul and consequently to express suspicion about death as determined by neurologic criteria [44]. Furthermore, when scripture is the sole moral authority, postures of ecclesiastical silence are interpreted as symbolizing de facto religious deference to fallible medical and scientific authority [45]. Relying on the same epistemic uncertainty as Catholic critics, Evangelical scholar Michael Munoz argues that a person determined dead by neurologic criteria is "dying" and will "eventually die." That is, neurologic criteria do not constitute the death of the

Table 2 Denominational statements on the determination of death or organ donation

Denomination	Statement on determination of death	Statement on organ donation	Congregational resources on organ donation
African Methodist Episcopal		Y	
Assemblies of God	Y	Y	Y
Christian Church (Disciples of Christ)		Y	
Christian Medical and Dental Association	Y	Y	
Church of Jesus Christ of Latter-Day Saints (Mormon)		Y	
Church of the Lutheran Brethren	Y		
Church of the Nazarene		Y	
Episcopal Church		Y	
Evangelical Lutheran Church in America		Y	Y
Lutheran Church-Missouri Synod		Y	Y
National Association of Evangelicals	Y		
Orthodox Church in America		Y	Y
Presbyterian Church USA	Y	Y	Y
Roman Catholic Church	Y	Y	Y
Southern Baptist Convention			Y
United Church of Christ			Y
United Methodist Church		Y	
Wesleyan Church		Y	
Wisconsin Evangelical Lutheran Synod	Y		

person. This interpretation leads Munoz to a "paradox" about whether there is a "biblical" warrant for organ donation [45]. The account also presents a metaphysical puzzle as it implies that ensoulment is conditioned by the application or discontinuation of technologies.

Medical Stewardship. Even though Scripture is the definitive word for human salvation, Protestant denominations commonly grant epistemic and moral authority to the medical profession regarding non-salvific matters of human health. As medicine is bestowed with a divine stewardship of caring for and healing of the human body, the Protestant Christian has a responsibility to interpret and weigh scientific evidence in ethical choices. In the theology of stewardship articulated by the Church of the Lutheran Brethren, though medical or philosophical accounts do not display the "true meaning of death" in Scripture, "Christians should regard the medical definition, 'brain death,' as the best available indicator of death" [46]. This stewardship accommodates shifts in Christian perspectives on neurologic criteria in coordination with advances in medical technology.

Dignity of the Person. The primary scriptural symbol expressing the dignity of the person is creation in "the image of God" (*imago Dei*). This symbol offers a scriptural basis for different views on the criteria for death as illustrated in three denominational resources. An interpretative essay on the Assemblies of God website contrasts the theological conviction of persons as *imago Dei* with bioethical definitions of personhood presumed in determinations of death by neurologic criteria [47]. In contrast, the Lutheran Church-Missouri Synod (LCMS) invokes the same symbol in expressing concern that exclusive reliance on cardiopulmonary criteria for death reduces the person to "animal and vital functions" and thereby fails to do justice to personal dignity. The LCMS contends that "the criterion of 'brain death' has contributed to a more constructive discussion" regarding end-of-life care [48]. The Christian Medical and Dental Association invokes the *imago Dei* symbol to support a direct equivalence between scriptural and scientific concepts of death: "The whole-brain definition and criterion of death is consistent with both the traditional concept of death and the Biblical definition of physical death" [49].

Dying Well and Opposition to Euthanasia. Many Protestant perspectives follow Hauerwas in situating the concept and criteria for death within narratives disclosing Christian values in dying well, particularly the sanctity and dignity of human life. These moral narratives invariably make a connection between the discourse on criteria for determining death and opposition to medical assisted suicide and euthanasia. Neurologic criteria per se are generally not considered to manifest a death-hastening judgment, but they are not viewed as "value-neutral" either. A website oriented by evangelical Protestant bioethics includes "brain death" in the terminology pertaining to "taking life," along with abortion, euthanasia, infanticide, and suicide [50].

Relational Life and Higher Brain Death. A central feature of the *imago Dei* symbol is that human persons are necessarily relational beings; relationship displays a distinctive character to human life. Conversely, the irreversible loss of relationship capacity can be a theological marker of death. This relational concept of death is associated in one congregational resource with a "medical judgment" of irretrievable loss of cerebral function [51]. The relational concept of death necessarily raises issues of human capacities conditioned by consciousness, often embedded in debates on "higher" brain death. Protestant scholars have generally criticized neocortical criteria for death by appealing to the concept of death: a neocortical account presupposes a mind/body dualism that neglects the integrated unity of the embodied person. Physician-ethicist Allen Roberts asserts: "For Christians to embrace a higher brain criterion for death requires the embrace of a lethal anthropological heresy" [52]. The dualistic anthropology erodes the sanctity and dignity of life and places vulnerable persons, such as those in a vegetative state/unaware-wakeful state (VS/UWS), in danger of being redefined as dead. However, the Protestant commitment to Christian freedom could support individual or communal advocacy of a higher brain standard for death. This perspective overlaps with distinctive claims in Orthodox Christian teaching.

4 Orthodox Perspectives

Orthodox Christianity similarly situates discourse on criteria for determining death within narrative accounts of human nature, the meanings of life, and concepts of death. Orthodox teaching interprets the nature of the human self through the *imago Dei* symbol, with the divine image most disclosed in relationships mirroring the mutuality of love in the divine Trinity. Humans are ensouled beings, and death represents a rupture of the unity and bond between soul and body. Orthodox ethicist Fr. Nikolaos Hatzinikolaou indicates that Orthodox Christians approach the "moment of biological death" with sentiments of "awe, respect, humbleness, and a deep sense of mystery." The characteristic of "mystery" precludes any ecclesiastical endorsement of specific criteria for determining death: "the Orthodox Christian Church avoids clear-cut statements that identify death with the cessation of the brain, cardiac, or any other function." Indeed, "the meticulous search for the exact moment of death has no special interest and meaning" in Orthodox tradition [53].

Orthodox teaching consequently expresses denominational, professional, and personal latitude regarding criteria for determining death. Stanley S. Harakas, the pioneering Orthodox scholar in medical ethics, maintained that Orthodox teaching could support both the circulatory-respiratory criterion and the higher-brain criterion for death: "Orthodox Christians once determined death by the cessation of heart and breathing activity. Recent Orthodox thinkers do not object to defining physical death in terms of brain death. Generally, the Orthodox recognize death as the cessation of higher human capacities concurrent with the demise of the cerebral cortex, even though lower brain stem activities may remain" [54].

This appeal to "higher" capacities is echoed by other Orthodox scholars in the context of an ethics of care for persons in a VS/UWS, with a shared concern to refrain from *overtreatment.* Fr. Hatzinikolaou contends that "in certain cases, when an individual has lost permanently his awareness, consciousness, cognition, volition, emotions, or any of the necessary brain functions, his body fails to manifest the soul, even if some of its functions are supported mechanically" [53]. Fr. John Breck argues that what is distinctively human about the gift of life, including the capacity for relationships, is dependent on cerebral cortical brain functions (not functions coordinated by the brain stem). Breck proposes that "in cases where accident or disease has irreversibly destroyed higher brain or neocortical activity, we must conclude the person in question is no longer alive" [55]. Breck acknowledges the inadequacy of clinical tests to reliably determine this concept of death; hence, what is theologically valid in principle is in practice constrained by the inherent mystery of death, which eludes precise medical judgment.

An interpretative document on the website of the Orthodox Church in America by Protodeacon Basil Andruchow, a member of the OCA's Medical Ethics Commission, likewise addresses the ethics of overtreatment of patients in a PS/UWS. Andruchow contends that brain activity within the cerebral cortex "defines the human condition," as it encompasses capacities for decision-making, organization, and speculative activities. He observes that "in the absence of any activity in this region, the attending physician(s) will declare the person as 'brain dead,'" and then offers an ecclesiastical normative position: "Our Church recognizes and agrees with this conclusion" [56]. While Andruchow's reasoning could be taken to entail redefining patients in VS/UWS as "dead" by the absence of cerebral cortical activity, its practical import is emphatically that organ support should be discontinued when the life force is no longer present in a biological organism.

The relational interpretation of the divine image in Orthodox teaching and scholarly focus on neocortical brain activity might imply the theological acceptability of higher brain criteria for death. However, Orthodox interpretations are focused on a practical context, *caring* for patients in a VS/UWS, not on a philosophical or conceptual argument stipulating that such persons are dead. Indeed, Allen Roberts contends that endorsement of a higher brain standard would be contrary to two millennia of Orthodox Christian teaching as reflected in Scripture, patristic teaching, ecumenical councils, and contemporary theologies [57]. The status of determining death by neurologic criteria invites continuing discussion in Orthodox Christianity.

5 Conclusion

Christian scholars, ecclesiastical leaders, and faith traditions have a legacy of informing and contributing to professional and policy discourse on determining death by neurologic criteria. This legacy of engagement is attributable to important theological and ethical interests, including the nature of the self, the relationship between religious thought and science, the integrity of organ donation, ethical end-of-life care, and moral authority in Christian life. The diversity of Christian perspectives on death by neurologic criteria is a function of both different communal interpretations of these interests and contemporary scientific controversies. The prevalence and intensity of this diversity may require legal accommodations for persons or traditions that hold conscientious objections to death by neurologic criteria.

References

1. Ad Hoc Committee of the Harvard Medical School. A definition of irreversible coma: report of the ad hoc Committee of the Harvard Medical School to Examine the Definition of Brain Death. JAMA. 1968;205:337–40.
2. Pius XII. Address to an international congress of anesthesiologists. Natl Catholic Bioethics Quar. 2002;2:309–14.
3. President's Commission for the Study of Ethical Problems in Medicine and Biomedical and Behavioral Research. Defining death: a report on the medical, legal, and ethical issues in the determination of death, vol. 11. Washington, DC: Government Printing Office; 1981. p. 43.
4. Vieth FJ, Fein JM, Tendler MD, et al. Brain death I: a status report on medical and ethical considerations. JAMA. 1977;238:1651–5.
5. John Paul II. Address to the 18th International Congress of the Transplantation Society [Internet]. Vatican.va. 2000. http://www.vatican.va/holy_ father/john_paul_ii/speeches/2000/jul-sep/documents/hf_jp-ii_spe_20000829_transplants_en.html. Accessed 1 July 2021.

6. Ramsey P. The patient as person: explorations in medical ethics. New Haven, CT: Yale University Press; 1970.
7. Bernat JL. The definition and criterion of death. Handb Clin Neurol. 2013;118:419–35.
8. May WF. Religious justifications for donating body parts. Hastings Cent Rep. 1985;15:38–42.
9. Alexander I. Humility before new scientific evidence: we no longer have moral certainty that 'brain death' is true death. Linacre Q. 2019;86:314–26.
10. Tonti-Filippini N. You only die twice: Augustine, Aquinas, the Council of Vienne, and death by the brain criterion. Communio. 2011;38:308–25.
11. Shewmon DA. You only die once: why brain death is not the death of a human being. Communio. 2012;39:422–94.
12. Aviv R. What does it mean to die? [internet]. The New Yorkercom, 2018; https://www.newyorker.com/magazine/2018/02/05/what-does-it-mean-to-die Accessed 15 Sept 2018.
13. Pontifical Academy of Sciences. Why the concept of brain death is valid as a definition of death. In: Sanchez Sarondo M, editor. The Signs of Death, The Proceedings of the Working Group of 11-12 September 2006, Scripta Varia, vol. 2007. Vatican City: The Pontifical Academy of Sciences. p. 110. http://www.casinapioiv.va/content/accademia/en/publications/scriptavaria/signsofdeath.html.
14. Moschella M, Condic ML. Symposium on the definition of death: summary statement. J Med Philos. 2016;41:351–61.
15. Jones DA. Loss of faith in brain death: catholic controversy over the determination of death by neurologic criteria. Clin Ethics. 2012;7:133–41.
16. Diamond EF. John Paul II and brain death. Natl Catholic Bioethics Quar. 2007;7:491–7.
17. Haas JM. Catholic teaching regarding the legitimacy of neurologic criteria for the determination of death. Natl Catholic Bioethics Quar. 2011;11:279–99.
18. Nguyen D. Pope John Paul II and the neurologic standard for the determination of death: a critical analysis of his address to the transplantation society. Linacre Q. 2017;84:155–86.
19. Brugger EC. Are brain dead individuals dead? Grounds for reasonable doubt. J Med Philos. 2016;41:329–50.
20. Ostertag K, Karches K. Brain death and the formation of moral conscience. Linacre Q. 2019;86:335–46.
21. Shewmon DA. Chronic "brain death": meta-analysis and conceptual consequences. Neurology. 1998;51:1538–45.
22. Shewmon DA. 'Brainstem death,' 'brain death' and death: A critical reevaluation of the purported equivalence. Issues Law Med. 1998;14:125–45.
23. President's Council on Bioethics. Controversies in the determination of death. Washington, DC; 2008. p. 37–40.
24. Nguyen D. Evolution of the criteria of 'brain death': a critical analysis based on scientific realism and Christian anthropology. Linacre Q. 2019;86:297–313.
25. Verheijde JL, Rady MY, Potts M. Neuroscience and brain death controversies: the elephant in the room. J Relig Health. 2018;57:1745–63.
26. Tonti-Filippini N. Religious and secular death: a parting of the ways. Bioethics. 2012;26:410–21.
27. Condic M. Determination of death: a scientific perspective on biological integration. J Med Philos. 2016;41:257–78.
28. Moschella M. Deconstructing the brain disconnection-brain destruction analogy and clarifying the rationale for the neurologic criterion of death. J Med Philos. 2016;41:279–99.
29. Francis, Pope. Address to the Italian Association for the Donation of Organs, Tissues, and Cells [Internet 2019]. Vatican.va. 2019. https://www.vatican.va/content/francesco/en/speeches/2019/april/documents/papa-francesco_20190413_donazione-organi.html. Accessed 10 Aug 2021.
30. John Paul II, Pope. Evangelium Vitae [Internet] 1995. https://www.vatican.va/content/john-paul-ii/en/encyclicals/documents/hf_jp-ii_enc_25031995_evangelium-vitae.html. Accessed 1 Aug 2021.
31. Pellegrino ED. Personal statement. In: Controversies in the determination of death. president's council on bioethics. Washington, DC; 2008. p. 107–19.

32. Shewmon DA. The dead donor role: lessons from linguistics. Kennedy Inst Ethics J. 2004;14:277–300.
33. Shewmon DA. Constructing the death elephant: a synthetic paradigm shift for the definition, criteria, and tests for death. J Med Philos. 2010;35:256–98.
34. Benedict XVI, Pope. Address to an International Congress organized by the Pontifical Academy for Life [Internet]. Vatican.va 2008. https://www.vatican.va/content/benedict-xvi/en/speeches/2008/november/documents/hf_ben-xvi_spe_20081107_acdlife.pdf. Accessed 31 Aug 2021.
35. Verheijde JL, Potts M. Commentary on the concept of brain death within the catholic bioethical framework. Christ Bioeth. 2010;16:246–56.
36. Olick RS, Braun EA, Potash J. Accommodating religious and moral objections to neurologic death. J Clin Ethics. 2009;20:183–91.
37. Campbell CS. Imposing death: religious witness on brain death. Hastings Cent Rep. 2018;48:S56–9.
38. Moschella M. Brain death and organ donation: a crisis of public trust. Christ Bioeth. 2018;24:133–50.
39. Ramsey P. Updating death. In: Cutler DR, editor. Updating life and death: essays in ethics and medicine. Boston, MA: Beacon Press; 1969. p. 31–54.
40. Hauerwas S. Suffering presence: theological reflections on medicine, the medically handicapped, and the church. Notre Dame, IN: University of Notre Dame Press; 1986. p. 87–99.
41. Christian Biowiki. Definition of death [Internet]. Christianbiowiki.org. http://christianbiowiki.org/wiki/index.php/Definition_of_Death. Accessed 25 July 2021.
42. United Network for Organ Sharing. Theological perspective on organ and tissue donation [Internet]. https://unos.org/transplant/facts/theological-perspective-on-organ-and-tissue-donation/. Accessed 15 Sept 2019.
43. Setta SM, Sheema SD. An explanation and analysis of how world religions formulate their ethical decisions on withdrawing treatment and determining death. Philos Ethics Humanit Med. 2015;10:1–22.
44. Wisconsin Evangelical Lutheran Synod. Life support and brain dead [Internet]. Wels.net. https://wels.net/faq/life-support-and-brain-dead/. Accessed 10 Aug 2021.
45. Munoz MG. Christian bioethics, brain death, and vital organ donation. Christ Bioeth. 2018;24:79–94.
46. Church of the Lutheran Brethren. Euthanasia and assisted suicide. Minneapolis, MN: Faith and Fellowship Publishing; 2010. p. 7.
47. Powell CMH. Being human: how should we define life and personhood? [Internet] News.ag. 2010. https://news.ag.org/en/Features/Being-Human-How-Should-We-Define-Life-and-Personhood. Accessed 1 Aug 2021.
48. Lutheran Church-Missouri Synod. Report on euthanasia with guiding principles [internet]. In: Abbott D, editor. The Lutheran tradition: religious beliefs and health care decisions, vol. 17. Chicago, IL: Park Ridge Center; 2002. https://www.advocatehealth.com/assets/documents/faith/lutheranfinal.pdf. Accessed 1 June 2021.
49. Christian Medical and Dental Association. Position Statements: Death [Internet]. Cmda.org. 2004. https://cmda.org/policy-issues-home/position-statements/. Accessed 22 Aug 2021.
50. Carter J. Basic bioethics: a glossary on taking life [Internet]. Erlc.com. 2018. https://erlc.com/resource-library/articles/basic-bioethics-a-glossary-on-taking-life/. Accessed 6 Aug 2021.
51. Presbyterian church, U. S. A. In life and in death we belong to god: euthanasia, assisted suicide, and end-of-life issues, a study guide [internet]. Presbyterianmission.org, vol. 42. Louisville, KY: Presbyterian Publishing; 1995. https://www.presbyterianmission.org/resource/paper-life-and-death-we-belong-god-euthanasia-assi/. Accessed 6 June 2021
52. Roberts AH II. The higher-brain concept of death: a Christian theological appraisal. Ethics Med. 2017;33:177–91.
53. Hatzinikolaou N. Prolonging life or hindering death? An orthodox perspective on death, dying and euthanasia. Christ Bioeth. 2003;9:187–209.

54. Harakas SS. The Orthodox Christian Tradition: Religious Beliefs and Healthcare Decisions [Internet]. Advocatehealth.com. Chicago, IL: Park Ridge Center; 1999. p. 10. https://www.advocatehealth.com/assets/documents/faith/orthodox_christian.pdf. Accessed 6 Aug 2021
55. Breck J. The sacred gift of life: orthodox Christianity and bioethics. Crestwood/New York: St. Vladimir's Seminary Press; 1997. p. 232–3.
56. Protodeacon Basil Andruchow. Medical bioethics: an orthodox Christian perspective for orthodox Christians [Internet]. OCAorg 2010. https://www.oca.org/parish-ministry/familylife/medical-bioethics-an-orthodox-christian-perspective-for-orthodox-christians. Accessed 10 Aug 2021.
57. Roberts AH II. Eastern orthodox views on a "higher-brain" death criterion: why theology must inform medical ethics. Int J Orthodox Theol. 2017;8:115–33.

Islamic Perspectives on Death by Neurologic Criteria

Aasim I. Padela and Rafaqat Rashid

Increased biomedical capacities have contributed to the medicalization of the dying process globally. For example, in the Middle East, societies which had little to no access to modern healthcare facilities a generation or two ago now have some of the most technologically advanced hospitals in their midst. This newfound access to the powers of modern medicine has certainly improved lives but also has fueled misgiven hopes in the restorative and/or curative powers of biotechnology when individuals near death. For example, experience suggests that the ability to maintain physiological functions via ventilators, vasopressors, and the like motivates family members to pursue aggressive treatment regimens for their loved ones despite diminished odds of success [1]. In this way, the newfound capacities of modern medicine impose upon the Islamic ethico-legal imperative of preserving life to generate greater numbers of Muslims meeting their death in the confines of sterile hospital wards rather than in the warmth of their own homes. Adding to this dynamic is the state of "brain death," which further confounds individuals because their loved ones appear to maintain the traditional markers of life, namely breathing and heartbeat, despite clinicians suggesting that human death has occurred. In response,

A. I. Padela (✉)
Department of Emergency Medicine, Medical College of Wisconsin, Milwaukee, WI, USA

Center for Bioethics and the Medical Humanities, Institute for Health and Equity, Medical College of Wisconsin, Milwaukee, WI, USA

Initiative on Islam and Medicine, Brookfield, WI, USA
e-mail: apadela@mcw.edu

R. Rashid
Al-Balagh Academy, Bradford, UK
e-mail: drrafaqat@albalaghacademy.com

A. Lewis, J. L. Bernat (eds.), *Death Determination by Neurologic Criteria*, Advances in Neuroethics, https://doi.org/10.1007/978-3-031-15947-3_26

some Muslims hold onto their loved ones by requesting continued life-sustaining measures citing religious values, dictates, and edicts that reject "brain death" [2–4]. [1]

This narrative is not unique to Muslims. Other religious communities and indeed even secular voices are critical of brain death, and many around the world decry the medicalization of death and dying. Indeed, this volume is full of narratives challenging the entity and the cultural shift in death and dying that undergirds it. Yet, there are nuances within Muslim responses that deserve dedicated attention. The plurality of religious views on brain death, the variability of Muslim state legislation on the matter, and the different sources of disquiet various Muslim stakeholders express, paint a picture worth examining. This chapter, accordingly, delves into both Islamic perspectives and Muslim experiences surrounding brain death. We begin with empirical insights into how Muslim clinicians, patients, religious leaders, and other stakeholders are challenged by brain death. Next, we detail the dominant Islamic juridical perspectives on brain death and describe their ungirding ethico-legal rationale while also critically appraising these views by pointing out their shortcomings and ambiguities. We end the chapter by proposing an approach to death by neurologic criteria which accounts for the Islamic plurality on the matter and addresses the disquiet Muslim patients, clinicians, and families have with brain death in clinical practice.

1 Muslim Disquiet with Brain Death

As biomedicine has made liminal states between traditional markers of life and death possible, and medicalization of the dying process has become the norm, Muslim clinicians, patients and their surrogate decision-makers, as well as religious scholars have been challenged by the ethical discourse and clinical practices surrounding brain death. Muslim clinicians analyze religious bioethics discourses seeking answers about their ethical duties surrounding this new form of death, Muslim patients and their caregivers wonder what sorts of decisions can be made once this state has been reached, and Islamic jurists debate whether brain death is a legitimate threshold for death within Islamic law. These engagements with brain death reveal substantial concerns about the entity and its associated clinical

[1] We have placed the term brain death in quotations to highlight that the term is a misnomer and controversial. "Brain death" is often used to denote that neurologic criteria for human death have been met. Death here refers to the death of the human being, not of the brain, because when neurologic criteria for death are met, the entirety of the brain may not have ceased functioning. Moreover, the term is a fact-value fusion. The medical fact that an individual who has met the neurological criteria for death will not be able to be revived to consciousness based on contemporary medical knowledge and biotechnology is fused with the value that such a state represents a life not worth living and/or maintaining. Both the medical facts here and the value attributed to it are contentious. Throughout the rest of the paper, quotation marks will not be used for the sake of maintaining flow; however the reader should hold these controversies in his/her mind.

practices. In what follows, we draw upon the empirical and academic bioethics literature to highlight this unease and confusion from various Muslim corners. [2]

1.1 Muslim Healthcare Providers

As a group, a significant proportion of Muslim healthcare providers who might be tasked with determining brain death, or certifying the death of their patient after the determination, are troubled by brain death. For example, a national survey of Muslim physicians in the United States (n = 255) reported that nearly half did not consider individuals determined to be dead by neurologic criteria to be dead legally, and half also did not consider them to be dead theologically. Moreover, respondents who were more religious had lower odds of believing brain death to signify the departure of one's soul from the body [5]. Relatedly, physicians who held religion to be an important part of their lives had greater odds of agreement with the belief that it is typically more ethically problematic to withdraw a life-sustaining treatment than to withhold it [6]. A smaller study of allied Muslim healthcare professionals (n = 91), including chaplains, found that half of participants felt that families should be given choice over whether brain death evaluations are performed because of religious and ethical conundrums associated with the determination [7].

As can be gleaned from the above, Muslim healthcare providers' unease with brain death occurs at both the conceptual and practical levels. Dr. Mohammed Rady, a critical care physician at Mayo Clinic represents one prominent voice among this group and levels his critique at the use of neurologic criteria to determine death. From a religious perspective, he contends that brain death cannot be equated with death because, according to him, the Qur'an and Prophetic traditions unequivocally characterize death as a single, irreversible event where the soul leaves the body [8, 9]. In his view, because individuals who are declared brain death retain somatic integration, either intrinsically or via supportive medical technology, such individuals cannot be considered dead because they do not meet the biological definition of death, and neither can be considered dead by religious criteria because the soul may still be attached to the body [9].

1.2 Islamic Jurists

Islamic jurists are similarly confounded by brain death, as they struggle to resolve its biomedical nature and its legitimacy as death according to the standards of Islamic law. While a fuller review of Islamic juridical perspectives will be provided in the next section, herein we share the critiques of a few prominent scholars to highlight points of contention.

[2] The review here is not a systematic review of the extant literature, rather we have handpicked certain studies to illustrate the phenomenon of Muslim stakeholder disquiet with brain death.

In response to the increasing calls for deceased organ donation programs and clarity over brain death in past few decades, Shaykh Muhammad Sayyid Tantawi, the former rector of Al-Azhar and grand Mufti of Egypt, declared the matter of ascertaining the occurrence of death to be a medical and not a religious affair [10]. On the other hand, Shaykh Tantawi's contemporary and colleague, Shaykh Ali Gomaa, also former Grand Mufti of Egypt, sees the issue not to be about applying the label of death to a physiological state, but about ontology and morality. He notes "it is not just a technical medical issue, it's also a human and moral issue... doctors cannot say it is only for them alone to decide. We [religious scholars] must get involved…the issue is not about definitions [of death], the issue is about uncovering the truth [reality] about something" [10], p. 73. The American Islamic jurist, Shaykh M. Amin Kholwadia also holds that physicians should not be given the authority to discern criteria for death; rather criteria must be based on religious sources. He further notes uncertainties surrounding the determination of brain death, which renders it insufficient grounds for determining human death in Islamic law [11]. Other jurists heatedly debate whether the brain is the "seat" of the soul, holding this to be a prerequisite for legitimating brain death within Islamic law [12–14].

These varied perspectives exemplify debates among Islamic jurists about the validity of neurologic criteria to determine death in Islamic law, as well as whether religious scholars or medical experts have the primary role in defining death. Generally speaking, scholars fall into one of two camps; some legists and juridical councils consider brain death to satisfy Islamic legal thresholds for declaring death and implicitly give medical science a role in determining death criteria, while others assert that the traditional criteria of cessation of heartbeat and breathing should be maintained as markers for human death as they are both biomedically and religiously sound [4, 15]. Critically, however there appears to be a near-consensus, at least among Sunni jurists, that when neurologic criteria for death are met, it is religiously permissible to withdraw and/or withhold life-sustaining treatment [16].

1.3 Muslim Patients and Their Surrogate Decision-Makers

Regrettably, there has been little empirical research on Muslim patient and caregiver attitudes towards brain death. The scant studies that exist suggest that these stakeholders wonder what brain death actually represents from a medical perspective, whether brain death represents true death in Islam, and whether withdrawing and withholding life support is religiously licit when neurologic criteria for death are met. Though systematic research remains wanting, our experience as religious advisors and ethics consultants [3] to Muslim patients and families concurs with these studies. Muslim community members are confused by the term "brain death." To

[3] Both authors have religious and medical training and have been serving as ad hoc religious bioethics consultants to the Muslim community in their respective nations. Moreover, AIP is also a trained clinical medical ethics consultant and as such provides formal ethics consulting to Muslim patients and families in hospitals around the US.

them, life is a miracle that pervades the entire body, and they find it odd to privilege one organ over others in defining the end of human life. Moreover, they wonder whether the healthcare system mislabels patients as "brain dead" when they are simply in a deep coma. Qur'anic narratives regarding the people of the cave who were thought to be dead but were simply in a state of suspended animation and deep sleep for hundreds of years gives credence to these concerns voiced by Muslim community members and religious scholars alike [17, p. 705, 18, pp. 346–347]. Moreover, the 'discovery' of a new physiological state close to death, termed a "minimally conscious state," fuels fears that one-day biomedicine will uncover their folly in labelling people dead who were yet living. These concerns feed into questions of moral duty as families and surrogate decision-makers are unsure whether they are called to live out the overarching objective of Islamic law, the preservation of life, *ḥifẓ al-nafs*, in advocating for continued medical treatment despite the determination of brain death, or whether they are religiously permitted to withdraw and/or withhold. This confluence of concerns about the medical reality of brain death alongside the moral duties owed to an individual declared brain dead has Muslim families and surrogate decision-makers reaching out to those with dual expertise in medicine and in Islamic law for guidance.

2 Islamic Juridical Views on Brain Death

All Islamic jurists agree that human death is marked by the soul leaving the body [19, p. 94, 20, p. 367, 21, p. 157]. While there is no ambiguity around this theological notion of death, there is no consensus on how the soul's departure ties in with determination of death. Are there assured physical correlates of this metaphysical event? By declaring brain death, is the physician attesting to the departure of the soul? Is death in Islamic law to be based on local custom or on expert testimony? These are but a few of the questions that Islamic jurists need to resolve when considering brain death.

Generally speaking, Islamic jurists have fallen into two camps regarding whether the soul's departure has physical signs. One camp considers the soul to be a fully immaterial entity and, as such, it is not possible to perceive how it is tied into the physical body when designating criteria for death. Because of this uncertainty, brain death is insufficient grounds for determining human death, rather classical criteria of cessation of heartbeat and breathing should be maintained as markers of human death. [4] The other camp holds that there are physical manifestations of the soul's departure, and these should be the basis of death criteria. On this basis, leading

[4] For some the cessation of breathing and heartbeat are indicants that the human body can no longer sustain its connection to the soul as the body is moving into a state of decomposition. Hence this camp infers the departure of the soul rather than identifying its signs. It is also important to note that breathing is closely related to the soul in Semitic languages as the two words share the same root letters, and hence the activity of breathing is often assumed to be related to the presence or a function of the soul. Finally, the cessation of heartbeat and breathing are held to be universally common criteria for death, rooted in human experience across cultures, not necessarily scripture.

Islamic jurists take three different approaches to designating death criteria, with each view tying into whether the soul's departure is associated/correlated with physical manifestations [22, pp. 280–283]. We describe these three views along with the main jurists and juridical bodies who align with each view below. The first camp, the 'traditional' one, affirms circulatory-respiratory criteria as indicants of human death, the second group accepts neurologic criteria as sufficient for human death determination, and the third group considers brain death to be an in-between state of unstable life where the human being has a moral status akin to a dead individual for some purposes and to a living individual for others.

2.1 The "Traditional" Camp: Brain Death Does Not Represent Human Death

Death according to this approach is the permanent, irreversible cessation of all biological functions that sustain a living organism. Islamic jurists who support this view resolve that there are no reliable physical correlates to the departure of the soul. Rather, they associate criteria for death to signs that the body has begun to decay irreversibly, and accordingly, the body can no longer carry out the commands of the soul, hence for all intents and purposes, the soul is assumed to be disconnected from the body and burial rituals commenced. [5] Historically, human death was determined through external examination for signs of rigor mortis and putrefaction. These somatic criteria are no longer practical because of the societal need for timely diagnosis in hospital and hospice settings. Those who are advocates of the traditional definition of death associate death with the irreversible loss of vital fluid flow as cessation of heart and lung function, determined by apnea and absence of pulse because after this point, the body will begin to decay irreversibly. Death is therefore associated with circulatory-respiratory collapse.

As a corollary, the patient is counted among the living so long as circulation and respiration (assisted or unassisted) is maintained. Consequently, brain death is unacceptable as a threshold for death declaration because circulatory-respiratory function continues, albeit with technological assistance [17, p. 718]. Furthermore, because the metaphysical occurrence, i.e., departure of the soul, has no definitive physical correlates, speculation about what happens when an individual has reached

---

[5] It bears mention that there are statements from the Prophet Muhammad that indicate that the soul's connection to its host body may not fully be severed upon death. For example, some narrations tell of the dead (or newly dead) being able to hear happenings around their host body or even perceive pain. The Prophet said, "When a dead servant is put into his grave (buried) and his fellows (relatives and those who took part in his burial) leave him, he hears the sound of their footsteps." [66] also some have interpreted that the dead feel pain from the prophetic tradition "Breaking the bone of a dead person is the same as breaking his bone when he is alive." [67, 68]. Other narrations discuss how the body will be resurrected such that the soul can reinhabit it, and as such suggesting a connection is maintained. "(No doubt), people will be resurrected on the Day of Judgment barefooted, naked and uncircumcised." Then, the Prophet recited (what means): {As We began the first creation, We (God) will repeat it.} [Q 21:104]." [66]

the physiological state of brain death is inadmissible for assigning 'new' death criteria [22, pp. 284–287]. Said another way, indicants such as breathing and circulation are definitive signs of a living person, while neurologic criteria for death are ambiguous as to whether they represent the soul's departure from the body.

Examples of juridical councils that fall into this include the Kuwaiti Awqaf Ministry, which in 1981 indirectly considered brain death impermissible, whilst adhering to the traditional approach by ruling that an individual remained alive if circulation and respiration continued, even if mechanically assisted [23, p. 154, 24, p. 41, 25]. Similarly, in 1987, most Islamic scholars of the Islamic Fiqh Academy of Muslim World League [IFA-MWL] held decree that brain death criteria were not equivalent to circulatory-respiratory criteria for human death and upheld the latter as in accordance with Islamic law [11, 26, p. 216, 27, 28, p. 8]. [6]

2.2 The Brain Death Is Human Death Camp

According to this view, the soul is the primary integrator of the human organism, and its metaphysical departure has physical correlates that are observed when someone is declared brain dead. Given that death is theologically tied to departure of the soul and that in the Islamic tradition the presence of the soul is what determines personhood, certain physical and mental abilities are markers of the soul's presence in the body. These include higher brain functioning related to sentience, *al-ḥiṣṣ*, and voluntary movement, *al-ḥarakat al-irādiyah*. The irreversible loss of these bodily capacities is associated with loss of critical functions of the brain, and signals the departure of the soul from the body [18, pp. 350–352, 29, pp. 656–658]. Additionally, the individual meeting neurologic criteria for death is equal in functionality to the individual meeting circulatory-respiratory criteria for death because neither

[6] Main Islamic bodies which adhere to this are:

- Kuwaiti Awqaf Ministry 1981, Lajnat al-Fatwā bi Wizārat al-Kuwaitiyyah, details can be found in [31, pp. 665–666].
- Some members of the IFA-MWL, 1987 tenth Session, were strict adherents to this view, however the final conclusion of IFA-MWL was according to the third camp, i.e., the 'brain death is unstable life' group (see later) [26, p. 216].
- Egyptian Awqaf Ministry, see [69, p. 3712].
- Other Islamic bodies include; Islamic Fiqh Academy India at the 16th Fiqhi Seminar 2007 in Azamgarh [70].

South Africa: Majlis al-Ulama in 1995 [71]. Proponents of this view include the Islamic scholars, Islamic legist Dr. Tawfīq al-Wāʿī from Kuwait, the Egyptian jurist Shaykh Badr al-Mutawallī ʿAbd al-Bāsiṭ, the Muftī of Tunis, Shaykh Muḥammad al-Mukhtār al-Salāmī, the great Saudi Islamic scholar Shaykh Muḥammad al-Mukhtār al-Shanqīṭī, the Grand Muftī of Syria Shaykh Ramaḍān al-Būṭī and many others, see: [22, p. 262, 72].

Among the Shīʿa Islamic legists; The Iranian Grand Ayatollahs, Mohammad Taqī Bahjat Fūmanī and Huṣayn Waḥīd Khurāsānī, and the Iraqi Grand Ayatollah al-Sayyid ʿAlī al-Ḥusaynī Sīstānī. Most of the Shīa clerics contend that brain death is insufficient to consider a person dead [73, pp. 95–96] physiological state retains capacity for the soul to perceive and act through the body. Consequently, both states are acceptable as death proper in Islamic law. [7]

One of the most influential bodies of Islamic bioethical deliberation, the Islamic Fiqh Academy (IFA-OIC) of the Organization of Islamic Cooperation adheres to this view. [8] The IFA-OIC in particular carries significant weight in medical and legal circles because it brings together jurists assigned to the council by their respective governments or though official recommendations of council members, includes medical experts, and has representation from jurists that span the various schools of Islamic theology and law [30]. In 1986, the IFA-OIC held that a person is pronounced legally dead and consequently, all dispositions of the Islamic law in case of death apply if one of the two following conditions has been established [28, 31, 32]: (i) there is total cessation of circulatory and respiratory functions, and doctors have ruled that such cessation is irreversible; (ii) there is total cessation of all cerebral functions and experienced specialized doctors have ruled that such cessation is irreversible and the brain has started to disintegrate [33]. Some Muslim countries, including Malaysia, Bangladesh, Turkey and Iran, have adopted these two standards for human death declaration which accords with the enactment of the Uniform Determination of Death Act (UDDA) in the United States [28, pp. 12–14, 34].

[7] The main Islamic bodies which adhere to this view are:

- IFA-OIC The Council of the International Islamic Fiqh Academy of the Organization of the Islamic Conference, holding its second session in Jeddah, Saudi Arabia on 10–16 Rabīᶜ al-Awwal 1406H (22–28 December 1985) was then reconvened, holding its third session in Amman, Hashemite Kingdom of Jordan, on 8–13 Ṣafar 1407 h (11–16 October 1986),
- Other Islamic bodies include European Council for Fatwa and Research (ECFR), South Africa: Majlis al-Shura al-Islami in 1994, United Kingdom: Muslim Law Council 1995 and Indonesia: Council of Ulama 1996 [71, 72].

Proponents of this view include the following Islamic scholars: the Syrian jurist, Shaykh Muṣtafā al-Zarqā', the Saudi jurist Shaykh Muḥammad Ibn Jubayr, the Jordanian Islamic legists, Dr. ᶜUmar Ibn Sulaymān al-Asqhar and Dr. Muḥammad Naᶜīm al-Yāsīn, the Egyptian Islamic scholar based in Doha, Dr. Yūsuf al-Qaraḍāwī and others, see: [22, p. 281]

Among the Shīᶜa Islamic legists, a number have delegated such determination to medical professionals, thereby accepting the brain death criterion eg. the grand Ayatollah of Iraq Moḥammad Isḥāq al-Fayāḍ and the Iranian grand Ayatollah Naṣir Makārim Shayrāzī. One very authoritative Shīᶜa cleric, grand Ayatollah Ḥusayn ᶜAlī Muntaẓrī considers death as being defined by the medical professionals. Interestingly the Parliament of Iran enacted a law on April 5, 2000 entitled: "Transplant of Organs from Deceased Patients or Patients with Evident Brain Death" The law was neither approved nor rejected by the Guardian Council and thus entered into force in accordance with Article 94 of the Constitution an Explanatory Note issued by the Legal Department of the Judiciary on 12 May 2008 which states that brain death is synonymous with death and entails all legal consequences of death [73, pp. 106–109]

[8] These organizations bring together medical scientists and Islamic legal scholars that are transnational in scope. They represent a plurality among the Islamic schools of law and theology. Islamic ethico-legal deliberation around bioethical challenges faced in the Muslim and non-Muslim world are addressed. As a result of this inclusivity, the verdicts issued by these organizations carry significant weight in medical and legal circles because these organizations are recognized as the forefront of Muslim efforts to address ethico-legal challenges brought forth by modern technological advances [4].

2.3 The Brain Death Is Unstable Life Camp

There are scholars who are, like the camp above, of the opinion that the presence of the soul is what determines personhood and certain physical and mental abilities are markers of the soul's presence in the body. Yet, they are not ready to discard traditional markers for human death for multiple reasons including the lack of certainty around neurologic criteria for death, and unease over the intrusion of medical specialists in determining new criteria for death that seem to go against commonplace understandings. Consequently, they do not equate brain death and circulatory-respiratory criteria for death proper in Islamic law. Rather, they consider brain death to represent a physiological state between life and death, where life support need not be continued [22, pp. 289–293, 35, pp. 668–669]. [9] The grounding for this in-between state is a construct within Islamic law, *al-ḥayāt ghayr al-mustaqirrah* (unstable life), which was traditionally used to resolve moral culpability for homicide in cases where the assaulted individual has already suffered an injury that may or may not be life threatening. Some jurists further suggest that organs may be procured after brain death is declared, while others do not [28, p. 6]. Human death is thus resolved into a functional legal construct with two different subtypes; the first subtype is associated with brain death, which allows for some moral rulings related to the death of the human being to apply, e.g. withdrawal of life support, while the second subtype is associated with death proper, *al-mawt al-ḥaqīqī*, where all rulings of death apply such as those related to burial, distribution of what is bequeathed and inheritance [22, pp. 289–293, 29, pp. 668–670, 35, pp. 668–671]. According to this camp, there are multiple different purposes that the pronouncement of human death serves, and different moral justifications for death behaviors. [10] Each of these purposes needs to be analyzed through the prism of Islamic law and then criteria can be adduced [36].

[9]Among the Shīʿa Islamic legists who hold a similar view, is grand Ayatollah Nasīr Mukārim Shayrāzī, who makes a distinction between the brain death criterion for the purpose of organ transplants, and the Sharī'a criteria of death for other legal or religious purposes such as power of attorney or burial [73, pp. 96–97].

[10]The term "death behaviors" has been borrowed from Dr. Robert Veatch who describes not only that some behaviors traditionally associated with death can be unbundled but also that other behaviors (including organ procurement) must continue to be associated with death [45]. (See Veatch 2005) Death behaviors in the Islamic tradition would include ritual acts/ practices which normally occur after death is announced, such as initiation of the three-day ritual mourning, ritual washing, *ghusl*, shrouding, *kafan*, funeral prayer, *janāzah*, distribution of inheritance, *wirātha*, burial, *dafan,* and all other associated actions after death. They also extend to those actions which are deemed permissible after death such as retrieving organs for organ transplantation and withdrawing life sustaining treatment.

The following Islamic bodies adhere to this approach; the Islamic Organization for Medical Sciences (IOMS) in a 1985 and 1996 meeting [11] and the Islamic Fiqh Academy of Muslim World League [IFA-MWL] in 1987 [28]. [12] It is claimed that this view is the dominant one in Muslim circles because 'many (Islamic scholars) concede that a brain-dead person is dying but will not accept that he is dead' [37].

3 A Critical Appraisal of Juridical Contentions Over Brain Death

Debates within Islamic juridical circles on death by neurologic criteria need both updating and deepening in light of recent biomedical knowledge. As highlighted below, there are conceptual as well as practical issues that limit the clinical applicability of some of the views [4]. Islamic jurists are required, as part of their research into a matter prior to issuing a ruling, to have an accurate understanding of the main issues involved, their social implications, and their legal relevance. Indeed, as the maxim goes, *al-ḥukmu ʿalā al-shayi' farʿun ʿan taṣawwurihi*, passing judgment on something is dependent on its proper conception [38, p. 314]. Towards that end, we will summarize the main points of contention within the juridical discourse over brain death. In our reading they relate to the following:

1. How does the metaphysical nature of death relate to the physical determination of it?
2. Which brain functions, when lost, signal human death?
3. What is the criterion for irreversibility to declare brain death?
4. What level of certainty is required to determine human death?

3.1 How Does the Metaphysical Nature of Death Relate to the Physical Determination of It?

We have noted above how views on this question generally inform juridical perspectives on neurologic criteria for death. In this section, we will focus on a couple of additional questions based on the juridical decrees noted above. To begin with, Islamic theologians, by and large, view the soul and body as separate entities. The Qur'ān and prophetic tradition relate a dualist conception of soul and body which is especially relevant to conceptions of death [39]. Common secular understanding

[11] The IOMS revisited the issue in 1996 after they sent three members to participate in an international bioethics conference. These members reported back to the IOMS, this time with some eminent Islamic Scholars attending the meeting including Shaykh Yūsuf Al-Qarḍāwī, Shaykh Khālid Al-Mathkūr, Professor of Islamic Law in Kuwait University, Dr. Ibrahīm ʿAlī Ḥasan, the Vice President of the High Government Council in Egypt, and Dr. ʿAbdullah Al ʿĪsa, Vice President of the High Court of Kuwait [33, 74].

[12] Tenth Session second Declaration of IFA-MWL, see [26].

advances a monist approach where the mind is simply a manifestation of the brain, and the body and mind are the same entity or single substance. In the monist view, the mind is an extension of the natural world, explainable in purely physical terms. The faculties of the mind, such as volition and sentience, are reduced to the physiological and not attributed to a soul. Whereas in Islam, these higher mental faculties are those attributed to the soul and not just the body.

The IOMS expressed the view that the determination and identification of the signs of death have always been a medical matter and accepted physician testimony to rule that cessation of brainstem functions reflects the death of the patient [29, p. 655, 659]. Similarly, the IFA-OIC declared brain death to meet the standards for legal death in Islam. These legal judgments have theological implications for a dead body in that it is considered to be one where there is a severed connection to the human soul [31]. Both rulings implicitly suggest that a brain-dead individual is one in whom there is no soul. Yet, at these council meetings, there was little discussion regarding questions that may ensue such as how the metaphysical truths about the soul associate with our biomedical understandings, and how Muslim theologians intend to tie vital functions of the brain to vital functions of the soul. At the IOMS in 1985, a few experts suggested that the brain was the seat of the soul, however it was decided to table such theological discussions to the future [18, pp. 350–352, 29, p. 655, 659].

The writings of classical jurists may provide some foundations for such deliberation. For example, the great Damascene jurist Ibn al-Qayyim al-Jawziyya (d. 751/1350) [13] discusses the human soul and its relationship to the body's functions in his discussion of the fetus. He argues that the fetus' life has two periods [40]:

1. *qabla-nafkh* (pre-ensoulment): where it is similar to plant life and the body grows and is nourished.
2. *baʿda-nafkh* (post-ensoulment): where it has the capacity for sentience, volition and voluntary motion since the human soul is now linked to the body.

[13] Ibn al-Qayyim al-Jawziyya was an important medieval Damascene Sunni Islamic jurisconsult, theologian, and spiritual writer. He belonged to the Hanbali school of Islamic jurisprudence, of which he is regarded as one of the most important jurists.

He thus asserts that bodily functions such as growth and nutrition can exist before ensoulment and without linkage of the soul to the body. [14] *Ipso facto*, if and when the soul leaves the body, these biological functions are not indicants of soul's presence, and as we know today, can be maintained through artificial means. Other theologian-jurists, such as Imam al-Ghazālī (d. 505/1111), [15] and metaphysicians, such as Ibn Sīnā (d. 427/1037), [16] comment on similar relationships where they attribute different types of functions to the developing fetus based on whether and when the human soul becomes linked to the physical body [41, 42]. These sorts of discussions suggest that the traditionalists need to explain why it is necessary for there to be permanent, irreversible cessation of *all* biological functions leading to irreversible bodily degeneration to declare death, when some classical scholars acknowledged that some biological functions occur independent of the soul.

Somewhat related is the practical concern that the IFA-OIC ruling requires brain degeneration for declaration of death. It is unclear why degeneration is part of the criteria when certain brain activity may occur without presence of the soul. Moreover, there is no similar certification of degeneration process pursued when circulatory-respiratory criteria are applied. This condition presents practical problems because brain death assessment protocols do not require verification of brain degeneration via biopsy or imaging, nor by assessing whether biochemical markers of brain degeneration are present in the bloodstream [43]. The criteria laid out by the IFA-OIC for acceptance of neurologic criteria for death as death in Islamic law

[14] al-Jawziyya further elaborates in his al-Tibyān: If it is said, does the fetus, *al-janīn*, before ensoulment have movement and perception or not. It will be said, that it has the movements of growth and nutrition just like plants, but these movements of growth and nutrition are not voluntary. When it is ensouled then voluntary sensory movements contribute to the movements of growth and nutrition [40]. Similarly, al-Jawziyya's teacher, the great Ḥanbalī polymath, theologian and judge, Ibn Taymiyya (d. 728/1328) asserts the same: Life is of two types; plant life, and animal life which is particular to sensory perception and voluntary movement, whereas plant life is that of growth and nutrition [75, p. 56]. The great Egyptian polymath and ḥadīth expert, Ibn Ḥajar al-ʿAsqalānī (d. 852/1449) claims that the first organ to develop is the liver because growth and nourishment arises from it which provides the strength to the body, some suggest that [the liver] is the requisite for the development of the natural system because growth is required first. Sensory perception or voluntary movement are not dependent on it, because at this stage it is like a plant. Rather, the capacity for sensory perception and voluntary, willed movement is associated with the soul (which comes after) [76, p. 482].

[15] Abū Ḥāmid Muḥammad ibn Muḥammad al-Ghazālī considered to be the *mujaddid*, reviver, of his age, a Sunnī, Shāfiʿī, Ashʿarī scholar, jurist, rationalist, and ṣūfī master of Persian descent. His book, the Iḥyāʾ ʿulūm al-dīn, The Revival of the Religious Sciences, was well received by Islamic scholars.

[16] Ibn Sīnā, often known in the West as Avicenna, was a Persian polymath who is regarded as one of the most significant physicians, astronomers, thinkers and writers of the Islamic Golden Age and the father of early modern medicine. He is arguably the most influential philosopher of the pre-modern era and was a Muslim Peripatetic philosopher influenced by Greek Aristotelian philosophy. Of the 450 works he is believed to have written, around 240 have survived, including 150 on philosophy and 40 on medicine. His most famous works are The Book of Healing (al-Shifā'), a philosophical and scientific encyclopedia, and The Canon of Medicine, a medical encyclopedia

appear to lack conceptual and clinical clarity, thus giving little guidance to Islamic theologians and Muslim clinicians.

3.2 Which Brain Functions, When Lost, Signal Human Death?

This question directly addresses the interface of biomedicine and religion. The first part of the question is attended to by the deliverables of neuroscience, for neuroscience would tell us not only what the physiological functions of brain matter are but also which of these functions are most critical to the manifestation of life in the human body. The latter half of the question relates to death, which is another matter altogether, and the crux of the matter this chapter is trying to shed light upon. Fatāwa, [17] such as those of the IFA-OIC, conceive of brain death as the vital functions of the brain having ceased, and when this happens legal death in Islam is fulfilled. However, their ruling does not explicitly denote which conception of brain death, brainstem or whole-brain, is aligned with Islam and accordingly which criteria are to be used in declaring human death [30].

Some argue that consciousness is an important determinant of an ensouled life. They point to classical *fiqh*, substantive law, literature which contains many cases where signs suggestive of consciousness are important determinants of an ensouled life, and the permanent loss of consciousness is *legal* death, *al-mawt al-hukmī*.[18] This understanding of permanent loss of consciousness as human death is deduced broadly from the descriptions of death asserted by jurists as permanent loss of willed (voluntary) action, *ḥarakat ikhtiyāri*, coherent speech, *nuṭq*, and sight, *ibṣār* [18, p. 352, 24, pp. 43–44, 29, pp. 656–660]. Since jurists back then differentiated between somatic signs of permanent loss of consciousness from those related to the decomposition of the human body, there is precedent to do so today. One could argue that permanent loss of consciousness could be a legitimate marker of death in Islamic law, standing alongside or replacing other indicants. More significantly, one could then generate a conception of brain death that is attached only to the capacity for consciousness, and consequently develop neurologic criteria for death that assess whether or not there remains capacity for consciousness. One of us (RR) makes this argument elsewhere [44].

[17] Fatāwā is plural for fatwā and is an authoritative, but nonbinding legal opinion or interpretation on a point of Islamic law given by a qualified legal scholar (known as a muftī) or collectively, comprising a number of Muslim scholars with an interdisciplinary team of biomedical scientists. A fatwa is usually issued in response to questions from individuals or Islamic courts.

[18] Classical Muslim jurists describe a state of permanent loss of voluntary movement, sentience and volition when a person has received injury after an assault or when they are in a state after a fatal illness. They allow for certain legal rulings which are associated with death enactments. If there is any movement, then this is described as being involuntary, not willed nor from conscious effort. This state is described to be synonymous to the normative state of *ḥarakāt al-madhbūḥ* (the involuntary reflexive movements seen after slaughtering an animal). In the state of *ḥarakāt al-madhbūḥ* the animal is legally treated as dead (*al-mawt al-ḥukmī*) [22, pp. 290–294, 77, p. 106, 78, p. 145]. For other examples and detail see: [79].

From the available juridical rulings, it remains unclear what activity or functions in the human brain is judged to be nonessential, not critical, or inconsequential and thus should not be assessed by a brain death evaluation, and is theologically and legally acceptable to persist after the declaration of death. [19] On a practical level, the whole-brain criterion requires permanent cessation of all brain functions for human death, yet many patients who are declared dead using clinical tests for this standard continue to have electroencephalographic activity and hypothalamic functioning [45, p. 482, 46, 47].

3.3 What Is the Criterion for Irreversibility to Declare Brain Death?

In Islam, human death is associated not only with the metaphysical departure of the soul from the body but also with changes in the physical body that lead to disintegration and decay. While the declaration of death occurs at a specific moment in time, one could argue that biologically, death signals the onset of a process of gradual decomposition and decay of the body. As noted previously, from a theological perspective, death refers to the separation of the soul from the body; however, the Islamic tradition is ambiguous as to whether there are physical correlates to the severance. Islamic jurists thus rely on the physical indicants that the body is moving towards permanent disintegration. A key aspect of these physical indicants is irreversibility, meaning that they must indicate that the onset of the process of disintegration and decay is irreversible.

The IFA-OIC explicitly states that Islamic law permits the declaration of death when all vital functions of the brain cease *irreversibly* [31]. Yet irreversibility is not defined. While the prognosis of those declared brain dead is very poor, as none of these individuals will regain consciousness despite continued life-sustaining treatment based on the current state of biomedicine, we do know that some brain functions may return. Neuroscientists note that brainstem reflexes may reappear after initial absence in brain-dead individuals, and a proportion of the brain may continue to function [48–50].

If irreversibility is a key aspect of what makes neurologic criteria for death acceptable in Islamic law, then it must be made clear what irreversibility indicates. Irreversibility is not a singular notion; rather, it can refer to many different states (as discussed elsewhere in this book). For example, it can refer to the state when functions of the brain will not naturally return, or when they cannot be reversed due to

[19] Elsewhere one of us (AIP) has argued that there are serious gaps in contemporary medical understanding and clinical diagnosis of brain death and its endorsement as human death in the Islamic faith. These gaps pertain to: (1) the retention of residual brain functions; (2) the recovery of some previously ceased brain functions; (3) the absence of whole brain degeneration and necrosis; and (4) the uncertainty of medical tests and bedside examination in determining this condition with reasonable accuracy [4].

limitations in our biomedical capacities, or alternatively when they will not be reversed because it is deemed *legally* or *morally* inappropriate to intervene [51].

If irreversibility means physiologic irreversibility, then one can understand why the IFA-OIC attached verification of the disintegration of the brain to their ruling. Physical signs of disintegration would attest that the absence of brain functions is irreversible. Irreversibility therefore equates with permanence because functions of the brain will not naturally return, nor can they be reversed due to limitations in our biomedical capacities. As noted above, some brain functions remain and others can return when a brain-dead patient is kept on life-sustaining instruments. If irreversibility means permanence, then these individuals cannot be said to be dead, but rather just imminently dying [50, 52].

If reversibility means the latter, then it is not whether the body or brain circulation and function can be resumed (because it can in some cases), but rather, whether the situation warrants reversal by clinicians, i.e., whether it will be. If it will not, then this state is also deemed permanent as it is morally or legally appropriate not to reverse.

Physiological death is not an event at some precise moment in time, but a process. The common contention revolves around the moment at which death can be declared. Islamic jurists need to be prescriptive about this juncture. Is death declaration justified in Islamic law when brain tissue begins to degenerate and brain functions *cannot* be reversed, or it is at a stage where physicians *will not* to try to reverse the clinical scenario because of various moral and legal considerations? Moreover, what level of certainty is required for physicians to make the diagnosis of brain death?

3.4 What Level of Certainty Is Required to Determine Human Death?

Detractors of brain death argue that brain death determinations are uncertain and, as such, insufficient as thresholds for death in Islamic law. Rather, the diagnosis is speculative, uncertain and doubtful [17, pp. 712–713, 18, pp. 349–350, 53, p532]. The Islamic ethico-legal maxim, certainty is not eliminated through doubt, *al-yaqīn la yuzūlu b'il-shakk*, would support the idea that brain death does not suffice for declaration of death. The patient is alive, and we are certain about this due to continued breathing and heartbeat, and death is uncertain. Other ethico-legal maxims, such as the original state remains on that what it was, *al-aṣlu biqā'a ma kāna ʿalā ma kāna*, can be used to further ground this view [54, p. 263]. *Istiṣḥāb*, presumptive continuity, a formal source in Islamic law, also suggests continuity of the original ruling of what is agreed upon by everyone, that the person is alive, and not upon the ruling in which there is dissension, i.e., that the person is dead [17, pp. 712–713, 18, pp. 347–348].

What is agreed upon is that the Qur'ān and prophetic tradition are not explicit in the criteria for death, and it is therefore left to *ijtihad,* juristic effort, to deduce rulings from these texts. Rulings derived from *ijtihād* are considered *ẓannī,* speculative or approximate, which is an epistemic category distinct from *qaṭʿī,* definitive or conclusive knowledge [55, p. 14]. Often jurists seek certainty, *al-yaqīn*, yet for legal purposes dominant probability, *ghalabat al-ẓann*, is considered sufficient to judge on a matter [29, pp. 645–659, 56, p. 155–170]. *Ghalabat al-ẓann* is formally defined as a situation where there is a possibility of two conclusions, but one is preferred above the other due to its higher likelihood based on formal, empirical, or other evidence [57, p. 4]. The difference between *yaqīn* and *ghalabat al-ẓann* is that the latter refers to an outcome that is preponderant when the remaining contrary outcome is minimized, very unlikely, and can be discarded, while the former requires proof that the other state is impossible [58, p. 144, 59, p. 77]. In other words, even though the remaining outcome can be dismissed on the basis that it is very unlikely, it cannot be excluded entirely, whereas for the epistemic level of certainty the remaining possibility must be excluded entirely.

Islamic juridical bodies stipulate *certainty* in determining brain death, but do not specify whether they mean true certainty *al-yaqīn* or *ghalabat al-ẓann*. If they mean the former, diagnostic testing can almost never conclusively judge a certain state is present with 100% positive predictive value. The threshold for certainty and declaration of death by neurologic criteria requires further elucidation.

4 Death in the Muslim Mind

Thus far, we have reviewed and critically appraised leading Islamic juridical perspectives on brain death and described how they relate to metaphysical understandings with biomedical approaches to the declaration of death. In this section, we propose an approach to brain death that accounts for Islamic plurality and addresses the disquiet Muslim patients, clinicians, and families have with brain death in clinical practice (see Table 1). All Islamic theologians and jurists, a large part of the Muslim public, agree upon the metaphysical or ontological representation of death as deduced from the Qur'an and Sunnah, that death represents the soul's functional separation from the human body. This theological notion or knowledge is given 'life' through human language, symbols and behaviors. For example, a semantical representation of death can be that the person has 'gone', 'departed this world', and

Table 1 Islamic views on death criteria and purposes

Death "type"	"Traditional" camp	"Brain death is human death" camp	"Brain death is unstable life" camp
Ontological death	CR	BD	CR
Semantic death	CR	BD	CR
Biomedical death			
• Withdrawal of life support	BD	BD	BD
• Retrieval of organs	CR	BD	Diversity of views
• Autopsy	CR	CR	CR
• Burial	CR	CR	CR

CR circulatory-respiratory criterion, *BD* brain death criterion 'moved on.' Muslim cultural behaviors representing death include the rapid burial of the body and various morning rituals. In this way, the theological understanding is given meaning in human language and culture.

More importantly, Islamic jurists glean that there are nearly universal semantic and behavioral practices that convey knowledge of, and thus signify, that death has occurred. While Muslims may bury their dead, other cultures cremate, yet all cultures across time conduct(ed) rituals that denotes a significant change has occurred such that the individual previously present in the body is no longer present. The point at which these rituals take place is when physical signs of bodily decomposition manifest. From our discussion above, it appears that many Islamic jurists, and a large proportion of the Muslim public, consider circulatory-respiratory cessation to be the point at which death should be declared because the path towards bodily decomposition has been set irreversibly.

What needs clarifying is the biomedical representation of death; a functional notion of death that serves normative purposes in clinical practice. Islam is predominantly a legal and normative tradition and Islamic law is the determining factor that adherents seek authority from for Islamic law represents a large part of Islamic morality [60]. Classical Muslim jurists thus looked at death with a normative lens seeking to address questions such as, "When should the waiting period between a future marriage, *ʿidda,* start for a wife whose husband has died?"; "When should the distribution of inheritance and property of the deceased take place?"; "When should the community obligated funeral prayer occur?"; "When should bodily burial take place?"; and the like. A description of death for Islamic legal purposes is normative in so far as it enables death behaviors which are ordained as communal or individual obligations. In clinical practice, the normative (read: moral) aspects of death relate to questions such as "When ought we discontinue life-sustaining treatments because the clinician's moral duty to restore health is no longer possible such that both the clinician, if Muslim, and the patient's surrogate decision-maker, if Muslim, will not be considered to be sinful?"

Another question relates to organ donation: "When is it religiously permissible to authorize the donation of one's organs without such donation being the proximate cause of the individual's death?" Withdrawing or withholding life support and/or organ donation serve moral purposes in biomedicine, and what we are asking Islamic jurists to do is to opine on whether and when the death behaviors carried out related to these purposes are religiously permitted. Said another way, not only is the biomedical representation of death about the biological or physiological point at which death can be determined or declared, but also from an Islamic ethico-legal perspective, this representation of death equates to moral end-points at which certain behaviors are permissible.

Given the medicalization of dying, there are multiple purposes for death declaration in hospital and hospice settings. Some of these relate to behaviors to be carried out by families/communities, clinicians, religious/spiritual support staff, and others. From an Islamic bioethical lens, we believe the following approach is most prudent, for it acknowledges a plurality of Islamic juridical views on brain death while demarcating moral boundaries for death behaviors that advance informed decision-making and are nearly universally acceptable to Islamic scholars and Muslim polities.

With respect to bodily burial and medical autopsy, [20] we hold that these actions should only occur after cessation of circulatory-respiratory activity, whether it be physiologically irreversible or judged to be permanent because important decision-makers (the patient or their surrogate decision-makers) agree not to attempt resuscitation after initial cessation, or because state statute prevents attempting resuscitation after initial cessation. Islamic jurists and Muslims uniformly agree the human body is on the path towards decomposition when the activities of circulation and respiration stop and thus death can be declared on the basis of this specific biomedical construction of death. Islamic jurists and theologians uniformly agree that the body cannot carry out the commands of the soul when it has reached the state of decomposition. As such, the metaphysical understanding of death also holds.

Furthermore, because there is a diversity of views on whether the soul's departure from the body has physical indicants, and if such signs exist whether we can ascertain them with certainty, our position adopts caution that we are not burying, or performing an autopsy upon those who may be still living. Additionally, the traditional criteria of circulatory-respiratory collapse are more appropriate to the moral purposes of burial and autopsy, because of the social and cross-cultural acceptability these markers possess. Moreover, they do not require specialist knowledge to ascertain; the common individual would be able to discern that a body is decomposing or that heartbeat and breathing have ceased. We may even reserve the language of 'death' as a customary semantic notion, *ʿurf*, for this standard of cardiopulmonary cessation.

Neurologic criteria for human death, on the other hand, suffice clinical moral purposes such as the obviation of duty to rescue by withdrawal and withholding life sustaining treatment for clinicians and families. Neurologic criteria are morally acceptable because continued life-sustaining measures are ineffective in restoring the health of the patient, and the harms of such sustaining compound. Indeed, clinical research demonstrates that individuals declared brain dead undergo cardiopulmonary cessation shortly thereafter [61], and the intensive treatments of ventilators, feeding tubes, medications and the like carry increasing risk of infection, decubitus ulcers, and otherwise violate bodily integrity and human sanctity [62]. Assuredly, clinical treatment has very little to offer patients in this state; we cannot restore brain function to a conscious state. Islamic jurists, by and large, consider it better to withdraw and/or withhold life-sustaining therapies on the brain-dead patient [16]. Indeed, there is no Islamic obligation to pursue continuing such measures at all.

With respect to organ donation, we advocate for plurality. For some jurists and patients, organ donation is only valid whilst living; hence organ donation after the declaration of death by neurologic death is not aligned. Others hold that organ

[20] It bears mention that autopsy requires explicit justification within the Islamic tradition. Tampering with the body after death is generally not permissible. Muslims have expressed their views on autopsy publically and internationally, and there have been claims of violation of the deceased, delays in burial, and nonconsideration of their religious beliefs. For more details see [80].

donation can take place after death, some restricting it to declaration of death after circulatory-respiratory arrest, and others permitting it when neurologic criteria are met [44, 63–65]. Procuring essential organs (i.e., those needed for life, such as the heart and/or both lungs) for transplantation may require the circulatory-respiratory criteria to be met according to some jurists and neurologic criteria to be met for others. Certainly, like their secular counterparts, Islamic jurists are cautious to not run afoul of the dead donor rule.

With respect to the brain-dead state, again there is little clinical treatment can offer because the patient has irretrievably lost the capacity for consciousness, and their essential organs have little or no 'value' for them. One could argue if an apparatus that replaces the functions of essential organs, e.g. a ventilator, can be turned off without moral sanction, then essential organs in the body should be made available for transplantation because they too can be 'turned off.' However, others may see this view as violating the sanctity of human body alongside its inviolability, *ḥurma and karāma*. Consequently, we suggest that neurologic criteria for death only be applied in cases where individuals have previously authorized organ donation. If these criteria for death and/or organ donation are not aligned with the Muslim patient or surrogate decision-maker's values, then traditional circulatory-respiratory criteria should be applied.

5 Conclusion

In summary, we believe that death is a moral affair. The purposes of death declaration need to be evaluated from a moral perspective, for the behaviors that are enacted after this declaration are moral in nature. Hence, the purposes of death declaration need to match up with the criteria that this declaration is based upon as well. We believe that determination of death by neurologic criteria, from an Islamic bioethical perspective, can nearly universally permit the withdrawal or withholding of life support, and in some cases/contexts organ donation. However, we do not believe that neurologic criteria for death should be enacted broadly for all legal and moral purposes.

References

1. Montgomery H, Grocott M, Mythen M. Critical care at the end of life: balancing technology with compassion and agreeing when to stop. Br J Anaesth. 2017;119(S1):i85–9.
2. Hamdy S. Not quite dead: why Egyptian doctors refuse the diagnosis of death by neurological criteria. Theor Med Bioeth. 2013 Apr;34(2):147–60. https://doi.org/10.1007/s11017-013-9245-5.
3. Rady MY, Verheijde JL. Legislative enforcement of nonconsensual determination of neurological (brain) death in Muslim patients: a violation of religious rights. J Relig Health. 2018;57(2):649–61.

4. Padela AI, Arozullah A, Moosa E. Brain death in Islamic ethico-legal deliberation: challenges for applied Islamic bioethics. Bioethics. 2013;27(3):132–9. https://doi.org/10.1111/j.1467-8519.2011.01935.x.
5. Popal S, Hall S, Padela AI. Muslim American physicians' views on brain death: findings from a national survey. Avicenna J Med. 2021;11(2):63.
6. Duivenbode R, Hall S, Padela AI. Assessing relationships between Muslim physicians' religiosity and end-of-life health-care attitudes and treatment recommendations: an exploratory national survey. Am J Hosp Palliat Med. 2019;36(9):780–8. https://doi.org/10.1177/1049909119833335.
7. Lewis A, Kitamura E, Padela AI. Allied Muslim healthcare professional perspectives on death by neurologic criteria. Neurocrit Care. 2020;33(2):347–57.
8. Rady MY, Verheijde JL. A response to the legitimacy of brain death in Islam. J Relig Health. 2016;55(4):1198–205. https://doi.org/10.1007/s10943-016-0221-z.
9. Rady MY, Verheijde JL. Brain death and the moral code of Islam. Chest. 2015;147(2):e69.
10. Hamdy S. Our bodies belong to god: organ transplants, Islam, and the struggle for human dignity in Egypt. Berkerley, CA: University of California Press; 2012.
11. Stodolsky MVY, Kholwadia MA. A Jurisprudential (Uṣūlī) Framework for cooperation between Muslim jurists and physicians and its application to the determination of death. In: Padela AI, editor. Medicine and shariah: a dialogue in Islamic bioethics. Indiana: University of Notre Dame Press; 2021.
12. Moosa E. Languages of change in Islamic law: redefining death in modernity. Islam Stud. 1999;38(3):305–42.
13. Ebrahim AFM. Islamic jurisprudence and the end of human life. Med Law. 1998;17:189.
14. al-Awadhi AE-RA. Nadwat al-ḥayāh al-insāniyyah bidāyatuhā wa nihāyatuhā fī al-mafhūm al-islāmī al-munaqidah, Majallat Majmaᶜ al-Fiqh al-Islāmī. Kuwait. 1985;3:2.
15. Al-Bar MA, Chamsi-Pasha H. Contemporary bioethics: Islamic perspective. London/Cham: Springer; 2015.
16. Mohiuddin A, Suleman M, Rasheed S, Padela AI. When can Muslims withdraw or withhold life support? A narrative review of Islamic juridical rulings. Glob Bioeth. 2020;31(1):29–46.
17. Wa'i T. Ḥaqīqat al-Mawt wa al-Ḥayāh fī al-Qur'ān wa al-Ahkām al-Sharīᶜah, Nadwa al-ḥayāh al-insāniyyah bidāyatuhā wa nihāyatuhā fī al-mafhūm al-islāmī, Majallat Majmaᶜ al-Fiqh al-Islāmī. 1986;3(2).
18. al Shinqīṭī, Muḥammad al-Mukhtār. Aḥkām al-jarāḥat al-ṭibbiyya wa al-āthār al-murattabah ᶜalayhā, 2nd Edition, Cairo: Maktabah al-Ṣaḥābah, 1994.
19. al Nawawī, al-Majmūᶜ sharḥ al-Muhadhdhab, ed. Maḥmūd al-Maṭrajī, 22. Beirut: Dār al-Fikr. 1996.
20. Ibn Qudāma, al-Mughnī, eds. ᶜAbd Allāh Ibn ᶜAbd al-Muḥsin al-Turkī and ᶜAbd al-Fattāḥ Muḥammad al-Hulw, 15 vols. Riyadh: Dār ᶜĀlam al-Kutub, 2007.
21. Niẓām al-Dīn et al., al-Fatāwā al-Hindiyya, ed. ᶜAbd al-Laṭīf Ḥasan ᶜAbd al-Raḥmān, 6 vols. Beirut: Dār al-Kutub al-ᶜIlmiyya; 2000.
22. al-Shuwayrakh, Sᶜad Ibn ᶜAbd al-ᶜAzīz, Mawt al-Dimāgh, Mujallat al-jamᶜīyyat al-fiqhiyyat al-saᶜaudiyya; 2011. http://ypeda.com/attachments/article/150/%D9%85%D9%88%D8%AA%20%D8%A7%D9%84%D8%AF%D9%85%D8%A7%D8%BA%20%D8%AF.%20%D8%B3%D8%B9%D8%AF%20%D8%A7%D9%84%D8%B4%D9%88%D9%8A%D8%B1%D8%AE.pdf. Accessed 11 July 2021.
23. al-ᶜUqaylī, ᶜAqīl, Ḥukm naql al-aᶜḍā' fī al-fiqh al-islāmī, First Edition, Cairo: Maktabah al-Saḥābah; 1990.
24. al-Ṭarīqī, ᶜAbd Allāh Ibn Muḥammad, Mawt al-Dimāgh, 1st edition; 2005.
25. Haque OS. Brain death and its entanglements: a redefinition of personhood for Islamic ethics. J Relig Ethics. 2008;36:13–36.
26. IFA-MWL, (Islamic Fiqh academy - Muslim world league) Rābiṭ al-ᶜĀlam al-Islāmī, Taqrīr al-wafāh Wa rafᶜ ajhazat al-inᶜāsh min jism al-insān, al-Majmaᶜ al-Fiqh al-Islāmī 10th session, Meccanica; 1987. https://ar.themwl.org/taxonomy/term/8

27. Al-Bar MA. Organ transplantation: a Sunni Islamic perspective. Saudi J Kidney Dis Transplant. 2012;23(4):817–22.
28. Grundmann J. Shariah, brain death, and organ transplantation: the context and effect of two Islamic legal decisions in the near and Middle East. Am J Islam Soc Sci. 2005;22(4):1–25.
29. Yāsīn MN. Nihāyat al-Insāniyyah fī ḍū' Ijtihādāt al-ʿUlamā' al-Muslimīn wa 'l-Muʿṭiyāt al-Ṭibbiyyah, Nadwa al-ḥayāh al-insāniyyah bidāyatuhā wa nihāyatuhā fī al-mafhūm al-islāmī al-munaqidah, Kuwait. Majallat Majmaʿ al-Fiqh al-Islāmī. 1985;3(2)
30. Moosa E. Brain death and organ transplantation – an Islamic opinion. S Afr Med J. 1993;83:385–6.
31. IFA-OIC, (Islamic Fiqh Academy- Organization of Islamic Cooperation) Munaẓẓamāt al-Taʿāwun al-Islāmī, Majallat Majmaʿ al-Fiqh al-Islāmī. https://iifa-aifi.org/en/32257.html. 1985 Jeddah see: Majallat Majmaʿ al-Fiqh al-Islāmī. 1985;2(1). 1986 Amman, Jordan see: ajhazat al-inʿāsh, Majallat Majmaʿ al-Fiqh al-Islāmī. 1986;3(2):809.
32. Chamsi-Pasha H, Albar MA. Do not resuscitate, brain death, and organ transplantation: Islamic perspective. Avicenna J Med. 2017;7(2):35–45. https://doi.org/10.4103/2231-0770.203608.
33. Al-Bar MA. Islamic ethics of organ transplantation and brain death. Saudi J Kidney Dis Transpl. 1996;7:109–14. https://www.sjkdt.org/text.asp?1996/7/2/109/39509
34. Wasay M, Mujtaba SB, Sheerani M, Barech S, Jooma R. A review of brain death protocols across the globe and need for brain death guideline for Pakistan. Pakistan J Neurologic Sci. 2020;15(2):Article 4. https://ecommons.aku.edu/pjns/vol15/iss2/4
35. al-Ashqar MS. Nihāyat al-ḥayā, nadwat al-ḥayāt al-insāniyyah bidāyatuhā wa nihāyatuhā fil al-mafhūm al-islāmī, Majallat Majmaʿ al-Fiqh al-Islāmī. Kuwait. 1985;3(2):661–71.
36. Padela AI. Muslim Disquiet over Brain-Death: Advancing Islamic Bioethics Discourses by Treating Death as a Social Construct that Aligns Purposes with Criteria and Ethical Behaviors. In: Ghaly, M., editor. End-of-Life Care, Dying and Death in the Islamic Moral Tradition. Netherlands: Brill. 2022
37. al-Mousawi M, et al. Views of Muslim scholars on organ donation and brain death. Transplant Proc. 1997;29:3217.
38. Ḥamwī S a-U. Ghamz ʿuyūn al-basā'ir fī sharḥ al-asbāh wa al-naẓā'ir, 4 Vols. Beirut: Dār al-Kutub al-ʿIlmiya; 1985.
39. Qazi F, Fette D, Jafri SS, Padela AI. Framing the mind–body problem in contemporary neuroscientific and sunni islamic theological discourse. New Bioethics. 2018;24(2):158–75. https://doi.org/10.1080/20502877.2018.1438835.
40. al-Jawziyya , Ibn al-Qayyim, al-Tibyān fī aqsām al-Qurʾān, ed. Fawwāz Aḥmad Zmirlī, Beirut: Dār al-Kitāb al-ʿArabī; 1994.
41. al-Ghazālī, Iḥyāʾ ʿulūm al-dīn, 6 vols. Damascus: Dār al-Fayḥāʾ; 2010.
42. Druart T-A. The human soul's individuation and its survival after the body's death: avicenna on the causal relation between body and soul. Arab Sci Philos. 2000;10(2):259–73.
43. AOMRC- A code of Practice for the Diagnosis and Confirmation of Brain Death [Internet], Academy of Medical Royal Colleges, aomrc.org.uk; 2008. https://www.aomrc.org.uk/wp-content/uploads/2016/04/Code_Practice_Confirmation_Diagnosis_Death_1008-4.pdf. Accessed 26 Dec 2021.
44. Rashid R. The Intersection between Science and Sunnī Theological and Legal Discourse in Defining Medical Death in Islam and Biomedicine. al-Akiti A. & Padela AI (editors), Springer 2022.
45. Potts M. A requiem for whole brain death. J Med Philos. 2001;26:479–92.
46. Veatch R. The impending collapse of the whole-brain definition of death. Hastings Cent Rep. 1993;23(4):18–24.
47. Nair-Collins M, Miller FG. Current practice diagnosing brain death is not consistent with legal statutes requiring the absence of all brain function. J Intensive Care Med 2022;37(2):153–156.
48. McCullagh P. Brain dead, brain absent, brain donors. New York: John Wiley and Sons; 1993.
49. Field DR, et al. Maternal brain death during pregnancy. Med Ethical Issues JAMA. 1988;260:816–22.

50. Joffe AR. Brain death is not death: a critique of the concept, criterion, and tests of brain death. Rev Neurosci. 2009;20:187–98.
51. Bernat JL. How the distinction between "irreversible" and "permanent" illuminates circulatory-respiratory death determination. J Med Philos. 2010;35(3):242–55.
52. Miller FG, Truog RD. Death, dying, and organ transplantation: reconstructing medical ethics at end of life. Oxford: Oxford University Press; 2012.
53. Abū Zayd, Bakr Ibn ʿAbd Allah, Ajhizat al-inʿāsh wa ḥaqīqat al-wafāt bayna 'l-fuqahā' wa 'l-attibā, Majallat Majmaʿ al-Fiqh al-Islāmī, Ammān. 1986;3:2.
54. Ibn Nujaym, Zain ud-Dīn bin Ibrāhīm (d. 970 A.H.), al-Āshbāh wa al-Naẓāir ʿalā Madhhab Abī Ḥanīfah al-Nuʿmān, Beirut: Dār al-Kutub al-ʿIlmiyya; 1999.
55. al-Āmidī, al-Iḥkām fī uṣūl al-aḥkām, ed. al-Sayyid al-Jumaylī, 4. Beirut: Dār al-Kitāb al-ʿArabī; 1984.
56. Majīd al-Mājid, ʿAbd al-Salām Ibrāhīm, Aḥkām al-ẓann wa alfāẓuhu wa aqsāmuhu fī sharī'at al-islāmiyyat, Mujallat al-abḥāth al-kulliyat al-tarbiyyati al-asāsiyati, 2007;7(2):151-170.
57. al-Saʿīdān, Walīd Ibn Rāshid, Risālat al-Iktifāʾ bi-al-ʿamal bi-ghalabat al-ẓann fī masāʾil al-fiqh, n.d.:4. http://www.saaid.net/book/open.php?cat=4&book=1406. Accessed 31 Aug 2019.
58. al-Jurjanī, Muʿjam al-taʿrifāt, Beirut: Dār al-Kutub al-ʿIlmiyya; 2002.
59. al-Zarkashī, al-Baḥr al-muḥīṭ fī uṣūl al-fiqh, ed. ʿAbd al-Qādir ʿAbd Allāh al-ʿĀnī, Kuwait: Wizārat al-Awqāf wa-al-Shuʾūn al-Islāmiyya; 1992.
60. Jackson SA. Shariah, democracy, and the modern nation-state: some reflections on Islam, popular rule, and pluralism. Fordham Internal Law J. 2003;27(1):Article 5.
61. McGee M, Gardiner D. Permanence can be defended. Bioethics. 2017;31(3):220–30.
62. Padela AI, Qureshi O. Islamic perspectives on clinical intervention near the end-of-life: we can but must we? Med Health Care Philos. 2017;20(4):545–59. https://doi.org/10.1007/s11019-016-9729-y.
63. Butt Zubair M, Organ donation and transplantation in Islam – an opinion, Institute of Islamic Jurisprudence, [Internet] organdonationnhsuk; 2020. https://www.organdonation.nhs.uk/helping-you-to-decide/your-faith-and-beliefs/islam/. Accessed 20 Dec 2021.
64. Ali M, Maravia U. Seven faces of a fatwa: organ transplantation and Islam. Religions. 2020 Feb;11(2):99.
65. Padela AI, Duivenbode R. The ethics of organ donation, donation after circulatory determination of death, and xenotransplantation from an Islamic perspective. Xenotransplantation. 2018;25(3):1–2.
66. Bukhārī, Muḥammad ibn Ismāʿīl. Sahih Bukhari. Karachi: Muhammad Sarid, 1966, (i) ḥadīth no. 1338, (ii) ḥadīth no. 6527.
67. Abū Dā'ūd Sulaymān ibn al-Ashʿath al-Sijistānī, and Ahmad Hasan.. Sunan abu dawud. Lahore: Sh. M. Ashraf. 1984 (ḥadīth 3207).
68. Ibn Mājah, Muḥammad ibn Yazīd, Injah al-hajah 'Abd al-Ghani al-Dihlawi, and Suyūṭī.. Sunan Ibn Majah Karachi: Vali Muhammad 1952 (ḥadīth 1616).
69. Egyptian Awqaf Ministry, al-Majlis al-ʿAʿlā li'l-Shu'ūn al-Islamiyyah bi Wizārat al-Awqāf al-Miṣriyyah, Dār al-Iftā' al-Miṣriyya, 2nd Edition al-Fatāwa al-Islamiyyah, Cairo: 1997.
70. IFA-India, Death, when it is to be declared & Artificial Respiratory Apparatus. Medical issues, Islamic Fiqh Academy India; 2007. www.ifa-india.org. http://www.ifa-india.org/pdfs/pdf-20210410012607.pdf. Accessed 3 Jan 2022.
71. Miller AC, Ziad-Miller A, Elamin EM. Brain death and Islam: the interface of religion, culture, history, law, and modern medicine. Chest. 2014;146(4):1092–101. https://doi.org/10.1378/chest.14-0130.
72. ʿAṭā Allāh MʿAM. Mawt al-dimāgh wa mā yataʿallaqu bihi min aḥkām darāsat fiqhiyyat maqārina, Majallat Kulliyat al-banāt al-islāmiyyah. Jāmiʿat al-Azhar. 2016;15:863–79.
73. Movassagh H. Human organ donations under the 'iranian model': a rewarding scheme for US regulatory reform? Indiana Health Law Rev. 2016;13(1):83–118. https://mckinneylaw.iu.edu/ihlr/pdf/vol13p82.pdf. Accessed 20 Dec 2021
74. IOMS, (The Islamic Organization of Medical Sciences), al-munaẓẓamat al-Islāmiyyah li'l-ʿulūm al-ṭibbiyyah. 1985 Kuwait, see: Nadwat al-ḥayāh al-insāniyyah bidāyatuhā wa

nihāyatuhā fī al-mafhūm al-islāmī al-munaqidah, Majallat Majmaʿ al-Fiqh al-Islāmī 1986;3(2). 1996 Kuwait, see: Nadwat al-taʿrīf al-ṭibbī li'l-mawt (Ru'yatu islāmiyyah li baʿḍ al-mushkilāt al-ṭibbiyyat al-muʿāṣarah), 915.
75. Ibn Taymiyya, Taqi al-Dīn (d. 728 A.H.), Majmūʿ al-Fatāwa, 21 Vols Dār al-Wafā'. 2001.
76. al-Asqalānī, Ibn Ḥajr, Fatḥ al-bārī, Riyāḍ: al-Maktabat al-Salafiyya. 2015.
77. al-Zarkashī, al-Manthūr fī al-qawāʿid, ed. Taysīr Fāʾiq Aḥmad Maḥmūd, 3, Kuwait: Wizārat al-Awqāf wa-al-Shuʾūn al-Islāmiyya; 1982.
78. al-Nawawī, Rawḍat al-ṭālibīn, ed. ʿĀdil Aḥmad ʿAbd al-Mawjūd and ʿAlī Muḥammad Muʿawwaḍ, 8. Riyadh: Dār al-ʿĀlam al-Kutub. 2003.
79. Qiyāsah, Nadā, al-Mawt al-Dimāghī bayn al-ṭibb wa al-dīn, Majalla Jāmiʿa Damashq l'il-ʿUlūm al-Iqtiṣādiyya wa al-Qānūniyya, Damascus; 2010;1:479–500. https://ketabpedia.com/%D8%AA%D8%AD%D9%85%D9%8A%D9%84/%D8%A7%D9%84%D9%85%D9%88%D8%AA-%D8%A7%D9%84%D8%AF%D9%85%D8%A7%D8%BA%D9%8A-%D8%A8%D9%8A%D9%86-%D8%A7%D9%84%D8%B7%D8%A8-%D9%88%D8%A7%D9%84%D8%AF%D9%8A%D9%86/. Accessed 12 Dec 2021.
80. Sajid MI. Autopsy in Islam: considerations for deceased muslims and their families currently and in the future. Am J Forensic Med Pathol. 2016;37(1):29–31. https://doi.org/10.1097/PAF.0000000000000207.

Jewish Perspectives on Death by Neurologic Criteria

Rabbi David Shabtai

1 Historical Perspective

Debates about determining death are not new to Judaism. In 1837, Rabbi Moshe Sofer (1762–1839), the universally recognized leading Torah scholar of his generation, dealt with a similar concern, albeit from the opposite perspective than the current moment [1]. Responding to what appeared to be an anonymous query, [1] the Pressburg-based Rabbi Sofer addressed the Duke of Mecklenberg-Schweren's recent decree that all corpses must be left unburied for 72 h following the declaration of death. The Duke was responding to a rash of reports of premature declarations of death which led to people being buried while still alive. He argued that determining death was demonstrably still an inexact science. So as to err on the side of preserving life, the Duke ordered the postponement of all burials for 3 days, to be able to ensure the detection of any inappropriately early declarations of death.

As a matter of Jewish law (Halakhah), the issue is far from clear. On the one hand, Jewish law demands a speedy burial. The Torah prohibits leaving a body in wait unless the delay directly enhances the deceased's honor (such as waiting a short while for her children or other dignitaries to arrive) [2, 3]. However, Jewish law also places a premium on life-saving, setting aside virtually all mitzvot (commandments) and prohibitions in efforts to save a life [4]. Rabbi Sofer's questioner reasoned that even while a speedy burial is preferred because the newly built-in delay was intended to save lives, Jewish law should not only allow for this delay but enthusiastically embrace it. Rabbi Sofer thought otherwise.

[1] Most of the responsa in the collection identify the questioner, although, in some instances, such as this specific query, Rabbi Sofer intentionally omits the name. Later scholars have identified the questioner as Rabbi Zevi Hirsch Chajes (1805–1855), who recorded his response to Rabbi Sofer's original letter in his work Darkei Hora'ah (Responsa Section, number 3).

R. D. Shabtai (✉)
Charles E. Schmidt College of Medicine, Florida Atlantic University, Boca Raton, FL, USA

A. Lewis, J. L. Bernat (eds.), *Death Determination by Neurologic Criteria*,
Advances in Neuroethics, https://doi.org/10.1007/978-3-031-15947-3_27

He argued that the Traditional halakhic determination of death stood firm and was unchanged by this recent decree, having been established and ensconced in the Talmudic literature. After analyzing the historical background to these criteria, Rabbi Sofer succinctly concluded that once a person is no longer moving, is not breathing, and his heartbeat has ceased, he is considered dead according to Jewish law. Relevant to the query he was asked, he ruled that once these criteria are met, burial must commence as soon as reasonably possible and fought vigorously against introducing any waiting period before a funeral. [2]

Effectively, what Rabbi Sofer was arguing was that determining death is not an exclusively scientific endeavor. While it indeed must deal with physiological realities, determining death is in essence a question of values and ethics, which he equated with Halakhah. Modern science can raise questions and concerns as to the proper facts and even devise new and improved tests to ascertain whether or not certain criteria have been met. But ultimately, defining death is an exercise in the application of values and ethical principles. The same must therefore apply to the criteria for determining death—the physiological representations of that definition.

While Rabbi Sofer looked to the Talmud to analyze the Halakhic perspective on determining death and his responsum itself became an essential component of the modern debate, both works predate the notion of brain death, the responsum by several centuries and the Talmud by over a millennium. That said, because the Talmud contains and forms the basic tenets and principles of Jewish law, opinions about the Jewish perspective on brain death (and all other matters) derive from them. The question becomes how to apply the ancient text and principles to the modern phenomenon.

2 The Evolution of Jewish Law

Over the last two centuries, various formal streams of Judaism have emerged. Among their differentiating characteristics is the question of fealty to the Talmudic textual tradition [5, 6]. Orthodox Judaism maintains the strongest reliance and adherence to Traditional precedent in all matters, not just determining death, and must navigate through its waters to deal with modern challenges. At the other end of the spectrum, Reconstructionist and Reform Jewish thought is generally more open to modern influence and analysis and less bound to the Tradition (sometimes describing it as having a "voice but not a veto" [7, 8]) and more freely and less hesitantly wholeheartedly embraced brain death as the modern equivalent of death [9, 10]. Even within each stream, there are varying opinions, some of which may be a function of the level of understanding and knowledge of each scholar of the

[2] For more resources on the historical debate about delaying burial in Jewish Law, see Altmann A. Moses Mendelssohn: A Biographical Study. Tuscaloosa, Ala.: University of Alabama Press; 1973. p. 288–293; Samet M. Leaving the Dead Overnight: A History of the Controversy on Establishing the Time of Death. Hadash Assur min ha-Torah: Chapters in the History of Orthodoxy. Jerusalem: Dinur Center and Carmel Publishing; 2005. p. 157–227

material, which varies quite significantly [11]. This chapter will focus on the Orthodox approach and specifically analyze the positions of those scholars who have spent significant efforts understanding the science and medicine of brain death.

2.1 History

Since the destruction of Jerusalem in 68 CE and the subsequent exile of the Jewish people, there has been no centralized authority of Jewish Law. Throughout the generations of the Jewish exile, each community needed to develop its own means of deciding policy for new and modern questions that arose.

For more complicated questions, those with far-reaching ramifications, or those that impacted significantly larger populations, opinions were solicited from greater scholars, often in the same region, but sometimes even further away. But practically speaking, each community came to recognize those scholars in their midst, who by virtue of their wisdom, knowledge, and piety, attained mastery of Torah and accepted them as their Posek (decisor of Jewish Law). Some scholars rose to such prominence so as to attain a much broader appeal across regions, cities, and countries. But even then, the system itself encourages constant and consistent questioning and challenging of assumptions, often leading to a spectrum of differing halakhic opinions. Meaning, that even while all participants acknowledge and abide by the "rules of the game," there is still ample leeway for differing positions to emerge. The same is true of brain death, where multiple, varied, and conflicting halakhic opinions exist. This is not a fault of the system, but rather a feature of the way Jewish law develops from the Talmud.

2.2 The Process of Jewish Law

Judaism believes that together with the Five Books of Moses divinely revealed on Mount Sinai some 3300 years ago, Moses also received a Divine oral tradition. This oral tradition compliments and supplements the written text, by expanding on various aspects, adding meaningful features, and more broadly, allowing for a dynamic, living process of Halakhah. For a variety of historical reasons, the basics of this tradition were eventually recorded in written form in the Mishnah, compiled circa 200 CE. A few centuries later came the Babylonian Talmud—the magnum opus of Jewish oral tradition (a parallel and somewhat less complete version was completed in Israel some hundred years prior). But while the Talmud is a book of law, it is most certainly not a law book. It contains many, many laws, but never lists them sequentially or organizes them in an exhaustive fashion. Instead, it is loosely structured based on the Mishnah, beginning each section discussing an aspect of the Mishnah, but then veering in many directions.

The Talmudic narrative takes the reader on a journey to observe the Talmudic laboratory in action. Ideas are presented and scholars offer their opinions with their respective rationales. The Talmud often proceeds in the manner of a narrated discussion, beginning with a statement or ruling of earlier scholars that is then analyzed, challenged, and interpreted. Discussion and debate ensue, with each side parrying his opponent's attack by seeking support from other halakhic sources, comparisons with other Talmudic disciplines, or demonstrating faults in his inquisitor's logic. Hypotheses are presented, challenged by evidence, and either ultimately withstand the attack, are rejected, or modified. The Talmud is more interested in the process itself than the normative conclusion, and, therefore, often leaves it out. It is the reason that Talmud is described as being learned as opposed to simply read, since it is rarely an intellectually passive endeavor. It is the later commentators, in various iterations through the generations who begin to parse out normative Halakhah, which itself is continually further debated.

3 Nature of the Brain Death Debate in Jewish Law

Halakhic perspectives on brain death span the gamut from complete acceptance to total rejection. These stances are not simply manifestations of a more liberal vs. a more conservative approach to either modern mores or to an acceptance of scientific advancement, but rather the outcomes of painstaking analysis, debate, and discussion.

Both positions start with a common halakhic axiom—that death is defined as the departure, separation, or exit of the soul from the body. But even while it is universally accepted, it is not terribly helpful in actually determining death since it describes a metaphysical reality that is not easily intellectually grasped or sensibly perceived. Practically, death is determined based on particular criteria, and it is these that are debated in Jewish literature. As the criteria are physiological representations of the definition in action, the debate is essentially about which physiological parameter most accurately captures the notion of the soul's departure.

Generally, halakhic positions on criteria for death fit into one of two categories: A. Death as the irreversible cessation of vital motion (defined as any spontaneous movement that contributes to the continued biological viability of the organism); B. Death as the irreversible cessation of spontaneous respiration. Applying these criteria practically, the first position rejects brain death as the death of the individual because of a brain dead patient's continued heartbeat, while the second embraces it, as the diagnosis of brain death necessitates a determination of the absence of any respiratory drive.

The debate centers on functional criteria for death, and even then, specifically around easily measurable physiological functions, namely, breathing (spontaneous respiration) or heartbeat (the most basic, but by no means sole determinant of vital motion). The reason that even in a debate about brain death, the actual functions of the brain are not the focus is a function of the nature of the adjudication of Jewish law.

Essentially, determining death is not a novel question of the twentieth century. From a halakhic perspective, it is about the modern application of previously established principles. Halakhic discussions of brain death therefore look to the Talmud and its ensuing literature to derive those very principles and distill their essential qualities such that they can be applied to modern phenomena. Because it necessarily flows from Talmudic sources, the debate is limited to those criteria that were not only known to the Talmudic sages but were practically applicable at the time. Advances in science and medicine can help us better understand those criteria better and more accurately explain the cellular and molecular mechanisms behind them, but cannot shift the goalposts.

For example, for those who accept the irreversible cessation of spontaneous respiration as the criterion for death, modern science has discovered that it is the cells of the medulla that control spontaneous respiration. When those cells no longer function, spontaneous respiration ceases; therefore, medullary function can act as a surrogate marker for death. Neurological function did not become a substitute criterion, but rather a more nuanced and precise method of ascertaining the already established functional criteria.

4 Talmudic Sources

There is no description of determining death under "normal" circumstances anywhere in ancient Jewish literature [12]. This makes a lot of sense. Historically, determining death was not usually difficult or fraught with any controversy. When somebody died naturally, all organ systems—including respiratory and circulatory—shut down within moments of each other. "Plain vanilla death" did not require extensive Talmudic analysis. Only in extreme circumstances was there any question as to whether a person was alive or dead. The Talmudic literature addresses only two such instances.

Before analyzing the Talmudic text, a quick tangent on Talmudic methodology is in order. In Talmudic discourse, it is common to find a recorded debate between two earlier scholars, often together with an accompanying statement about the parameters of the disagreement. However, as it originated as an oral Tradition, even while the debate itself was acknowledged and (usually) accurately reported, the context, premises, and assumptions were not always faithfully maintained. In most cases, the specific detailed topic under debate and the varying perspectives (the superficial levels of the argument) are accurately relayed. This presentation leads many Talmudic analyses to begin by proposing a potential mechanism for interpreting the debate, only to be subsequently challenged and questioned by other similar cases that may turn on those same issues.

The first is a case of a person found under a collapsed building on Shabbat (the day of rest) who appears dead, but because of the circumstances, a visual inspection is insufficient to make a definite determination [13]. The issue at hand is the violation of Shabbat restrictions that may be necessary to save the victim's life. The Talmud suggests that if there is any chance that the person may be alive or that resuscitative efforts might be successful, virtually all halakhic restrictions are set aside for the purposes of life-saving. However, should the person be ascertained to be dead, all efforts must cease. While care for the dead is a paramount value in Judaism, Shabbat restrictions take precedence, and therefore no preparations for burial may ensue until after Shabbat has ended (Saturday night).

The Talmud records a debate as to how to ascertain that a person has in fact died [14]. In the context of a building collapse where rescuers are removing debris to hopefully uncover victims, the Talmud asks how much of the body must be uncovered before determining that the person has died. Two opinions are cited, somewhat cryptically: the first suggests that once the nose is uncovered, death can be ascertained, while the second argues that death cannot be determined until the chest is revealed. (In this instance, the Talmud does not ascribe attribution to each particular stance, but simply records a tradition that such a debate exists.) Clearly not focusing on particular bodily landmarks, but on physiological functions, the first opinion argues that a lack of breathing (determined by checking the nose) determines death, while the second opinion argues that cessation of cardiac activity is necessary. As presented, the two opinions seem to argue about the halakhic criteria for death: respiratory vs. cardiac, with practical relevance to the halakhic stance on brain death. For a variety of reasons, normative Halakhah sides with the first opinion that requires checking the nose and seemingly arguing for an exclusively respiratory criterion for death [15].

However, as the Talmud is wont to do, it proceeds to question the premises of the aforementioned argument.

Technically, the disagreement was not about the actual criteria for death, but rather the location of the body that must be checked after being uncovered from under the rubble. Meaning, these are not meant to be "spot checks" on various body parts, but rather an argument of how much rubble needs to be removed before an accurate determination of death can be made. The Talmudic sage Rav Pappa points out that the orientation of the victim, in relation to how the rescue efforts are proceeding, is a very relevant factor to this debate. In fact, Rav Pappa posits that the two opinions only argue when the victim is uncovered feet first. However, should the victim be uncovered head first, both opinions would agree that checking the nose is sufficient [14].

Rashi [Rabbi Shlomo Yitzhaki (1040–1105)], one of the earliest, most prolific, and influential medieval Talmudic commentators) explains that when finding a victim's feet first, the rescue efforts of removing rubble proceed toward the head [16]. As such, during the course of removing the rubble, the chest is exposed before the head and the presence of a heartbeat can be ascertained before even exposing the head and reaching the nose. Seemingly, Rav Pappa is arguing that after uncovering the chest, the first opinion demands continuing the rescue efforts until exposing the nose to check for breathing, while the second opinion suffices with checking for a heartbeat.

4.1 Opposing Brain Death as Death

Rashi is bothered by an obvious problem. If the two opinions are arguing about the criteria for death, it means that when uncovering the victim feet first and reaching the chest, the second opinion argues that a negative cardiac exam is sufficient to determine death. After all, this opinion argues that determining death depends on the cessation of cardiac activity. But then why would the first opinion demand continuing the rescue effort to uncover the nose? If it has already been established that the victim's heart is no longer beating, what use is there in checking for breathing? Without a heartbeat, breathing is not possible, leaving no reason at all to check the nose. Why then does the first opinion disagree? What are they really arguing about?

Sensitive to this point, Rashi explains that the key to understanding this Talmudic debate hinges upon precision in terminology. Had Rashi had the luxury of our modern lexicon, he likely would have explained that the original debate was never about the criteria for death, but rather about the tests that could sufficiently ascertain that the criteria had been met. But even while lacking that specific language, Rashi essentially proposes that very idea [16].

He describes the case as one where the victim is found feet first under a pile of rubble with rescue efforts of removing debris slowly working toward the head. Once the chest is uncovered, the rescuers perform a cardiac exam. The question the two Talmudic opinions debated was whether a negative cardiac exam is sufficiently specific to verify that the heartbeat has indeed irreversibly stopped. And as Rashi notes, the real question was not about a "simple" or "standard" cardiac exam, but one performed while the victim was still mostly covered by rubble in the context of a building collapse [16]. The second opinion argues that the inability to detect a heartbeat is sufficient to declare death. Effectively, this position believes that a negative cardiac exam—a test for death—is sensitive enough to ascertain that the heart is no longer beating—a criterion for death. But even while the second opinion trusts that not auscultating or not palpating a heartbeat sufficiently establishes that the heart is no longer beating, the first opinion is not as sure. [3]

Importantly, it is not that this first opinion debates the notion that a lack of a heartbeat denotes death, but simply questions whether or not the lack of a heartbeat has been demonstrably established. As Rashi notes, "it is possible to not observe any signs of life in the heart but observe them in the nostrils" [16]. There are various reasons that sometimes—particularly under the circumstances described by the Talmud—a heartbeat cannot be detected even though in actuality, it is still beating. The test—be it auscultation or palpation—is simply insufficiently specific to rule out a possible misdiagnosis. This opinion therefore demands an additional test, a check for breathing, before definitively determining death. Even while either exam may not be sufficiently specific on their own, taken together, the first Talmudic opinion is satisfied that the criteria—assessing for the absence of cardiac activity—can be determined.

This is entirely in concert with understanding the terminology in the modern discourse of determining death. A definition of death is a philosophical or other value-based notion as to the essence or of how a dead person fundamentally differs from the living. The criteria for death are physiological representations of a definition and are therefore, in essence, also philosophical or value-based constructs. Describing criteria for death necessitates an intimate understanding of physiology,

[3] Rabbi Tzvi Ashkenazi [1656–1718] is one of the earliest scholars to interpret Rashi in this manner. See his Responsa Hakham Tzvi (1712). Amsterdam. no. 77.

but only inasmuch as science is the fabric upon which the definition can manifest itself. In the Jewish Tradition, both the definition and criteria for death are determined by Jewish law. Representing a system of values, ethics, and morals, Jewish law posits an approach for death that reflects these fundamental elements. But not tests.

Understood properly, tests are wholly different in that they do not represent a stance taken on a matter of values and morals, but rather answer a simple question: have the criteria been met? A test answers a question of fact—are the physiological parameters set out by the accepted criteria (whatever they might be and each system unto itself) definitively manifest in this particular patient? The answer should be a simple yes or no (any ambiguity would mean that the criteria have not been definitively met). As assessments of facts, tests should utilize the most sensitive, specific, and updated equipment, technology, and skills available. The fact that the Talmud describes a simple cardiac exam without even describing the details of the exam is highly irrelevant. Even though the Talmud is interested in all aspects of the tripartite determination, it is only the definition and the criteria that have any eternal value. The tests described merely reflected the state of medicine of that ancient era and, as such, should be appropriately updated with each advancement in diagnostic assessment [17, 18].

As an aside, it is vital to note that all commentators take it as a given that any and all criteria for death—be they what they may—must include the notion of irreversibility. In Talmudic times, and for much of world history thereafter, once a person's heart had indeed stopped, there was no possibility for reversal. The knowledge and skills necessary to restart a heart and perpetuate life would take several centuries to even begin to emerge in rudimentary form. The scientific advances of the last century have changed this entire field. As such, modern Halakhists point out that nowadays, no determination of death should be made on the basis of these crude assessments. Instead, any and all modern tools should be employed not only to detect a heartbeat but also to resuscitate a patient whenever possible [19].

Taking all of this into consideration, the fact that normative Halakhah sides with the position that requires checking the nose is not taking a stand on the issue of identifying criteria for death at all because that was not what their debate was about. This position, much like the second, believes that the criteria for death must include the cessation of the heartbeat; it is only taking a position on the necessary and sufficient test to establish those criteria. Understood in this light, there is no Talmudic opinion that supports a solely respiratory criterion for death. The question was only about the specificity of the then available diagnostic assessments, which, as noted, must be updated in accordance with medical advances.

4.2 Accepting Brain Death as Death

The modern Halakhic perspective that argues for a solely respiratory-based criterion for death generally takes one of two forms. The first tries to reread this Talmudic passage and particularly focuses on the universally agreed upon conclusion that a determination of death requires "checking the nose" [20, 21]. From the Halakhic system's perspective, Rashi's comments cannot simply be ignored. As such, even while Rashi—as presented previously—seems to reinterpret this understanding to reflect a debate about tests for death and not criteria, some try to read Rashi in a different light. Considering that, for most of Jewish history, determining death was not a controversial issue, there is not a great deal of literature analyzing this particular comment, allowing for greater leeway in the modern interpretation.

A second approach relates this debate to larger scientific themes found in the Talmud. In so doing, it can both accept Rashi's insistence on focusing on the heartbeat and yet still argue for a solely respiratory criterion for death [12]. As presented, the working assumption was that the criteria for death were largely a spiritual concern, and, therefore, the cessation of the heartbeat (in the context of and as part of the cessation of all vital motion) somehow reflects the departure of the soul. However, the opposing perspective argues that even the cessation of the heartbeat reflects an essentially respiratory criterion for death.

As is well known, the ancients often described the heart as a respiratory organ, for which inhalation cooled down its activity and exhalation allowed the newly heated air to escape [22]. The heart was only formally established as an essentially circulatory organ during the time of William Harvey (1578–1657). As such, perhaps Rashi and even the Talmud intended on focusing exclusively on respiration as the functional criterion for death and simply used the example of the heart as what-they-understood-to-be a respiratory organ. According to this approach, any and all Talmudic and medieval references to the cessation of the heartbeat as part of the criteria for death is actually arguing for a respiratory criterion for death, with the cessation of heartbeat merely a surrogate marker for the irreversible cessation of respiration. The opposing perspective would respond that although this understanding of the Talmud and Rashi's comments is certainly possible, it is just as equally, if not more, likely that Rashi simply meant what he said. Meaning, that he was describing the accepted Tradition that the halakhic criterion for death is the irreversible cessation of all vital motion, including the heartbeat.

5 Second Talmudic Source

The second explicit source dealing with determining death also describes particularly exigent circumstances. In Jewish law, humans and certain creatures become ritually impure upon their death. This metaphysical reality has practical, real-world consequences inasmuch as people who come into contact with those corpses may not enter the Holy Temple or eat certain ritual foods and male Kohanim (descendants of Aharon the Priest) are forbidden from contacting human remains.

The Mishnah explains that ritual defilement only devolves upon a corpse once the soul has departed. So long as a person is alive, he or she is free from corpse impurity. The Mishnah continues, "If their heads were severed, even if they still wiggle about, they become impure (immediately), much like the severed tail of a lizard continues to wiggle" [23] Recognizing that decapitation is thankfully something with which few have experience, the Mishnah compares it to a more commonly observed phenomenon. Much as it is clear that the severed tail of the lizard is not living, despite its movement, so too a body's movement after decapitation is not indicative of life.

Moving on immediately to other topics, the Mishnah does not offer any further analysis and this is a section of the Mishnah for which there is no corresponding Talmudic tractate. For ancient readers of the Mishnah, there was likely little to discuss, given that there was no question as to why the decapitated person was dead. While the practical ruling of the Mishnah stands today as well (with the irreversibility caveat mentioned above, which as of yet is not yet in the realm of medical possibility), the reasons behind the ruling have become potentially far more relevant. Given the possibility to tease out various physiological functions from each other, it is important to question the mechanism through which the Mishnah posits to determine death in the case of decapitation, which may shed light on the larger question of determining death.

For those Halakhists who adopt a respiratory criterion for death, the issue is quite straightforward. Severing the head from the lungs of a person renders them irreversibly incapable of breathing and as such, dead [24]. The issue is far less clear for the opposing perspective. While breathing immediately stops upon decapitation, the heart still continues to beat for a short while. If the criteria for death require the cessation of all bodily motion, the continued heartbeat should render the person still alive—if even for a few more moments—and yet the Mishnah declares him dead immediately.

There are two ways to resolve this challenge: 1. Posit that there are two alternate pathways for determining death; 2. Reevaluate and refine the original position to include decapitation. Famously, Rabbi Shlomo Zalman Auerbach (1910–1995) argued that Halakhah acknowledged two tracks for determining death: A. Irreversible cessation of vital motion; B. Decapitation [25]. He resolved the contradiction by explaining that each discussion was describing a different criterion for death. He then went further and equated decapitation with the death and liquefaction of all brain cells even while the head itself might still be attached [26]. Applying this principle, he was initially led to believe that this was indeed an accurate description of a "brain dead" brain and as such would qualify as halakhic death. However, when learning that vast parts of the brain were still structurally intact, he retracted this erroneous application and based on his own principles ruled a brain dead person to be halakhically alive [25].

A far more common resolution to this conundrum was hinted at earlier. As Rabbi J. David Bleich (b. 1936) explains, there is nothing particularly special about the heart in terms of Halakhah's determination of death [27]. Instead, Rashi's (and others') focus on heartbeat is simply an example of "vital motion"—the irreversible cessation of which constitutes death. Vital motion is defined as any spontaneous bodily movement that contributes to, or is reflective of, the continued viability of the organism. Breathing and heartbeat are the most obvious examples but it also includes talking and coordinated motion. In most "natural" deaths, the heartbeat is often the last vestige of vital motion to be physiologically manifest and upon its irreversible cessation, the person is declared dead.

A close reading of Rashi reveals that this too was his actual intent. In introducing the discussion, Rashi describes the victim as lying motionless ("like a rock") and it is then that there is a debate as to how to ascertain his death [16]. There is no visible motion, for if there was, it would be clear that the person was still alive. For this otherwise motionless victim, the only question is whether some other, more subtle vital motion exists that is not readily perceptible. So long as any vital motion continues, the person is alive—which is why the rescuers look to detect a heartbeat—an otherwise, not readily perceptible vital motion and sign of continued life [28].

The reason the Mishnah declared a decapitated person to be dead immediately upon decapitation despite the short-lived continued heartbeat is because that heartbeat is not considered vital motion. Because it is not the continued heartbeat per se that determines life, but rather only the heart's contribution to the overall vital motion that matters, if the heart would beat in such a manner that was not deemed vital, its continued heartbeat would not be a hindrance to the determination of death. That is exactly what happens in decapitation. Once the head is severed, the circulatory system is rendered instantly useless. The heartbeat only serves to promote speedier exsanguination by continuing to pump blood toward the gaping hole out of the body. This certainly does not contribute to the continued viability of the organism and is therefore not considered "vital motion." As such, it is not a sign of life and the person is therefore understandably considered dead [28].

This is likely why the Mishnah immediately appends the phrase "even if they still wiggle about" to its description of decapitation. The reason is that, under "normal" circumstances, movement is generally considered a sign of life. The case of decapitation is different because all of the perceived motion—whether of the limbs or the heart—no longer contribute to the continued viability of the organism and therefore are irrelevant (and not considered an impediment) to determining death. Considering that brain dead patients all continue to experience significant vital motion—at the very least with regard to heartbeat—this perspective considers them to be alive.

6 Conclusion

These are two of the earliest sources marshaled in the current halakhic brain death debate. Modern Poskim (Jewish legal scholars) must integrate these Talmudic elements together with the layered commentary of medieval and later scholars as well as the rich response literature throughout the generations to arrive at a coherent and cohesive perspective to be brought to bear on the current scientific understanding of physiology. The halakhic principles derived are part of the eternal truth, representing a modern manifestation of the Divine oral tradition and the process is therefore not taken lightly.

The intricacies of the Halakhic system often support varied perspectives, even among world renown and recognized scholars, and particularly as regards modern phenomena, unknown to previous generations. The question of brain death as the death of the individual fits well into this halakhic paradigm. While the overwhelming majority of Poskim recognize brain dead patients as alive, a not-insignificant minority consider them dead. Practically speaking, this has serious ramifications for questions of continued care, allocation of resources, and organ donation. Because many of these issues are not individual in nature, but affect society, it is important that societal decisions ensure that individual, deeply held philosophical and religious truths are respected and supported.

References

1. Sofer M. Responsa Hatam Sofer. Pressburg, Hungary; 1855. no. 338.
2. Deuteronomy 21:23.
3. Shulhan Arukh, Yoreh De'ah 357:1.
4. Talmud Yoma 85b; Shulhan Arukh, Orah Hayyim 329:1.
5. Dorff E. The Jewish Tradition: Religious Beliefs and Healthcare Decisions [Internet]. Advocatehealth.com. 2002 [cited 26 December 2021]. https://www.advocatehealth.com/assets/documents/faith/jewish4.pdf.
6. Kaplan M. Not so random thoughts. New York: The Reconstructionist Press; 1966. p. 263.
7. Steinberg A. Entzyklopedia Hilkhatit Refu'it. Jerusalem, 2006: v. 6, s.v. rega ha-mavet, col. 834.
8. MRR 188–197 - Central Conference of American Rabbis [Internet]. Reform Responsa. 1983 [cited 26 December 2021]. https://www.ccarnet.org/ccar-responsa/mrr-188-197/.
9. Teutsch D. Bioethics: reinvigorating the practice of contemporary Jewish ethics. Wyncote, PA: Reconstructionist Rabbinical College Press; 2005.
10. Lewis A. A survey of multidenominational rabbis on death by neurologic criteria. Neurocrit Care. 2019;31(2):411–8.
11. Steinberg A. Determining the moment of death and heart transplants in: Keviat Rega ha-Mavet. Jerusalem: Schlesinger Institute; 2007.
12. Talmud Yoma 83a.
13. Talmud Yoma 85a.
14. Tendler M. Cessation of brain function: ethical implications in terminal care and organ transplant. N Y Acad Sci. 1978;315:394–7.
15. Yitzhaki, S. Commentary to the Talmud, Yoma 85a, s.v. hakhi garsinan.
16. Rabbi Shlomo Zalman Auerbach cited in Abraham A. Nishmat Avraham, vol. 1. 2nd ed. Jerusalem: Schlesinger Institute; 2014. p. 502.
17. Yosef O. Livyat Hen Jerusalem; 1986. p. 150.
18. Feinstein M. Iggerot Moshe, Yoreh Deah 2. New York; 1973. no.146.
19. Tendler M. Halakhic death means brain death. Jewish Rev. 1990: 6–8
20. Friedman JS. Guidance, not governance: Rabbi Solomon B. Freehof and reform responsa. Cincinnati, OH: Union College Press; 2013.
21. Bolli R. William Harvey and the discovery of the circulation of the blood. Circ Res. 2019;124:1300–2.
22. Mishnah Ohalot 1:6.
23. Veith F, Tendler M. Brain death. II. A status report of legal considerations. JAMA. 1977;238:1744–8.
24. Avraham AS. Nishmat Avraham. 2nd ed. Jerusalem: Schlesinger Institute; 2007.
25. Leizerson SB. Mishnat Hayyei Sha'ah. Jerusalem; 1994.

26. Bleich JD. Of cerebral, respiratory and cardiac death. In: Contemporary halakhic problems. Hoboken, NJ: Ktav; 1995. p. 316–50.
27. Maimonides, Laws of the Sabbath 2:19; Shulhan Arukh, Orah Hayyim 329:4.
28. Bleich JD. BiNetivot HaHalakhah, Volume 3. N Y; 2001. p. 119–121.

Part VI

Ethical and Social Issues

Public Views on Death by Neurologic Criteria

Antonio Ríos and Pedro R. Gutiérrez

In the past, the moment of death was marked by the cessation of heartbeat and breathing. As a result of technological advances during the 1950s and 1960s, the Harvard Medical School Ad Hoc Committee proposed the first standards for death by neurologic criteria in 1968 [1].

Fifty years later, death by neurologic criteria is now considered the legal equivalent of death by circulatory-respiratory criteria, synonymous with the death of a person [2], and its incorporation in many legal systems around the world has transformed end-of-life care and organ transplantation [3].

However, there is a key difference between death by neurologic criteria and death by circulatory-respiratory criteria: a person who is dead by neurologic criteria can be sustained by a machine and lacks the traditional visual signs of death. Because of this, although death by neurologic criteria is conceptually well-founded and generally accepted, debate persists as to whether it is *really* equivalent to "traditional" death by circulatory-respiratory criteria [3]. The acceptance of death by neurologic criteria has been debated in many forums where different problems and

A. Ríos (✉)
Instituto Murciano de Investigación Bio-Sanitaria (IMIB-Arrixaca), Universitario Virgen de la Arrixaca, Murcia, Spain

Department of Surgery, Pediatrics, Obstetrics and Gynecology, University of Murcia, Murcia, Spain

Proyecto Colaborativo Internacional Donante (International Collaborative Donor Project), Murcia, Spain
e-mail: arzrios@um.es

P. R. Gutiérrez
Proyecto Colaborativo Internacional Donante (International Collaborative Donor Project), Murcia, Spain

Departament of Urology, Complejo Hospitalario Universitario de Canarias, Tenerife, Spain

Departament of Surgery, University of La Laguna, Tenerife, Spain
e-mail: prgti@ull.edu.es

A. Lewis, J. L. Bernat (eds.), *Death Determination by Neurologic Criteria*, Advances in Neuroethics, https://doi.org/10.1007/978-3-031-15947-3_28

varying solutions have been raised [4–6], as discussed throughout this book. There is criticism of the concept of death by neurologic criteria in both the medical and philosophical-ethical communities. This stems, in part, from the fact that it is difficult to accept a concept of death that abandons the traditional view of visibly defined death by circulatory-respiratory criteria. Further, philosophical-ethical objections to the concept of death by neurologic criteria center on the concern that it is merely a pragmatic way to free patients and their families from the burden of an indefinitely prolonged coma, reallocate medical resources, and increase the supply of organs for donation. Another ethical point of view is that while it may be convenient to redefine death, termination of "organ support" could mean ending a human life for utilitarian reasons.

Because death by neurologic criteria is a social construct and views on when life ends vary (Fig. 1), it will always be subject to criticism [7, 8]. However, it is necessary to recognize that controversies about death have medical, legal, social, cultural, and economic implications. Although discussions about death by neurologic criteria generally are not held in an open, clear, and transparent way in front of the eyes of society [9], these debates can impact public perspectives, leading to doubt and fear about death by neurologic criteria. Because it is important for the public and clinicians to agree on whether an intubated person who has a catastrophic brain injury is alive or dead [10], this chapter will examine public views on death by neurologic criteria.

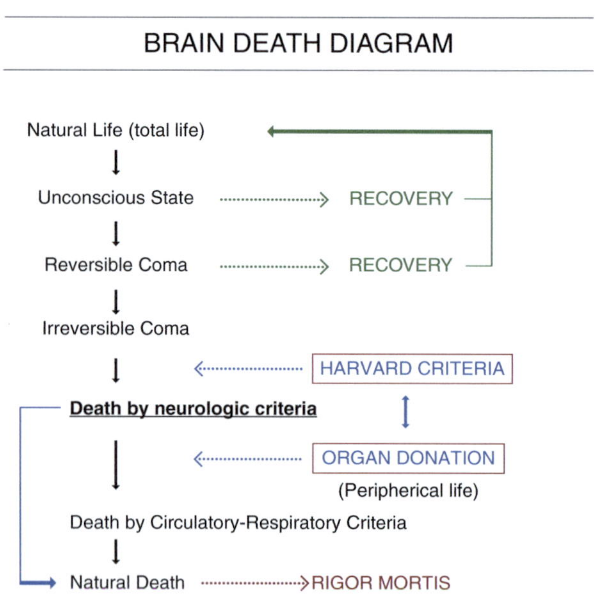

Fig. 1 Diagram of the states from natural life to natural death, where brain death means no central life, which provides limited time for peripheral organs to be donated and transplanted to allow others to continue living

1 Medical and Legal Acceptance of Death by Neurologic Criteria Around the World

Although the concept of death by neurologic criteria is medically and legally accepted throughout much of the world, medical standards and laws about the declaration of death are inconsistent, as discussed elsewhere in this book [11, 12]. The content of a given country's standards and laws about death declaration, and this variability itself, can impact public views on death by neurologic criteria [12]. However, there are numerous other factors that contribute to public perspectives about death by neurologic criteria.

2 Sources of Public Knowledge About Death by Neurologic Criteria

Public awareness about death by neurologic criteria is largely based on traditional or social media (television, radio, internet, etc.) [7]. Although this information is often inaccurate, the coverage of medical and health issues in the media "educates" the public and affects their perceptions [13]. An analysis of the mainstream social media coverage about death by neurologic criteria revealed that erroneous information was presented in 72% of stories [14]. The most common errors were the imprecise use of medical terms and the misrepresentation of death by neurologic criteria as a state of life [14]. Further, less than 4% of the articles explained the medical conditions required to declare death by neurologic criteria [14, 15]. Similarly, the portrayal of death by neurologic criteria in film and television is misleading; in a review of 24 productions that addressed death by neurologic criteria, two clinicians felt that only 13% provided the public with a complete and accurate understanding of death by neurologic criteria [16].

3 How Much Does the Public Know About Death by Neurologic Criteria?

There are few studies on public knowledge about death by neurologic criteria. In a survey of 1351 Ohio residents, Siminoff et al. found that although more than 98% of respondents had heard of the term "brain death," two-thirds did not know that this was a form of legal death and more than half were not capable of differentiating coma from death [17]. In the review of 43 articles with over 18,000 participants, Shah et al. found that there is a general lack of understanding of the biological and legal facts about death by neurologic criteria [18]. Even the families of patients declared dead by neurologic criteria have a dearth of knowledge about this topic. Siminoff et al. found that only 28% of 403 families in Southwest Pennsylvania and Northeast Ohio were able to provide a completely correct definition [19]. In a survey of knowledge about death by neurologic criteria amongst Latin American immigrants in Spain, only 25% of respondents were aware of the concept of death by neurologic criteria [20]. Contrastingly, in a survey of the native Spanish population, 51% of respondents knew about death by neurologic criteria [21].

4 The Difference Between Knowledge and Acceptance of Death by Neurologic Criteria

With these data in mind, one could conclude that public rejection or uncertainty about death by neurologic criteria is attributed to a lack of knowledge that could or should be addressed by higher education. This type of assumption is well-recognized in public policy and healthcare debates as the "knowledge deficit model." However, this is a problematic assumption because it fails to see that differences of opinion may represent genuine differences in sociocultural values [22]. There is considerable data to suggest that while knowledge and education may predict the strength of attitudes toward scientific matters, the positivity of attitudes is poorly correlated with knowledge [23].

5 What Does the Public Think About Death by Neurologic Criteria?

In considering the available data on public views about death by neurologic criteria, it is necessary to recognize there are a number of fundamental limitations. There is a small number of studies that address this topic, and there is marked heterogeneity across studies. Study cohorts are generally very small and restricted to professional, cultural, or regional populations [22]. Therefore, they are unable to provide a complete view of public perception at the population level [22, 24]. Of course, there are always biases inherent in all questionnaire-based studies [25]. However, this issue is magnified because each study uses a unique questionnaire, hampering the comparison of results and making it challenging to know whether the differences between populations are due to the subjects themselves or to the questionnaire [26, 27].

In a scoping review of the literature, Skowronski et al. identified 32 studies that analyze attitudes towards death by neurologic criteria [22]. The vast majority of these studies focus on the views of healthcare professionals and university students; only six are population studies. In most of the 32 studies, around 75% of respondents accepted death by neurologic criteria as the death of the person. The authors conclude that a "considerable proportion of people, including healthcare professionals, have doubts about the medical and ethical validity of modern determinations of death".

The few studies on public perceptions of death by neurologic criteria reveal mixed views. A 1995 survey of public attitudes about death and organ donation in Spain demonstrated that 70–80% of respondents were unaware and/or fearful of the concept of death by neurologic criteria [28]. In the aforementioned study by Siminoff et al. surveying 1351 Ohio residents on their knowledge and beliefs about death by neurologic criteria, 40% believed a patient declared dead by neurologic criteria was dead, 43% considered them to be "as good as dead," and 16% felt they were alive [17, 29]. In a survey of 136 patients hospitalized in Brazil, Teixeira et al. found that 77% believed a patient declared dead by neurologic criteria had the "potential to live" [30].

There is scant empirical data on public perspectives about death by neurologic criteria outside of Europe and the Americas. However, the survey by Roels et al. of international healthcare professionals on attitudes about death is revealing [31]. Support for the statement, "Brain death is a valid determination of death," was highest in Norway (95%) and lowest in Japan (36%). Despite the fact that the world is becoming more globalized, it is important to recognize that the public perception of death is linked to culture. There is a greater cultural reluctance to accept the concept of death by neurologic criteria in Eastern cultures compared to Western cultures, particularly in Asian societies based on Confucianism like Japan, China, and Korea [32]. This is discussed in detail elsewhere in this book.

There have not been any studies in Africa on public perspectives about death by neurologic criteria. However, surveys of emigrants to Europe demonstrate that only 21% of respondents accept death by neurologic criteria as the death of a person [33]. Cultural views of death by neurologic criteria in Africa are discussed in detail elsewhere in this book.

In summary, although there are studies that address the public perspective on death by neurologic criteria, additional high-quality studies are needed to examine both factual knowledge and value-based attitudes internationally.

6 Impact of the Public Perception of Death by Neurologic Criteria on Organ Donation and Transplantation

While organ donation and transplantation are outside the scope of this book, it is worth noting that most studies report a close relationship between knowledge and acceptance of death by neurologic criteria and attitude towards organ donation and transplantation [21, 34, 35]. Thus, those who understand the concept of death by neurologic criteria are more in favor of organ donation compared to those who do not understand it. In general, based on this association, it has been suggested that reinforcing knowledge and acceptance of death by neurologic criteria could contribute to improved attitudes towards organ donation and transplantation.

7 The Role of the Healthcare Profession in Educating the Public About Death by Neurologic Criteria

Since the establishment of the Hippocratic tradition, one responsibility of healthcare professionals has been the promotion of health and preventive care [36]. In a statement supported by the American Academy of Neurology, American Academy of Pediatrics, American College of Chest Physicians, American College of Radiology, American Neurological Association, American Society of Neuroradiology, Child Neurology Society, and Neurocritical Care Society, it was suggested that healthcare professionals need to develop and promote educational initiatives for the public about death by neurologic criteria [36, 37]. Public confusion about death by neurologic criteria can contribute to mistrust and decrease confidence in the ability of healthcare professionals to accurately determine death [36, 37].

8 Conclusion

We examined public perspectives about death by neurologic criteria. Although there are studies that address this topic, it is clear that additional research is needed. Nonetheless, healthcare professionals should be mindful that public knowledge about death by neurologic criteria is lacking and that perspectives vary based on cultural beliefs.

References

1. Report of the ad hoc committee of the Harvard Medical School. A definition of irreversible coma. JAMA. 1968:85–8.
2. Magnus DC, Wilfond BS, Caplan AL. Accepting brain death. N Engl J Med. 2014;370:891–4.
3. International figures on organ, tissue & haematopoietic stem cell donation & transplantation activities. Documents produced by the council of Europe European Committee (Partial Agreement) on Organ Transplantation (CD-P-TO). Year 2020. Dominguez-Gil B editor. Newsletter Transplant 2021; 26 (Monographic volume). ISSN: 2171–4118. http://www.ont.es/publicaciones/Documents/V4%202021.pdf. Accessed 12 Marc 2022.
4. Greer DM. Determination of brain death. N Engl J Med. 2021;385:2554–61.
5. O'Keeffe FJ, Mendz GL. Diagnosing death 50 years after the Harvard brain death report. New Bioeth. 2021;27(1):46–64.
6. Shewmon DA. The brain and somatic integration: insights into the standard biological rationale for equating "brain death" with death. J Med Philos. 2001;26:457–78.
7. Busl KM. What does the public need to know about brain death? AMA J Ethics. 2020;22:E1047–54.
8. Truog RD. Is it time to abandon brain death? Hast Cent Rep. 1997;27:29–37.
9. Lewis A. Contentious ethical and legal aspects of determination of brain death. Semin Neurol. 2018;38:576–82.
10. Shewmon DA. The brain and somatic integartion: insights into standard biological rationale for equating "brain death" with death. J Med Philos. 2021;26:457–78.
11. Wijdicks EF. Brain death worldwide: accepted fact but no global consensus in diagnostic criteria. Neurology. 2002;58:20–5.
12. Greer DM, Shemie SD, Lewis A, et al. Determination of brain death/death by neurologic criteria: the world brain death project. JAMA. 2020;324:1078–97.
13. Larsson A, Oxman AD, Carling C, Herrin J. Medical messages in the media—barriers and solutions to improving medical journalism. Health Expect. 2003;6:323–31.

14. Lewis A, Lord AS, Czeisler BM, Caplan A. Public education and misinformation on brain death in mainstream media. Clin Transpl. 2016;30:1082–9.
15. Daoust A, Racine E. Depictions of "brain death" in the media: medical and ethical implications. J Med Ethics. 2014;40:253–9.
16. Lewis A, Weaver J, Caplan A. Portrayal of brain death in film and television. Am J Transplant. 2017;17:761–9.
17. Siminoff LA, Burant C, Youngner SJ. Death and organ procurement: public beliefs and attitudes. Soc Sci Med. 2004;59:2325–34.
18. Seema KS, Kenneth K, Franklin GM. A narrative review of the empirical evidence on public accitudes on brain death vital organ transpalntation: the need for better data to inform policy. J Med Ethics. 2015;41:291–6.
19. Simininoff LA, Mercer MB, Arnold R. Families's understanding of brain death. Prog Transplant. 2003;13:218–24.
20. Ríos A, López-Navas A, Navalón JC, et al. The Latin-American population in Spain and organ donation. Attitude toward deceased organ donation and organ donation rates. Transplant Int. 2015;28:437–47.
21. Conesa C, Ríos A, Ramírez P, et al. Estudio multivariante de los factores psicosociales que influyen en la actitud poblacional hacia la donación de órganos. [Multivariate study of the psychosocial factors affecting public attitude towards organ donation]. Nefrologia. 2005;25:684–97.
22. Skowronski G, Ramnani A, Walton-Sonda D, et al. A scoping review of the perceptions of death in the context of organ donation and transplantation. BMC Med Ethics. 2021;22:167.
23. Allum N, Sturgis P, Tabourazi D, Brunton-Smith I. Science knowledge and attitudes across cultures: a meta-analysis. Public Underst Sci. 2008;17:35–54.
24. Ríos Zambudio A, López-Navas A, Ayala-García M, et al. Knowledge of the brain death concept by personnel in Spanish and Latin-American healthcare centers. Int J Artif Organs. 2014;37:336–46.
25. Choi BC, Pak AW. A catalog of biases in questionnaires. Prev Chronic Dis. 2005;2:A13.
26. Ríos A, López Navas A, de Francisco C, et al. Psychometric characteristics of the attitude questionnaire toward the donation of organs for transplant (PCID-DTO-RIOS). Transplant Proc. 2018;50:345–9.
27. Ríos ZA. Proyecto Colaborativo Internacional Donante (International Colaborative Donor Projet). Cir Esp. 2018;96:69–75.
28. Martínez JM, Martín A, López Jorge S. La opinión pública española ante la donación y el trasplante de órganos. [Spanish public opinion concerning organ donation and transplantation]. Med Clin. 1995;105:401–6.
29. Siminoff LA, Burant C, Youngner SJ. Death and organ procurement: public beliefs and attitudes. Kennedy Inst Ethics J. 2004;14:217–34.
30. Teixeira RKC, Gonçalves TB, Silva JA. Is the intention to donate organs influenced by the public's understanding of brain death? Rev Bras Ter Intensiva. 2012;24:258–62.
31. Roels L, Spaight C, Smits J, Cohen B. Critical care staffs' attitudes, confidence levels and educational needs correlate with countries' donation rates: data from the donor action database. Transpl Int. 2010;23:842–50.
32. Terunuma Y, Mathis BJ. Cultural sensitivity in brain death determination: a necessity in end-of-life decisions in Japan. BMC Med Ethics. 2021;22(1):58.
33. Actitud hacia la donación de órganos para trasplante de la población africana residente en España. Estudio Nacional Español Estratificado. Doctorando: Joaquín Carrillo Director: Antonio Ríos Escuela Internacional de Doctorado Universidad de Murcia (Spain) Lecture: 19–2-2021.
34. Ríos A, Cascales P, Martínez L, et al. Emigration from the British Isles to South-Eastern Spain: a study of attitudes toward organ donation. Am J Transplant. 2007;7:2020–30.

35. Irving MJ, Tong A, Jan S, et al. Factors that influence the decision to be an organ donor: a systematic review of the qualitative literature. Nephrol Dial Transplant. 2012;27:2526–33.
36. Rubin MA. Guidance for physicians who wish to influence policy development on determination of death by neurologic criteria. AMA J Ethics. 2020;22:E1033–7.
37. Lewis A, Bernat JL, Blosser S, et al. An interdisciplinary response to contemporary concerns about brain death determination. Neurology. 2018;90:423–6.

Cultural Considerations in the Declaration of Death by Neurologic Criteria in Asia

Bryan J. Mathis, Yuri Terunuma, and Yuji Hiramatsu

1 Introduction

Asia is comprised of over 40 countries and houses nearly 60% of the world's population. Although there are common Confucian underpinnings across the continent, the myriad cultures that coexist in modern Asia are heterogenous in technology and Westernization. In the context of globalization, Westernization within Asian societies raises post-industrial issues for which traditional culture has no easy answer. Brain death, a relatively new concept first introduced to Asia from the West, is no exception. The questions that must be asked are "Should Asia employ the same criteria to determine death as the West?" and "Can Asia employ the same criteria to determine death as the West?" To address these questions, we discuss the common fundamental characteristics of Asian cultures, major religions, and values endemic to Asia with the aim of delineating the necessity, presence, and dissemination of the concept of brain death within the Asian cultural milieu.

While the western world debates the merits of whole-brain versus brainstem criteria [1], in Eastern and Southeastern Asia, the concept of brain death does not even exist unless organ donation is considered. In some cases, like in India and China, this is attributed to resource limitations [2]. In Japan, South Korea, Malaysia, and Thailand, there are guidelines for the evaluation of brain death, but these

Bryan J. Mathis and Yuri Terunuma contributed equally.

B. J. Mathis (✉) · Y. Hiramatsu
International Medical Center, University of Tsukuba Hospital, Ibaraki, Japan
e-mail: bmathis@md.tsukuba.ac.jp; yuji3@md.tsukuba.ac.jp

Y. Terunuma
School of Medicine, University of Tsukuba, Ibaraki, Japan

A. Lewis, J. L. Bernat (eds.), *Death Determination by Neurologic Criteria*, Advances in Neuroethics, https://doi.org/10.1007/978-3-031-15947-3_29

guidelines are highly selective (e.g., granting the ability to diagnose brain death only to physicians of certain specialties, requiring ancillary tests, and a legal guardian to proceed with diagnosis, etc.) [3–6]. Comparisons of brain death in Asia are clouded by the fact that brain death determination is limited to considerations of organ donation, coupled with a patchwork of these rigorous guidelines. Here, we focus on the cultural contention related to the determination of death by neurologic criteria in Eastern and Southeastern Asia.

2 Example: The Progression of Thought Regarding Brain Death in Japan

In her 2002 foundational work, *Twice Dead*, Dr. Margaret Lock contrasts the relatively facile Western commodification of human organs with the more complex Japanese view on organ donation as viewed through the prism of Buddhism, Shintoism, Confucianism, and traditional culture [7]. Through a series of interviews, case studies, research reports, and media analysis, the thoughts of doctors, patients, donor families, and recipients reveal Japanese brain death laws to be a poor fit for the cultural and religious expectations of the patients they purport to serve. Concerns about the "unnatural" nature of dying in a hospital while the organs of a loved one are given to unknown strangers, worries that medical care will be prematurely withdrawn to preserve organs, and the redefinition of death as neurologically centered when bodies are visibly breathing or functioning have not been allayed by medical science or legislative initiatives. This lack of progress in public acceptance, even while organ donation is built into the driver licensing system (similar to the US), stems from multiple causes both common to Asia and unique to Japan. The next sections will detail the roots of these issues from religious, technological, and cultural perspectives.

3 Brain Death-Associated Factors, Organ Donation, and Statistics in Asia

Mortality, judging from reported statistics, is greatly heterogeneous across Asia. The most common causes of death differ depending on the degree of industrialization in each region: injuries, infectious diseases, and pregnancy/childbirth-related causes account for a large proportion of adult deaths in more agrarian Southeast Asia, while cancer deaths and suicide are more common in industrial East Asia [8–10]. Similarly, the reported incidence of brain death varies greatly due to different degrees of acceptance and implementation of the use of neurologic criteria, especially with regard to any cause of irreversible brain injury, to declare death. Japan and South Korea, which have laws that accept brain death after any irreversible brain injury and coma as legal forms of death, have national organizations that collect and report data on the number of brain-dead donor transplants or deaths that could have otherwise been recorded as brain deaths if organ donation was conducted [11, 12]. In contrast, even with a traumatic brain injury prevalence of 644 per 100,000 people, Vietnam does not publicize brain death statistics or existing laws that determine brain death (at least in English), in line with many countries in Pacific Asia that lack protocols or laws for determining/reporting brain death [13–15]. Such a lack of consistency in laws and reporting across the entire continent complicate efforts to collect data on specific causes and dispensation of neurological injuries.

Brain death statistics, when reported at all, are often specific to organ donation (the number of organ donations from brain-dead donors). This is a distinguishing characteristic of brain death in several Asian countries, largely due to the fact that brain death there often exists solely in the context of organ donation. But, even in countries with such organ donation laws, such as Vietnam, the aforementioned lack of published statistics makes it difficult to determine whether organ donations are tied to brain death. Additionally, organ donation may be complicated by demographic factors such as income, education, and ethnicity. A study in the higher-income Asian country of Singapore found that 85% of 799 participants would agree to donate the organs of family members, but this was highly correlated with education and knowledge of the finality of brain death [16]. However, since even a single objection by a family member can halt the process in Japan or China, larger or richer countries still face obstacles to organ donation and, therefore, the determination of brain death necessary to start the donation process. Additionally, ethical factors, obfuscation of actual incidence, and evidence of organ procurement from prisoners of conscience in countries like China may raise fear among the populace, casting organ donation in the light of political punishment and reducing participation [17].

3.1 Eastern Asia Traumatic Brain Injury Statistics and Laws on Brain Death or Organ Donation

Most countries in East Asia, such as Japan and Korea, have established national registries to record new cases of traumatic brain injury, which is a leading cause of brain death, second only to strokes [12, 18]. However, these databanks are not comprehensive; many only include a fraction of the hospitals that participate in data registration. Additionally, there have been cases of missing information and incomplete reporting [19]. Thus, traumatic brain injury incidence is underreported and current statistics are largely dependent on population-based epidemiological studies, regional initiatives, or national health insurance claims instead of nationwide reports [20–22]. Further development in national TBI incidence reporting is warranted to ascertain the true incidence of traumatic brain injury in these countries. Tables 1 and 2 list available traumatic brain injury statistics and the legal status of brain death/organ donation legislation in China, Japan, North Korea, South Korea, Taiwan, Hong Kong, and Macao.

Table 1 Incidence of traumatic brain injuries in East Asia (China, Japan, North Korea, South Korea, Taiwan, Hong Kong, and Macao).

1. Incidence of traumatic brain injury (TBI)				
	Name of TBI Databank (where available)	Incidence	Average annual incidence rate	Source
China	Chinese Head Trauma Data Bank (No national TBI incidence reported)	770,060–890,990 cases (1983–1985 estimate)	55.4–64.1 cases per 100,000 population (1983–1985 estimate) 442.4 cases per 100,000 population (2020 survey estimate)	[22, 86, 87]
Japan	Japan Neurotrauma Data Bank (not comprehensive; only includes data of **severe** TBI) Japan Trauma Data Bank (not all hospitals participate in data registration)	95,484 cases (2004–2018 cumulative)	No official data	[88, 89]
North Korea	Unknown	No official data	No official data	None found
South Korea	Korean Neuro-Trauma Data Bank System (not all hospitals participate)	247,989 cases (2017)	484.1 per 100,000 population	[21, 90]
Taiwan	Head Injury Registry in Taiwan (not all hospitals participate)	99,391 cases (2007–2008 cumulative)	No official data 49,695.5 Total cases (approx. 216.1 per 100,000 population) as calculated from incidence	[91, 92]
Hong Kong	No TBI-specific data bank Each of the 5 designated trauma centers has its own trauma registry ➞ Data from all 5 registries are annually merged for administrative purposes	No official data	No official data	[93]
Macao	Unknown	No official data	No official data	None found

Table 2 Data on brain death/organ donation statistics in East Asia (China, Japan, North Korea, South Korea, Taiwan, Hong Kong, and Macao)

	Presence of laws related to brain death or organ donation	Country-level reporting authority	Statistics from national reports	Statistics from literature reports
China	No laws Living-related donors only [94]	The China Organ Transplant Response System (COTRS)	Inaccessible/No English data available	
Japan	Organ Transplant Law (1997)	Japan Organ Transplant Network (JOTN)	68 brain-dead donors total = 0.54 brain-dead donors per million population (2020) [12]	
North Korea	Unknown	Unknown	Unknown	
South Korea	Internal Organs Transplant Act (2000) [95]	Korean Network for Organ Sharing (KONOS)	No English data available	2426 bases of potential brain death (2018) = 47 per million population [11] 449 diagnosed brain-dead donors total = 8.7 per million population [11]
Taiwan	Human Organ Transplant Act (1987) [95] Living-related donors only [94]	Taiwan Organ Registry and Sharing Center	Inaccessible	12.3 deceased organ donations per million population (2016) [96]
Hong Kong	Human Organ Transplant Ordinance enacted 1995 [97] Guidelines (Expired 2009) [98] Living-related donors only [94]	Department of Health (Organ donation statistics)	No English data available	6.1 deceased donors per million population (2013) [97] 80–120 certified brain deaths in 2015 [99]
Macao	Brain Death Standards and Regulations (2016)	Unknown	No English data available	

3.2 Southeastern Asia TBI Statistics and Laws on Brain Death or Organ Donation

Traumatic brain injury is common in Southeast Asia due to a rapid increase in motorcycle usage and lack of helmet mandates [23, 24]. Despite a high frequency, no Southeast Asian country has a dedicated database for traumatic brain injury cases. However, Southeast Asian countries are believed to have some of the highest traumatic brain injury mortality rates in the world due in part to limited resources, facilities, and expertise in neurocritical care that complicate treatment [13, 25]. Efforts to collect data from these countries are also hampered by a lack of electronic medical records [26]. Tables 3 and 4 list available traumatic brain injury statistics

Table 3 Incidence of traumatic brain injuries in Southeast Asia (Cambodia, Laos, Malaysia, Myanmar, Thailand, Vietnam, Brunei, East Timor, Indonesia, Philippines, Singapore)

	Name of TBI Databank (where available)	Incidence	Average annual incidence rate	Source
Brunei	No	1535 (2016)	384 per 100,000 population	[13]
Cambodia	No	1200 cases (Nov 2013–Mar 2016 Cumulative incidence at a single institute)	No English data available	[100]
Indonesia	Riskesdas (Basic Health Research) Not specific to TBI	TBI prevalence is 8.2% (2013) [101]	No English data available	[102–104] (single institute reports)
Laos	No	No English data available	No English data available	None found
Malaysia	National Trauma Database Malaysia (Not specific to TBI; only 8 participating hospitals; underreporting of trauma cases has been recognized)	742 pediatric cases (2010, regional)	32 pediatric cases per 100,000 population (2010, regional)	[105, 106]
Myanmar	No	No English data available	No English data available	[26]
Philippines	Online National Electronic Injury Surveillance System (not specific to TBI)	1729 cases (2014)	No English data available	[107]
Singapore	National Trauma Registry, Singapore (not specific to TBI)	780 cases (2004–2011) at a single institute	No English data available	[108]
Thailand	No	No English data available	No English data available	[109]
Vietnam	No	No English data available	No English data available	None found
East Timor	No	No English data available	No English data available	None found

Table 4 Data on brain death/organ donation statistics in Southeast Asia (Cambodia, Laos, Malaysia, Myanmar, Thailand, Vietnam, Brunei, East Timor, Indonesia, Philippines, Singapore)

	Presence of laws related to brain death or organ donation	Country-level reporting authority	Statistics from national reports	Statistics from literature reports
Brunei	No	Unknown	No English data available	
Cambodia	No	Unknown	No English data available	
Indonesia	Only living donors are allowed [94]	Unknown	No English data available	
Laos	Unknown	Unknown	No English data available	
Malaysia	Criteria/guidelines exist, but no substantial laws [5, 110]	National Transplant Resource Center	105 diagnosed cases of brain death (2015) 32 brain-dead donors (2015) = 1.06 per million population (12th report of national transplant registry 2015)	122 diagnosed cases of brain death (2018–2019 cumulative) [111]
Myanmar	Body Organ Donation Law [112]	Unknown	No English data available	
Philippines	Organ Donation Act of 1991 (Republic Act No. 7170, Amended No.7885) [113]	Human Organ Preservation Effort (HOPE) of National Kidney and Transplant Institute (NKTI) Founded 1983	No English data available	437 referrals for potential deceased organ donors (2009–2012 cumulative) 104 actual deceased donors (2009–2012 cumulative) [114]
Singapore	Human Organ Transplant Act from 1987 [115]	National Organ Transplant Unit of the Ministry of Health	No data for brain death or brain-dead donors	26 deceased organ donors (2007) include both cardiac and brain death donors [116]
Thailand	Brain death criteria established by Medical Council in 1989 [4] No specific laws, but basic principles are outlined by MC and Red Cross [117]	The Thai Red Cross Organ Donation Centre Founded 1994	No English data available	791 deceased donors (2007) [118] 3.33 deceased donors per million population (2016) [119]

(continued)

Table 4 (continued)

	Presence of laws related to brain death or organ donation	Country-level reporting authority	Statistics from national reports	Statistics from literature reports
Vietnam	Donors must be ≥18 years old [120] Law on donation, removal, and transplantation of human tissues and organs and donation and recovery of cadavers established 2006 [121]	No, only self-reporting from select transplant centers [119]	No English data available	
East Timor	No	Unknown	No English data available	

and the legal status of brain death/organ donation legislation in Cambodia, Laos, Malaysia, Myanmar, Thailand, Vietnam, Brunei, East Timor, Indonesia, Philippines, and Singapore.

4 Rapid Technological Advancement in Asia Vs. Slow Cultural Change

Historically, rapid technological advancements in East and South Asian countries have been fueled by a drive to upgrade their economies to keep up with the competitive international market [27]. However, even as technology and industrialization flow from the West, delayed regulatory policies in the face of rapid technological infusions may complicate efforts to fully integrate the indirect Westernization that results from such importation [28].

Unlike industrialization and economics, cultural change in Asia occurs at a slower rate. Socioreligious beliefs are incorporated into the languages themselves. For example, the Japanese language has Confucian and traditional Chinese medicine concepts such as "気", or ki/chi, as a part of daily greetings like "o genki desu ka?" (Are you well?), which directly translates to "how solid is your ki?", or "Kimochi ga warui" (I feel bad), which directly translates to "the ki I have is bad." Importantly, these concepts are in the "onyomi" or "Chinese" forms, while words borrowed from the West, such as medical terms, are usually converted to katakana, a script specifically used for nonnative Japanese words. This gives rise to the *Asian medical paradox*: even within imported Western medical concepts, such as psychosomatic therapy, some Shintoism has been found useful, but most Western medicine is kept separate both culturally and linguistically from native socioreligious beliefs [29]. As an example, in Japan, which adopted Western medicine during the Meiji Era (1868–1912) and spread it to South Korea and Taiwan, physicians are trained exclusively in Western techniques and certification has no relationship with traditional Chinese medicine even though most Japanese believe that the spirit (ki) is diffused throughout the organs versus the Western ideal of the brain as the seat of the mind/soul (Fig. 1) [7, 30, 31]. This means that Japanese and Korean doctors are trained to ask direct questions about symptoms and not the traditional Chinese medicine-laced general questions about ki. Therefore, the importation of specific medical protocols, technologies, or procedures from the West does not mean the automatic integration of Western mores or general culture into the tapestry of Asian society. Thus, in cases of death by neurologic criteria, the culturally accepted definition of death as a holistic process overrides the current Western medical definition of a brain-dead body as a simple source of functional organs [7].

It is perhaps due to the mysteries surrounding medicine, health, and life (versus the assimilation of more concrete industrialization principles needed for an export economy) that the Asian medical paradox occurs, namely, that slow religious and cultural changes conflict with Western medical training founded on cultural concepts not yet integrated into Eastern societies. Thus, doctors in Asia must walk a very fine line between cultural/familial expectations of medical care (mostly traditional Chinese medicine based) and their exclusively Western training [7, 32]. Efforts to transition Asian countries from an ancient Neo-Confucianism (960–1279 AD) to a modern, New Confucianism that attempts to blend classic concepts with the challenges of modern life have failed as, despite attacks by Maoism, Socialism, Communism, Democracy, and Capitalism, China, Japan, Korea, and other Asian countries still cling tightly to these classic ideas [33, 34].

Despite the abundance of Western-trained physicians, resistance remains; a survey of 476 Chinese healthcare providers in three separate facilities found that only about half actually recognized the concept of brain death regardless of official recognition [35]. This confirms that challenges remain in the changing of culture needed to drive brain death legislation and also that reliance on technology to deliver information about it may not be as effective in Asia as in the West.

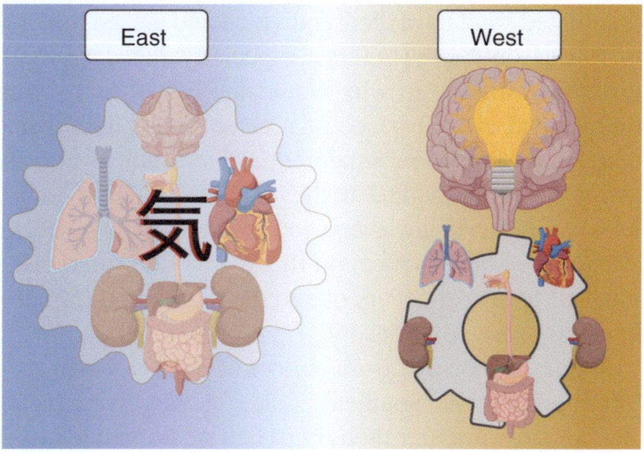

Fig. 1 In Asia, the life force (ki 気) is thought to be diffused throughout the entire body, permeating all organs, while, in the West, the brain alone is the seat of the mind, soul, and reason. Organs are relegated to mechanical functions only. (Created by BioRender.com)

5 Culture, Religion, and Law

Laws are the formal codification of values, religion, culture, and mores that serve the perceived needs of a group, community, or nation. Particularly in the Muslim-dominant countries of Malaysia and Indonesia or Buddhist-dominated Thailand, laws based on religious restrictions can shape the direction of popular thought on brain death and emic analyses of the intersection of religion, culture, and law in Asia have found that religion is a crucible from which the popular support of actions and thoughts regarding interpersonal relationships, including end-of-life issues, emerges [36]. Categorization of intangible mores, often based on animist or spiritualist traditional religions which derived such concepts by melding with Confucianism, is thus complicated by the clashing of diverse cultures within the same country (as seen in the US fight between the medical establishment and religious authorities over abortion), but this discord is also seen in ethnically and culturally homogenous countries such as China or Korea, where less variance from the accepted is found, even when technology is assimilated [37–39]. Medical guidelines are no different, codifying best practices as determined by the outcome and cultural suitability, and Asia has operated under the rules of traditional Chinese medicine for millennia.

Of some importance in the concept of brain death is the traditional Chinese medicine definition of death itself, as this is defined by heart or respiratory-related issues associated with death and, coupled with the focus on organ donation after brain death, can scare people into fearing that they will "wake up" during an organ transplant or receive substandard care [7, 40, 41]. Additionally, the Western concept that brain-dead individuals are "gone" and therefore are not accorded the dignity of either the living (no choice in their own care) or the dead (since Confucian funeral rites cannot take place until after heart death) may be repulsive to many Asians, even with the advent of rarely discussed end-of-life directives [7, 16, 42]. Also, native religions (of which there are many, such as Shintoism) often require an intact body for funeral rites, similar to Rabbinic laws against the desecration of corpses [43, 44]. Why then would a relative willingly allow a close family member to be "mutilated" and have their ki divided and placed into a stranger's body so that someone else could live? Asians, especially Japanese, do not feel comfortable giving gifts to strangers, so why would the life of that stranger be worth more than the dignity and eternal peace of a close family member [7]? The strong collectivity concerning family in Asian culture does not rest on the same bedrock as Western law (i.e., the individual is the smallest divisible component of society), and thus, why would Asian law even need to reflect such Western mores? Can Asia rely on the West to be 100% correct on this matter and intervene with reparations if social or medical upheaval occurs due to the dramatic differences in human rights that exist in some Asian countries [45]?

Since laws evolve from collective culture, when they do not exist with regard to brain death (as in Southeast Asia), cases will be handled in a manner compatible with protean popular thought in lieu of formalized rules [15, 46]. Additionally, the ruling bodies of almost every Asian country are comprised of the elderly, who, in spite of being more likely to require the exercise of such laws, act as gatekeepers to the problem by not culturally/legally constructing, or at least facilitating, frameworks upon which brain death can be decided [47]. Several possibilities exist for this lack of progress, especially in East Asia. One powerful theory is the deterministic conundrum, in which either technological forces drive societal change or societal change drives the development of new technologies to suit it [48]. Traditional media communication may be instrumental in either of these scenarios in the West, disseminating updates on social progression, highlighting areas of needed change for mass opinion shifting, and adopting new technology to facilitate the mission of change. Traditional Western thinking often sees media, especially online media, as a key driving force for societal change via technology, but in Asia, the common Confucian background does not drive societal innovation as governments/elite usually own or control it and studies have shown that the expression of minority opinions (the basis of societal change) is limited [49, 50]. In this way, Asia seems to be stagnating in the realm of brain death law as media outlets are not focusing on such problems to drive change.

Another area in which brain death law is held back is the focus by governments on current care strategies. Building digital databases, reducing disparities in healthcare resources, enacting patient information/privacy laws, handling chronic diseases, dealing with aging societies, and managing the medical records of socialized medical societies require an enormous commitment of organizational resources that are often not available for patient-level care [51, 52]. This is seen in Japan, where the country graduates sufficient neurosurgeons but requires them to manage patients from pre-surgery through discharge, increasing workloads and draining energy needed to address brain death itself [53]. Even when Japanese neurologists could diagnose brain death, Lock found that they often said nothing for fear of unduly influencing families towards donation [7]. This interpersonal stress on doctors may also add to the collective strain on the medical system. Brain death legislation in Japan, enacted only in the 1990s and with a limited capacity for organ donation, has not seen progress to parity with the West's rather rapid acceptance of the commodification of the human body [7]. This may suggest that Japan, a society still culturally constrained by the concept of traditional Chinese medicine death, the requirement of Shinto for an intact body, and little media attention to the issue, considers current legislation "good enough" to keep up appearances with the West while wrestling with larger issues such as its rapidly aging society [7, 54]. This is evident in the concept that the Asian definition of death was settled thousands of years prior and does not need revisiting since it follows the "Mandate of Heaven," while, in the West, death is seen as a defeat or failure of medicine [55]. Meanwhile, in Asian countries with younger demographics, end-of-life issues such as brain death are likely of less concern to centrally planned economies than employment, medical care for workers, or resource management [56].

Another possibility of note is the concept of healthcare fragmentation, in which the provision of medical care becomes fragmented by the frequent comorbidities seen in elderly patients (heart disease, cancer, diabetes, etc.), requiring patients to have multiple specialist providers. In these cases, brain death legislation may be complicated by the need to address diverse areas of medicine simultaneously, e.g., a brain condition for which brain death laws may be needed could be complicated by comorbidities, such as diabetes or heart disease [56]. Do those conditions need to be stabilized first before brain death can be accurately judged, or would medical care in these cases be best rationed for the purpose of saving organs over lives? The inextricable link between culture, religion, and law in Asia, coupled with the non-iconoclastic role of media, means that the popular opinion needed to drive change towards Western ideals of sufficient brain death definitions may not happen in the immediate future.

5.1 Patriarchy and Cultural Change in Asia (a.k.a. Don't Rock the Boat!)

Native religions of Asia, often spiritual or animist in nature (derived from Confucianism), are the bedrock of culture and rarely change [39, 57]. However, since changes in laws begin with changes in culture, dynamism within cultural propagation (moving away from religion) would be expected to play an important role in the process of enhancing brain death legislation. However, the views on brain death, even in technologically advanced countries like Singapore, have not changed regardless of efforts by legislation or awareness [41]. The tight social control of popular thought in Asian countries causative for this effect may be the concept of filial piety that stems from the Confucian concepts of inheritance, divine right, and the ordered, patriarchal relationships that are "proper under Heaven" [58]. Of the five major, defined interpersonal relationships (ruler-subject, father-son, elder brother-younger brother, husband-wife, friend-friend), the father-son relationship (indicative of familial patriarchy) may have the strongest influence on Asian thought. Based on the concepts of humanistic reciprocity and group identity, this established model for conflict resolution is based on the idea that the greater/older will be benevolent to the lesser/younger while the lesser/younger reciprocates with unwavering loyalty, a stricture that prohibits direct defiance and introduces the idea of indirect communication to indicate dissatisfaction with leadership [59].

Thus, Asian countries are tightly knit societies that would see defiance or questioning from youth as an affront to the established order, and social punishment, such as ostracism, could occur. This cultural rigidity has been reported as causative for alexithymia, or repression of true emotion to fit in, and this is engendered in the general Asian concept of "face" or specific Japanese concept of "tatemae/honne" (façade vs. true feelings) [37, 60, 61]. Additionally, the emphasis on age as a hallmark of leadership is reflected in the two Confucian relationships most normally encountered in families (father-son, husband-wife) and, as seen in a study where the elderly can be thought of as both "meriting pity and compassion" and threatening in status due to perceived social power, the permission of the elderly (who most likely believe in traditional Chinese medicine principles and Confucianism-based traditional religions) is required to generate cultural change [39, 58]. In this way, unlike the West, in which the generative power of youth instigates progressive ideas, Eastern sociocultural power flows from old to young and, thus, media influence focused on youth (i.e., social media) is unlikely to have any real cultural effect unless significant numbers of elderly participate.

Even within the patriarchal family structure, basic Confucian concepts are usually taught to children with all available resources, including through the use of money, as found in a study where Chinese or Vietnamese parents used allowances to teach family culture, Korean parents used allowances to build family trust, and Japanese parents used allowances as a pressure relief valve from Confucian obligations [62]. Additionally, as a result of the group being the smallest Confucian social unit, conflict avoidance/mediation, peace, harmony, and intrasocial stability are key parts of Asian culture that do not permit the iconoclastic generation of progressive ideas even in the medical realm [34, 63]. Consequently, the anticipation of benevolence from the elderly (who should possess wisdom and understanding according to Confucianism) may cause families to expect (1) that doctors should decide what is best while the doctors themselves are beholden to family expectations often based on traditional Chinese medicine principles and (2) that, if the "benevolent" elderly leaders in charge are not concerned with the brain death concept, then it is not significant. Meanwhile, the elderly rely on the youth for support in their old age and give approval based on support given in concert with the ideas of Confucianism and following tradition. Therefore, an intergenerational "soft" conflict is delineated; each side tries to keep harmonious or "amicable" relationships while neither is willing to wage a harder conflict against commonly accepted principles, progressing to detached or ambivalent feelings rather than straining the family structure (especially, as Nauck found, in the case of young women vs. parents) [64]. In this way, the rigid, age-focused, and patriarchal structure of Asian societies and religions may interfere with efforts to enact significant legislation with regard to brain death.

5.2 Westernization of Youth and Social Media

Social media, driven by consensus-focused design, tends to dilute opinions to an acceptable least common denominator: Ideas and content are graded (via "likes") and discussed in real time, arriving at some "final" judgement within hours, days, or weeks. The rapid nature through which any issue can be sorted may not be sufficient for topics of such gravitas as brain death. Additionally, social media is designed around the collection of personal information for ad revenue or data brokering, and thus, topics or content that draws a lot of views and discussions is preferentially selected by systems that track and communicate positive audience engagement to content providers. In this case, topics that are already commonly accepted among Asian youth (such as the existing Confucian or religious ideas around death) would not be as controversial and would therefore not draw views or engagement. Also, as social media is dominated by youth who use it to connect with peers and seek information, topics not interesting to that demographic may never be discussed [65, 66]. Indeed, reports on social media tend to focus on monetization and electronic word of mouth, usually around merchandise (real or digital), and the manipulation of opinions about those products through influencers [67]. Thus, even when groups around medical issues form, they tend to be "tribal" and have discussions that do not bleed into the popular sphere [68]. For these reasons, coupled with the ossified, patriarchal social structure, social media may not be as reliable a driver of changes to legislation (through cultural progressivism) as in the West due to a lack of interest in topics that are already well known or are not of interest to youth.

But, if the youth are gradually being Westernized through social media, won't they become amenable to more Western ideas (such as brain death)? Multiple reports from the English as a Second Language field have confirmed that, unlike decades past, modern youth/students study foreign languages for personal and professional enrichment (instrumental) rather than the ability to seamlessly join a particular culture associated with that language (integrative) [69–72]. Westernization is therefore seen as a way to improve life through technology versus generating social change by the integration of Western ideas or concepts. In this manner, schools play a powerful role in resisting the definition of foreign words, concepts, ideas, religions, and cultures as something to be enmeshed and submerged in. Rather, the excluded/included group concept inherent to Confucianism is used to segregate a foreign language (and its associated culture) as an "outside" subject in contrast to that native country's language, history, social studies, etc. ("inside" subjects) [73, 74]. With this in mind, any educational initiative to change opinions on brain death would need to first penetrate the group membership barrier and become an "inside" subject.

5.3 Confrontational and Context Communications in Cultural Change

A reliance on harmony would indicate that the Asian communication paradigm is dominated by a lack of conflict at both interpersonal and intergroup levels. This means not only that, when concepts or feelings are expressed, indirect methods of communication are crucial for context but also that complex topics (such as brain death) cannot be openly discussed. Additionally, there is a concept, expressed in the Japanese word "言霊" (kotodama), or the belief that saying something will make it come true, that prevents frank discussions of brain death to ward off the chances of it happening [75]. Moreover, some religious mythologies of Asia based on Confucianism (including Shinto, animist, and ancestral worship styles) expect that spiritual connections remain after death and people who suffer a "bad" death (unnatural; defined by mutilation or dying in a hospital) may have their spirits (reikon) wander about, unable to fully pass out of the material world and become venerated ancestral spirits [7, 39, 76]. Thus, if a decision on brain death according to Western principles was made regarding a loved one, nascent guilt over a potential "wrong" decision could affect family relations as well as open the family up to perceived bad luck and a loss of ancestral protection [7].

A final concept in any form of communication in Asia is the idea of concerted operation or feelings, namely, the idea that everyone thinks the same way so much communication can remain unspoken. In Japan, the concept of "以心伝心"

(ishindenshin), or telepathy, is shared with Confucian ideals [77]. If everyone is adhering properly to the correct Confucian relationships, then all communication flows effortlessly without the need for verbalization or direct challenge. This is why Asian countries still maintain an element of uniqueness over their Confucian foundations; each subculture of the heterogenous "Asian" culture is homogenous within itself, permitting such effortless communication, and so only shallower pop culture tends to be shared with outside cultures [78]. Therefore, on a national level, telepathic connectivity is required while, thanks to the concepts of included/excluded culture and face, sharing this "inside" thought with other countries is socially forbidden [79]. As such, assimilation of Western ideals in one Asian country does not necessarily result in a domino effect of mass acceptance in others.

5.4 Summary

It is, therefore, within this context of emotional self-repression, religiocultural rigidity, and intercultural isolationism that communication on neurological death in Asia must occur. A summary of such differences between the East and the West is available in Fig. 2. Since, as previously explained, social and mass media are not the conduits of progressive cultural change in Asia, what then is the mechanism by which communication and change on this issue can occur? One possible solution is the concept of cultural borrowing in which Asia waits for the West to arrive at the correct answer before integrating that into its own structures. A good example of this is the Westphalian political order of sovereignty (in which each state is the supreme power in its territory) that has been assimilated into Asian culture in steps since the nineteenth century [80]. Another prime example can be found in the material/financial/cultural Western trappings that have been integrated into Asia in diverse imitative forms following the importation of economic and political ideas (e.g., katakana loan words in Japan, smartphone payments in China, Western-style hotel rooms in Bangkok, Western luxury brands in Seoul, etc.). These amenities are

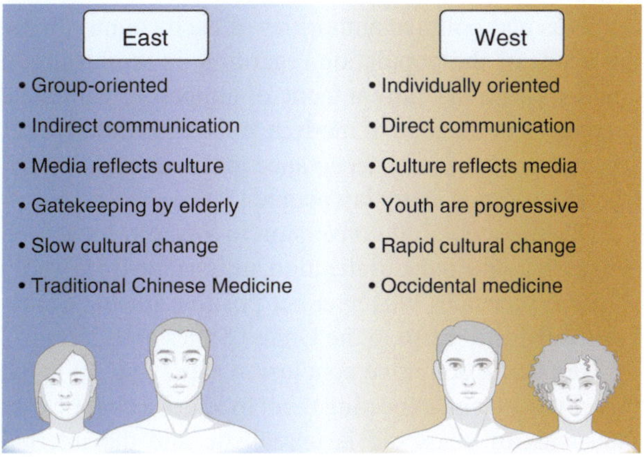

Fig. 2 A summary of Eastern vs. Western cultural factors that influence brain death legislation. (Created by BioRender.com)

not solely for tourist use and, since they do not upset the traditional Confucian/religious order, have been readily assimilated into Asian societies. Even the concept of abortion, the cause of tremendous social upheaval in the United States, is fairly routine and uncontroversial in Japan, indicating that birth issues are of much less importance than death (possibly since death is believed to be the transition to eternal ancestral spirithood) [7]. Asians have become more familiar with Western lifestyles as they achieve industrial and trade parity with Western nations; they seem to have, in effect, absorbed some of the surface concepts from some financial and industrial protocols, but only those that they believe are the "correct" answers to national wealth building and collective prosperity. In China, the Western concept of wealth, as exhibited in ostentatious displays of material goods, especially jewelry, clothing, and cars, resulted in the merging of happiness and success in the popular perception, while centrality, or the idea of the role that possessions play in one's life (a concept influenced by cultural/religious definitions of value), did not change [81]. This drive to acquire wealth and external displays of material power has resulted in the pursuit of industrial and export-driven success on national levels as China, Japan, Vietnam, and South Korea (plus the ASEAN economic bloc) are tying their industries into supplier-manufacturer-exporter networks to dominate world exports [82]. As such, it is clear that, when Asia judges something Western to be beneficial, it can rapidly and successfully integrate that into itself as long as no cultural or religious conflicts exist and that any such conflicts remain the primary obstacle to successful sociocultural integration.

Thus, in a similar fashion, the drive to achieve Western success on Asian terms could be exploited to change the Asian perception of brain death as richer Asian nations (e.g., Japan) come to terms with rapidly aging populations for which the industrialized destruction of closely knit family groups through migration to cities during boom times creates complications regarding end-of-life care [83]. However, in Southeast Asia, the extended family system has not changed much, and the elderly are often still cohabiting with their children and grandchildren [84]. As such, there is no one-size-fits-all solution since Asian countries vary in ethnic, socioeconomic, industrial, and cultural diversity. What is therefore needed is a national-level conversation in which Western principles are debated on an abstract level by the religious and political authorities of each country, with details hashed out based on the needs of the population regarding cultural, religious, and popular needs. This contrasts with the current tactic of tampering with the definition of death in Asia to keep up with the idea of modern Westernization, resulting in measures not likely to see widespread social acceptance and pushback from both gatekeepers of tradition (the elderly) and everyday people [7].

Unfortunately, full conversion to Western thought on this issue relies on a consensus that, like industrialization, the change from traditional principles and required spiritual sacrifices will be a net positive for the entire country. With debates over brain death still fresh in the United States and Europe, Asians cannot view the West and see that a definitive conclusion satisfying all parties has been reached [32]. In this environment, expecting them to adopt a chaotic disruption to a working formula thousands of years old is unrealistic and the internal conflict it would cause is anathema to the harmony-based foundational cultures of Asia. It is for these reasons that some of the more industrialized nations, such as Japan, South Korea, or China, will need to take the lead in formulating compromises to laws and principles on brain death diagnosis that make them "Asiacentric" and more palatable to other Asian communities and nations within the pan-Confucian region [85].

References

1. Dhanwate AD. Brainstem death: a comprehensive review in Indian perspective. Indian J Crit Care Med. 2014;18(9):596–605.
2. Ding Z-Y, Zhang Q, Wu J-W, Yang Z-H, Zhao X-Q. A comparison of brain death criteria between China and the United States. Chin Med J. 2015;128(21):2896–901.
3. Kim M-I, Oh J, Cho WH, et al. Causes of failure during the management process from identification of brain-dead potential organ donors to actual donation in Korea: a 5-year data analysis (2012-2016). J Korean Med Sci. 2018;33(50):e326.
4. Tungsiripat R, Tangcharoensathien V. Regulation of organ transplantation in Thailand: does it work? Rachata Tungsiripat and Viroj Tangcharoensathien. Health Economics & Financing Programme. 2003.
5. Malaysian Society of Neurosciences. Consensus Statement on Brain Death 2003. 2003 p. 24.
6. Yokota H. Noushi Hantei (Brain Death Judgement) Tokyo; 2010.
7. Lock M. In: Borofsky R, Scheper-Hughes N, Borgois P, Schneider N (editors). Twice dead. 1st ed. Berkeley, CA: University of California Press; 2002. p. 429.
8. Mathers CD, Boerma T, Ma FD. Global and regional causes of death. Br Med Bull. 2009;92(1):7–32.
9. Chan-on C, Sarwal MM. A comprehensive analysis of the current status and unmet needs in kidney transplantation in Southeast Asia. Front Med. 2017;4:1–14.
10. Institute for Health Metrics and Evaluation. GBD Compare: VizHub 2021.
11. Cho W-H. Organ donation in Korea in 2018 and an introduction of the Korea national organ donation system. Korean J Transplant. 2019;33(4):83.
12. Japan Organ Transplant Network. JOTNW Data 2020 (Japanese).pdf. 2020.
13. James SL, Bannick MS, Montjoy-Venning WC, et al. Global, regional, and national burden of traumatic brain injury and spinal cord injury, 1990-2016: a systematic analysis for the global burden of disease study 2016. Lancet Neurol. 2019;18(1):56–87.
14. Wahlster S, Wijdicks EFM, Patel PV, et al. Brain death declaration: practices and perceptions worldwide. Neurology. 2015;84(18):1870–9.
15. Lewis A, Liebman J, Bakkar A, et al. Determination of brain death/death by neurologic criteria in countries in Asia and the Pacific. J Clin Neurol. 2020;16(3):480.
16. Muthiah MD, Chua MSH, Griva K, et al. A multiethnic Asian perspective of presumed consent for organ donation: a population-based perception study. Front Public Heal. 2021;9:1451.
17. Sharif A. Why journals should stop publishing transplantation research from China [internet]. Br Med J. 2021:2. https://blogs.bmj.com/bmj/2021/04/13/why-journals-should-stop-publishing-transplantation-research-from-china/
18. Kim K, Son Y, Jung S, et al. Changing cause of mortality in general population and changing source of brain death in Korea. Transplantation. 2020;104(S3):S277.
19. Tohira H, Matsuoka T, Watanabe H, Ueno M. Characteristics of missing data of the Japan trauma data Bank. Nihon Kyukyu Igakukai Zasshi. 2011;22(4):147–55.
20. Chiu WT, Huang SJ, Tsai SH, et al. The impact of time, legislation, and geography on the epidemiology of traumatic brain injury. J Clin Neurosci. 2007;14(10):930–5.
21. Kim HK, Leigh JH, Lee YS, et al. Decreasing incidence and mortality in traumatic brain injury in Korea, 2008–2017: a population-based longitudinal study. Int J Environ Res Public Health. 2020;17(17):1–13.

22. Jiang J-Y, Gao G-Y, Feng J-F, et al. Traumatic brain injury in China. Lancet Neurol. 2019;18(3):286–95.
23. Hoang HTM, Pham TL, Vo TTN, et al. The costs of traumatic brain injury due to motorcycle accidents in Hanoi. Vietnam Cost Eff Resour Alloc. 2008;6:1–7.
24. Gupta S, Klaric K, Sam N, et al. Impact of helmet use on traumatic brain injury from road traffic accidents in Cambodia. Traffic Inj Prev. 2018;19(1):66–70.
25. Dewan MC, Rattani A, Gupta S, et al. Estimating the global incidence of traumatic brain injury. J Neurosurg. 2019;130(4):1080–97.
26. Rock JP, Prentiss T, Mo SM, et al. Traumatic brain injury in Myanmar: preliminary results and development of an adjunct electronic medical record. World Neurosurg. 2020;140:e260–5.
27. Li B, Piachaud D. Technological innovations and social development in Asia. J Asian Public Policy. 2019;12(1):1–14.
28. Chou Y, Wang SB, Lin Y. Long-term care and technological innovation: the application and policy development of care robots in Taiwan. J Asian Public Policy. 2019;12(1):104–23.
29. Nakao M, Ohara C. The perspective of psychosomatic medicine on the effect of religion on the mind-body relationship in Japan. J Relig Health. 2014;53(1):46–55.
30. Onishi H. History of Japanese medical education. Korean J Med Educ. 2018;30(4):283–94.
31. Nomura O, Onishi H, Park YS, et al. Predictors of performance on the pediatric board certification examination. BMC Med Educ. 2021;21(1):122.
32. Terunuma Y, Mathis BJ. Cultural sensitivity in brain death determination: a necessity in end - of - life decisions in Japan. BMC Med Ethics. 2021;22(58):1–6.
33. Madsen R. Confucianism: ethical uniformity and diversity. In: Sullivan WM, Kymlicka W, editors. The globalization of ethics. Cambridge: Cambridge University Press; 2009. p. 117–33.
34. Solé-Farràs J. Harmony in Contemporary New Confucianism and in Socialism with Chinese Characteristics. China Media Res. 2008;4(4):14–24.
35. Yang Q, Fan Y, Cheng Q, Li X, Khoshnood K, Miller G. Acceptance in theory but not practice – Chinese medical providers' perception of brain death. Neuroethics. 2015;8(3):299–313.
36. Harding A. Law, religion, and constitutionalism in Asia. Asian J Comp Law. 2018;13(2):227–32.
37. Han E, Paik C. Ethnic integration and development in China. World Dev. 2017;93:31–42.
38. Fedorenko O. The insiders and outsiders of Korean culture guest Editor's note: the insiders and outsiders of Korean culture. Acta Koreana. 2021;21(2):7–16.
39. Ogura K. 8. Animism and spiritualism: the two origins of life in Confucianism. In: Confucianisms for a changing world cultural order. Honolulu: University of Hawaii Press; 2017. p. 112–27.
40. Hsieh J-G, Wang Y-W. Application of signs of dying identified in traditional Chinese, Tibetan, and modern Western medicine in terminal care. Tzu Chi Med J. 2012;24(1):12–5.
41. Liu CW, Yeo C, Lu ZB, et al. Brain death in Asia: do public views still influence organ donation in the 21st century? Transplantation. 2019;103(4):755–63.
42. Martina D, Lin C-P, Kristanti MS, et al. Advance care planning in Asia: a systematic narrative review of healthcare professionals' knowledge, attitude, and experience. J Am Med Dir Assoc. 2021;22(2):349.e1–349.e28.
43. Oliver M, Ahmed A, Woywodt A. Donating in good faith or getting into trouble religion and organ donation revisited. World J Transplant. 2012;2(5):69–73.
44. Feld J, Sherbin P, Cole E. Barriers to organ donation in the Jewish community. J Transpl Coord. 1998;8(1):19–24.
45. Dellapenna J, D'Amato A, Kausikan B. An Asian approach to human rights. Proc Annu Meet (American Soc Int Law). 1995;89:146–52.
46. Ming Y, Tu B, Zhuang Q. Current situation of organ donation in China. In: Organ Donation and Transplantation - Current Status and Future Challenges. InTech; 2018.
47. Oku A, Ichimura E, Tsukamoto M. Aging population in Asian countries - Lessons from Japanese experiences. PRI Discussion Paper Series 2017;17(A-12).

48. Danowski JA, Park HW. East Asian communication technology use and cultural values. J Contemp East Asia. 2020;19(1):43–58.
49. Shen F, Wang N, Guo Z, Guo L. Online network size, efficacy, and opinion expression: assessing the impacts of internet use in China. Int J Public Opin Res. 2009;21(4):451–76.
50. Osman A, Subhani MI, Hasan SA. Asian ascendancy: media in the age of globalization. Springerplus. 2013;2:646.
51. Chongsuvivatwong V, Phua KH, Yap MT, et al. Health and health-care systems in Southeast Asia: diversity and transitions. Lancet (London, England). 2011;377(9763):429–37.
52. Nageshwaran G, Harris RC, El Guerche-Seblain C. Review of the role of big data and digital technologies in controlling COVID-19 in Asia: public health interest vs. privacy. Digit Heal. 2021;7:20552076211002 9.
53. Ferraris KP, Matsumura H, Wardhana DPW, et al. The state of neurosurgical training and education in East Asia: analysis and strategy development for this frontier of the world. Neurosurg Focus. 2020;48(3):E7.
54. Yang Q, Miller G. East–West differences in perception of brain death. J Bioeth Inq. 2015;12(2):211–25.
55. Lee SK. East Asian attitudes toward death- a search for the ways to help east Asian elderly dying in contemporary America. Perm J. 2009;13(3):55–60.
56. He AJ, Tang VFY. Integration of health services for the elderly in Asia: a scoping review of Hong Kong, Singapore, Malaysia, Indonesia. Health Policy (New York). 2021;125(3):351–62.
57. Croucher SM, Zeng C, Rahmani D, Sommier M. Religion, culture, and communication. In: Oxford research encyclopedia of communication. New York: Oxford University Press; 2017.
58. Vauclair CM, Hanke K, Huang LL, Abrams D. Are Asian cultures really less ageist than Western ones? It depends on the questions asked. Int J Psychol. 2017;52(2):136–44.
59. Yum JO. The impact of Confucianism on interpersonal relationships and communication patterns in East Asia. Commun Monogr. 1988;55(4):374–88.
60. Lo C. Cultural values and Alexithymia. SAGE Open. 2014;4(4):2158244014555117.
61. Kim JY, Nam SH. The concept and dynamics of face: implications for organizational behavior in Asia. Organ Sci. 1998;9(4):522–34.
62. Takahashi N, Yamamoto T, Takeo K, et al. East Asian children and money as a cultural tool: dialectically understanding different cultures. Jpn Psychol Res. 2016;58(1):14–27.
63. Kim T, Wang C, Kondo M, Kim T. Conflict management styles: the differences among the Chinese, Japanese, and Koreans. Int J Confl Manag. 2007;18(1):23–41.
64. Nauck B. Affection and conflict in intergenerational relationships of women in sixteen areas in Asia, Africa, Europe, and America. Comp Popul Stud. 2014;39(4)
65. Plaisime M, Robertson-James C, Mejia L, et al. Social media and teens: a needs assessment exploring the potential role of social media in promoting health. Soc Media + Soc. 2020;6(1):2056305119886 02.
66. Abi-Jaoude E, Naylor KT, Pignatiello A. Smartphones, social media use and youth mental health. Can Med Assoc J. 2020;192(6):E136–41.
67. Dwivedi YK, Ismagilova E, Hughes DL, et al. Setting the future of digital and social media marketing research: perspectives and research propositions. Int J Inf Manag. 2021;59:102168.
68. Rolls K, Hansen M, Jackson D, Elliott D. How health care professionals use social media to create virtual communities: an integrative review. J Med Internet Res. 2016;18(6):e166.
69. Osterman GL. Experiences of Japanese university students' willingness to speak english in class: a multiple case study. SAGE Open. 2014;4(3):2158244014543779.
70. Seki A, La GJ. Motivational and attitudinal factors influencing English study of Japanese tertiary level students. Commun Sci. 2006;29:47–66.
71. Yashima T. Orientations and motivation in foreign language learning: a study of Japanese college students. Japan Assoc Coll English Teach Bull. 2000;31:121–33.
72. Adachi R. Motivation of English learning and intercultural communication: a case of Japanese college students. J Sch Foreign Lang Nagoya Univ Foreign Stud. 2009;37:119–43.

73. Guo Y, Guo S, Yochim L, Liu X. Internationalization of Chinese higher education: is it westernization? J Stud Int Educ. 2022;26(4):436–53.
74. Nagai M. Chapter II. Westernization and Japanization: The early Meiji transformation of education. In: Tradition and modernization in Japanese culture. Princeton, NJ: Princeton University Press; 2015. p. 35–76.
75. Hara K. The word is the thing: the "Kotodama" belief in Japanese communication: part II. ETC A Rev Gen Semant. 2002;58(4):408–19.
76. Hirai N. Traditional Cultures and Modernization: Several Problems in the Case of Japan [Internet]. Institute for Japanese Culture and Classics, Kokugakuin University. 1999 [cited 2021 Dec 3]. p. 3. https://www2.kokugakuin.ac.jp/ijcc/wp/cimac/hirai.html
77. Zhang YB, Zhang YB. Asian communication modes. In: Donsbach W, editor. The Blackwell International Encyclopedia of Communication. Blackwell Publishing; 2008. p. 775–9. Publisher's official version: http://www.communicationencyclopedia.com/public/tocnode?id=g.
78. Kim M-S. Intercultural communication in Asia: current state and future prospects. Asian J Commun. 2010;20(2):166–80.
79. Han KH. The feeling of "face" in Confucian society: from a perspective of psychosocial equilibrium. Front Psychol. 2016;7:1–9.
80. Park S-H. Changing definitions of sovereignty in nineteenth-century East Asia: Japan and Korea between China and the West. J East Asian Stud. 2013;13(2):281–307.
81. Liao J, Wang L. The structure of the Chinese material value scale: an eastern cultural view. Front Psychol. 2017:8, 1852.
82. Peng D. The changing nature of East Asia as an economic region. Pac Aff. 2000;73(2):171.
83. Kato A. The Japanese family system: change, continuity, and regionality over the twentieth century. Mpidr Work Pap. 2013;49:48.
84. Yeung WJJ, Desai S, Jones GW. Families in Southeast and South Asia. Annu Rev Sociol. 2018;44:469–95.
85. Swinbanks D. Japan reaches a compromise on organ transplants. Nature. 1997;387(6636):835.
86. Maas AIR, Menon DK, David Adelson PD, et al. Traumatic brain injury: integrated approaches to improve prevention, clinical care, and research. Lancet Neurol. 2017;16(12):987–1048.
87. Sun D, Jiang B, Ru X, Sun H, Fu J, Wang L, et al. Prevalence and altered causes of traumatic brain injury in China: a Nationwide survey in 2013. Neuroepidemiology. 2020;54(2):106–13.
88. Hosomi S, Kitamura T, Sobue T, Ogura H, Shimazu T. Survival outcomes after traumatic brain injury during national academic meeting days in Japan. Sci Rep. 2021;11(1):1–7.
89. Suzuki M, Ono J, Ogawa T, Suehiro E. Japan neurotrauma data bank (JNTDB): the past, the present and the future. Japan J Neurosurg. 2014;23(12):934–41.
90. Eom KS, Seo BR, Rim BC, et al. Epidemiology and outcomes of traumatic brain injury in elderly population: a multicenter analysis using Korean neuro-trauma data bank system 2010-2014. J Korean Neurosurg Soc. 2019;62(2):243–55.
91. Hsu IL, Li CY, Chu DC, Chien LC. An epidemiological analysis of head injuries in Taiwan. Int J Environ Res Public Health. 2018;15(11):2457.
92. Chiu WT, Yeh KH, Li YC, et al. Traumatic brain injury registry in Taiwan. Neurol Res. 1997;19(3):261–4.
93. Lai CY, Maegele M, Yeung JHH, et al. Major trauma care in Hong Kong and Germany: a trauma registry data benchmark study. Eur J Trauma Emerg Surg. 2021;47(5):1581–90.
94. Jingwei AH, Yu-Hung AL, Ching L. Living organ transplantation policy transition in Asia: towards adaptive policy changes. Glob Heal Gov. 2010;3(2):1–4.
95. The Republic of Korea. Internal Organs Transplant Act. 11976 South Korea; 2013 p. 6.
96. Wang TH, Lee PC, Chiang YJ. Taiwan's organ donation and transplantation: observation from national registry point of view. J Formos Med Assoc. 2017;116(9):649–51.
97. Cheung CY, Pong ML, Au Yeung SF, Chau KF. Factors affecting the deceased organ donation rate in the Chinese community: an audit of hospital medical records in Hong Kong. Hong Kong Med J. 2016;22(6):570–5.

98. Buckley T, So HY, Wong KK, Yan WW. The Hong Kong Society of Critical Care Medicine Position Statement Guidelines on Certification of Brain Death [Internet]. The Hong Kong Society of Critical Care Medicine. 2009 [cited 2021 Dec 7]. https://hksccm.org/index.php/professional/30-expired-guidelines/46-guidelines-on-certification-of-brain-death
99. Legislative Council Secretariat. Research Brief: Organ donation in Hong Kong. 2016.
100. Peeters S, Blaine C, Vycheth I, et al. Epidemiology of traumatic brain injuries at a major government Hospital in Cambodia. World Neurosurg. 2017;97:580–9.
101. Tandean S, Japardi J, Kollins F, Loe ML. Epidemiology of traumatic brain injury in neurosurgery Department of Tertiary Referral Hospital at North Sumatera, Indonesia. Medicinus. 2020;7(5):146.
102. Faried A, Bachani AM, Sendjaja AN, Hung YW, Arifin MZ. Characteristics of moderate and severe traumatic brain injury of motorcycle crashes in Bandung, Indonesia. World Neurosurg. 2017;100:195–200.
103. Niryana IW, Jorden IW, Darmawan R, Widyadharma IPE. Characteristics of traumatic brain injury in Sanglah Hospital, Bali, Indonesia: a retrospective study. Biomed Pharmacol J. 2020;13(3):1431–7.
104. Rosyidi RM, Priyanto B, Laraswati NKP, et al. Characteristics and clinical outcome of traumatic brain injury in Lombok, Indonesia. Interdiscip Neurosurg Adv Tech Case Manag. 2019;18(February):100470.
105. Arulsamy A, Shaikh MF. Current status of traumatic brain injury research in Malaysia: a systematic review. Neurosci Res Notes. 2020;3(4):1–21.
106. Tay EL, Lee SWH, Jamaluddin SF, Tam CL, Wong CP. The epidemiology of childhood brain injury in the state of Selangor and Federal Territory of Kuala Lumpur, Malaysia. BMC Pediatr. 2016;16(1):1–6.
107. Rivera AS, Lam HY, MacAlino JU. Epidemiology of injuries in the Philippines: an analysis of secondary data. Acta Med Philipp. 2018;52(2):180–6.
108. Wee JZ, Yang YRJ, Lee QYR, Cao K, Chong CT. Demographic profile and extent of healthcare resource utilisation of patients with severe traumatic brain injury: still a major public health problem. Singap Med J. 2016;57(9):491–6.
109. Damkliang J, Considine J, Kent B. Thai emergency nurses' management of patients with severe traumatic brain injury: comparison of knowledge and clinical management with best available evidence. Australas Emerg Nurs J. 2013;16(4):127–35.
110. Malaysian Medical Council. Guideline of the Malaysian Medical Council: Brain Death 2006.
111. Attamimi ZE, Awang H, Mansor HS. Challenges and barriers of brain death. Int J Public Heal Clin Sci. 2021;8(4):87–98.
112. The Union of Myanmar State Peace and Development Council. No Title. 1/2004 Myanmar; 2004 p. 4.
113. Republic of the Philippines Department of Health. No Title [Internet]. Report. https://doh.gov.ph/node/15149
114. Suguitan GA, Cabanayan-Casasola CB, Danguilan RA, Jaro JMA. Outcome of referrals for deceased organ donation to the government organ procurement organization. Transplant Proc. 2014;46(4):1074–6.
115. Tan G. Singapore Human Organ Transplants Act (HOTA) [Internet]. Personal Page. 2012 [cited 2021 Dec 7]. https://www.geraldtan.com/medaffairs/hota.html
116. Kwek TK, Lew TWK, Hui LT, Kong S. The transplantable organ shortage in Singapore - has implementation of presumed consent to organ donation made a difference? Ann Acad Med Singap. 2009;38(4):346–53.
117. Martphol T. Alternative choice of organ donation in Thailand: a study of opt-out and mandated choice systems. Thammasat Bus Law J. 2020;10:49–58.
118. Nivatvongs S, Dhitavat V, Jungsangasom A, et al. Thirteen years of the Thai red cross organ donation Centre. Transplant Proc. 2008;40(7):2091–4.
119. Pinmongkol C, Sroyson S, Dhitavat V, et al. Success in organ donation program in Thailand. Transplantation. 2017;101(August):S101.

120. Tran SH, Du TT, Thai SM, et al. Current status of organ donation for transplant in Vietnam. Transplantation. 2018;32(15):2264–6.
121. The Socialist Republic of Vietnam On donation, removal and transplantation of human tissues and organs and donation and recovery of cadavers. Vietnam; 2006.

Cultural Considerations in the Declaration of Death by Neurologic Criteria in Africa

Wangari Waweru-Siika, Dilraj Singh Sokhi, and Violet Naanyu

"Wherever the African is there is religion [1]."

Who exactly is "the African" and is there such a thing as "African culture"[2]? The North African is remarkably different from the South African and the East African from the West African. Even within regions and countries, incredible cultural diversity exists [2–4]. Culture, that complex whole which encompasses knowledge, beliefs, morals, laws, and customs, includes habits acquired by virtue of belonging to a given society [5]. But are African people more similar than we are different when it comes to death [2, 6]?

In this chapter, we discuss the African perception of death and dying and unravel the ways this influences the discussion about the declaration of death by neurologic criteria. We explore areas of controversy, identify potential challenges to its widespread acceptance and offer suggestions for the future.

1 Declaration of Death by Neurologic Criteria in Africa

The concept of death by neurologic criteria has not gained much traction in Africa, perhaps because it is not discussed with African philosophy in mind [7]. Of the 54 countries on the continent, only a handful have legal frameworks or medical

W. Waweru-Siika (✉)
Department of Anesthesia, Aga Khan University, Nairobi, Kenya
e-mail: wangari.siika@aku.edu

D. S. Sokhi
Adult Clinical Neurology, Department of Medicine, Aga Khan University, Nairobi, Kenya
e-mail: dilraj.sokhi@aku.edu

V. Naanyu
Department of Sociology Psychology and Anthropology, School of Arts and Social Sciences, Moi University, Eldoret, Kenya
e-mail: vnaanyu@ampath.or.ke

A. Lewis, J. L. Bernat (eds.), *Death Determination by Neurologic Criteria*, Advances in Neuroethics, https://doi.org/10.1007/978-3-031-15947-3_30

standards that accept death by neurologic criteria as death and support the withdrawal of intensive care under such circumstances [3, 8–10]. Declaration of death by neurologic criteria is nevertheless widely performed by clinicians in Africa, despite the absence of laws to support this practice.

However, in our experience, the absence of national laws or standards has led to significant variation in the quality, performance, and interpretation of brainstem function testing and the steps taken after the declaration of death by neurologic criteria. Even within a given country, the accepted interval between two independent tests of brainstem function may vary between and within institutions. In Kenya, for instance, following the confirmation that irreversible loss of brainstem function has occurred, most clinicians institute what is referred to as "minimal support," a poorly defined process in which ventilator settings are modified [3]. As a result, the changes made by some clinicians lead to circulatory-respiratory arrest within a matter of minutes; in others, the patient remains ventilated for several more hours or even days. Where families refuse to accept a declaration of death by neurologic criteria, the patient's status is marked as "full code" and aggressive critical care support continues to be provided. Critical care support continues until circulatory-respiratory arrest occurs naturally, irrespective of whether the process takes days or weeks.

In the absence of a legal framework to support the withdrawal of intensive care after the declaration of death by neurologic criteria, some healthcare facilities in Kenya have developed internal policies that guide the signing of a "Change of Goals of Care" order. This order allows clinicians to document that further critical care interventions will not be performed after the declaration of death by neurologic criteria or in other circumstances of perceived medical futility. This plan follows a discussion with the patient's next of kin or legal representative, to ensure concurrence and avoid litigation. Most other health facilities, however, lack such policies. Nevertheless, extubation is never performed, even in the face of an uncontested declaration of death by neurologic criteria.

2 Areas of Controversy

2.1 Death as Ceasing to Be

While medicine defines death as the end of life [11], death in many African cultures is viewed as a transition from one part of the cycle of life to another [1, 2, 4, 6, 12]. One does not cease to exist, but rather moves from the state of a physical being to a spiritual one [6, 13, 14]. To die means to enter the ancestral world, a world with which the living interact [4, 14].

Culture and religion or spirituality in Africa are inextricably linked [1, 15]. While a significant proportion of Africans are either Christian or Muslim, many subscribe to their traditional beliefs, openly or in private [6, 12, 15, 16]. Accordingly, giving the dead a good sendoff is key to keeping the spirits of the departed happy as they are considered by some to be more powerful than the living [14]. For these reasons, the way the dead and dying are treated is of utmost importance as it may have far-reaching consequences for those left behind. Organ donation in Africa in the context of death by neurologic criteria is therefore shunned because it is thought that it may lead to a very unhappy ancestor: one who arrives in the afterlife without all essential organs [4]. Organ donation rates are thus low amongst Africans, negatively impacting those in need of a donor [17].

2.2 Speaking of Death

For many Africans, death is feared and is neither openly discussed nor even contemplated [4, 18, 19], because "to speak of death is to invite it" [20]. The inevitability of death amongst the aged nevertheless is accepted. The aged are seen as having lived a full life and are therefore ready to move on to the afterlife. The gradual process of dying, no matter how long it takes, is used to set one's affairs in order. This is the definition of a "good death" [14, 19] although this view varies across cultures, even within the geographical confines of a single nation [21]. Dying young or suddenly is inevitably seen as having been brought on by evil forces [4, 22]. Declaring death by neurologic criteria in Africa is therefore likely to be more acceptable for elderly members of families; amongst children and those who suffer sudden, catastrophic illnesses or injuries, it is more likely to be stigmatized, with the family considered to have been "bewitched" [23].

2.3 Culture, Religion, and the Dying

Where death by neurologic criteria has been declared, family concurrence is requested to avoid further critical care interventions, but religion often becomes a stumbling block. A significant proportion of Africans are Christian or Muslim [24]. Even in the face of medical futility, many are encouraged to ignore science, to have faith, and to continue to "expect a miracle." When asked to accept a declaration of death by neurologic criteria, many are conflicted, feeling that in doing so they will have given up on their loved one, and that they will have refused to allow a miracle to occur. Presenting a declaration of death by neurologic criteria in Africa therefore has the potential to cause conflict within families and between families and clinicians, particularly when the more influential in the family are unwilling to accept any outcome other than complete recovery.

In addition to the impact of religion, views are shaped by the influence of traditional African beliefs: no relative wants to be associated with discussions about withholding or withdrawing intensive care for fear that they will be blamed for that person's death, and that curses will befall them and their offspring. Similarly, in Islam, one of the fastest growing religions on the continent, life must be preserved at all costs, raising contentious issues when discussing death by neurologic criteria (as reviewed in detail elsewhere in this book) [25].

3 Solutions for the Future: Education, the Law, or Both?

3.1 The Role of Education

In considering the approach towards educating the general public on the declaration of death by neurologic criteria, the African scenario is complicated by the admixture of Western education, the influence of world religions, and the broad range of deeply rooted traditional African beliefs [18]. Even amongst some clinicians, these factors influence their approach to the concept of death by neurologic criteria and how—or indeed if—to bring it up with families. Many clinicians in Africa struggle with the concept of death by neurologic criteria and find it difficult to have conversations with families about it. This barrier will certainly continue for the foreseeable future.

Demystifying the concept of death by neurologic criteria for patients and their families is further compounded by widespread neurophobia—the fear of, or reluctance to understand, the complexity of the nervous system and its disorders—amongst healthcare providers. This fear is highly prevalent in Africa [26]. Whilst the neurologist, neurosurgeon, or intensivist is well-placed to talk to families about the dying brain [27], there are far too few specialists in Africa to expect this conversation to become routine practice. Tackling neurophobia amongst healthcare workers in Africa is therefore a big challenge. Despite the term being coined almost three decades ago, little has changed to overcome neurophobia in Africa due to the marked lack of specialists in the neuroscience fields on the continent [26].

Instilling confidence in medical students, nurse trainees, and even graduates that they too are the custodians of brain health would go a long way to mitigate fears, misconceptions, and stigma in the general population at grass-root levels. Capacity-building geared towards increasing the neuroscience healthcare workforce in all cadres of mainstream and allied services will allow for more sustainable change in shifting the current landscape of brain disease, including death by neurologic criteria.

Further, the timing and technique for communication with families also matter. Immediately before and following a declaration of death by neurologic criteria, dialogue with families is crucial to help them understand and accept their loved one's state [10, 28]. The use of alternative phrases such as "irreversible brain failure" during such conversations has been suggested to help families come to terms with the reality, given that conceptually, critical care units exist primarily to support organ failure [10].

3.2 Policy Makers and the Law

A legislative framework to guide the declaration of death by neurologic criteria is urgently required.

Given the limited critical care capacity in Africa, continuing organ support for patients declared dead, or believed to be dead, by neurologic criteria can have a significant knock-on effect: (1) Critically ill patients who would benefit from a critical care admission are denied the service if the few available beds are occupied by patients who are dead by neurologic criteria; (2) in fee-paying facilities, family resources are depleted as these patients continue to receive futile yet expensive critical care indefinitely; and (3) staff working in these critical care units suffer the moral distress of caring for patients they know will not benefit from any further critical care interventions.

Critical care resource limitation in many African countries demands that clear policies be enacted to guide who should receive aggressive interventions and to define those circumstances where continuation of support is futile. Providing these scarce resources to a person who is dead by neurologic criteria generates an opportunity cost by denying a potential rescue if, instead, the resources had been allocated to them. Further, accepting death by neurologic criteria would help facilitate organ donation that could save additional lives.

4 Conclusion

Death by neurologic criteria, as a concept, is difficult to advance in Africa, where there is little dialogue about death and where laws and medical standards to guide the conversation are nearly nonexistent. Despite this absence, clinicians in Africa need to navigate this situation to minimize (or ideally avoid) the provision of futile interventions and to ensure that the limited available critical care services are provided to patients who are likely to benefit. Acknowledging African cultural perceptions of death and dying is key to advancing the discourse about death by neurologic criteria on the continent. Critical care resource limitation, public education that acknowledges religious and cultural diversity, and possibly even organ donation, are powerful drivers that we hope will guide future policy creation in Africa.

References

1. Mbiti JS. African religions & philosophy. 2nd rev. and enl. ed. African religions and philosophy. Oxford: Heinemann; 1990.
2. Baloyi L, Makobe-Rabothata M. The African conception of death: a cultural implication. In: Meiring LTBJD, Van de Vijver FJR, Idemoudia ES, Gabrenya Jr WK, editors. Toward sustainable development through nurturing diversity: Proceedings from the 21st International Congress of the International Association for Cross-Cultural Psychology; 2014.
3. Waweru-Siika W, Clement ME, Lukoko L, et al. Brain death determination: the imperative for policy and legal initiatives in Sub-Saharan Africa. Glob Public Health. 2017;12(5):589–600.
4. Ekore RI, Lanre-Abass B. African cultural concept of death and the idea of advance care directives. Indian J Palliat Care. 2016;22(4):369–72.
5. Tylor EBHHC. Primitive culture : researches into the development of mythology, philosophy, religion, language, art, and custom. London; New York: John Murray/G.P. Putnam's Sons; 1920.
6. Asuquo O. A rationalization of an African concept of life, death and the hereafter. Am J Soc Manage Sci. 2011;2:171–5.

7. Gardiner D, Shemie S, Manara A, Opdam H. International perspective on the diagnosis of death. Br J Anaesth. 2012;108(Suppl1):i14–28.
8. Lewis A, Kumpfbeck A, Liebman J, et al. Barriers to the use of neurologic criteria to declare death in Africa. Am J Hosp Palliat Care. 2022;39(2):243–9.
9. Hardcastle TC. Diagnosis of death guidelines for South Africa – timely and necessary. South Afr J Crit Care. 2021;37(1):6–7.
10. Prin M, Quinsey C, Kadyaudzu C, Hadar E, Charles A. Brain death in low-income countries: a report from Malawi. Trop Dr. 2019;49(2):107–12.
11. World Health O. Clinical criteria for the determination of death, WHO technical expert consultation, WHO headquarters, Geneva, Switzerland, 22–23 September 2014. Geneva: World Health Organization; 2017.
12. Eyetsemitan F. Cultural interpretation of dying and death in a non-western society: the case of nigeria. online readings in psychology and culture. Online Readings in Psychology and Culture. 2002;3(2):1.
13. Meister CV, Copan P. The Routledge companion to philosophy of religion. 2nd ed. Companion to philosophy of religion, Routledge. Abingdon, Oxon/New York; 2013.
14. Selin H, Rakoff RM. Death across cultures. Science across cultures: the history of non-Western science. Cham: Springer International Publishing AG; 2019.
15. Chitando E. Religion in Africa: confessions of an animist, written by Orobator, Agbonkhianmeghe E. J Relig Afr. 2021;49(2):217–8.
16. Aderibigbe IS. Religious traditions in Africa: an overview of origins, basic beliefs, and practices. In: Aderibigbe IS, Medine CMJ, editors. Contemporary perspectives on religions in Africa and the African diaspora. New York: Palgrave Macmillan US; 2015. p. 7–29.
17. Etheredge HR. Assessing global organ donation policies: opt-in vs opt-out. Risk Manag Healthc Policy. 2021;14:1985–98.
18. Weru J. What Kenya needs to do to end the taboo of talking about 'end of life care'. The Conversation; 2017.
19. Nasimango E. Quality of Death and Dying: What matters? An African perspective. eHospice 2019(Africa edition1).
20. Love KR, Karin E, Morogo D, et al. "To speak of death is to invite it": provider perceptions of palliative care for cardiovascular patients in Western Kenya. J Pain Symptom Manag. 2020;60(4):717–24.
21. Gafaar TO, Pesambili M, Henke O, et al. Good death: an exploratory study on perceptions and attitudes of patients, relatives, and healthcare providers, in northern Tanzania. PLoS One. 2020;15(7):e0233494.
22. Umoh D. Death is not natural: the African story. J Relig Soc. 2012;14
23. Fottrell E, Tollman S, Byass P, Golooba-Mutebi F, Kahn K. The epidemiology of 'bewitchment' as a lay-reported cause of death in rural South Africa. J Epidemiol Community Health. 2012;66(8):704–9.
24. Saleh M. African countries with the highest share of Christians as of 2019. Statista 2021.
25. Lewis A. Contentious ethical and legal aspects of determination of brain death. Semin Neurol. 2018;38(5):576–82.
26. McDonough A, Chishimba L, Chomba M, Zimba S, Mwendaweli N, Asukile M, et al. Neurophobia in Africa: survey responses from fifteen African countries. J Neurol Sci. 2022;434:120161.
27. Kondziella D. The neurology of death and the dying brain: a pictorial essay. Front Neurol. 2020;11:736.
28. Ali SK, Ochola A, Juma F, Daroowalla F. Using CIRUPA to help foster communication with families about brain death in sub-Saharan Africa. Indian J Palliat Care. 2019;25(1):162–4.

The Argument for Personal Choice in Determining Death

Christos Lazaridis and Lainie Friedman Ross

1 The Fourfold Scheme of Defining Death

In 1972, Capron and Kass asserted that determining whether and when a person has died entails both medical and metaphysical/normative components. They began with several assumptions. First, death is a discrete unitary concept. Second, the basic concept of defining death incorporates both clinical and philosophical factors, the former best enumerated and determined by physicians, the latter by a broader group of stakeholders. Third, although a unitary concept, its determination need not be [1].

In evaluating the desirability of legislation to "define death," Capron and Kass described four distinct levels of "definitions" that would give substance to the formal notion: (1) the *basic concept or idea*; (2) *general physiological standards*; (3) *operational criteria*; and (4) *specific tests or procedures* [1].

The *basic concept* refers to the philosophical grounding required to understand how the term "death" is employed and to what it refers. It is a metaphysical question about the essential nature of the entity whose death is in question. Although science may inform certain views about human ontology and place plausibility constraints, the question cannot be settled in strictly scientific or medical terms [2]. For our purposes, it is a question of human or personal ontology. In contrast, some (like

C. Lazaridis
Departments of Neurology and Neurosurgery, University of Chicago Medical Center, Chicago, IL, USA

MacLean Center for Clinical Medical Ethics, University of Chicago, IL, Chicago, USA
e-mail: lazaridis@uchicago.edu

L. F. Ross (✉)
Departments of Pediatrics, Medicine, Surgery, Institute for Translational Medicine, University of Chicago, Chicago, IL, USA

MacLean Center for Clinical Medical Ethics, University of Chicago, IL, Chicago, USA
e-mail: lross@uchicago.edu

A. Lewis, J. L. Bernat (eds.), *Death Determination by Neurologic Criteria*, Advances in Neuroethics, https://doi.org/10.1007/978-3-031-15947-3_31

Nair-Collins) argue that "since we are a part of the natural world, our lives and deaths can and should be explained in the same terms as we explain the rest of the natural world" [3]. What underlies this normative claim is the metaphysical view of animalism—that we, human beings, share the same essential nature as the rest of the natural world. While we respect Nair-Collin's perspective, it is a metaphysical view that we do not share.[1] Thus, the basic concept of death incorporates a degree of personal choice about what it means to be a living person, resulting in different people adopting very different metaphysical perspectives.

The second level, *general physiological standards*, refers to the choice to define death in terms of organ systems, physiological functions, or recognizable human activities, capacities, and conditions. As Capron and Kass note, the definition moves to a level which is more medico-technical, but not wholly so, and philosophical issues persist. Importantly, the choice of a standard depends on a corresponding death concept. The third level discusses the *operational criterion or set of criteria* required to satisfy the chosen general physiological standard and is mainly technical in nature. Finally, *tests* are required to provide empirical evidence that the criterion or criteria of death have been met. These are medical tests based on clinical examination, laboratory, and radiographic investigations.

The distinctions drawn by Capron and Kass' fourfold analysis help explain the policy decision by Capron and Kass to select the general physiological standards as the level at which legislation regarding the definition of death should focus. They reject the level of a *basic concept* because it would either impose a particular metaphysical view or not provide adequate guidance. They also reject legislation at the *operational criteria* level because of the need for flexibility in the face of medical advances, and they reject legislation at the specific test level due to its technical nature, which is best left to physicians. As such, in their law review article, Capron and Kass support legislation at the level of *general physiological standards* (level 2), which they understood to include "irreversible cessation of spontaneous respiratory and/or circulatory functions," "irreversible loss of spontaneous brain functions," and "irreversible loss of the ability to respond or communicate," or "some combination of these" [1, pp. 102–3].[2] A non-exhaustive range of plausible standards from recent literature include (a) the irreversible cessation of the integrated functioning of the organism as a whole, such that the organism no longer has the

[1] We do not deny that besides whatever we may essentially be, we also are biologic organisms or animals. For a book-length discussion, see E.T. Olson, *What Are We? A Study in Personal Ontology* (New York: Oxford University Press, 2007). Rather we focus on the normative nature of the claim; even if indeed we share the same essential nature with other living entities or animals, that does not settle the issue in terms of the policies we adopt in determining when life starts or ends. This may be better understood by seeing that a different set of considerations bears on the appropriateness of initiating death behaviors for a human person vs. a mouse. This is not a matter of speciesism; it is merely a recognition of the different kinds of lives that are available to these creatures.

[2] We will continue using the term "irreversible" as it is more commonly employed in this context, although we are sympathetic to the suggestion that the term "permanent" is more appropriate. Nevertheless, we have elected to leave this debate for another time as it is not relevant to our focus on choice.

capacity to restore homeostasis and thereby resist entropy (somatic or organismic standard) [4]; (b) the irreversible cessation of all functions of the entire brain, including the brainstem (whole-brain standard) [5]; (c) the irreversible cessation of all functions of the brainstem (brainstem standard) [6]; and (d) the irreversible cessation of personhood or the loss of the capacity of consciousness (higher-brain or consciousness-based standard) [7–9].[3]

2 The Uniform Determination of Death Act

In his capacity as the executive director of the President's Commission for the Study of Ethical Problems in Medicine and Biomedical and Behavioral Research from 1979 to 1983, Capron worked on developing statutory standards for the definition of death. In 1980, the President's Commission drafted the Uniform Determination of Death Act (UDDA), which read, "An individual who has sustained either (1) irreversible cessation of circulatory and respiratory functions, or (2) irreversible cessation of all functions of the entire brain, including the brain stem, is dead. A determination of death must be made in accordance with accepted medical standards"[10]. One year later, the President's Commission published its report "Defining Death: A Report on the Medical, Legal, and Ethical Issues in the Determination of Death," which included this model legislation [10]. The UDDA was approved by the National Conference of Commissioners on Uniform State Laws, the American Medical Association, and the American Bar Association. These groups endorsed the model legislation and recommended that it be adopted in all jurisdictions in the United States.

The model legislation was designed to declare a person dead if they meet either physiological standard. Current controversies about the UDDA include (1) whether the determination of death by neurologic criteria as currently performed in accordance with the American Academy of Neurology (AAN) testing guidelines is consistent with the whole-brain standard [11]; (2) whether higher brain death should be a valid alternative; and (3) whether individual choice, at the level of physiologic standards, should be available in the determination of death.

[3]We base our discussion on the aforementioned fourfold scheme, although more recent scholars like Bernat and colleagues propose a threefold approach which includes (1) the definition of death, (2) the medical criterion for determining that death has occurred, and (3) the tests to prove that the criterion has been satisfied. Bernat and colleagues understand death as an event that separates the process of dying from the process of disintegration: "the permanent cessation of functioning of the organism as a whole." This commitment to defining death in reference to the organism-as-a-whole originates, according to these authors, from an effort to maintain an understanding of death as it is used in everyday language. But this definition interprets death purely as a biological concept, in contradiction with their claim that the definition is a philosophical task. For this reason, we find that the fourfold scheme allows for a better discernment of the nature of the ontological commitments and clinical questions that arise in the debate about how death ought to be determined.

2.1 Controversies with the UDDA: Is Determination of Death by Neurologic Criteria Using the AAN Testing Guidelines Consistent with the Whole-Brain Standard?

A crucial requirement for any physiological standard is that there are testable criteria which consistently and reliably determine whether a person has died. Determining whether the AAN testing guidelines are consistent with the whole-brain standard requires consideration of what counts as "all functions of the entire brain" (vs. some specific set of functions). The AAN endorses the position that whole brain death can be assessed by the lack of capacity for consciousness, the inability to breathe spontaneously, and the lack of brainstem reflexes. The AAN also endorses the belief that preserved neuroendocrine function may be present despite irreversible injury of the cerebral hemispheres and brainstem, and that this is not inconsistent with the whole brain standard as stated in the UDDA [12].

Several critics maintain that the presence of neuroendocrine function is not consistent with the whole brain standard as enshrined in the current form of the law (as discussed elsewhere in this book). These critics require that either the law be changed to meet practice or neuroendocrine function be tested as part of a determination of death [13–15]. There are other versions of the neurologic standard such as those used in Canada and the United Kingdom which focus on brainstem function and the capacity for consciousness, avoiding the aforementioned issue of preserved neuroendocrine function, but these criteria are not widely supported in the US [16].

2.2 Controversies with the UDDA: Higher Brain

Understanding our essential nature as embodied minds, or human persons, engenders a different neurologic standard, also known as the higher brain standard. The criterion that follows is irreversible coma or irreversible loss of the capacity of consciousness. Regardless of what one thinks about the philosophical merits of this proposal, the requirement for consistent and reliable tests poses significant problems. There is a growing literature on the discovery of covert consciousness in individuals thought to have irreversible loss of the capacity for consciousness, and high rates of misdiagnosing minimally conscious patients as being in a vegetative state/unaware-wakeful state (VS/UWS) [17–19]. It has become evident that behavior is an unreliable surrogate for consciousness, and bedside neurologic examination can misclassify patients with cognitive-motor dissociation (this refers to patients with preserved brainstem function who are behaviorally classified as in a coma or in a VS/UWS; [17]). One could consider advanced neuroimaging (e.g., rest and task-based fMRI) or neurophysiology testing to probe for the presence of consciousness in behaviorally unresponsive patients [18, 19], but these techniques can have high false negative rates and, more importantly, do not provide information about the capacity for consciousness [20]. Although we lean to respecting an individual's preferences for different neurologic standards, the lack of operational criteria and tests that are necessary and sufficient to determine higher brain function means that the current AAN criteria constitute the safest place to draw the line on determining death, even for higher-brain advocates.[4]

2.3 Controversies with the UDDA: Is It Necessary to Accept Both Standards as Valid Means to Determine Death?

The third concern with the UDDA is whether one must accept both physiological standards as valid means to determine death. Scholars like Truog and Miller argue that human beings are essentially biologic organisms such that human death ought to be identically defined and determined as for other biologic organisms or animals; the standard that follows is the irreversible loss of homeostatic integration of the organism-as-a-whole. They deny that the death of the brain signifies the death of the human being [21]. Others, like Veatch and Lizza, understand our essential nature as persons or embodied minds, and not organisms; such a stance motivates a neurologic standard where the irreversible loss of consciousness is a necessary and sufficient criterion of death [7, 22], and cardiorespiratory death is not death per se, but a determination of impending irreversible loss of consciousness.

We accept that people reasoning sincerely and to the best of their abilities may disagree about which physiological standard is necessary to define death. Rather, if one accepts that the determination of death (at least at level 1) is not purely biological but incorporates philosophical values, then it is not clear why one must accept the public policy that asserts that one can be determined to be dead by either standard. In a pluralistic liberal society, reasonable disagreement ought not be resolved by fiat [23–25]. Rather, the foundation of public policy in a liberal society is the principle of respect for persons (and respect for their personal values and beliefs) which requires respecting their right to choose by which standard(s) their death may be determined.

While thirty-eight states have adopted the UDDA verbatim (or with very similar wording), several states have opted to allow patients and their families a greater degree of choice [26]. New Jersey's statute is most noticeable for offering choice through a "conscience clause" which stipulates that if an authorized physician has reason to believe that an individual has religious opposition to the determination of death based on neurologic criteria, then death can only be pronounced on the basis of circulatory-respiratory criteria [27]. New York also respects some degree of choice by requiring that all hospitals have a policy for "reasonable accommodation" of such religious or moral opposition [28].

Why should one allow for choice in the determination of death, as has been legislated in New Jersey and New York and in some other states to a more limited degree? While both somatic and neurologic standards can ground a determination of

[4] A possible response from higher-brain proponents could be that a well-informed person should be free to take their chances and draw the line at a certain level of brain injury short of meeting whole brain or brain stem criteria on the basis that they discount the prospect of minimal consciousness. This is a proposal that may merit more work and development.

death (level 2), the reason to permit choice is that requiring a person or a person's family to accept both pathways is to impose a metaphysical conception of death (level 1). For those whose beliefs reject death by neurologic criteria, choice is necessary in order not to impose a particular ontology [29].

One important difference between the New Jersey and New York statutes is the question of what reasonable justifications are to demand that the state respect one's decision for death to be determined by only a circulatory-respiratory standard. While personal choice in the determination of death has precedent in several state statutes out of deference to the first amendment's rights regarding religious freedoms, there is no principled reason to privilege religion in such a matter, and no justification for why secular reasons should not be admissible. The Supreme Court was called on to interpret the exemption for conscientious objection to the military draft and its relation to the first amendment [30]. The Court ruled: "What is necessary for a registrant's conscientious objection to all war to be 'religious' within the meaning of 6(j) is that this opposition to war stems from the registrant's moral, ethical, or religious beliefs about what is right and wrong and that these beliefs be held with the strength of traditional religious convictions." Similarly, other sincerely held beliefs and values could lead one to reject death by neurologic criteria, and we support a broader exemption in all states.

We recognize the two distinct standards delineated in the UDDA as two ways to determine death at the level of physiological standards. However, we believe that all states should allow for exemptions for individuals who reject the neurologic criteria.

3 Implementation

How should a conscience clause—an exemption for those who reject the neurologic criteria—be operationalized? We favor the idea of a preselected opt-out as an opportunity to explicitly reject one of the two standards and their corresponding criteria. This could take a similar form to what is currently done for registering organ donation preferences. The major advantage of this approach is that it removes decision-making at the time of illness when the very high stakes and emotional overload can make such a decision extremely difficult. It also avoids situations where third-party interests may derail the process and potentially not serve the individuals' wishes or preferences. On the other hand, registering an opt-out choice has its own problems including that it requires an active choice on the part of the individual, a well-informed presentation of the options, and a high enough level of health literacy in order to comprehend the options and their implications.

3.1 Consent for Testing

Another way to reject a standard or a criterion is to object to the performance of the necessary testing that would confirm the criterion. A recent debate has centered on the question of whether consent is necessary before one performs an apnea test for the determination of death by neurologic criteria [31]. The issue with the apnea test is that it is not risk-free, and several commentators have suggested that apnea testing, like any other procedure performed on a patient, should require the consent of the patient or his or her surrogate [32, 33]. Others disagree, maintaining that the risks are exaggerated and that the test is necessary for the determination of death, and thus not optional or a subject of consent [34]. Although we do not necessarily advocate for consent, we nevertheless argue here for *choice*, and so we think that a family objecting to death determination should be treated with respect, and at least an attempt towards accommodation of their view [35], unless the patient, when competent, had expressed a specific preference.

4 Objections

There are three objections in recent literature to the idea of choice. One is that choice will lead to *policy chaos*. The second is that rejection of the neurologic standard will have important *deleterious consequences* for organ transplantation and other scarce resources. The third poses the problem of opting out of the circulatory-respiratory standard.

4.1 Policy Chaos

We have said above that level 1 is the level at which one philosophically, metaphysically, and potentially spiritually, defines human death. An obvious problem is that we can imagine views that do not generate plausible standards or criteria, and are altogether untestable. However, we are not advocating for unregulated choice. As mentioned earlier, the determination of death in modern human societies has to be medicalized and constrained by medical science and technology. Although one is free to hold personal idiosyncratic views about the concept and meaning of death, they still have to conform to plausible and testable standards and criteria. In fact, we have made the modest proposal of not introducing any additional or novel operational criteria or tests, merely to provide choice among the existing ones.

A related concern is that choice could lead to a situation in which there are two identical patients side-by-side in hospital beds, one treated as alive and the other pronounced dead [15]. This would have implications for health insurance, life insurance, and inheritance. But this issue already occurs because surrogates are empowered to choose what health care their loved ones receive. In some cases, the surrogate demands to do everything, and in other families, the surrogate asks for withdrawal of treatment, resulting in one patient being maintained and the other being allowed to die. These contradictory outcomes are a reasonable cost to pay in order to treat all patients according to their own wishes.

Furthermore, if one were to judge from a state like New Jersey that has a conscience clause, there is no evidence that implementing a system of choice as pertains to the determination of death leads to a large number of refusals or strains scarce resources, or that policy chaos ensues. Accommodating opt-out from death by neurologic criteria could impose extra costs on health insurance, but there is no evidence that the social costs would be significant. Because such cases would probably be rare, the cost to insurers would be minimal. That cost should be seen as worthwhile if it protects autonomous choice and respects reasonable personal belief systems. If objections to the use of neurologic criteria to determine death were to become frequent, then one could imagine that health insurance policies could be written to state that usual coverage only pays for treatment until an individual is dead according to uniform standards as delineated by the UDDA but that individuals who choose to be determined to be dead only by circulatory-respiratory criteria could seek out supplemental health insurance, plan to self-pay, or seek charity care from communities that share their views.

4.2 Consequences for Scarce Resources

Patients declared dead by neurologic criteria remain the main source of transplantable organs [36]. Although donation after death by circulatory-respiratory criteria is gaining in numbers, it currently provides less organs and carries a higher risk for ischemic time, resulting in worse organ quality [37]. To date, there is no evidence that many people would choose to opt out of brain-based determinations. However, if a large number of people were to opt out of brain-based determinations, this could have a negative impact on organs available for transplantation. While this would harm the 100,000 people who are on organ waitlists, that does not justify forcing patients and their families to adopt a certain physiological standard for how death is determined. On the contrary, we believe that there is the potential to do serious harm by refusing to respect opt-outs in terms of the loss of public trust regarding medicine, organ donation, and transplantation. Instead, the medical community would need to balance the national expansion of conscience clause exemptions with the development of innovative educational campaigns about the various ways to determine death and the value of organ transplantation. Allowing individuals to accept or refuse alternative definitions of death may increase public confidence in the organ procurement system and promote voluntary organ donation.

4.3 Circulatory Opt-Out

Daniel Sulmasy poses the following challenge: "Could a family demand indefinite cardiopulmonary resuscitation? Or could they demand life support with a left ventricular assist device and extracorporeal membrane oxygenation for a loved one who meets the cardiopulmonary definition of death on the supposition that he would not accept that he is dead?" [15]. This challenge seems to ignore the fact that the purpose of continued cardiorespiratory function is to maintain brain function. Families may, and at times do, request the continuation or escalation of extraordinary means of circulatory support, but if such interventions are considered clinically inappropriate, they may not be granted [38].

Table 1 offers a summary of the main arguments, objections, and replies.

Table 1 Argument supporting choice, objections, and replies

Argument	Objections	Replies
Defining death entails metaphysical commitments open to reasonable disagreement that should not be resolved by fiat; choice is a requirement of respect for persons	Death is a common term in everyday use and refers to the cessation of organismal homeostatic integration	1. The standard of organismal integration only obtains under a human ontology view known as animalism 2. Ontological and biological reasoning may not settle the normative issue of determining death 3. Death of the brain is not required for homeostatic integration
	A potentially large number of metaphysical, religious, or spiritual views could be proposed	1. People are entitled to their personal understandings of death; nevertheless, its determination still has to conform to scientifically plausible and testable standards and criteria 2. The most common objection is to determination by neurologic criteria; determination by cardiorespiratory criteria is widely accepted
	Policy and implementation chaos may ensue, leading to differential treatment of patients with similar clinical characteristics	1. No empirical evidence for chaos in states with conscience clauses that permit exemptions to the neurological determination of death such as in NJ and NY
		2. Differential end-of-life treatment is dictated by individualized patient preferences and this is common practice
	Deleterious consequences for organ transplantation and the allocation of scarce resources	1. Determination of death should remain a separate issue 2. No evidence for large-scale rejection of the neurological standard 3. Potential for negative backlash from failures to respect refusals 4. Avoids the more complicated proposal of abandoning the dead donor rule

5 Conclusions

The medical definition of human death cannot rest on contested metaphysics or unmeasurable standards; it should, rather, be regarded as a plausible and widely accepted social construct that conforms to the best available and pragmatic medical science and practice. The standard(s) and criteria should be transparent, consistent, reliable, and publicly justifiable and allow for the accommodation of reasonable choice. The law in its current form treats death as a unitary phenomenon with two sets of physiological standards: one based on circulatory-respiratory cessation and one based on cessation of brain function. We have argued that these criteria carry built-in assumptions that may not be acceptable to all. The justification for allowing choice to decide by which physiological standards and criteria one may be determined to be dead rests on the foundational principle of respect for persons.

References

1. Capron AM, Kass LR. A statutory definition of the standards for determining human death: an appraisal and a proposal. Univ Pennsylvania Law Rev. 1972;121(1):87–118.
2. Veatch RM. Controversies in defining death: a case for choice. Theor Med Bioeth. 2019;40(5):381–401.
3. Nair-Collins M. We Die When Entropy Overwhelms Homeostasis. In: Cholbi M, Timmerman T, editors. Chapter in exploring the philosophy of death and dying: classical and contemporary perspectives. 1st ed. Abingdon: Routledge; 2020.
4. Bernat JL, Culver CM, Gert B. On the definition and criterion of death. Ann Intern Med. 1981;94(3):389–94.
5. President's Council on Bioethics. Controversies in the determination of death: a white paper by the President's Council on Bioethics. Washington, DC: President's Council on Bioethics; 2008.
6. Academy of Royal Medical Colleges. A code of practice for the diagnosis and confirmation of death. London: Academy of Royal Medical Colleges; 2008. p. 1–42.
7. Veatch R. Death, dying, and the biological revolution. New Haven, CT: Yale University Press; 1976.
8. Green MB, Wikler D. Brain death and personal identity. Philos Public Aff. 1980;9(2):105–33.
9. Veatch RM, Ross LF. Defining death: The case for choice. Washington, DC: Georgetown University Press; 2016.
10. The National Conference of Commissioners on Uniform State Laws. Uniform Determination of Death Act. Approved by all the states July 26–August 1, 1980. http://www.lchc.ucsd.edu/cogn_150/Readings/death_act.pdf. Accessed September 2022.
11. Wijdicks EFM, Varelas PN, Gronseth GS, et al. Evidence-based guideline update: determining brain death in adults: report of the Quality Standards Subcommittee of the American Academy of Neurology. Neurology. 2010;74(23):1911–8.
12. Russell JA, Epstein LG, Greer DM, et al. Brain death, the determination of brain death, and member guidance for brain death accommodation requests: AAN position statement. Neurology. 2019;92(6):304.
13. Nair-Collins M, Joffe AR. Hypothalamic function in patients diagnosed as brain dead and its practical consequences. Handb Clin Neurol. 2021;182:433–46.
14. Lewis A, Bonnie RJ, Pope T. It's time to revise the uniform determination of death act. Ann Intern Med. 2020;172(2):143–4.
15. Sulmasy DP. Whole-brain death and integration: realigning the ontological concept with clinical diagnostic tests. Theor Med Bioeth. 2019;40(5):455–81.

16. Wijdicks EF. The transatlantic divide over brain death determination and the debate. Brain. 2012;135(Pt 4):1321–31.
17. Schnakers C, Vanhaudenhuyse A, Giacino J, et al. Diagnostic accuracy of the vegetative and minimally conscious state: clinical consensus versus standardized neurobehavioral assessment. BMC Neurol. 2009;9:35.
18. Monti MM, Vanhaudenhuyse A, Coleman MR, et al. Willful modulation of brain activity in disorders of consciousness. N Engl J Med. 2010;362(7):579–89.
19. Cruse D, Chennu S, Chatelle C, et al. Bedside detection of awareness in the vegetative state: a cohort study. Lancet. 2011;378(9809):2088–94.
20. Peterson A, Cruse D, Naci L, et al. Risk, diagnostic error, and the clinical science of consciousness. Neuroimage Clin. 2015;7:588–97.
21. Truog RD, Miller FG. Changing the conversation about brain death. Am J Bioeth. 2014;14(8):9–14.
22. Lizza JP. Where's Waldo? The 'decapitation gambit' and the definition of death. J Med Ethics. 2011;37(12):743–6.
23. Larmore C. What is political philosophy? vol. 5. Princeton, NJ: Princeton University Press; 2020. p. 151.
24. Lazaridis C, Johnson LSM. Determination of brain death. JAMA. 2021;325(5):492.
25. Lazaridis C. Defining death: reasonableness and legitimacy. J Clin Ethics. 2021;32(2):109–13.
26. Nikas NT, Bordlee DC, Moreira M. Determination of death and the dead donor rule: a survey of the current law on brain death. J Med Philos. 2016;41(3):237–56.
27. NJ Rev. Stat § 26:6A-5, 2013.
28. New York Codes, Rules and Regulations. Title: Section 400.16 - Determination of death, 10 NYCRR 400.16.
29. Ross LF. Respecting choice in definitions of death. Hastings Cent Rep. 2018;48(Suppl 4):S53–5.
30. Welsh v. United States 398 US 333. 1970
31. Berkowitz I, Garrett JR. Legal and ethical considerations for requiring consent for apnea testing in brain death determination. Am J Bioeth. 2020;20(6):4–16.
32. Joffe AR. The apnea test: requiring consent for a test that is a self-fulfilling prophecy, not fit for purpose, and always confounded? Am J Bioeth. 2020;20(6):42–4.
33. Paquette E, Frader J, Shah SC, Tasker R, Truog R. Beyond the apnea test: an argument to broaden the requirement for consent to the entire brain death evaluation. Am J Bioeth. 2020;20(6):17–9.
34. Bhagat D, Lewis A. The case against solicitation of consent for apnea testing. Am J Bioeth. 2020;20(6):20–2.
35. Lazaridis C. Accommodating apnea testing not death determination refusal. Am J Bioeth. 2020;20(6):47–9.
36. Global Observatory on Organ Donation and Transplantation. http://www.transplant-observatory.org. Accessed Sep 2022.
37. Domínguez-Gil B, Ascher N, Capron AM, et al. Expanding controlled donation after the circulatory determination of death: statement from an international collaborative. Intensive Care Med. 2021;47(3):265–81.
38. Bosslet GT, Pope TM, Rubenfeld GD, et al. Responding to requests for potentially inappropriate treatments in intensive care units. Am J Respir Crit Care Med. 2015;191(11):1318–30.

The Distinction Between *Determination* of Death by Neurologic Criteria and *Declaration* of Death

Robert C. Tasker

The origins and meanings of words are found in standard dictionaries of etymology [1, 2], and my précis for the words *determination* and *declaration* is:

> ***Determination:*** In the Latin origin, *de-* means completely or thoroughly, and *terminare* means terminate. The derivative words are *determinare* and *determinatio*, meaning limit or fix. Subsequent language iterations are *determiner* (old French) and *determine* (Middle English). In modern usage, *determine* refers to "the process of ascertaining something exactly."
>
> ***Declaration:*** In its Latin origin, *clarus* or *clarare* means clear or make clear. The derivative Latin words are *declarare* and *declaratio*, and the Middle English word is *declare*. In modern usage, *declare* refers to "the process of a formal announcement or explicit statement on the beginning of a state or condition."

These two words—*determination* and *declaration*—are now embedded in medical narratives about the deaths of individuals in an irreversible and unresponsive coma with apnea and no brainstem reflexes, i.e., death by neurologic criteria. These terms were used in the 1968 report of the Ad Hoc *Committee of the Harvard Medical School to examine the definition of death* [3]. At that time, *determination* was used to describe aspects of the diagnosis of death by neurologic criteria (i.e., the process of ascertaining something completely), and *declaration* was used to describe a legal proclamation of death or the assignment of a person as now dead (i.e., the explicit statement on the beginning of a state or condition).

Over 50 years later, in the 2020 report from the World Brain Death Project, these same terms are used, but with subtle differences in the language [4]. *Death determination* is defined as "processes and tests required to diagnose death in accordance

R. C. Tasker (✉)
Department of Anesthesiology, Critical Care and Pain Medicine, Boston Children's Hospital, Boston, MA, USA

Selwyn College, Cambridge University, Cambridge, UK
e-mail: robert.tasker@childrens.harvard.edu

A. Lewis, J. L. Bernat (eds.), *Death Determination by Neurologic Criteria*, Advances in Neuroethics, https://doi.org/10.1007/978-3-031-15947-3_32

with established criteria." In this publication, *determination* appears to be considered synonymous with words such as "diagnosis" or "definition." *Death declaration* is defined as "the point at which a health professional, having determined that an individual is dead, formally states this finding." *Declaration*, or action *to declare*, is implied as something that comes after or follows a diagnosis of death by neurologic criteria. For example, see Table 3 and Supplement 12, which cover the issue of managing requests to forgo an evaluation or continue organ support after death by neurologic criteria [4]. However, in the Supplement, the word *declare* is used repeatedly in the context of religion and/or culture, and there are instances when *declare* is used interchangeably with *determine*, e.g., "criteria to determine death" and "criteria to declare death." There are also times when *declare* appears to be a word that is used when referring to public understanding or a cultural perspective on diagnosing death.

Use of these words—*determination* and *declaration*—can be followed through the various official standards and texts referring to death by neurologic criteria between 1968 and 2020. Of note, the words featured in the 1981 President's Commission for the Study of Ethical Problems in Medicine and Biomedical and Behavioral Research on *"Defining death: a report on the medical, legal, and ethical issues in the determination of death"* [5]. For example, the following two quotes indicate that *determination* and *declaration* not only are only distinct terms but also describe different activities that may or may not occur at the same time. These quotes also illustrate the possibility of a separation between the activity of *determination* and the activity of *declaration*:

> *Determinations* of death must be made in a consistent and evenhanded fashion, but the statute does not preclude flexibility in responding to individual circumstances after determination has been made… (page 43)

> …physicians would *declare* dead those bodies…in the absence of all brain functions. Nonetheless, people who believe…would not be forced by the statute to… (page 43)

1 Semantics or a Necessary Distinction Between Determination and Declaration?

In most instances, the logical or lexical distinction between words about death by neurologic criteria like *determination* and *declaration* may not matter. In contemporary practice, there is a standard clinical and legal procedure for death by neurologic criteria that is followed simultaneously [4], with a seamless combination of the clinical aspects of *determination* and the consequent legal responsibility of *declaration*. The progression in this sequence with *declaration* following *determination* was highlighted in Henry K. Beecher's 1968 special article on ethical problems in the "hopelessly unconscious patient" [6]:

> When is death? …the exact moment of death is not useful…What usually matters is not the time of death, but the time when a physician undertook *to declare* the patient dead…

Of note, recommendations about the time of death in children and adults after defining (or the *Determination*) of death by neurologic criteria may make the use of the term *declaration* of death redundant or implied. For example, in the 2011 Guidelines for the Determination of Brain Death in Infants and Children [7], two clinical evaluations are used for death by neurologic criteria. The first evaluation determines that the child has met the conditions necessary to undergo evaluation for death by neurologic criteria. The second evaluation, performed by a different physician, confirms that the child has fulfilled the conditions required for death by neurologic criteria. Death is declared after the second neurologic evaluation and apnea test confirm an unchanged and irreversible condition. The time of death is taken as the time of the second evaluation, or completion of an ancillary study.

The 2020 World Brain Death Project recommendations and suggestions for the time of death include, for a single evaluation, the time that the partial pressure of carbon dioxide in arterial blood ($PaCO_2$) reaches the target during the apnea test; for two evaluations, the time that the second evaluation is completed; and, for conditions that require ancillary testing, the time that the tests are interpreted and documented [4]. That is, there is no distinction between the time of *determination* of death by neurologic criteria and the *declaration* of death when only one clinical evaluation is required, as in adults. The apnea test and the time that $PaCO_2$ reaches the necessary target signifies not only the completed *determination* of death by neurologic criteria but also the *declaration* of a new state, death. Issuing a death certificate, registering death, and/or referral to the medical examiner/coroner are actions that follow, but the word *declaration* may be redundant. However, in infants and children (or those adults requiring a second evaluation or ancillary testing), the process of *determination* of death by neurologic criteria may have been started in an individual even though one apnea test has been carried out, but the individual is not dead, and *declaration* of death cannot occur. Here, there is no redundancy in the distinct terms.

Taking all the above usages together, there is, on the one hand, a workable framework at the end-of-life for usual practice within hospitals. However, on the other hand, there are also instances when there is objection to death by neurologic criteria [8]. Here, the distinction between *determination* of death by neurologic criteria and *declaration* of death may be a useful medicolegal construction that provides sanctuary (in space or time) while adjustments are made to accommodate a family's needs and/or conscientious objection; that is, so-called "reasonable accommodation."

2 "Reasonable Accommodation" and th e *Determination-Declaration* Construct

The 1981 President's Commission for the Study of Ethical Problems in Medicine and Biomedical and Behavioral Research in *Defining death: a report on the medical, legal and ethical issues in the determination of death* introduced the idea or possibility of "reasonable accommodation" being offered to families once a *determination* of death had been made [5]. It is interesting that the Commission appeared to consider this support as a legal matter with "no personal discretion as to the fact of death" and occurring without invoking a *declaration* of death. For example:

> …as a legal matter…no personal discretion as to the fact of death…room remains for reasonable accommodation of personal beliefs regarding the actions to be taken once a *determination* of death has been made. (page 81)

Declaration of death with continued organ support using mechanical ventilation has consequences, beyond simply issuing a legal certificate of death (see Supplement 11 of the 2020 World Brain Death Project [4]). There are many stakeholders, and in a privatized healthcare system financed by individuals and their insurers, the question is "who pays?" since health insurance policies do not ordinarily cover care provided after the *determination* of death [9]. For example, in the case of a neurosurgical intensive care patient, GF, in the state of New York discussed by Olick et al. [10], hospital costs were considered in relation to the timing of end-of-life procedures:

> …who bore the cost of care during the period of accommodation was not an issue in GF's case. Because GF's spouse objected to confirmation of brain death, the medical team delayed completing the hospital's protocol for determining death. In the meantime, since death had not been declared, GF's insurance covered the cost of care…

Other cases, perspectives, and discussions about reasonable accommodation in death by neurologic criteria are well reported in the literature [11–16] and discussed in detail elsewhere in this book. In addition, we also have the views of physicians in general [17], the American Academy of Neurology [18], and the World Brain Death Project [4]. However, the optimal clinical approach when supporting families with a belief that differs from the prevailing medical professional and legal opinions is far from clear, and the debate continues [8, 19–21].

Thus, while accepting the principle of *reasonable accommodation* around the time of death by neurologic criteria, the practicalities of unifying guidance [18] and approach to timing and duration [9] are yet to be agreed upon by the wider professional community involved with evaluation for death by neurologic criteria (see Supplements 11 and 12 of the 2020 World Brain Death Project [4]). That said, it is clear there is the historical precedent of using an apparent distinction between the process of *determination* of death by neurologic criteria and the act of *declaration* of death to enable continued coverage of healthcare expenditure for ongoing organ support. But how long is "reasonable" and who decides on the outcome?

In the American Academy of Neurology 2019 position statement on "Brain death, the determination of brain death, and member guidance for brain death accommodation requests," the emphasis appears to be on the procedure for the diagnosis of death by neurologic criteria [18]. In fact, the word *determination* (or derivatives) is used a dozen times. In contrast, the word *declaration* is not used at all. This lack of use of the word *declaration* could reflect its redundancy, as described above. Alternatively, the words *determination* and *declaration* may be used interchangeably because completion of the *determination* of death by neurologic criteria and the timing of *declaration* of death are, in most instances in adult individuals, the same and the latter is implied by the former. For example, the position statement describes the state of death using other terms for the onset of something different, e.g., "newly dead" (p. 231), "premortal wishes of the brain-dead patient" (p232). Issues such as death certification, death registration, and ongoing health insurance are not discussed.

3 Organ Support After Death and Postmortem Pregnancy

An exceptional circumstance regarding "reasonable accommodation" occurs when death by neurologic criteria has been determined in a pregnant woman and the decision is made to continue organ support (i.e., support to maintain the function of the heart and lungs) to enable fetal development to attain extrauterine viability. This situation is discussed in detail elsewhere in this book, but a recent systematic review (using medical literature electronic databases from inception to April 2020) identified and summarized findings from 35 cases of death by neurologic criteria in pregnancy in which maternal organ support was continued to optimize perinatal outcome [22]. The mean gestational age at the time of maternal death by neurologic criteria was 20 weeks, and maternal organ support was continued for a mean duration of 7 weeks, resulting in 27/35 (77%) live neonatal births. Of note, a case report from the United States specified that organ support was continued for 90 days after "brain death testing was performed, and the pregnant patient was declared brain-dead" [23]. Most recently, organ support with veno-venous extracorporeal membrane oxygenation was used in a pregnant woman after death by neurologic criteria from 22 to 26 weeks of gestation [24].

In general intensive care unit practice, organ support after *determination* of death by neurologic criteria and *declaration* of death commonly occurs and may last a few days in organ donors before organs are removed. However, the examples of the continuation of organ support for pregnant women for weeks to months after the *declaration* of death indicate a status beyond being dead. For example, a child who is born via posthumous birth—as in birth after the mother has been declared dead—may be treated as though they were living at the death of a parent [25], and the mother may be included in the birth certificate, but there is a substantial financial burden for the family or mother's estate [26–28].

Together, these examples of the *determination* of death by neurologic criteria and *declaration* of death in the extraordinary circumstances leading to continued organ support demonstrate nuances to the terms that in other situations would be considered as being final. Organ support can be continued for months after death by neurologic criteria, and *determination* of death by neurologic criteria does not equate with imminent disintegration of body systems. Furthermore, *declaration* of death despite ongoing provision of organ support may impact legal status, and there can be costs not otherwise attributed to death.

4 Is the Distinction Between *Determination* of Death by Neurologic Criteria and *Declaration* of Death Useful?

The language of death, particularly death by neurologic criteria, is fraught with problems. At present, there is no consensus on how words such as *determination* and *declaration* should be used, despite their distinct meanings and derivation [1, 2]. Language evolves, and who is to say if the meaning of these words used by contemporary thinkers and frontline expositors and practitioners is incorrect, or inexact? And does it matter? I believe it does and these are my arguments for and against the placement of these distinct terms:

1. In the adult individual undergoing one clinical evaluation for the *determination* of death by neurologic criteria with the subsequent *declaration* of death, there is no distinction between *determination* and *declaration.* The latter is contemporaneous with the former at the time that $PaCO_2$ reaches the target during the apnea test.
2. In the infant, child, or adult individual requiring more than one clinical evaluation or ancillary testing for the *determination* of death by neurologic criteria, there is a potential distinction between the *determination* of death by neurologic criteria and *declaration* of death that can be exploited in some instances. For example, when there is objection to death by neurologic criteria or there is a need for reasonable accommodation in the process leading to death, there is a potential place of sanctuary for the family in this distinction.
3. The extraordinary circumstances of prolonged organ support after the *declaration* of death show us that during postmortem pregnancy, this new state can, potentially, accrue responsibilities and costs not otherwise attributed to being in the process of *determination* of death by neurologic criteria, but not yet in the state of *declaration* of death.

References

1. Onions CT, Friedrichsen GWS, Burchfield RW, editors. The Oxford Dictionary of English Etymology. 1st ed. Oxford: Oxford University Press; 1966.
2. Barnhart RK. In: Steinmetz, editor. The Chambers Dictionary of Etymology. 1st ed. London: Chambers Publishing Limited; 1988.
3. Report of the Ad Hoc Committee of the Harvard Medical School to Examine the Definition of Brain Death. A definition of irreversible coma. JAMA. 1968;205:337–40.
4. Greer DM, Shemie SD, Lewis A, et al. Determination of brain death/death by neurologic criteria: the world brain death project. JAMA. 2020;324:1078–97.
5. President's Commission for the Study of Ethical Problems in Medicine and Biomedical and Behavioral Research. Defining death: a report on the medical, legal and ethical issues in the determination of death. U.S. government printing Office; 1981.
6. Beecher HK. Ethical problems created by the hopelessly unconscious patient. N Engl J Med. 1968;278:1425–30.

7. Nakagawa TA, Ashwal S, Mathur M, et al. Clinical report – guidelines for the determination of brain death in infants and children: an update of the 1987 task force recommendations. Pediatrics. 2011;128:e720–40.
8. Truog RD, Morrison W, Kirschen M. What should we do when families refuse testing for brain death? AMA J Ethics. 2020;22:E986–94.
9. Truog RD. Brain death – too flawed to endure, too ingrained to abandon. J Law Med Ethics. 2007;35:273–81.
10. Olick RS, Braun EA, Potash J. Accommodating religious and moral objections to neurological death. J Clin Ethics. 2009;20:183–91.
11. Smith ML, Flamm AL. Accommodating religious beliefs in the ICU: a narrative account of a disputed death. Narrat Inq Bioeth. 2011;1:55–64.
12. Flamm AL, Smith ML, Mayer PA. Family members' requests to extend physiologic support after declaration of brain death: a case series analysis and proposed guidelines for clinical management. J Clin Ethics. 2014;25:222–37.
13. Bliss SE, Macauley RC. The least bad option: unilateral extubation after declaration of death by neurological criteria. J Clin Ethics. 2015;26:260–5.
14. Johnson LS. The case of reasonable accommodation of conscientious objections to declarations of brain death. J Bioeth Inq. 2016;13:105–15.
15. Miller G. Re-examining the origin and application of death by neurological criteria (DDNC): a commentary on "the case for reasonable accommodation of conscientious objections to declarations of brain death" by L Syd M Johnson. J Bioeth Inq. 2016;13:27–9.
16. Gabbay E, Fins JJ. Go in peace: brain death, reasonable accommodation and Jewish mourning rituals. J Relig Health. 2019;58:1672–86.
17. Lewis A, Varelas P, Greer D. Prolonging support after brain death: when families ask for more. Neurocrit Care. 2016;24:481–7.
18. Russell JA, Epstein LG, Greer DM, et al. Brain death, the determination of brain death, and member guidance for brain death accommodation requests: AAN position statement. Neurology. 2019;92(5):228-32.
19. McEvoy MJ, Scott MJ, Sawyer KE. Requests for accommodation in brain death cases: emerging role for pediatric palliative care. J Pain Symptom Manag. 2021;62:1319–24.
20. Joffe AR, Hansen G, Tibballs J. The world brain death project: the more you say it does not make it true. J Clin Ethics. 2021;32:97–108.
21. Lazaridis C. Defining death: reasonableness and legitimacy. J Clin Ethics. 2021;32:109–13.
22. Dodaro MG, Seidenari A, Marino IR, et al. Brain death in pregnancy: a systematic review focusing on perinatal outcomes. Am J Obstet Gynecol. 2021;224:445–69.
23. Holliday S, Magnuson-Woodward B. Somatic support following cardiac arrest for 90 days leading to a healthy baby boy: a case report. Heart Lung. 2017;46:397–400.
24. Salazar L, Arora L, Botia M, et al. Somatic support with veno-venous ECMO in a pregnant woman with brain death: a case report. ASAIO J. 2022;68:e16–8.
25. Sekino RH. Posthumous conception: the birth of a new class: Woodward vs. Commissioner of Social Security. Boston Univ J Sci Technol Law. 2002;8:362–73.
26. Spike JP. Pregnancy, brain death, and posthumous motherhood: a provisional policy proposal. Am J Bioeth. 2014;14:48–50.
27. Mayo TW. Brain-dead and pregnant in Texas. Am J Bioeth. 2014;14:15–8.
28. Barr JJ. When death is not the end: continuing somatic care during postmortem pregnancy. Linacre Q. 2019;86:275–82.

Why Families Object to Declaration of Death by Neurologic Criteria

Aleksandra E. Olszewski and Erin Talati Paquette

In 1968, after almost a century of unclear and conflicting discussions regarding the concept and determination of death, a committee at Harvard Medical School defined "irreversible coma as a new criterion for death" [1, 2]. They described the primary purposes of delineating this new criterion for death as being twofold: (1) to reduce the use of scarce intensive care resources, as well as to reduce the burden on families of those who suffer irreversible brain damage, and (2) to increase the organ donor pool [1, 2]. In 1981, to support the clinical description of the determination of death by neurologic criteria, the Uniform Determination of Death Act (UDDA) proposed two legal standards for death: irreversible cessation of circulatory and respiratory functions or irreversible cessation of all functions of the entire brain, including the brainstem [3, 4]. The UDDA was promulgated by the Uniform Law Commission and adopted by each state in some form [5, 6].

Despite this law, there have been many challenges to the determination of death by neurologic criteria conceptually, procedurally, practically, and legally, as discussed throughout this book [7]. There are vast differences across the world and within the United States in how death by neurologic criteria is determined and declared [8, 9]. Among the most controversial are differences in the evaluation for death by neurologic criteria, the need for consent to perform the evaluation, and the management of objections to the determination and declaration of death by neurologic criteria [10]. Legally, there are differences across states as well—the language

A. E. Olszewski
Department of Pediatrics, Division of Critical Care, Northwestern University Feinberg School of Medicine, Ann and Robert H. Lurie Children's Hospital, Chicago, IL, USA
e-mail: aolszewski@luriechildrens.org

E. T. Paquette (✉)
Department of Pediatrics, Division of Critical Care, Northwestern University Feinberg School of Medicine, Ann and Robert H. Lurie Children's Hospital, Northwestern University Pritzker School of Law (Courtesy), Chicago, IL, USA
e-mail: epaquette@luriechildrens.org

A. Lewis, J. L. Bernat (eds.), *Death Determination by Neurologic Criteria*, Advances in Neuroethics, https://doi.org/10.1007/978-3-031-15947-3_33

of the UDDA is used verbatim by 36 states and the District of Columbia, but 14 states use different wording, and only a few, like New Jersey and New York, legally address accommodation of religious objections to death by neurologic criteria [2].

Cases involving death by neurologic criteria are complex and challenging, particularly when families object to the determination or declaration of death [11]. Given that some clinicians, ethicists, and philosophers question and debate the concept of death by neurologic criteria, [11] one can only imagine how challenging and confusing it must be for patients' families to understand the concept and its application, particularly when faced with the need for urgent and traumatic decision-making. Unsurprisingly, objections to the declaration of death by neurologic criteria and the subsequent discontinuation of organ support are increasingly common in both clinical and legal settings [12, 13]. One study showed that half of the neurologists have been asked to continue organ support after declaring patients dead by neurologic criteria [10]. Despite these data, there are only limited studies about public perceptions of death by neurologic criteria [5]. There is even less in the literature about family objections to declaration of death by neurologic criteria and how clinicians respond to these objections. Here, we will discuss contemporary legal cases that highlight key factors influencing family objections to determination and declaration of death by neurologic criteria, analyze empirical data about the most commonly known reasons for objections, and share ideas for the management of objections.

1 Cases Highlighting Objections to Declaration of Death by Neurologic Criteria

In a recent review of court cases involving disputes regarding death by neurologic criteria, 30 unique cases were identified [14]. The majority of cases occurred after 2010, highlighting the fact that conflicts about death by neurologic criteria are increasing. Religion was a factor in nearly all of the cases. Many involved disputes about whether death by neurologic criteria was true death; in some, it was argued that discontinuation of organ support was homicide. The cases touched on alleged violations of the right to privacy over one's body as guaranteed by the Constitution, as well as the Americans with Disabilities Act and the Rehabilitation Act, which prohibit discrimination on the basis of disability, including severe neurologic impairment. Several prominent cases have had the potential to influence subsequent objections and policy development.

The case of Aiden Hailu involved a 20-year-old woman in Nevada whose father objected to the discontinuation of her organ support after she was determined to be dead by neurologic criteria [15]. Questions were raised about whether prior electroencephalographic activity confounded the determination. While the district court found that the hospital's application of the clinical standards described by the American Academy of Neurology to determine death by neurologic criteria was sufficient, the Supreme Court of Nevada rejected this finding, noting that such standards could only be dispositive if uniformly applied across the states. Ms. Hailu eventually met circulatory-respiratory criteria for death, making the facts of the case no longer relevant. However, following this case, the Nevada legislature revised the state's determination of death act to explicitly indicate that the American Academy of Neurology standard was sufficient for the determination of death by neurologic criteria [15].

Two prominent California cases—those of Jahi McMath and Israel Stinson—challenged California's death statute, and were eventually brought to the ninth Circuit Court of Appeals for consideration. Jahi McMath was a 13-year-old girl who underwent tonsillectomy and adenoidectomy. She experienced post-operative bleeding complications, leading to a circulatory-respiratory arrest and severe hypoxic-ischemic brain injury. She was eventually determined to be dead by neurologic criteria. Her family objected to this determination, citing a violation of their religious beliefs. After she was declared dead, her family moved her on ventilatory support to a facility in New Jersey, a state where objections to death by neurologic criteria on religious grounds are permitted. For more than 4 years, Jahi's organs were supported by technology, until she eventually met the criteria for circulatory-respiratory death due to complications from liver failure. Jahi's case brought national attention to the issue of death by neurologic criteria [16].

In a separate California case, Israel Stinson, a 2-year-old boy, suffered a circulatory-respiratory arrest following a severe asthma attack [17]. He was initially declared dead by neurologic criteria, but his parents disagreed and transferred him to Guatemala for further evaluation. He was subsequently brought back to another California hospital, but that hospital also determined him to be dead by neurologic criteria. Israel's parents filed suit to enjoin the hospital from removing Israel from the ventilator, but the court disagreed and the ventilator was ultimately discontinued. His family continued to pursue legal remedies, bringing suit to argue that the California death statute was unconstitutional. Despite their best efforts, the case was dismissed by the lower courts in California because the family was not able to demonstrate the foundational element of standing (having the right to bring suit). The court never ruled on the constitutional arguments brought forth in the case.

These cases demonstrate a lack of uniformity in views on death by neurologic criteria. Importantly, in each of them, the patients and their families were of a minority race, raising the concern that minority viewpoints are prejudicially rejected or ignored with no accommodation to objections to the determination of death by neurologic criteria. In the next section, we consider the reasons why families object to the determination and declaration of death by neurologic criteria.

2 Why Do Families Object to the Determination and Declaration of Death by Neurologic Criteria?

There are few studies on clinician perspectives about the reasons why families object to death by neurologic criteria, and no studies summarize the perspectives of families themselves. Here, we summarize clinician views on why families object to the determination and declaration of death by neurologic criteria, recognizing that further study is needed on the views of families themselves.

Families object to the determination and declaration of death by neurologic criteria for a variety of complex reasons. In a recent study surveying pediatric neurologist and intensivist perspectives, participants were asked about their experiences with objections to determination and declaration of death by neurologic criteria [18, 19]. The top reasons for objections were "waiting on a miracle" ($n = 56$, 70%), not wanting to "give up" ($n = 54$, 67%), believing reports of patients "waking up" after determination of death by neurologic criteria ($n = 32$, 40%), not believing in death by neurologic criteria ($n = 31$, 39%), and having religious objections ($n = 28$, 35%). In studies surveying physicians practicing with adults, families were reported to object to discontinuation of organ support after death by neurologic criteria for similar reasons: belief that the patient would regain neurologic function; lack of acceptance that a person whose heart is beating could be dead; a desire to continue receiving Social Security benefits; and religious beliefs [10, 20].

Families object to the determination and declaration of death by neurologic criteria at different points in a patient's course. Figure 1 depicts the more common steps involved in the determination and declaration, though the process may not always appear as linearly as depicted. Of note, at any point in the process, a family may request that testing not be performed, or support not be withdrawn. Interviews of twelve pediatric neurologists and critical care providers demonstrated that the majority of family objections occurred between the first and second evaluation for determination of death by neurologic criteria, and requests ranged from asking for a brief delay prior to testing to refusing the evaluation for a prolonged period of time [19]. In another study, 48% of neurologists reported they had encountered families

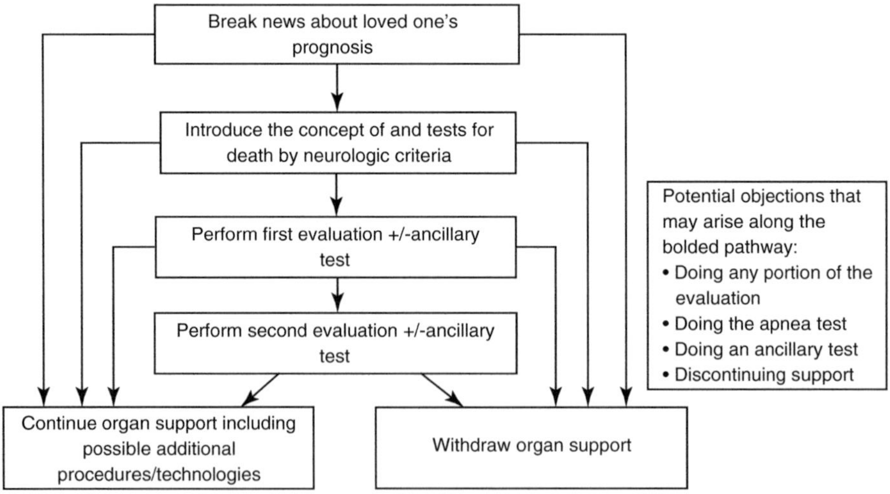

Fig. 1 The typical stages of discussing, determining, and declaring death by neurologic criteria, with potential outcomes of continuation or withdrawal of organ support. Family objection or request for accommodation may occur at many stages in the process

who requested prolonged support after the determination of death by neurologic criteria [10]. Time is often key to objections—family members may ask for time to accommodate needs related to the dying process or end-of-life care planning, time for anticipatory grief, and time for "a miracle" to occur. Below, we discuss in greater detail each of the more commonly described reasons for family objections.

2.1 Belief that Only Cessation of Circulatory-Respiratory Function Can Signify Death

In cases involving death by neurologic criteria, families are asked to reckon with the confusing and painful idea that their family member has a beating heart, but is dead [5, 11, 21]. In a qualitative study, family members described how challenging it is to accept death by neurologic criteria when faced with their loved ones' beating hearts, growing beards, warm bodies, and perspiration [22]. In a recent review of court cases, likewise, common family perceptions of death by neurologic criteria included beliefs that recovery was possible; the presence of a heartbeat signifies life; if patients were truly dead, their bodies would decay quickly; patients declared dead by neurologic criteria could feel pain; and death by neurologic criteria and coma are the same thing [14]. Understandably, the idea that a loved one transitions from being alive to being dead with no visible change in their condition (they continue to look the same as other ill people in an intensive care unit) would be confusing to accept for many [23]. Importantly, for some who believe death occurs when the heart ceases to beat, they understand death by neurologic criteria to be a step towards death, rather than death itself. Therefore, its determination and declaration may be viewed as killing their loved one [5, 24].

Death by neurologic criteria is a relatively new concept, and it is unlike what many expect when they imagine or experience death. Whether death by neurologic criteria should actually constitute death remains debated among scholars, so it is unsurprising that there are varying views on it among the public [4, 5, 25, 26]. The meaning of death, and its experience, can be incredibly personal. Even more tangible and concrete aspects of it, like the time of death, can differ among family members of the same patient based on their individual perceptions and experience. Even in cases of death experienced in a more traditional or expected way, denial is often an aspect of the grieving process, and this may be confounded by grappling with understanding death by neurologic criteria. Separating expected denial from confusion that deserves additional clarity from the medical team can be a challenging balance, as clinicians strive to support families through this process [5, 13]. Finally, religion plays a factor in how families understand and experience death, as described in detail below. In court cases, the belief that life persists as long as the heart is beating was often connected to religious beliefs, across many different religions [5].

2.2 Religious Objections to the Idea of Death by Neurologic Criteria

In the few studies that surveyed and interviewed clinicians, religion rose to the top as a reason for family objection to death by neurologic criteria [27, 28]. Likewise, in reviewing court cases on the topic, religion was a factor in over half of the cases [18, 29]. In interviews with pediatric neurologists and intensivists, religion was the most frequently cited reason for objection [18]. In this same study, surveying a broader group of clinicians, religiosity was associated with objection to the declaration of death by neurologic criteria. Furthermore, family-clinician religiosity discordance was associated with religious objections [18]. The association of religiosity with objections was not isolated to any specific religion—refusals of determination and declaration of death by neurologic criteria were described in families practicing Christianity, Buddhism, Catholicism, Judaism, and Islam. Key religious beliefs that were felt to impact objections were the belief that (1) life continues as long as the heart is beating, (2) withdrawal of support may be viewed as murder, and (3) miraculous recovery is possible.

2.3 International Differences in Belief in the Concept of Death by Neurologic Criteria

While approaches to determination and declaration of death by neurologic criteria and laws about the management of objections to death by neurologic criteria vary across the United States, this variability is heightened when considering approaches and laws internationally [6, 30–32]. In discussing why families might object to death by neurologic criteria, it is important to recognize international differences in the concept. First, death by neurologic criteria is not uniformly accepted across the world. Second, not all countries have standards to determine death by neurologic criteria. Though the World Brain Death Project argues that most countries have a process for death by neurologic criteria, this majority does not equal uniformity [33]. Among countries that have a standard for the determination of death by neurologic criteria, there is considerable variability among practices [30–32]. This suggests that beliefs are not uniform even among those who may agree with having a process for determination in place. There may be very real philosophical and religious differences that will not allow uniformity, and these differences may impact individual patients and families, especially when the family's culture is not the same as that of the clinical team.

2.4 Not Wanting to Give Up

Although death by neurologic criteria is widely accepted to denote permanent brain injury, it is not necessarily understood that way by the public [5]. In studies conducted in Southern Iran, India, and Sweden, among families who refused organ donation after death by neurologic criteria, a large portion of them made this decision because they believed future recovery was possible [34–38]. Though findings vary, a significant portion of families faced with death by neurologic criteria and decisions about organ donation believed their loved one had the potential to recover [35, 39–41]. In one study, 52% of family members who refused organ donation after death by neurologic criteria believed their family members could recover, while an additional 14% were not sure [41]. In another study, less than 5% of those who refused organ donation after death by neurologic criteria believed that patients who met the neurologic criteria for death were unable to recover; only 20% of those who agreed to organ donation believed their loved one was unable to recover [40]. In a study surveying pediatric neurologists and intensivists, the belief that allowing determination of death by neurologic criteria signified giving up and investment in "doing everything" were key themes in family member objections [19].

2.5 The Influences of Trauma, Distrust, and Bias

Some have described that a family's previous experience with healthcare systems, particularly errors, incorrect diagnoses or prognoses, or difficulties in communication, may lessen their trust in the clinical team in scenarios involving the determination of death by neurologic criteria [11, 13]. Prior experiences influence how families perceive information from the clinical team, how they behave, and how they make decisions on behalf of family members [11].

Exposure to prior and current trauma is shaped by social and structural health determinants. Known groups that are at higher risk of cumulative trauma include racial/ethnic minorities, those facing socioeconomic hardships, and those with intellectual or developmental disabilities [42, 43]. In addition, experiencing racism and discrimination is associated with a higher incidence of traumatic stress symptoms [11, 43]. Individuals from these groups are more likely to experience bias in the healthcare system, which may further influence their experiences in healthcare settings [44–47]. The race/ethnicity of the family was associated with requests to allow time for others to arrive before the declaration of death by neurologic criteria, particularly among Black, mixed/other, and Hispanic race/ethnicity [19]. Low socio-economic status was correlated with needing more time to grieve. Lower education was associated with reports of individuals "waking up." Though previous, current, and cumulative exposure to trauma and bias are not always necessarily factors in family objections, there are complex intersecting social and structural factors that may correlate with and influence family objections. These interactions bear further study to better understand whether such connections indeed exist and, if so, their mechanisms to potentially mitigate perpetuating additional trauma.

In addition to personal prior or current experiences with healthcare, families are influenced by stories from known contacts and the media. Families increasingly engage with social media and news to inform their understanding and decision-making. This can be particularly challenging when terms such as "coma" and "brain death" are used interchangeably, and families can access news stories, movies, or books about people who "wake up" after being "brain dead" [48, 49]. Misinformation about the determination and declaration of death by neurologic criteria, especially relating to organ donation, is widespread in media coverage [50]. A narrative review of public perceptions of death by neurologic criteria showed the public generally does not understand legal aspects of death by neurologic criteria, uncontested medical facts about death by neurologic criteria, and that organs are procured from patients who are dead by neurologic criteria while their hearts are still beating [5].

One study demonstrated that family mistrust of the clinical team was associated with waiting for a miracle, not wanting to give up, and not believing in death by neurologic criteria [18]. In interviews, clinicians further described that challenges with communication about the evaluation for death by neurologic criteria or prognosis contributed to objections [19]. Some felt that confidence and competence in performing the evaluation could lead to family objections.

One specific form of mistrust that can shape objections is the belief that medical teams determine death by neurologic criteria simply to obtain organs [5, 39, 41]. The majority of the public does not understand how organ donation in the setting of death by neurologic criteria works. Most do not understand that organs are procured before mechanical support is removed, while the heart is still beating [5]. Learning this during a traumatic and urgent time may influence how families feel about organ donation and declaration of death by neurologic criteria. Families who object to organ donation may be more likely to object to the declaration of death by neurologic criteria [34–36]. Misconceptions may also be influenced by prominent court cases and media coverage. Because historically and currently, death by neurologic criteria determination and declaration is linked to organ donation, it is critical that clinical teams approach discussions about death by neurologic criteria with sensitivity, honesty, and openness to family perspectives, and do so separately from discussions about organ donation.

3 Responding to Objections

As shown in Fig. 1, family objections may occur at different points during the determination and declaration of death by the neurologic criteria process. Objections to the declaration of death by neurologic criteria encompass objections to the determination of death by neurologic criteria.

Prior to the determination of death by neurologic criteria, obtaining consent from families for the determination is one way to deal with potential objections after declaration. Even among family members who may accept that meaningful recovery is not possible, being faced with a decision to withdraw organ support, or to perform an evaluation for death, is incredibly challenging and devastating at best [23]. It may be particularly challenging for parents of children and young patients to make the decision to perform an evaluation for determination of death or withdraw support, which may be why some of the most prominent cases in the law and media are pediatric ones [13]. Some ethicists and clinicians argue that it is unethical to task family members with these decisions, because they are impossible decisions for anyone, and because there is significant legal or medical ambiguity [23]. Others argue that, despite the risk that families will request to withhold the evaluation, in order to respect diverse viewpoints about death by neurologic criteria, it is ethical and legally permissible to obtain consent for determination [10, 51, 52].

Mataya et al. found that 64% of pediatric neurologists and intensivists would not obtain permission to determine death by neurologic criteria, indicating that 36% would [53]. Reasons for obtaining permission included that it is the last decision a family may make for their loved one, there is a duty to obtain permission, and that the imperfect definition of death by neurologic criteria requires discussion with families. In clinician interviews, conducted as a subset of the prior study, respondents were asked whether they sought permission prior to determining death by neurologic criteria [19]. Many expressed that they felt permission was unnecessary and that it could create challenges if families expected it; others argued that seeking consent is an important aspect of communication and relationship-building with families. Lewis et al. found that, although 78% of neurologists did not agree that there was a need for family members to be asked to give consent prior to performing the evaluation for determination of death by neurologic criteria, 42% felt that permission should be obtained prior to discontinuation of organ support after determination or declaration [10]. This disagreement among experts highlights the lack of clarity on the topic, and the disparate approaches used [23].

After the determination and before the declaration of death by neurologic criteria, responding to objections is often referred to as accommodation, or "a medically unnecessary period of time between the determination that a patient meets death by neurologic criteria and the discontinuation of cardiopulmonary sustaining therapies" [11]. Four states have legal accommodation exceptions, but questions about determining reasonable accommodations and managing these objections exist in all states. Arguments against accommodation include concerns about futility, inappropriate use of resources, and a desire to avoid participating in actions that may not be in the patient's best interest [27]. Arguments for accommodation include respecting family requests and care goals [54]. In a survey of neurologists, 70% believed that there are times when organ support ought to be continued after declaration, and 95% believed there are times when it should be permitted, with the most common reason being the avoidance of legal challenges [10]. Though the majority of respondents would accommodate families with time, few (14%) would start additional therapies such as nutrition, fluid, vasopressors, antibiotics, etc.

Cases involving death by neurologic criteria and requests for accommodation can be incredibly challenging, as clinical teams are tasked with balancing patient care, family support and education, appropriate use of resources, and, increasingly, worries about negative press or legal challenges. Clinicians have commented that management of objections to the determination of death by neurologic criteria causes moral distress, distracts from care for other patients, and creates personal strain on individual clinicians and healthcare teams [19, 53].

Because there are multiple and complex reasons for families to object to the determination and declaration of death by neurologic criteria, responding to an objection ought to include an individualized approach that considers each family's concerns and needs. Open communication and negotiation with families, while continuously responding to new data and circumstances, are critically important. Also useful is the employment of second opinions and the involvement of a supportive multidisciplinary team [11]. Resources described as helpful include palliative care, ethics, other subspecialty consultants, hospital legal, hospital administration, nursing colleagues, and spiritual support.

Currently, approaches are markedly varied, often not because of individualized family needs, but because of differing clinician attitudes, lack of education, and lack of standards or guidance. Indeed, one respondent in the study of Paquette et al. commented that approaches to consent in these circumstances are not taught in medical training [19]. Surveyed clinicians noted that legal guidance or laws are lacking, there are rarely internal policies, and when there are, they are unhelpful or not followed. State laws vary regarding whether family requests for accommodation or objection to testing must be honored [55]. Clinicians describe a need for resources, guidance, and support in addressing family requests for accommodation [19]. Even when policies may support clinicians in avoiding the provision of "futile" care, concerns about supporting families and avoiding bad press or legal action affect how clinicians approach these scenarios, especially in the context of existing legal cases and the media coverage of them.

In interviews with clinicians, several potential approaches have been proposed [19]. Often, an approach of waiting until the family is "ready" has been utilized, though clinicians described a wide range of approaches and outcomes [23]. When family objections are accommodated, they sometimes come to agree with the withdrawal of organ support. Other times, they did not, and clinical teams facilitated the placement of tracheostomy and gastrostomy tube, followed by transfer home, internationally, or to another state where refusal of determination of death by neurologic criteria is allowed. Some clinical teams accommodated aspects of the family's requests but placed restrictions, with the refusal to perform additional interventions and the use of a do-not-resuscitate order.

Anecdotally, families who request accommodation are more often than not merely asking for time to process and plan [11, 13, 23]. As stated by Dr. Robert Truog, "Most families don't want to sustain the life of a loved one who will never wake up, any more than clinicians want to participate in care that is essentially futile" [23]. Thus, often, giving them time resolves potential conflicts without the need for confrontation [11, 13, 23]. When this is not the case, some guidance does exist for clinical teams to support families through the determination and declaration of death by neurologic criteria. Some strategies for working with families during these conflicts are listed in Fig. 2. Of note, some of these may be more useful early in the process, while others may be more useful later in the process. Approaches should be targeted at the specific timing and specific concerns of each family.

First, clinical teams should be transparent in discussing the evaluation for death by neurologic criteria and offer services for anticipatory grief as well as suggestions for the family to gather to have time with the patient early on in the process. Next, when objections do occur, clinical teams can reframe objections as family requests for accommodation, striving to understand what it is that a family is asking. Clinical

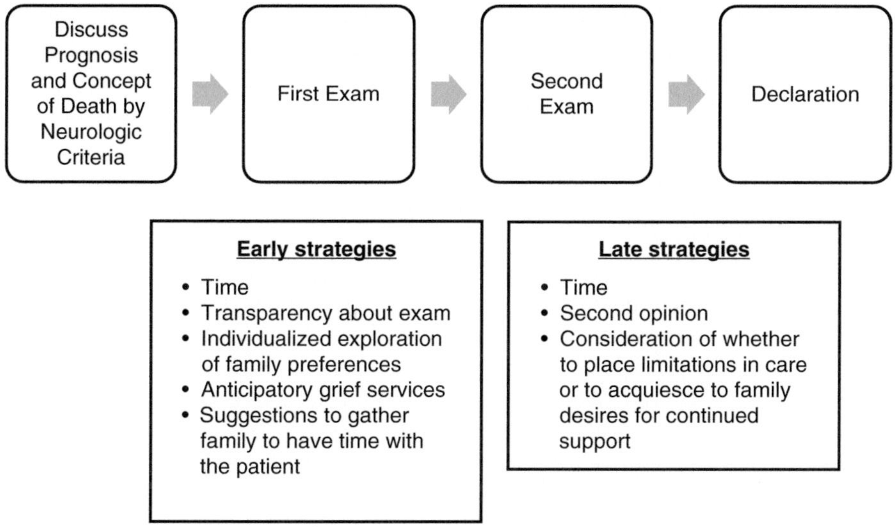

Fig. 2 Strategies for addressing objections or requests for accommodation in the determination and declaration of death by neurologic criteria. Some strategies may be more useful early on, while others may be more useful later on in the process

teams ought to seek to understand not only the reasons for a family's request but also whether they are objecting to the determination of death by neurologic criteria itself, the evaluation or a component of the evaluation, or withdrawal of support regardless of the circumstances. Knowing how complicated the terminology is and how much misinformation exists, clinical teams can explore family understanding, working to develop a shared language. Next, they can compassionately explore family reasons for their requests. In order to do justice to this task, clinical teams ought to employ a trauma-informed approach that focuses on family strengths and protective factors while working to create a safe environment, minimizing new trauma and triggers of past trauma [11, 43]. As argued by McEvoy et al., "Recognizing, normalizing, and supporting families through this natural stress response can be critical to developing an effective therapeutic relationship" [11]. When the request is due to disagreement that death by neurologic criteria is death, some argue that clinical teams should consider accommodation for this belief [56].

Later in the process, it is important for clinical teams to recognize and discuss whether second opinions may be useful in partnering with the family and to consider whether the team will place limitations on treatment or will acquiesce to family desires for continued support. There are multiple decisions within this larger one, regarding what types of support and interventions clinical teams may perform, requiring a nuanced and detailed multidisciplinary discussion.

Finally, at all steps in the process, clinical teams must recognize that individual bias and institutional discriminatory practices can influence how clinicians value a patient's life, how clinicians behave, the language they use, and what they offer families [11, 57]. Clinicians may be at particular risk of acting in biased manners, given that death by neurologic criteria determinations and declarations often represent urgent, high-risk, complicated, and poorly understood situations that lack standardized approaches and frameworks. Recognizing these factors, reflecting individually, and working to address biases are important components of managing these conflicts.

4 Conclusion

As controversies about determination and declaration of death by neurologic criteria have increased in recent years and as these controversies have become more publicized, family refusals of determination or declaration of death by neurologic criteria have also become increasingly common. Refusals, or requests for accommodation, can happen at different stages in the patient's care, and for a variety of reasons. Existing studies describing family reasons for refusals and requests for accommodation use interviews and surveys of clinical teams, but more research is needed to better understand family perspectives directly. Because the process is complex, a clinical team should begin by seeking to understand what a family is refusing or requesting, and then seek to understand the reasons why, in order to partner with families in moving through these processes.

References

1. Beecher HK. Report of the ad hoc committee of the Harvard Medical School to examine the definition of brain death: the definition of irreversible coma. Transplantation. 1969;7(3):204.
2. Spinello IM. Brain death determination. J Intensive Care Med. 2015;30(6):326–37.
3. President's Commission for the Study of Ethical Problems in Medicine, Biomedical, Behavioral Research. Deciding to forego life-sustaining treatment: a report on the ethical, medical, and legal issues in treatment decisions. United States; 1983.
4. Bernat JL, Culver CM, Gert B. On the definition and criterion of death. Ann Intern Med. 1981;94(3):389–94.
5. Shah SK, Kasper K, Miller FG. A narrative review of the empirical evidence on public attitudes on brain death and vital organ transplantation: the need for better data to inform policy. J Med Ethics. 2015;41(4):291–6.
6. Lewis A, Cahn-Fuller K, Caplan A. Shouldn't dead be dead?: the search for a uniform definition of death. J Law Med Ethics. 2017;45(1):112–28.
7. Wahlster S, Wijdicks EF, Patel PV, et al. Brain death declaration: practices and perceptions worldwide. Neurology. 2015;84(18):1870–9.
8. Greer DM, Wang HH, Robinson JD, et al. Variability of brain death policies in the United States. JAMA Neurol. 2016;73(2):213–8.
9. Powner DJ, Hernandez M, Rives TE. Variability among hospital policies for determining brain death in adults. Crit Care Med. 2004;32(6):1284–8.
10. Lewis A, Adams N, Varelas P, Greer D, Caplan A. Organ support after death by neurologic criteria: results of a survey of US neurologists. Neurology. 2016;87(8):827–34.
11. McEvoy MJ, Scott MJ, Sawyer KE. Requests for accommodation in brain death cases: emerging role for pediatric palliative care. J Pain Symptom Manag. 2021;62(6):1319–24.
12. Kirschen MP, Lewis A, Rubin M, Kurtz P, Greer DM. New perspectives on brain death. J Neurol Neurosurg Psychiatry. 2021;92(3):255–62.

13. Scott M. Jahi McMath: lessons learned. Pediatrics. 2020;146(Supplement_1):S81–5.
14. Paquette E, Pinto V, Renteria VA, Frader J. 544: legal cases challenging death by neurologic criteria add insight into family perspectives. Crit Care Med. 2022;50(1):263.
15. Cerminara KL. Rip currents: rough water for end of life decision making. J Health Care L Policy. 2018;21:59.
16. Schmidt S. Jahi McMath, the Calif. girl in life-support controversy, is now dead. Washington Post 2018 Jun 29.
17. Fonseca v. Smith, No. 2:16-cv-00889-KJM-EFB | Casetext Search + Citator. https://casetext.com/case/fonseca-v-smith. Accessed 15 Feb 2022.
18. Paquette E, Frader J, Ross L. 531: sociodemographic characteristics and reasons for refusals of determination of brain death. Crit Care Med. 2021;49(1):258.
19. Paquette E, Chavez J, Ross L Frader J. Refusals of the determination of death by neurologic criteria: a mixed methods study of provider perspectives on refusals cases. Manuscript submitted.
20. van Beinum A, Healey A, Chandler J, et al. Requests for somatic support after neurologic death determination: Canadian physician experiences. Can J Anesth. 2021;68(3):293–314.
21. Sque M, Payne SA. Dissonant loss: the experiences of donor relatives. Soc Sci Med. 1996;43(9):1359–70.
22. Pelletier M. The organ donor family members' perception of stressful situations during the organ donation experience. J Adv Nurs. 1992;17(1):90–7.
23. Truog RD, Morrison W, Kirschen M. What should we do when families refuse testing for brain death? AMA J Ethics. 2020;22(12):986–94.
24. Long T, Sque M, Addington-Hall J. Conflict rationalisation: how family members cope with a diagnosis of brain stem death. Soc Sci Med. 2008;67(2):253–61.
25. Shewmon DA. Brainstem death, brain death and death: a critical re-evaluation of the purported equivalence. Issues Law Med. 1998;14:125.
26. Bernat JL. A defense of the whole-brain concept of death. Hastings Cent Rep. 1998;28(2):14–23.
27. Lewis A, Varelas P, Greer D. Prolonging support after brain death: when families ask for more. Neurocrit Care. 2016;24(3):481–7.
28. Smith ML, Flamm AL. Accommodating religious beliefs in the ICU: a narrative account of a disputed death. Narrat Inq Bioeth. 2011;1(1):55–64.
29. Burkle CM, Schipper AM, Wijdicks EF. Brain death and the courts. Neurology. 2011;76(9):837–41.
30. Lewis A, Bakkar A, Kreiger-Benson E, et al. Determination of death by neurologic criteria around the world. Neurology. 2020;95(3):e299–309.
31. Lewis A, Kreiger-Benson E, Kumpfbeck A, et al. Determination of death by neurologic criteria in Latin American and Caribbean countries. Clin Neurol Neurosurg. 2020;197:105953.
32. Lewis A, Liebman J, Bakkar A, et al. Determination of brain death/death by neurologic criteria in countries in Asia and the pacific. J Clin Neurol (Seoul, Korea). 2020;16(3):480.
33. Greer DM, Shemie SD, Lewis A, et al. Determination of brain death/death by neurologic criteria: the world brain death project. JAMA. 2020;324(11):1078–97.
34. Broumand M, Parsapoor A, Asghari F. Public opinion of organ donation: a survey in Iran. Clin Transpl. 2012;26(5):E500–4.
35. Dehghani SM, Gholami S, Bahador A, et al. Causes of organ donation refusal in southern Iran. Transplant Proc. 2011;43(2):410–1.
36. Seth AK, Nambiar P, Joshi A, et al. First prospective study on brain stem death and attitudes toward organ donation in India. Liver Transpl. 2009;15(11):1443–7.
37. Sanner M. Attitudes toward organ donation and transplantation: a model for understanding reactions to medical procedures after death. Soc Sci Med. 1994;38(8):1141–52.
38. Jasper JD, Harris RJ, Lee BC, Miller KE. Organ donation terminology: are we communicating life or death? Health Psychol. 1991;10(1):34.
39. Siminoff LA, Burant C, Youngner SJ. Death and organ procurement: public beliefs and attitudes. Soc Sci Med. 2004;59(11):2325–34.
40. DuBois JM, Anderson EE. Attitudes toward death criteria and organ donation among healthcare personnel and the general public. Prog Transplant. 2006;16(1):65–73.

41. Franz HG, DeJong W, Wolfe SM, et al. Explaining brain death: a critical feature of the donation process. J Transpl Coord. 1997;7(1):14–21.
42. Bourgois P, Holmes SM, Sue K, Quesada J. Structural vulnerability: operationalizing the concept to address health disparities in clinical care. Acad Med. 2017;92(3):299.
43. Goddard A. Adverse childhood experiences and trauma-informed care. J Pediatr Health Care. 2021;35(2):145–55.
44. Oliver MN, Wells KM, Joy-Gaba JA, Hawkins CB, Nosek BA. Do physicians' implicit views of African Americans affect clinical decision making? J Am Board Fam Med. 2014;27(2):177–88.
45. Trent M, Dooley DG, Dougé J, et al. The impact of racism on child and adolescent health. Pediatrics. 2019;144(2).
46. Hall WJ, Chapman MV, Lee KM, et al. Implicit racial/ethnic bias among health care professionals and its influence on health care outcomes: a systematic review. Am J Public Health. 2015;105(12):e60–76.
47. Dehon E, Weiss N, Jones J, et al. A systematic review of the impact of physician implicit racial bias on clinical decision making. Acad Emerg Med. 2017;24(8):895–904.
48. Goldschmidt D. Jahi McMath, California teen at center of brain-death controversy, has died. CNN. 2018;29
49. Slomski A. Another "brain dead" patient wakes up just in time. New York: Medscape; 2018.
50. Lewis A, Weaver J, Caplan A. Portrayal of brain death in film and television. Am J Transplant. 2017;17(3):761–9.
51. Paquette E, Frader J, Shah S, Tasker RC, Truog R. Beyond the apnea test: an argument to broaden the requirement for consent to the entire brain death evaluation. Am J Bioeth. 2020;20(6):17–9.
52. Leemputte M, Paquette E. Consent for conducting evaluations to determine death by neurologic criteria: a legally permissible and ethically required approach to addressing current controversies. Curr Pediatr Rep. 2019;7(4):152–62.
53. Mataya L, Ross LF, Ghavam A, Paquette ET. Pediatric intensivist and pediatric neurologist perspectives and practices on death by neurologic criteria. J Clin Ethics. 2021;32(3):195–205.
54. Truog RD, Tasker RC. Counterpoint: should informed consent be required for apnea testing in patients with suspected brain death? Yes Chest. 2017;152(4):702–4.
55. Pope TM. Brain death forsaken: growing conflict and new legal challenges. J Legal Med. 2017;37(3–4):265–324.
56. Veatch RM, Ross LF. Defining death: the case for choice. Georgetown University Press; 2016.
57. FitzGerald C, Hurst S. Implicit bias in healthcare professionals: a systematic review. BMC Med Ethics. 2017;18(1):1–8.

Arguments Favoring Continuation of "Organ Support" when Families Object to Declaration of Death by Neurologic Criteria

L. Syd M Johnson

Death by neurologic criteria occurs infrequently, accounting for approximately 2% of hospital deaths in the United States [1] and 2–3% in the European Union [2]. Despite this relative rarity, the number and frequency of objections and legal challenges to declarations of death by neurologic criteria are increasing, especially in the aftermath of the highly publicized 2013 case of Jahi McMath, a 13-year-old girl who was declared dead in California after experiencing post-surgical bleeding and circulatory-respiratory arrest.[1] There are several foci of ethical concern when families and surrogates object to the determination or declaration of death by neurologic criteria. This chapter examines those concerns, focusing on the unshared metaphysical, and moral commitments implied by neurologic criteria for death that give rise to distrust and injustice. These concerns provide compelling ethical reasons for recognizing reasonable objections. Reasonable objections are those that accept an already socially, medically, and legally recognized alternative, namely, death by circulatory-respiratory criteria.

[1] Jahi's mother denied that her daughter was dead, but accepted, consistent with her religious beliefs death by circulatory-respiratory criteria as a legitimate definition of death. After being relocated to New Jersey, a state with an expansive religious exemption in its law, Jahi "statutorily resurrected" [3] and lived for nearly 4 years. Shewmon has argued that there is evidence that Jahi recovered brain function consistent with a diagnosis of minimally conscious state [3].

L. S. M Johnson (✉)
Center for Bioethics and Humanities, SUNY Upstate Medical University, Syracuse, NY, USA
e-mail: johnsols@upstate.edu

A. Lewis, J. L. Bernat (eds.), *Death Determination by Neurologic Criteria*, Advances in Neuroethics, https://doi.org/10.1007/978-3-031-15947-3_34

1 Death by Neurologic Criteria as a Metaphysical Stance

The medicolegal language of death by neurologic criteria tacitly endorses a metaphysical position that persons are reducible to functioning brains by defining death in terms of the brain and brain functions, or other functions of the body that can be mediated by the brain (such as respiration) [4, 5]. In 1968, the Ad Hoc Committee of the Harvard Medical School claimed that our ancient understanding of death was always based on the understanding that the death of the brain was what counted as death. They arrived at this conclusion by pointing out that the brain inevitably dies when the heart and respiration stop [6]. The Harvard Committee claimed that the semblance of life achieved through modern medicine is but an illusion, mistaking signs of life for life itself. Arguably, they made the mirror image of that mistake, mistaking signs of death for death itself. The Harvard Committee also concluded that the permanently nonfunctioning brain is "for all practical purposes dead" [6]. This conclusion begs the question, however, for it very much depends on how death is defined and whether death "for all practical purposes" is actually death.

Certainly long before humans knew about brains, they knew death. Neurologic death is decidedly not how death was traditionally understood. Perhaps those earlier views were unscientific or prescientific, but some of those notions were not wrong. The body in rigor mortis is without question and scientifically confirmable as dead, and so is the decaying body. The body with prolonged cessation of circulatory and respiratory activity is dead. These ways of recognizing death are almost universally agreed upon across cultures and across religious and spiritual traditions. Today, a body with a severely damaged brain can continue to live in traditionally understood ways—breathing, with a beating heart—given medical support. If this is but an illusion of life, it is one that has been long recognized and is still accepted.

In the United States, the 1981 President's Commission said, "Even if life continues in individual cells or organs, life of the organism as a whole requires complex integration, and without the latter, a person cannot properly be regarded as alive" [7]. What did they mean by *person*? They rejected the possibility of defining the essence of persons and personhood, noting the lack of agreement among philosophers, physicians, and the general public. In the United Kingdom, the Academy of Royal Medical Colleges went where the President's Commission feared to tread: "Death entails the irreversible loss of those essential characteristics which are necessary to the existence of a living human person and, thus, the definition of death should be regarded as the irreversible loss of the capacity for consciousness, combined with irreversible loss of the capacity to breathe" [4].

The claim of the Academy is that there are two capacities that are jointly necessary characteristics of a living human *person*. Both of these capacities are lost, as it happens, when there is injury to the brainstem, specifically to the reticular formation. The relevant parts of "living human person" are obviously "living" and "person" since membership in the human species is presumably resistant to damage to the reticular formation. This is, then, explicitly a definition of the death of the *person*, for which the anatomical proxy is irreversible damage to the reticular formation. Since the capacity to breathe is not essential to being alive or being a person, the implication of the Academy's position is that personhood is reducible to consciousness (or consciousness as it can be clinically defined, as awareness and arousal). In attempting to define the *death* of a person, the Academy implies a conceptually confused definition of personhood. Personhood is not a biological or medical concept, but rather a moral (and sometimes legal) concept.[2]

To its credit, the academy is very clear about what is to be medically determined. It doesn't claim biological equivalence between death by neurologic criteria and death by circulatory-respiratory criteria. And it is clear that the death of the brainstem is the death of a *person*. Although theirs is a controversial definition of death, it is one that is accepted by many people (including physicians) who would apply it to themselves. It aligns with a belief—not only controversial but also acceptable to many people—that the permanent loss of consciousness is analogous to the loss or death of personhood. We should be careful, however, about what we actually mean when we say that loss of consciousness is analogous to the death of persons. Some, like the Academy, take it to mean "literally dead." Some, like the Harvard Committee, take it to mean "for all practical purposes dead" [6] or as good as dead. And some mean it in a prescriptive way, implying that chronically unconscious individuals "should be treated *as if* they are dead," that is, not be medically treated at all, and sent to the morgue. But analogies, however apt they might be, are not medical, scientific, or legal facts. Unconscious apneic individuals are not literally dead. They are not decaying corpses. Are they as good as dead? Should they be treated as if they are dead? Those are important ethical questions, but they are not medical questions.

The *death of the person* is implicit in the determination and declaration of death by neurologic criteria. Death by neurologic criteria was meant to solve the problems created when a patient could be kept alive despite the fact that recovery of the injured brain and its associated functions (including consciousness, interaction, thought, movement about the world) were no longer thought to be possible. That this is viewed as a problem implies an ideological commitment, a value judgment about what makes human existence valuable and what makes a human being a person. But medicine and the law are silent on what makes human existence valuable. It's not the capacity to breathe, obviously. The capacity for consciousness is widely thought to be an essential component of a life with subjective value, and many people agree that without consciousness, life cannot have value. *What makes human*

[2]There is considerable philosophical controversy about the definition of personhood as a moral concept (see, e.g., [8], Chap. 7). In legal settings, personhood is frequently understood to designate those beings with sufficiently high moral standing to be subjects of justice and to warrant protection under the law [9]. As a legal concept, personhood has often been defined on an ad hoc basis, as previously excluded humans are admitted into the category of legal persons. For example, the pathbreaking legal case *Somerset v Stewart* (1772) in England established the personhood of an enslaved Black man [10, 11]; in the United States, the abused and neglected child Mary Ellen McCormack was recognized to be a person, and not the property of her parents, in an 1874 legal case brought by Henry Bergh, the founder of the American Society for the Prevention of Cruelty to Animals [12]. More recently, a number of legal challenges have been brought attempting to establish the legal personhood of some nonhuman animals [9].

existence valuable? is a question of vital ethical, cultural, spiritual, and personal importance, but it is not a medical or legal question.

2 Synecdoche as an Ethical Stance: Is It Just "Organ Support"?

The most plausible justification for having a concept of death by neurologic criteria is that devastating brain injuries resulting in the irreversible and chronic loss of consciousness result in the death of the *moral person,* the person who is the subject of moral concern [13]. Biology, medicine, and the law do not tell us what moral persons are, or when personhood begins and ends, and there is no biological, medical, or legal fact of the matter. The idea that persons (in the metaphysical and moral sense) can die while their physical bodies remain alive makes intuitive sense and is acceptable to many people, for many cultural and spiritual traditions contain an implicit or explicit dualism in which body and soul can part ways. But the declaration that *this moral or metaphysical person has shuffled off this mortal coil* is not a medical diagnosis nor a matter for the law.

After a declaration of death by neurologic criteria, what was once called life-sustaining treatment for a patient becomes "organ support," reinforcing the (unscientific) dualist notion that there is no longer a person who must be treated as a patient, but only a material body, a collection of living organs. This deployment of synecdoche, in which a part represents the whole, is fundamentally dehumanizing and takes a metaphysical stance regarding what persons are, and an ethical stance concerning which humans have value.[3] It shifts the locus of ethical concern from the individual patient—now a former person/patient—to the needs of others, such as potential organ recipients who might benefit from the act of "organ support." Within medicine and medical ethics, there has long been recognition of the need for patient-centered care to counter the tendency to view patients with complex lives and needs as merely a disease, dysfunction, or diagnosis [15]. Yet, in death by neurologic criteria, we see synecdoche in medicine reach its apogee, in the diagnostic transformation of "life support" into "organ support," and the conversion of patients (and persons) into mere vessels for organs. It instrumentalizes patients in a way that would be frankly objectionable if they were considered persons. It is not surprising, then, that surrogates and families might object to the instrumentalization, dehumanization, and downgrading of their loved ones into a collection of organs when their values and beliefs are inconsistent with viewing persons as reducible to brains-that-support-consciousness.

[3] In other contexts, such as bioethical discussions of prenatal screening, the use of synecdoche to view a possible person solely in terms of a potential disability (or anomalous gene or chromosome) has been criticized as similarly dehumanizing and morally problematic [14].

3 Distrust

Death by neurologic criteria is a locus of controversy and distrust because of its implications for those persons declared dead. The ethical implications of a medico-legal declaration of the point at which a human being ceases to be a person with rights are many and significant. Laws have coercive force, and compelling a definition of personhood that is incompatible with an individual's moral, philosophical, cultural, social, and spiritual beliefs is coercive in a way that can easily be perceived as oppressive, or as legally sanctioned medical neglect. When Jahi McMath was declared legally dead, for example, Oakland Children's Hospital was no longer obligated to provide medical treatment under California law, except for court-ordered ventilation. She was not fed or medicated for the next 3 weeks while her case was litigated [3]. When the hospital released her to her family, they sent her to the morgue, still on a ventilator.

It is not surprising that death by neurologic criteria is contested when it is viewed as a way of determining what is only implicit in the law—the transformation of a person with rights, a subject of justice, into a nonperson who no longer has rights. It is not surprising that it engenders distrust because the law, and other human institutions, have historically and frequently been morally wrong in their recognition of who is and is not a person with rights.

Conflicts over death by neurologic criteria frequently involve patients from historically marginalized and exploited communities, revealing another source of distrust.[4] The families of these patients commonly express religious or cultural objections to death by neurologic criteria, as well as concerns fueled by distrust that include perceptions of a medical rush to judgment, negligent, indifferent, or openly hostile treatment of their loved ones, and pressure to donate organs. Jahi McMath's family have described the lack of compassion and respect shown to them by healthcare staff and the hospital [17]. Elijah Smith's parents objected to the determination of death by neurologic criteria and refused consent for organ donation. They argued that they wanted to give their son more time to recover and that he did not understand that organ donation occurred while the heart continued to beat. Lifeline Ohio, an organ procurement organization, obtained a court order for organ procurement. Smith's parents believe he was killed for his organs [18]. The experiences of these families must be viewed in the context of a long history of discrimination and racism, both in societal institutions and in healthcare, that has created persistent racial disparities in access to healthcare and in objective health outcomes [19–21]. These phenomena are experienced by minority and immigrant communities worldwide, resulting in distrust of healthcare systems and healthcare workers [22]. Imposing an

[4] Some recent cases have involved the families of Jahi McMath (2013), Israel Stinson (2016), Aden Hailu (2015), Elijah Smith (2013), Michael Todd (2006), Areen Chakrabarti (2018), Jayden Auyeung (2018), Cho Fook Cheng (2006), and Tara Hawkins-Bottoms (2004) in the United States, Taquisha McKitty (2018) in Canada, and Child A (2015) in the UK (for background on these and many other cases, see Pope [16])

unshared view of death on families struggling with a patient's devastating brain injury does not encourage trust.

Institutional policies for managing so-called futile treatment, which may be invoked in disputes over death by neurologic criteria, can also target and burden marginalized and vulnerable patients and their families in their implementation, by ignoring well-documented intergroup differences and preferences concerning end-of-life care and the withdrawal of life-sustaining treatment [23–25].[5] The intention of enacting policies may be benign: to prevent bias and avoid conflicts through facially neutral policies that appear to treat everyone the same. However, when racial, ethnic, cultural, and religious minorities are harmed more frequently and disproportionately by policies implemented on the bodies of their children and loved ones, those policies do not effectively prevent bias and are not neutral [27]. Moreover, the *appearance of bias*, particularly concerning a matter as emotional and culturally laden with meaning and import as death, further erodes trust in persons and communities that already have reasons to be distrustful of healthcare providers.

The legal declaration of death means that someone is no longer a person with a right to treatment. It means their surrogates no longer have the right to make decisions concerning treatment. In some of the most contentious cases, it means that organs are procured from those individuals without the consent of their families [18]. When families believe that their loved ones are killed by an involuntary withdrawal of life-sustaining treatment or by the removal of their organs, the most extreme implication of the collusion of law and medicine in death by neurologic criteria is one that ironically circles back to one of the motives for inventing it, which was to avoid the charge of medical homicide.

The issue we have been evading, the bush we have been beating around for decades now when it comes to death by neurologic criteria, is that it does not merely harmonize modern medicine, biological facts, and the law. Death by neurologic criteria imposes a particular moral and metaphysical viewpoint that is at odds with what many people from many cultural, social, and spiritual traditions believe. Generally, when unconventional and minority beliefs and values conflict with medical orthodoxies, we are *permissive* regarding the authority of patients and their surrogates to decide what kind of life is valuable and what is in their interests. But legal statutes and medical codes of practice concerning death by neurologic criteria run counter to that permissiveness. They're prescriptive and paternalistic and have been used coercively to impose a moral worldview about which humans are valuable and worth treating [28].

---

[5] Research shows that the families of patients of color tend to decide to withdraw life-sustaining treatment later than white families. Rubin et al. found a one-day difference between decisions to withdraw life-sustaining treatment after severe neurological injuries made by families of patients of color and families of white patients, with no difference in mortality [26]; Hornor et al. found a greater time difference—4 days versus 15 days—in decisions to withdraw ventilation in intensive care, with African-American families most likely to wait the longest [24]. As is often the case when difficult decisions concerning treatment withdrawal are made, tincture of time can be an effective remedy.

The President's Commission argued against adopting a standard of "higher brain death," or death as the loss of consciousness and cognition. "On a matter so fundamental to a society's sense of itself—touching deeply held personal and religious beliefs—and so final for the individuals involved, one would desire much greater consensus than now exists before taking the major step of radically revising the concept of death" [7]. They revised it anyway with the concept of whole brain death, but the sentiment, and the moral and social commitments, were right. In pluralistic, multicultural, democratic societies, transparent public engagement and a pluralistic consensus—and not medicolegal cultural imperialism—are necessary to maintain trust and to respect the values and autonomy of patients and their surrogates.

4 Justice and Equal Treatment

The effects of practices and policies concerning death by neurologic criteria implicate issues of justice that go well beyond simplistic calculations of costs, benefits, and the allocation of medical resources, or what is allowed by the law. Indeed, justice in this context would argue against using clinical and practice guidelines, policies, or the letter of the law coercively even when doing so appears to treat everyone equally [28–30]. The coercive force of laws, policies, and guidelines is specifically brought to bear on those who resist death by neurologic criteria, and not on those who accept it. In practice, they treat alike only those who agree that death by neurologic criteria is death. Those who deny that it is death are likely to be subjected to the involuntary withdrawal of life-sustaining treatment and, in some cases, procurement of organs against the wishes of the family or surrogate [18].

While death by circulatory-respiratory criteria is accepted scientifically, medically, legally, and across nearly all cultures and religious/spiritual traditions, death by neurologic criteria has long been contentious and its conceptual soundness has been challenged on biological and philosophical grounds [13, 31–35]. "Death by neurologic criteria is death" is not a matter of scientific fact like "The Earth orbits the Sun," but has been accurately described instead as a "legal fiction" [36, 37]. It has social and legal value as a way of declaring death and facilitating organ donation. But not all legal fictions are equally benign in their intrusions on vitally important aspects of human life. Those that define the legal age of majority, for example, are relatively benign, necessary, and fictive, and avoid the practically impossible task of deciding who is adult enough to be allowed to do things like get married, enlist in the military, vote, and consent to medical treatment. Treating death by neurologic criteria as legal death is an example of what Shah and Miller call a *status legal fiction*, one that draws an analogy between two clearly different concepts, in that it "allows us to treat persons who are not dead as if they are dead" [38]. This is a legal fiction that intrudes on significant social, cultural, spiritual, and personal aspects of human life. It alters the moral/legal rights of the individual determined to be dead; it severs important social connections and relationships; it precipitates a host of social and legal practices. It does not do so to extend rights or protect the affected individual from serious harm.[6] Indeed, the ethical worry is that the legal fiction and force of death by neurologic criteria causes harm to the individual declared dead, transforming them from a person with moral and legal rights into a fictive ex-person, a mere vessel for living organs, or a corpse. It is a legal fiction that imposes and enforces a particular moral worldview in a way that is onerous when brought to bear on those who deny it. The burdens are neither benign nor equally distributed, and are thus unjust.

5 The Case for Reasonable Objections and Choice Concerning Continuation of Treatment

Societies and people need to know who is actually dead, and not who is merely as good as dead. We need to know because there are traditions, practices, and legal actions that occur when we *know* someone has died. Thus, *Who is dead?* has long been socially and culturally defined. The relatively recent designation of death by neurologic criteria as another way of defining death has been endorsed by many as a necessary and useful approach that solves the practical and ethical problems that have emerged alongside advancements in medicine. That the matter of death became more complex and fit less easily into traditional understandings as a result of medical interventions does not make *What is death?* and *Who is dead?* medical questions. Medicine can confirm when death has occurred according to some accepted criteria (e.g., when the heart stops beating, or when the brainstem is severely damaged). The controversy over death by neurologic criteria is about the proper criteria for death, whether what is taken to be a sign of death—the loss of some or all functions of the brain—is actually the death of a person.[7] It is important to acknowledge the instrumental value of death by neurologic criteria, which includes being transparent about its status as a legal fiction. But it is even more important that it not be brandished to instrumentalize persons, to treat them as organ donors who exist only to serve others, or as corpses to whom we owe nothing [40].

Worldwide, across individuals, societies, cultures, religions, and spiritual traditions, death by circulatory-respiratory criteria is accepted. Some societies accept

[6] It might be argued that it does harm a patient to provide futile treatment, to maintain biological life when consciousness and brain function are forever gone. Much hinges on what is meant by harm and whether those considered dead according to neurologic criteria are still moral subjects who *can* be harmed. To conclude that they can't be harmed because they are already dead begs the question, and also ignores the harm to their families and loved ones. To conclude that they are harmed by treatment because they are dead is confusing at best.

[7] There are important questions about whether whole brain death, a formulation of death by neurologic criteria which requires loss of function of the entire brain, including the brainstem, is consistent with current clinical guidelines [5], or whether the loss of function of the entire brain is medically diagnosable at all [34]. At present, the tests used to confirm whole brain death cannot establish that all functions of the entire brain have been irreversibly lost [3, 39].

only death by circulatory-respiratory criteria.[8] Others, like the United States, accept death by circulatory-respiratory *or* neurologic criteria. There are compelling ethical reasons to endorse the disjunctive *or*, and recognize the validity of objections to death by neurologic criteria given that death by circulatory-respiratory criteria is a reasonable, medically valid, and almost universally accepted alternative [30, 42]. Allowing individuals and their surrogates to choose continuation or withdrawal of treatment and choose the death that aligns with reasonable, medically and socially accepted beliefs also aligns with core ethical values in medicine, including respecting persons and respecting patient and surrogate autonomy. It answers the demands of justice. It fosters trust and patient-centered care [27], and treats grieving and struggling families with compassion as they make the most weighty, consequential, highly emotional, and momentous decisions many will ever face.

References

1. Seifi A, Lacci JV, Godoy DA. Incidence of brain death in the United States. Clin Neurol Neurosurg. 2020;195:105885.
2. Council of Europe. Organ shortage: current status and strategies for improvement of organ donation-A European consensus document 2013. https://www.edqm.eu/medias/fichiers/Organ_shortagecurrent_status_and_strategies_for_improvement_of_organ_donation_A_European_consensus_document.pdf
3. Shewmon DA, Salamon N. The extraordinary case of Jahi McMath. Perspect Biol Med. 2021;64(4):457–78.
4. Academy of Royal Medical Colleges. A code of practice for the diagnosis and confirmation of death. UK: Academy of Medical Royal Colleges; 2008.
5. Wijdicks EFM, Varelas PN, Gronseth GS, Greer DM. Evidence-based guideline update: determining brain death in adults. Report of the Quality Standards Subcommittee of the American Academy of Neurology. Neurology. 2010;74(23):1911–8.
6. Beecher H, Adams R, Barger C, et al. A definition of irreversible coma: report of the Ad Hoc Committee of the Harvard Medical School to examine the definition of brain death. JAMA. 1968;205(6):337–40.
7. President's Commission for the Study of Ethical Problems in Medicine and Biomedical and Behavioral Research. Defining death: a report on the medical, legal and ethical issues in the determination of death. Washington DC: President's Commission for the Study of Ethical Problems in Medicine and Biomedical and Behavioral Research; 1981.
8. Johnson LSM. The ethics of uncertainty: entangled ethical and epistemic risks in disorders of consciousness. New York: Oxford University Press; 2022.
9. Andrews K, Comstock G, Crozier G, et al. Chimpanzee rights: the philosophers' brief. New York: Routledge; 2019.
10. Wise SM. Though the heavens may fall: the landmark trial that led to the end of human slavery. Cambridge, MA: Da Capo Press; 2005.
11. Lord Mansfield. Somerset v Stewart 98 ER 499.98 ER 499. England 1772.
12. Markel H. Case shined first light on abuse of children. The New York Times; December 14, 2009.

[8] Japan, after much social and political debate, arrived at an interesting compromise that appears to transparently acknowledge death by neurologic criteria as a legal fiction. Japan accepts circulatory-respiratory criteria for death. Death by neurologic criteria is accepted as death only for purposes of organ donation and only with the explicit consent of the donor [41].

13. Veatch RM. The death of whole-brain death: the plague of the disaggregators, somaticists, and mentalists. J Med Philos. 2005;30(4):353–78.
14. Parens E, Asch A. Prenatal testing and disability rights. Washington, DC: Georgetown University Press; 2000.
15. Epstein RM, Street RL. The values and value of patient-centered care. Ann Fam Med. 2011;9(2):100–3.
16. Pope T. Brain death. http://www.thaddeuspope.com/braindeath.html Accessed 19 Mar 2020.
17. Aviv R. What does it mean to die? The New Yorker January 28, 2018.
18. Manning A. Family loses fight to keep son's organs from donation. The Columbus Dispatch Columbus Ohio July 11, 2013. https://www.dispatch.com/story/news/crime/2013/07/11/family-loses-fight-to-keep/24114454007/
19. Gordon HS, Street RL Jr, Sharf BF, Souchek J. Racial differences in doctors' information-giving and patients' participation. Cancer. 2006;107(6):1313–20.
20. Sorkin DH, Ngo-Metzger Q, De Alba I. Racial/ethnic discrimination in health care: impact on perceived quality of care. J Gen Intern Med. 2010;25(5):390–6.
21. Smedley BD, Stith AY, Nelson AR. Institute of Medicine. Unequal treatment: confronting racial and ethnic disparities in health care. Washington, DC: National Academies Press; 2003.
22. Alpers L-M. Distrust and patients in intercultural healthcare: A qualitative interview study. Nurs Ethics. 2018;25(3):313–23.
23. Jacobs EA, Rolle I, Ferrans CE, Whitaker EE, Warnecke RB. Understanding African Americans' views of the trustworthiness of physicians. J Gen Intern Med. 2006;21(6):642.
24. Hornor MA, Byrne JP, Engelhardt KE, Nathens AB. Examining racial disparities in the time to withdrawal of life-sustaining treatment in trauma. J Trauma Acute Care Surg. 2018;84(4):590–7.
25. Degenholtz HB, Thomas SB, Miller MJ. Race and the intensive care unit: disparities and preferences for end-of-life care. Crit Care Med. 2003;31(5):S373–S8.
26. Rubin MA, Dhar R, Diringer MN. Racial differences in withdrawal of mechanical ventilation do not alter mortality in neurologically injured patients. J Crit Care. 2014;29(1):49–53.
27. Johnson LSM. Restoring trust and requiring consent in death by neurological criteria. Am J Bioeth. 2020;20(6):33–5.
28. Choong KA, Rady MY. Re A (a child) and the United Kingdom Code of Practice for the Diagnosis and Confirmation of Death: Should a secular construct of death override religious values in a pluralistic society? HEC Forum. 2018:71–89.
29. Rady MY, Verheijde JL. Legislative enforcement of nonconsensual determination of neurological (brain) death in muslim patients: a violation of religious rights. J Relig Health. 2018;57:649–61.
30. Johnson LSM. The case for reasonable accommodation of conscientious objections to declarations of brain death. J Bioethic Inqu. 2016;13(1):105–15.
31. Shewmon DA. Chronic "brain death": meta-analysis and conceptual consequences. Neurology. 1998;51(6):1538–45.
32. Shewmon DA. Brain death: a conclusion in search of a justification. Hastings Cent Rep. 2018;48(S4):S22–S5.
33. Miller FG, Truog RD. The incoherence of determining death by neurological criteria: a commentary on Controversies in the determination of death, a white paper by the President's Council on Bioethics. Kennedy Inst Ethics J. 2009;19(2):185–93.
34. Joffe AR, Khaira G, de Caen AR. The intractable problems with brain death and possible solutions. Philos Ethics Humanit Med. 2021;16(1):11.
35. Nair-Collins M, Miller FG. Do the 'brain dead' merely appear to be alive? J Med Ethics. 2017;43(11):747–53.
36. Shah SK. Rethinking brain death as a legal fiction: is the terminology the problem? Hastings Cent Rep. 2018;48:S49–52.
37. Truog RD, Miller FG. Changing the conversation about brain death. Am J Bioeth. 2014;14(8):9–14.

38. Shah SK, Miller FG. Can we handle the truth? Legal fictions in the determination of death. Am J Law Med. 2010;36(4):540–85.
39. Latorre JGS, Schmidt EB, Greer DM. Another pitfall in brain death diagnosis: return of cerebral function after determination of brain death by both clinical and radionuclide cerebral perfusion imaging. Neurocrit Care. 2020:1–7.
40. Johnson LSM. A legal fiction with real consequences. Am J Bioeth. 2014;14(8):34–6.
41. Morioka M. Reconsidering brain death: a lesson from Japan's fifteen years of experience. Hastings Cent Rep. 2001;31(4):41–6.
42. Veatch RM. Controversies in defining death: a case for choice. Theor Med Bioeth. 2019;40(5):381–401.

Arguments Opposing Continuation of Organ Support When Families Object to Declaration of Death by Neurological Criteria

Adam Omelianchuk and David Magnus

There are two main reasons why a patient or family may object to a declaration of death by neurologic criteria. The family may believe that the declaration is premature and continue to hope for recovery. Or the patient and family may believe that life continues because the heart still beats and that "life" in this state is worth preserving. Of course, these reasons can coexist. Either way, the implication is that families wish for supportive measures to continue until hope for recovery or circulatory-respiratory arrest.

What these measures support is a vexed question. For those who object to the declaration, it is the life of the patient, so "life support" is at stake. For those making the declaration, it is the functional integrity of the body's organ systems below the neck, so "organ support" is at stake. Normally, these measures are provided for transplant purposes or for a limited time of "reasonable accommodation" so the family can gather at the bedside. On rare occasions, they are provided for the safe delivery of a fetus. They may even be provided to give family members a specified amount of time to come to accept that there is no hope for recovery. In each case, the time of support is defined and justified by the benefits. If none of these purposes are in view, and supportive measures are requested for "life-preservation" purposes, then the time of support is indefinite, requiring what Flamm and colleagues call "indefinite accommodation" [1]. Is indefinite accommodation ever reasonable accommodation? What does the team owe, if anything, to those who request that supportive measures be provided indefinitely?

Those sympathetic to principled or religious objections to death by neurologic criteria are not automatically committed to providing indefinite accommodation.

A. Omelianchuk
Center for Medical Ethics and Health Policy, Baylor College of Medicine, Houston, TX, USA
e-mail: adam.omelianchuk@bcm.edu

D. Magnus (✉)
Stanford Center for Biomedical Ethics, Stanford, CA, USA
e-mail: dmagnus@stanford.edu

A. Lewis, J. L. Bernat (eds.), *Death Determination by Neurologic Criteria*,
Advances in Neuroethics, https://doi.org/10.1007/978-3-031-15947-3_35

One could offer to withdraw the ventilator and declare death on the basis of circulatory-respiratory failure. While many would probably recommend withdrawing support, those inclined to defer to patient autonomy will have strong, if not decisive, reasons to honor a family's request [2]. Some go so far as to argue that the full battery of resuscitation measures in addition to ongoing supportive measures and critical care medicine should be offered; to withhold any of these things is to express "arbitrary power" over the patient [3], p. 682. Preserving patient trust and avoiding paternalism or cultural imperialism are the purported benefits of such an approach [4]; we call this the preservation approach.

We believe the preservation approach goes too far. Instead, we argue in support of the limitation approach that indefinite accommodation should not be given because (1) there is no party to which a justifiable medical benefit accrues, (2) it undermines the professional integrity of the providers involved and is unfair to the wider community, (3) it gives the family false hope for recovery and prolongs their grieving process, and (4) it produces societal confusion by permitting a "negotiated and inconsistent standard of death" [5]. These reasons are implied by any valid death declaration. For example, they are sufficient to reject requests for interventions from a patient or family who believes death can be declared only after every possible measure to support "circulatory functioning" has failed.[1] Therefore, if neurologic criteria are on par with circulatory-respiratory criteria for determining death, and there is no morally relevant difference between withdrawing and withholding medical treatment, and it is unreasonable to provide extracorporeal membrane oxygenation (ECMO) *only* for the sake of helping patients avoid pending circulatory-respiratory arrest, then withdrawing supportive measures from patients declared dead on the basis of neurologic criteria is on par with not offering ECMO when the patient can neither survive outside of the intensive care unit nor appreciate the benefits of being alive.

The crucial question, then, is whether the neurologic criteria used to determine death are valid. As important as that question is, it is not the focus of our inquiry, and we will assume for the sake of argument that they are. Our aim is to support the key premise of the limitation approach: whatever the rights of the dead or of those related to the dead may be, they are not so strong as to generate duties in anyone to provide the deceased with organ support indefinitely, and one ought to refrain from providing it because the burdens of doing so are disproportionate to the benefits. If this is the case, then indefinite accommodations are not reasonable accommodations and supportive measures should be discontinued when the time of reasonable accommodation is over.

[1] One of us consulted on a case in which a surrogate requested that their obtunded loved one, who was "hours-to-days" away from heart failure, be put on extracorporeal membrane oxygenation for the sole purpose of extending her life. Facts about asystole, auto-resuscitation, or hypotension did not matter to this surrogate.

1 Preliminaries

It is important to situate the case against the preservation approach within the context of societal policy. Disputes over death declarations go beyond a conflict between families and physicians; rather, they are disputes over the criteria for death a given community recognizes. Supporters of the preservation approach argue that it is reasonable to dispute the validity of neurologic criteria for death and point to communities or whole societies that either reject them (e.g., some Orthodox Jews) or have been reluctant to embrace them (e.g., Japan) [6]. The preservation approach is on firmer ground when the validity of neurologic criteria is not recognized, or an exemption from their otherwise recognized validity is available [7]. However, the vast majority of jurisdictions have decided that neurologic criteria for death are on par with circulatory-respiratory criteria for death.

Despite this broad consensus, there is evidence suggesting that some clinicians still feel tempted to take the preservation approach for whatever reason despite practicing in jurisdictions where there are no exemptions. A survey of pediatric critical care clinicians showed that 8% of the respondents in such jurisdictions would continue organ support indefinitely and only 7% would discontinue it without family consent [8]. A similar survey of neurologists showed that 48% of respondents would continue it to avoid legal action [9]. We will argue that these clinicians have no obligation to continue indefinite supportive measures.

2 There Is No Party to Which a Justifiable Medical Benefit Accrues

To declare death is to make a claim about the subject of life being lost. Thus, to continue providing someone with organ support after they were truthfully declared dead is to maintain a certain condition of their remains for some purpose; the measures used cannot, in principle, maintain their life because the subject of life has passed away. The best reasons to continue organ support involve securing benefits to third parties, such as the potential recipient(s) of their vital organs, their family members needing time to gather at the bedside, or criminal investigators who need time to gather evidence against their assailant. Perhaps a 24-to-48-h period of "reasonable accommodation" ought to be offered so that the deceased's loved ones can come to terms with the devastating news. Objections against instrumentalizing the deceased in these ways do not apply, because the body is no longer to be identified with the person themselves, but only with their remains.

This is not to say that someone's remains merit no respect or that their premortem wishes about what to do with them should not be honored. We should honor the deceased's wishes regarding their postmortem goals if they are reasonable. Donating their organs and receiving a ceremonial rite after death are good examples. The type of "benefit" at stake is an act of respect that accrues to a person's life history, which is the "narrative," "legacy," or "memory" of the person which deserves a measure of respect. If someone has died, but previously made their wishes to be an organ donor and have a proper burial known, then given that these things are reasonable, they ought to be accommodated to respect their life history. Failing to do so is disrespectful.

Now what if someone did not wish to be an organ donor, and they denied the validity of neurologic criteria for death because of religious reasons and desired to be kept on "life support" as long as possible, complete with access to surgical interventions for a tracheostomy and gastrointestinal feeding, dialysis, vasopressor medications, antibiotics, blood products, and a full code? Should these preferences be respected? We think not, because these preferences are unreasonable. This does not mean that the rejection of death by neurologic criteria is itself unreasonable. Rather, it means no one is obligated to provide any of the requested measures to help someone avoid death by circulatory-respiratory criteria for as long as possible. Nor is anyone obligated to believe that neurologic criteria for death are valid only if the patient says so, as if the validity of death criteria were dependent on the controlling principles of individual relativism. Legitimate goals of medicine, such as preserving life, curing disease, restoring baseline functioning, or providing comfort, cannot be achieved because a living patient is not there to receive them. The benefits attained through respecting someone's preferences are disproportionate to the burdens imposed on the living tasked with providing them, namely the medical teams, the institution, and the wider community (see below).

What about the benefits of continued organ support to third parties, such as the parents of a young child? The most excruciating cases of death by neurologic criteria involve pediatric patients. Children are not supposed to die, especially in tragic ways that can lead to a determination of death by neurologic criteria. What's more, family-centered medicine is permitted, if not required, in the pediatric and neonatal context [10]. Pediatric intensivists should be permitted to offer some definite time of organ support if, in their clinical judgment, it would benefit the family of the deceased (see below). Nonetheless, a limiting principle remains: benefits to families should be measured against the burdens to the team, the institution, and the wider community (again, see below). Additionally, the decedent's body demands its own form of respect. Unlike the adult case in which someone autonomously expresses wishes to receive supportive measures until their heart fails, pediatric and neonatal patients are (almost) never in a position to do this; thus, their remains should not be medically manipulated indefinitely. The respect owed to the newly dead demands a dignified dispossession of the body, not ongoing medical interference that can achieve no medical goal. Supportive measures, if they are to be given at all, ought to be time-limited for a specific goal. The problem with indefinite accommodation as permitted by the preservation approach is that it effectively nullifies any assessment of the burdens imposed by the interventions and effectively communicates to the family that they are allowed to drive care however they wish no matter the cost.

3 The Preservation Approach Undermines the Professional Integrity of the Clinical Teams Involved and Is Unfair to the Wider Community

So far, we have talked about the nature of the benefits at stake in the preservation approach and how they are disproportionate to the burdens. Just what are the burdens? Before answering this question, we must address how the burden-benefit analysis should be understood. Normally, it is limited to measuring how benefits and burdens affect the patient, not the clinical teams, their institution, or the wider community. Even if a treatment is expensive and requires many medical resources, it is wrong to classify these facts as "burdens" if they do not adversely affect the patient. A modest benefit may be worth pursuing if the benefits outweigh the burdens the treatment would impose on the patient. Moreover, patients and their families should have a significant measure of authority in determining what benefits are worth pursuing or what burdens are worth avoiding.

Nonetheless, if there is no patient to be benefited from after a valid declaration of death (see above), then the benefit-burden analysis of what to do next cannot be constrained by patient welfare.[2] The circle of concern automatically extends to other considerations such as clinician welfare and fair resource allocation; therefore the "burden-benefit" analysis becomes more like a traditional cost-benefit analysis that accounts for the burdens and benefits to everyone involved. True, the family may have some claims as a potential beneficiary of ongoing interventions on the deceased, but these claims are not rooted in the patient's therapeutic relationship with the team. When the patient becomes a decedent, the primary party to which the team relates in the therapeutic relationship is no more, and the decision-making responsibilities shift to a posthumous context governed by different relationships. The relationship with the family is ethically significant and is essential to settling a crucial question: what must be done for the body to be properly dispossessed or used in accordance with the patient's antemortem wishes? Any answer to this question is subject to a wider analysis of the benefits and burdens [10].

We rejected a patient's preference to be given indefinite accommodation because it is unreasonable, and it is unreasonable because it undermines the professional integrity of the clinical teams involved and is unfair to the wider community. Again, this judgment does not entail that someone's rejection of death by neurologic criteria is itself unreasonable. What is unreasonable is the expectation that everyone else involved must act *as if* death by neurologic criteria is invalid and provide medical treatment for a body that would otherwise be volunteered for organ donation or sent to the morgue. The clinical team's sense of moral coherence is undermined when

[2] Indeed, this is acknowledged by some who believe that neurological criteria fail to indicate biological death. For example, Miller and Truog write, "With the occurrence of total brain failure, the person who inhabited the body whose biological functions are being sustained by mechanical ventilation has ceased to exist. This drastic change in status provides a good reason to claim that no harm or wrong is done by stopping mechanical ventilation. Because the former person who occupied this body no longer exists, no interests of that person are set back by stopping the ventilator" ([11], p. 190).

they have to treat similar cases in *radically* dissimilar ways for no other reason than someone's expressed a preference that they do so. No one should blame the team for experiencing profound cognitive dissonance resulting from the responsibility to help a "brain dead" body "survive" with a full code in one case, and the responsibility to help the organ procurement organization determine if it should be subjected to the eviscerating acts of transplant surgery in another. The problem isn't just one of "moral distress" that sensitive clinicians might or might not experience when providing indefinite accommodation, but of a fundamental confusion over their roles and responsibilities as health care providers. Why should one body be treated in ways that reflect a lack of moral status when we volunteer it for transplant surgery while the other is treated in ways that reflect the sort of moral status that belongs to every other neurocritical care patient on the service if both bodies are in the same condition?

One should be forgiven for being skeptical of the claim that autonomous choice makes all the difference. Either neurologic criteria for death are valid, or they are not; either they draw the line at which constitutional rights are lost or they do not; either they distinguish organ procurement from euthanasia, or they do not; either they make the difference between murder and assault, or they do not; either they mark the point when health insurance coverage stops, or they do not. Saying their validity varies with the vicissitudes of patient autonomy is to say they are not valid at all, but mere political tools unmoored from the physical facts of death. As such, they can be suspended or implemented whether one likes them or not, regardless of (1) the coherence of medical standards of care, (2) the welfare of those who would benefit from intensive care resources while waiting in the emergency department, and (3) the financial costs incurred by the health care system which the wider community is responsible for supporting in one way or another. The fact that most patients would not choose prolonged ventilation and other forms of support after a determination of death by neurologic criteria is no objection to the demands of a fundamental principle of justice that *like cases must be treated alike*. In this case, it is the difference between what we owe the living and what we owe the dead. It is not that some amount of indefinite accommodation pursuant to the wishes of the dead would at some point cause *too much* unfairness to the living. It's that *any* amount of indefinite accommodation of the wishes of the dead is intolerably unfair to the living.

4 The Preservation Approach Gives the Family False Hope for Recovery and Prolongs Their Grieving Process

Patients who are declared dead by neurologic criteria are in a strange state. The body appears asleep, is warm to the touch, and shows no signs of decay; yet, it displays no openness to the world, no response to its environment, no felt sense of its needs. Only the modern intensive care unit with its ever-improving "life-sustaining technology" could produce such a phenomenon. The circumstances that lead to this condition are often sudden and tragic. Readily accessible and medically ill-informed stories proliferate in the media about "brain dead" patients recovering [12]. Distrust in clinicians is common and worries that they might be giving up too early on a loved one are understandable. As already mentioned, there is some evidence to suggest that supportive measures may be continued for a very limited amount of time (e.g., 24–48 h) to help families come to terms with the sudden fact that their loved one is gone. As Flamm and colleagues write:

> The most common motivation in finite-goal accommodation was to enable family members to arrive to view and/or say "goodbye" to their loved one while the body resembled its still-living form and to perceive that the patient had passed away peacefully and was surrounded by family. By maintaining physiological support, the clinical team provided the practical means of achieving this goal. Most clinicians acknowledged the ethic of holistic and patient-centered care that encompasses compassion for family members and could support the goal of a more consoling experience for those grieving and experiencing loss [1, p. 230].

Providing "organ support" to achieve these goals, especially in pediatric cases, is surely permissible.

Yet, providing indefinite accommodation sustains the belief that recovery is possible and that grieving is inappropriate. Why would anyone believe otherwise if the team and the institution agree to provide it? After all, the medical resources needed to deliver it are costly. While compassion may motivate a team to offer indefinite accommodation in the hope that the family may come around, the family should be forgiven for interpreting the supportive measures as "life support." Even those who reject the validity of neurologic criteria for death know that the "brain dead" state is not practically sustainable. A deep storyline often missed in the case of Jahi McMath is the family's tireless effort to give her the best opportunity to recover brain function. For many families, the preservation of life functions as the foundation for their hope—the possibility of a "miracle," however remote. Moreover, families are not likely to change their perspective if their requests for indefinite accommodation arise from religious reasons; it would be futile, if not deceptive, to offer "organ support" for the purpose of getting the family to "come around" and withdraw what they believe is "life support" on their own.

One of the intangible costs of the preservation approach is that it undermines the family's sense of closure. The grieving process is either delayed or stretches on from the time neurologic death is declared to the time the patient "dies on the vent" with little hope for change in between. An interesting turning point in the McMath case was when a scan revealed that her brain had not liquified, which gave the family hope. Yet before the scan, her grandmother was quoted as saying, "If her brain is jelly, we are going to have to accept that. I don't think people should live on that way. If they're gone, they're gone" [13]. The ambivalence of these statements reflects the strange state of the body that is determined dead by neurologic criteria. Jahi's grandmother believed that Jahi could "live on" in one sense but be "gone" in another, and that supportive measures should not continue if she were "gone." Note the consistency between rejecting the death declaration and knowing that their loved one has crossed into the shadowlands where people "pass away." In this view, the "brain dead" state marks the beginning of a fatal process, which has already eliminated the most primitive forms of consciousness from the human life-form. The supportive measures at issue can only delay the inevitable collapse of any remaining physiological functioning, which would rapidly occur if the measures were withdrawn. In the rare case when this process takes years to terminate in heart failure, it is years of the patient being bound to a ventilator, totally unable to appreciate the benefits of being alive while the families agonize over medical decisions involving varying levels of care and code status. Families who take on the fight to keep their loved ones "alive" in this state risk (1) locking themselves into a treatment course they will feel unable to stop, closed off from the solace that closure would bring, perhaps through meaningful grieving rituals [14], and (2) becoming bound to persistent feelings of guilt, ambivalence, and indecision [15].

5 The Preservation Approach Produces Societal Confusion by Permitting a Negotiated and Inconsistent Standard of Death

The preservation approach is motivated by the values of a liberal, pluralistic society as reflected in the majority ruling of *Planned Parenthood v. Casey* [16]:

> At the heart of liberty is the right to define one's own concept of existence, of meaning, of the universe, and of the mystery of human life. Beliefs about these matters could not define the attributes of personhood were they formed under compulsion of the State.

Although written in the context of abortion, the enumerated liberties extend to end-of-life issues, as reflected in a public statement issued by the McMath family while Jahi was still in the California hospital where she was declared dead:

> We have our strong religious convictions and set of beliefs and we believe that, in this country, a parent has the right to make decisions concerning the existence of their child: not a doctor who looks only at lines on a paper, or reads the cold black and white words on a law that says "brain dead" and definitely not a doctor who runs the facility that caused the brain death in the first place [17].

At issue is a simple argument that rests on two premises: (1) The choice of criteria for determining death can only be settled by controversial philosophical premises that go beyond the biological sciences; (2) if this is so, then individuals are permitted to make an autonomous choice between reasonable options for determining their deaths. As usually qualified, the "reasonable options" are limited to three: circulatory-respiratory criteria, the currently accepted "whole brain" criterion for death by neurologic criteria, and the so-called "higher-brain" criterion for death by neurologic criteria [18]. The higher-brain criterion, however, has never been validated and is not legal anywhere in the world. The case for choice effectively terminates in an opt-out policy permitting refusal of neurologic criteria for death, the validity of which is otherwise presumed.

The case for choice is flawed because both premises are false. First, no one gets to opt out of death by circulatory-respiratory criteria, because the complete and permanent loss of circulation and respiration imposes a limit on choice that is well within the realm of biology. Second, assuming for the sake of argument that the choice of using neurologic criteria for death can only be settled by controversial philosophical premises that go beyond the biological sciences, it does not follow that the choice should be left solely up to the individual or family. The choice to recognize the validity of neurologic criteria can legitimately be made by lawmakers within a jurisdiction in consultation with specialists, scientists, and citizens. Indeed, this is precisely what happened throughout the United States and other parts of the world. It is no accident either because there are compelling government interests at stake: being able to determine when a marriage ends, when a murder occurs, when constitutional rights are lost, when health insurance ends, when life insurance pays out, when an autopsy can be requested, and when vital organs can be donated (see above).

While the argument for neurologic criteria for death has always been disputed, it has proved hard to overturn. The reasons for this involve the stubborn fact that nothing can be done to change the "brain dead" condition. The clinical standards for the neurologic determination of death have stood the test of time.

Moreover, even if one believed that the definition of death, though clinically well-defined and supported and legally well-established and entrenched, was completely conventionally defined, it does not follow that the legal line should not be respected. Surely the line between adulthood and childhood at the age of 18 in most states in the United States is a convention at best, and arbitrary at worst. Yet it is a line that requires respect. A mature 17-year-old cannot vote in presidential elections—even if their religion tells them they are adults.

The truth is that the plausibility of the preservation approach depends on the view that the legal, social, and medical lines drawn by neurologic criteria for death should be ignored and that "futility" is in the eye of the beholder [19]. In this view, neurologic criteria cannot be used to determine "true death"—you might be mostly dead or as good as dead or legally dead, but true death is to be associated with corpses: complete stillness, paleness of the skin, and coldness to the touch. Of course, no one waits to declare death after such features become noticeable, and the Preservation Approach's defenders do not insist on it. They might believe that neurologic criteria for death are a useful legal fiction for facilitating organ donation [20]. Yet, the end result of such a distinction between "true death" and "pseudo-death" is confusion, which allows there to be different definitions of death for different purposes, both of which are inconsistent with one another. One of those purposes is to secure a peculiar right, the right to die on the ventilator. Denying that right in the name of "futility" is not adequate because the preservation approach assumes that the goals of care are to be determined solely by patients and families and that they have a right to pursue whatever treatment options are available to any other neurocritical care patient. Failing to do this, they argue, would be tyrannical, paternalistic, and imperialistic. Yet, this is true only if patient autonomy matters most in biomedical ethics; yet that is a questionable assumption.

6 Conclusion

Our goal has been to explain why the limitation approach should be preferred to the preservation approach in response to those who object to a declaration of death by neurologic criteria. We offered four reasons against offering indefinite accommodation: (1) there is no party to which a justifiable medical benefit accrues, (2) it undermines the professional integrity of the clinical teams involved and is unfair to the wider community, (3) it gives the family false hope for recovery and prolongs their grieving process, and (4) it produces societal confusion by permitting a negotiated and inconsistent standard of death. We understand that the limitation approach assumes something in dispute—the validity of neurologic criteria for death. Yet, this assumption was made to prove conditional: if neurologic criteria for death are valid, then the limitation approach ought to be followed and the preservation approach ought to be rejected. The authority these criteria have over our lives depends on their validity and on how communities and jurisdictions adjudicate them. Neither medical professionals nor patients and their families have the authority to wield them however they see fit.

References

1. Flamm AL, Smith ML, Mayer PA. Family members' requests to extend physiologic support after declaration of brain death: a case series analysis and proposed guidelines for clinical management. J Clin Ethics. 2014;25:222–37.
2. Burt RA. The medical futility debate: patient choice, physician obligation, and end-of-life care. J Palliat Med. 2002;5:249–54.
3. Brown SD. Is there a place for CPR and sustained physiological support in brain-dead non-donors? J Med Ethics. 2017;43:679–83.
4. Johnson LSM. Death by neurological criteria: expert definitions and lay misgivings. QJM. 2017;110:267–70.
5. Russell JA, Epstein LG, Greer DM, et al. Brain death, the determination of brain death, and member guidance for brain death accommodation requests: AAN position statement. Neurology. 2019;92:228–32.
6. Johnson LSM. The case for reasonable accommodation of conscientious objections to declarations of brain death. Bioeth Inq. 2016;13:105–15.
7. Olick RS. Brain death, religious freedom, and public policy: New Jersey's landmark legislative initiative. Kennedy Inst Ethics J. 1991;1:275–88.
8. Lewis A, Adams N, Chopra A, Kirschen MP. Organ support after death by neurologic criteria in pediatric patients. Crit Care Med. 2017;45:e916–24.
9. Lewis A, Adams N, Varelas P, Greer D, Caplan A. Organ support after death by neurologic criteria: results of a survey of US neurologists. Neurology. 2016;87:827–34.
10. Olson NW. Why should medical care be family-centered?: understanding ethical responsibilities for patients' family members. Kennedy Inst Ethics J. 2019;29:159–85.
11. Miller FG, Truog RD. The incoherence of determining death by neurological criteria: a commentary on controversies in the determination of death, a white paper by the President's council on bioethics. Kennedy Inst Ethics J. 2009;19:185–93.
12. Busl KM. What does the public need to know about brain death? AMA J Ethics. 2020;22:1047–54.
13. Aviv R. What does it mean to die? The New Yorker 2018.

14. Gabbay E, Fins JJ. Go in peace: brain death, reasonable accommodation and jewish mourning rituals. J Relig Health. 2019;58:1672–86.
15. Paris JJ, Cummings BM, Moore MP. "Brain death," "dead," and parental denial: the case of Jahi McMath. Camb Q Healthc Ethics. 2014;23:371–82.
16. Kennedy A. Planned Parenthood of southeastern PA. v. Casey; 1992.
17. Gafni M. Jahi McMath: family, attorney release letter addressing critics. The Mercury News 2013.
18. Veatch RM. Controversies in defining death: a case for choice. Theor Med Bioeth. 2019;40:381–401.
19. Nair-Collins M. Laying futility to rest. J Med Philos. 2015;40:554–83.
20. Shah SK, Miller FG. Can we handle the truth? Legal fictions in the determination of death. Am J Law Med. 2010;36:540–85.

Index

A. Lewis, J. L. Bernat (eds.), *Death Determination by Neurologic Criteria*, Advances in Neuroethics, https://doi.org/10.1007/978-3-031-15947-3

MIX
Papier aus verantwortungsvollen Quellen
Paper from responsible sources
FSC® C105338

If you have any concerns about our products,
you can contact us on
ProductSafety@springernature.com

In case Publisher is established outside the EU,
the EU authorized representative is:
Springer Nature Customer Service Center GmbH
Europaplatz 3, 69115 Heidelberg, Germany

Printed by Libri Plureos GmbH
in Hamburg, Germany